T0181863

Lecture Notes in Computer Science   12867

Advanced Research in Computing and Software Science

Subline of Lecture Notes in Computer Science

More information about this subseries at http://www.springer.com/series/7407

Evripidis Bampis · Aris Pagourtzis (Eds.)

Fundamentals of Computation Theory

23rd International Symposium, FCT 2021
Athens, Greece, September 12–15, 2021
Proceedings

 Springer

Editors
Evripidis Bampis 🔟
Sorbonne University
Paris, France

Aris Pagourtzis 🔟
National Technical University of Athens
Athens, Greece

ISSN 0302-9743   ISSN 1611-3349 (electronic)
Lecture Notes in Computer Science
ISBN 978-3-030-86592-4   ISBN 978-3-030-86593-1   (eBook)
https://doi.org/10.1007/978-3-030-86593-1

LNCS Sublibrary: SL1 – Theoretical Computer Science and General Issues

This Springer imprint is published by the registered company Springer Nature Switzerland AG
The registered company address is: Gewerbestrasse 11, 6330 Cham, Switzerland

Preface

The 23rd International Symposium on Fundamentals of Computation Theory (FCT 2021) was hosted virtually by the National Technical University of Athens due to the COVID-19 pandemic during September 12–15, 2021. The Symposium on Fundamentals of Computation Theory (FCT) was established in 1977 for researchers interested in all aspects of theoretical computer science and in particular algorithms, complexity, and formal and logical methods. FCT is a biennial conference. Previous symposia have been held in Poznan (Poland, 1977), Wendisch-Rietz (Germany, 1979), Szeged (Hungary, 1981), Borgholm (Sweden, 1983), Cottbus (Germany, 1985), Kazan (Russia, 1987), Szeged (Hungary, 1989), Gosen-Berlin (Germany, 1991), Szeged (Hungary, 1993), Dresden (Germany, 1995), Krakow (Poland, 1997), Iasi (Romania, 1999), Riga (Latvia, 2001), Malmö (Sweden, 2003), Lübeck (Germany, 2005), Budapest (Hungary, 2007), Wroclaw (Poland, 2009), Oslo (Norway, 2011), Liverpool (UK, 2013), Gdansk (Poland, 2015), Bordeaux (France, 2017), and Copenhagen (Denmark, 2019).

The Program Committee (PC) of FCT 2021 received 94 submissions. Each submission was reviewed by at least three PC members and some trusted external reviewers, and evaluated on its quality, originality, and relevance to the symposium. The PC selected 30 papers for presentation at the conference and inclusion in these proceedings.

Four invited talks were given at FCT 2021 by Constantinos Daskalakis (Massachusetts Institute of Technology, USA), Daniel Marx (Max Planck Institute for Informatics, Germany), Claire Mathieu (CNRS and University of Paris, France), and Nobuko Yoshida (Imperial College, UK). David Richerby (University of Essex, UK) offered an invited tutorial.

This volume contains, in addition to the 30 accepted regular papers, the papers of the invited talks of Claire Mathieu and Nobuko Yoshida, the abstracts of the invited talks of Constantinos Daskalakis and Daniel Marx, and the abstract of the invited tutorial of David Richerby.

The Program Committee selected one contribution for the best paper award and two contributions for the best student paper awards, all sponsored by Springer:

- The best paper award went to Marc Neveling, Jörg Rothe, and Robin Weishaupt for their paper "The Possible Winner Problem with Uncertain Weights Revisited."
- Two papers shared the best student paper award: (a) "Faster FPT Algorithms for Deletion to Pairs of Graph Classes" by Ashwin Jacob, Diptapriyo Majumdar, and Venkatesh Raman, and (b) "On Finding Separators in Temporal Split and Permutation Graphs" by Nicolas Maack, Hendrik Molter, Rolf Niedermeier, and Malte Renken.

We thank the Steering Committee and its chair, Marek Karpinski, for giving us the opportunity to serve as the program chairs of FCT 2021, and for trusting us with the responsibilities of selecting the Program Committee, the conference program, and publications.

We would like to thank all the authors who responded to the call for papers, the invited speakers, the members of the Program Committee, and the external reviewers for their diligent work in evaluating the submissions and for their contributions to the electronic discussions. We would also like to thank the members of the Organizing Committee and the members of the Local Arrangements team for the great job they have done; special thanks go to Dimitris Fotakis, Ioanna Protekdikou, and Antonis Antonopoulos.

We would like to thank Springer for publishing the proceedings of FCT 2021 in their ARCoSS/LNCS series and for their sponsoring of the best paper awards. We are thankful to the members of the Editorial Board of *Lecture Notes in Computer Science* and the editors at Springer for their help throughout the publication process. We also acknowledge support from the Institute of Communication and Computer Systems of the School of Electrical and Computer Engineering of the National Technical University of Athens, towards covering teleconference expenses and registration costs for a number of students. Sponsors that provided support after the preparation of these proceedings appear on the webpage of the conference: https://www.corelab.ntua.gr/fct2021/.

The EasyChair conference system was used to manage the electronic submissions, the review process, and the electronic Program Committee discussions. It made our task much easier.

This volume is dedicated to the fond memory of our friend and colleague Yannis Manoussakis, Professor at University of Paris-Saclay, France. Yannis, a specialist in graph theory, unexpectedly passed away earlier this year in his beloved hometown on Crete. We will always remember him for his open heart and his great passion for theoretical computer science.

July 2021

Evripidis Bampis
Aris Pagourtzis

Organization

Steering Committee

Bogdan Chlebus	Augusta University, USA
Marek Karpinski (Chair)	University of Bonn, Germany
Andrzej Lingas	Lund University, Sweden
Miklos Santha	CNRS and University Paris Diderot, France
Eli Upfal	Brown University, USA

Program Committee

Evripidis Bampis (Co-chair)	Sorbonne University, France
Petra Berenbrink	University of Hamburg, Germany
Arnaud Casteigts	University of Bordeaux, France
Marek Chrobak	University of California, Riverside, USA
Hans van Ditmarsch	CNRS and University of Lorraine, France
Thomas Erlebach	University of Leicester, UK
Bruno Escoffier	Sorbonne University, France
Henning Fernau	University of Trier, Germany
Dimitris Fotakis	National Technical University of Athens, Greece
Pierre Fraigniaud	CNRS and University of Paris, France
Leszek Gasieniec	University of Liverpool, UK, and Augusta University, USA
Laurent Gourves	CNRS and Paris Dauphine University, France
Giuseppe F. Italiano	LUISS Guido Carli University, Italy
Ralf Klasing	CNRS and University of Bordeaux, France
Alexander Kononov	Sobolev Institute of Mathematics and Novosibirsk State University, Russia
Antonin Kucera	Masaryk University, Czech Republic
Dietrich Kuske	Technische Universität Ilmenau, Germany
Nikos Leonardos	National and Kapodistrian University of Athens, Greece
Minming Li	City University of Hong Kong, Hong Kong
Zsuzsanna Liptak	University of Verona, Italy
Giorgio Lucarelli	University of Lorraine, France
Vangelis Markakis	Athens University of Economics and Business, Greece
Nicole Megow	University of Bremen, Germany
Andrzej Murawski	University of Oxford, UK
Aris Pagourtzis (Co-chair)	National Technical University of Athens, Greece
Charis Papadopoulos	University of Ioannina, Greece
Igor Potapov	University of Liverpool, UK
Tomasz Radzik	King's College London, UK

Maria Serna	Universitat Politecnica de Catalunya, Spain
Hadas Shachnai	Technion, Israel
Vorapong Suppakitpaisarn	University of Tokyo, Japan
Nikos Tzevelekos	Queen Mary University of London, UK
Guochuan Zhang	Zhejiang University, China

Organizing Committee

Dimitris Fotakis (Co-chair)	National Technical University of Athens, Greece
Nikos Leonardos	National and Kapodistrian University of Athens, Greece
Thanasis Lianeas	National Technical University of Athens, Greece
Aris Pagourtzis (Co-chair)	National Technical University of Athens, Greece

Additional Reviewers

Faisal Abu-Khzam
Ioannis Anagnostides
Antonis Antonopoulos
Andrei Asinowski
Max Bannach
Rémy Belmonte
Nathalie Bertrand
René Van Bevern
Therese Biedl
Felix Biermeier
Davide Bilò
Ahmad Biniaz
Johanna Björklund
Benedikt Bollig
Michaël Cadilhac
Olivier Carton
Armando Castaneda
Pyrros Chaidos
Sankardeep Chakraborty
Pierre Charbit
Hunter Chase
Vincent Chau
Leroy Chew
Dmitry Chistikov
Dimitrios Christou
Ferdinando Cicalese
Florence Clerc

Bruno Courcelle
Geoffrey Cruttwell
Dominik D. Freydenberger
Clément Dallard
Minati De
Yichao Duan
Swan Dubois
Pavlos Efraimidis
Matthias Englert
Leah Epstein
Vincent Fagnon
Qilong Feng
Irene Finocchi
Florent Foucaud
Shayan Garani
Ran Gelles
Marios Georgiou
Archontia Giannopoulou
Andreas Göbel
Stefan Göller
Radu Grigore
Nathan Grosshans
Hermann Gruber
Longkun Guo
Siddharth Gupta
Anthony Guttmann
Christoph Haase

Nicolas Wieseke

Kyrill Winkler

Petra Wolf

Lirong Xia

Chenyang Xu

Kuan Yang

Wei Yu

Tom van der Zanden

Jingru Zhang

Peng Zhang

Ruilong Zhang

Xu Zijian

Plenary Talks

Min-Max Optimization: From von Neumann to Deep Learning Plenary Talks

Constantinos Daskalakis

Massachusetts Institute of Technology, Cambridge, MA, USA

Abstract. Deep Learning applications, such as Generative Adversarial Networks and other adversarial training frameworks, motivate min-maximization of nonconvex-nonconcave objectives. Unlike their convex-concave counterparts, however, for which a multitude of equilibrium computation methods are available, nonconvex-nonconcave objectives pose significant optimization challenges. Gradient-descent based methods commonly fail to identify equilibria, and even computing local approximate equilibria has remained daunting. We shed light on this challenge through a combination of complexity-theoretic, game-theoretic and topological techniques, presenting obstacles and opportunities for Deep Learning and Game Theory going forward.

(This talk is based on joint works with Noah Golowich, Stratis Skoulakis and Manolis Zampetakis)

Tight Complexity Results for Algorithms Using Tree Decompositions

Dániel Marx

CISPA Helmholtz Center for Information Security, Saarbrücken, Germany

Abstract. It is well known that hard algorithmic problems on graphs are easier to solve if we are given a low-width tree composition of the input graph. For many problems, if a tree decomposition of width k is available, algorithms with running time of the form f(k)*poly(n) are known; that is, the problem is fixed-parameter tractable (FPT) parameterized by the width of the given decomposition. But what is the best possible function f(k) in such an algorithm? In the past decade, a series of new upper and lower bounds gave us a tight understanding of this question for particular problems. The talk will give a survey of these results and some new developments.

The Complexity of Counting Problems (Tutorial)

David Richerby

University of Essex, Colchester, UK

Abstract. Every computational decision problem ("Is there an X?") has a natural counting variant ("How many X's are there?"). More generally, computing weighted sums such as integrals, expectations and partition functions in statistical physics can also be seen as counting problems.

This tutorial will give an introduction to the complexity of solving counting problems, both exactly and approximately. I will focus on variants of constraint satisfaction problems. These are powerful enough to naturally express many important problems, but also being restricted enough to allow their computational complexity to be classified completely and elegantly. No prior knowledge of counting problems will be assumed.

Contents

Invited Papers

Two-Sided Matching Markets
with Strongly Correlated Preferences

Hugo Gimbert[1], Claire Mathieu[2(✉)], and Simon Mauras[3]

[1] CNRS, LaBRI, Bordeaux, France
hugo.gimbert@cnrs.fr
[2] CNRS, IRIF, Paris, France
Claire.Mathieu@irif.fr
[3] Université de Paris, IRIF, Paris, France
simon.mauras@irif.fr

Abstract. Stable matching in a community consisting of men and women is a classical combinatorial problem that has been the subject of intense theoretical and empirical study since its introduction in 1962 in a seminal paper by Gale and Shapley, who designed the celebrated "deferred acceptance" algorithm for the problem.

In the input, each participant ranks participants of the opposite type, so the input consists of a collection of permutations, representing the preference lists. A bipartite matching is unstable if some man-woman pair is blocking: both strictly prefer each other to their partner in the matching. Stability is an important economics concept in matching markets from the viewpoint of manipulability. The unicity of a stable matching implies non-manipulability, and near-unicity implies limited manipulability, thus these are mathematical properties related to the quality of stable matching algorithms.

This paper is a theoretical study of the effect of correlations on approximate manipulability of stable matching algorithms. Our approach is to go beyond worst case, assuming that some of the input preference lists are drawn from a distribution. Approximate manipulability is approached from several angles: when all stable partners of a person have approximately the same rank; or when most persons have a unique stable partner.

1 Introduction

In the classical stable matching problem, a certain community consists of men and women (all heterosexual and monogamous) where each person ranks those of the opposite sex in accordance with his or her preferences for a marriage partner (possibly declaring some matches as unacceptable). Our objective is to marry off the members of the community in such a way that the established matching is *stable, i.e.* such that there is no *blocking pair*. A man and a woman who are not married to each other form a blocking pair if they prefer each other to their mates.

In their seminal paper, Gale and Shapley [11] designed the *men-proposing deferred acceptance* procedure, where men propose while women disposes. This

© Springer Nature Switzerland AG 2021
E. Bampis and A. Pagourtzis (Eds.): FCT 2021, LNCS 12867, pp. 3–17, 2021.
https://doi.org/10.1007/978-3-030-86593-1_1

algorithm always outputs a matching which is stable, optimal for men and pessimal for women (in terms of rank of each person's partner). By symmetry, there also exists a women-optimal/men-pessimal stable matching. Gale and Shapley's original motivation was the assignment of students to colleges, a setting to which the algorithm and results extend, and their approach was successfully implemented in many matching markets; see for example [1,2,8,29].

However, there exists instances where the men-optimal and women-optimal stable matchings are different, and even extreme cases of instances in which every man/woman pair belongs to some stable matching. This raises the question of which matching to choose [14,15] and of possible strategic behavior [9,10,28]. More precisely, if a woman lies about her preference list, this gives rise to new stable matchings, where she will be no better off than she would be in the true women-optimal matching. Thus, a woman can only gain from strategic manipulation up to the maximum difference between her best and worst partners in stable matchings. By symmetry, this also implies that the men proposing deferred acceptance procedure is strategy-proof for men (as they will get their best possible partner by telling the truth).

Fortunately, there is empirical evidence that in many instances, in practice the stable matching is essentially unique (a phenomenon often referred to as "core-convergence"); see for example [6,16,23,29]. One of the empirical explanations for core-convergence given by Roth and Peranson in [29] is that the preference lists are correlated: *"One factor that strongly influences the size of the set of stable matchings is the correlation of preferences among programs and among applicants. When preferences are highly correlated (i.e., when similar programs tend to agree which are the most desirable applicants, and applicants tend to agree which are the most desirable programs), the set of stable matchings is small."*

Following that direction of enquiry, we study the core-convergence phenomenon, in a model where preferences are stochastic. When preferences of women are strongly correlated, Theorem 1 shows that the expected difference of rank between each woman's worst and best stable partner is a constant, hence the incentives to manipulate are limited. If additionally the preferences of men are uncorrelated, Theorem 2 shows that most women have a unique stable partner, and therefore have no incentives to manipulate.

1.1 Definitions and Main Theorems

Matchings. Let $\mathcal{M} = \{m_1, \ldots, m_M\}$ be a set of M men, $\mathcal{W} = \{w_1, \ldots, w_W\}$ be a set of W women, and $N = \min(M, W)$. In a matching, each person is either single, or matched with someone of the opposite sex. Formally, we see a matching as a function $\mu : \mathcal{M} \cup \mathcal{W} \to \mathcal{M} \cup \mathcal{W}$, which is self-inverse ($\mu^2 = \mathrm{Id}$), where each man m is paired either with a woman or himself ($\mu(m) \in \mathcal{W} \cup \{m\}$), and symmetrically, each woman w is paired with a man or herself ($\mu(w) \in \mathcal{M} \cup \{w\}$).

Preference Lists. Each person declares which members of the opposite sex they find acceptable, then gives a strictly ordered preference list of those members.

Preference lists are *complete* when no one is declared unacceptable. Formally, we represent the preference list of a man m as a total order $\succ_m$ over $\mathcal{W} \cup \{m\}$, where $w \succ_m m$ means that man m finds woman w acceptable, and $w \succ_m w'$ means that man m prefers woman w to woman w'. Similarly we define the preference list $\succ_w$ of woman w.

Stability. A man-woman pair (m, w) is blocking a matching μ when $m \succ_w \mu(w)$ and $w \succ_m \mu(m)$. Abusing notations, observe that μ matches a person p with an unacceptable partner when p would prefer to remain single, that is when the pair (p, p) is blocking. A matching with no blocking pair is stable. A stable pair is a pair which belongs to at least one stable matching.

Random Preferences. We consider a model where each person's set of acceptable partners is deterministic, and orderings of acceptable partners are drawn independently from *regular* distributions. When unspecified, someone's acceptable partners and/or their ordering is *adversarial*, that is chosen by an adversary who knows the input model but does not know the outcome of the random coin flips.

Definition 1 (Regular distribution). *A distribution of preferences lists is* **regular** *when for every sequence of acceptable partners* $a_1, \ldots, a_k$ *we have* $\mathbb{P}[a_1 \succ a_2 \mid a_2 \succ \cdots \succ a_k] \leq \mathbb{P}[a_1 \succ a_2]$.

Intuitively, knowing that a_2 is ranked well only decreases the probability that a_1 beats a_2. Most probability distributions that have been studied are regular. In particular, sorting acceptable partners by scores (drawn independently from distributions on $\mathbb{R}$), yields a regular distribution. As an example of regular distribution, we study popularity preferences, introduced by Immorlica and Madhian [17].

Definition 2 (Popularity preferences). *When a woman w has* popularity *preferences, she gives a positive popularity* $\mathcal{D}_w(m)$ *to each acceptable partner m. We see* $\mathcal{D}_w$ *as a distribution over her acceptable partners, scaled so that it sums to 1. She uses this distribution to draw her favourite partner, then her second favourite, and so on until her least favourite partner.*

The following Theorem shows that under some assumptions every woman gives approximately the same rank to all of her stable partners.

Theorem 1. *Assume that each woman independently draws her preference list from a regular distribution. The men's preference lists are arbitrary. Let u_k be an upper bound on the odds that man m_{i+k} is ranked before man m_i:*

$$\forall k \geq 1, \quad u_k = \max_{w,i} \left\{ \frac{\mathbb{P}[m_{i+k} \succ_w m_i]}{\mathbb{P}[m_i \succ_w m_{i+k}]} \;\middle|\; w \text{ finds both } m_i \text{ and } m_{i+k} \text{ acceptable} \right\}$$

Then for each woman with at least one stable partner, in expectation all of her stable partners are ranked within $(1 + 2\exp(\sum_{k \geq 1} k u_k)) \sum_{k \geq 1} k^2 u_k$ *of one another in her preference list.*

Theorem 1 is most relevant when the women's preference lists are strongly correlated, that is, when every woman's preference list is "close" to a single ranking $m_1 \succ m_2 \succ \ldots \succ m_M$. This closeness is measured by the odds that in some ranking, some man is ranked ahead of a man who, in the ranking $m_1 \succ m_2 \succ \ldots \succ m_M$, would be k slots ahead of him.

We detail below three examples of applications, where the expected difference of ranks between each woman's best and worst partners is $O(1)$, and thus her incentives to misreport her preferences are limited.

– *Identical preferences.* If all women rank their acceptable partners using a master list $m_1 \succ m_2 \succ \cdots \succ m_M$, then all u_k's are equal to 0. Then Theorem 1 states that each woman has a unique stable husband, a well-known result for this type of instances.
– *Preferences from identical popularities.* Assume that women have popularity preferences (Definition 2) and that each woman gives man m_i popularity 2^{-i}. Then $u_k = 2^{-k}$ and the expected rank difference is at most $O(1)$.
– *Preferences from correlated utilities.* Assume that women have similar preferences: each woman w gives man m_i a score that is the sum of a common value i and an idiosyncratic value η_i^w which is normally distributed with mean 0 and variance σ^2; she then sorts men by increasing scores. Then $u_k \leq \max_{w,i} \{2 \cdot \mathbb{P}[\eta_i^w - \eta_{i+k}^w > k]\} \leq 2e^{-(k/2\sigma)^2}$ and the expected rank difference, by a short calculation, is at most $4\sqrt{\pi}\sigma^3(1 + 2e^{4\sigma^2}) = O(1)$.

A stronger notion of approximate incentive compatibility is near-unicity of a stable matching, meaning that most persons have either no or one unique stable partner, and thus have no incentive to misreport their preferences. When does that hold? One answer is given by Theorem 2.

Theorem 2. *Assume that each woman independently draws her preference list from a regular distribution. Let u_k be an upper bound on the odds that man m_{i+k} is ranked before man m_i:*

$$\forall k \geq 1, \quad u_k = \max_{w,i}\left\{ \frac{\mathbb{P}[m_{i+k} \succ_w m_i]}{\mathbb{P}[m_i \succ_w m_{i+k}]} \,\middle|\, w \text{ finds both } m_i \text{ and } m_{i+k} \text{ acceptable}\right\}$$

Further assume that all preferences are complete, that $u_k = \exp(-\Omega(k))$, and that men have uniformly random preferences. Then, in expectation the fraction of persons who have multiple stable partners converges to 0.

Notice that in the three examples of Theorem 1, the sequence $(u_k)_{k \geq 1}$ is exponentially decreasing. The assumptions of Theorem 2 are minimal in the sense that removing one would bring us back to a case where a constant fraction of woman have multiple stable partners.

– *Preference lists of women.* If we remove the assumption that u_k is exponentially decreasing, the conclusion no longer holds: consider a balanced market balanced ($M = W$) in which both men and women have complete uniformly random preferences; then most women have $\sim \ln N$ stable husbands [19,25].

- *Preference lists of men.* Assume that men have random preference built as follows: starting from the ordering $w_1, w_2, \ldots, w_M$, each pair (w_{2i-1}, w_{2i}) is swapped with probability $1/2$, for all i. A symmetric definition for women's preferences satisfy the hypothesis of Theorem 2, with $u_1 = 1$ and $u_k = 0$ for all $k \geq 2$. Then there is a $1/8$ probability that men m_{2i-1} and m_{2i} are both stable partners of women w_{2i-1} and w_{2i}, for all i, hence a constant expected fraction of persons with multiple stable partners.
- *Incomplete preferences.* Consider a market divided into groups of size 4 of the form $\{m_{2i-1}, m_{2i}, w_{2i-1}, w_{2i}\}$, where a man and a woman are mutually acceptable if they belong to the same group. Once again, with constant probability, m_{2i-1} and m_{2i} are both stable partners of women w_{2i-1} and w_{2i}.

1.2   Related Work

Analyzing instances that are less far-fetched than in the worst case is the motivation underlying the model of stochastically generated preference lists. A series of papers [19, 22, 24–26] study the model where N men and N women have complete uniformly random preferences. Asymptotically, and in expectation, the total number of stable matchings is $\sim e^{-1} N \ln N$, in which a fixed woman has $\sim \ln N$ stable husbands, where her best stable husband has rank $\sim \ln N$ and her worst stable husband has rank $\sim N/\ln N$.

The first theoretical explanations of the "core-convergence" phenomenon where given in [17] and [4], in variations of the standard uniform model. Immorlica and Mahdian [17] consider the case where men have constant size random preferences (truncated popularity preferences). Ashlagi, Kanoria and Leshno [4], consider slightly unbalanced matching markets $(M < W)$. Both articles prove that the fraction of persons with several stable partners tends to 0 as the market grows large. Theorem 2 and its proof incorporate ideas from those two papers.

Beyond strong "core-convergence", where most agents have a unique stable partner, one can bound the utility gain by manipulating a stable mechanism. Lee [21] considers a model with random cardinal utilities, and shows that agents receive almost the same utility in all stable matchings. Kanoria, Min and Qian [18], and Ashlagi, Braverman, Thomas and Zhao [3] study the rank of each person's partner, under the men and women optimal stable matchings, as a function of the market imbalance and the size of preference lists [18], or as a function of each person's (bounded) popularity [3]. Theorem 1 can be compared with such results.

Beyond one-to-one matchings, school choice is an example of many-to-one markets. Kojima and Pathak [20] generalize results from [17] and prove that most schools have no incentives to manipulate. Azevedo and Leshno [5] show that large markets converge to a unique stable matching in a model with a continuum of students. To counter balance those findings, Biró, Hassidim, Romm and Shorer [7], and Rheingans-Yoo [27] argue that socioeconomic status and geographic preferences might undermine core-convergence, thus some incentives remain in such markets.

2 Strongly Correlated Preferences: Proof of Theorem 1

Theorem 1. *Assume that each woman independently draws her preference list from a regular distribution. The men's preference lists are arbitrary. Let u_k be an upper bound on the odds that man m_{i+k} is ranked before man m_i:*

$$\forall k \geq 1, \quad u_k - \max_{w,i} \left\{ \frac{\mathbb{P}[m_{i+k} \succ_w m_i]}{\mathbb{P}[m_i \succ_w m_{i+k}]} \,\middle|\, w \text{ finds both } m_i \text{ and } m_{i+k} \text{ acceptable} \right\}$$

Then for each woman with at least one stable partner, in expectation all of her stable partners are ranked within $(1 + 2\exp(\sum_{k \geq 1} k u_k)) \sum_{k \geq 1} k^2 u_k$ of one another in her preference list.

In Subsect. 2.1, we define a partition of stable matching instances into *blocks*. For strongly correlated instances, blocks provide the structural insight to start the analysis: in Lemma 3, we use them to upper-bound the difference of ranks between a woman's worst and best stable partners by the sum of (1) the number x of men coming from other blocks and who are placed between stable husbands in the woman's preference list, and (2) the block size.

The analysis requires a delicate handling of conditional probabilities. In Subsect. 2.2, we explain how to condition on the men-optimal stable matching, when preferences are random.

Subsection 2.3 analyzes (1). The men involved are out of place compared to their position in the ranking $m_1 \succ \ldots \succ m_M$, and the odds of such events can be bounded, thanks to the assumption that distributions of preferences are regular. Our main technical lemma there is Lemma 4.

Subsection 2.4 analyzes (2), the block size by first giving a simple greedy algorithm (Algorithm 2) to compute a block. Each of the two limits of a block is computed by a sequence of "jumps", so the total distance traveled is a sum of jumps which, thanks to Lemma 4 again, can be stochastically dominated by a sum X of independent random variables (see Lemma 7); thus it all reduces to analyzing X, a simple mathematical exercise (Lemma 8).

Finally, Subsect. 2.5 combines the Lemmas previously established to prove Theorem 1.

Our analysis builds on Theorems 1 and 2, two fundamental and well-known results.

Theorem 1 (Adapted from [11]). *Algorithm 1 outputs a stable matching $\mu_{\mathcal{M}}$ in which every man (resp. woman) has his best (resp. her worst) stable partner. Symmetrically, there exists a stable matching $\mu_{\mathcal{W}}$ in which every woman (resp. man) has her best (resp. his worst) stable partner.*

Theorem 2 (Adapted from [12]). *Each person is either matched in all stable matchings, or single in all stable matchings. In particular, a woman is matched in all stable matchings if and only if she received at least one acceptable proposal during Algorithm 1.*

Algorithm 1. Men Proposing Deferred Acceptance.

Input: Preferences of men $(\succ_m)_{m \in M}$ and women $(\succ_w)_{w \in \mathcal{W}}$.

Initialization: Start with an empty matching μ.

While a man m is single and has not proposed to every woman he finds acceptable, **do**

$\quad$ m proposes to his favorite woman w he has not proposed to yet.

$\quad$ **If** m is w's favorite acceptable man among all proposals she received,

$\quad\quad$ w accepts m's proposal, and rejects her previous husband if she was married.

Output: Resulting matching.

2.1 Separators and Blocks

In this subsection, we define the block structure underlying our analysis.

Definition 3 (separator). *A separator is a set $S \subseteq \mathcal{M}$ of men such that in the men-optimal stable matching $\mu_{\mathcal{M}}$, each woman married to a man in S prefers him to all men outside S:*

$$\forall w \in \mu_{\mathcal{M}}(S) \cap \mathcal{W}, \quad \forall m \in \mathcal{M} \setminus S, \quad \mu_{\mathcal{M}}(w) \succ_w m$$

Lemma 1. *Given a separator $S \subseteq \mathcal{M}$, every stable matching matches S to the same set of women.*

Proof. Let $w \in \mu_{\mathcal{M}}(S)$ and let m be the partner of w in some stable matching. Since $\mu_{\mathcal{M}}$ is the woman-pessimal stable matching by Theorem 1, w prefers m to $\mu_{\mathcal{M}}(w)$. By definition of separators, that implies that $m \in S$. Hence, in every stable matching μ, women of $\mu_{\mathcal{M}}(S)$ are matched to men in S. By a cardinality argument, men of S are matched by μ to $\mu_{\mathcal{M}}(S)$.

Definition 4 (prefix separator, block). *A prefix separator is a separator S such that $S = \{m_1, m_2, \ldots, m_t\}$ for some $0 \leq t \leq N$. Given a collection of $b+1$ prefix separators $S_i = \{m_1, \ldots, m_{t_i}\}$ with $0 = t_0 < t_1 < \cdots < t_b = N$, the i-th block is the set $B_i = S_{t_i} \setminus S_{t_{i-1}}$ with $1 \leq i \leq b$.*

Abusing notations, we will denote S as the prefix separator t and B as the block $(t_{i-1}, t_i]$.

Lemma 2. *Given a block $B \subseteq \mathcal{M}$, every stable matching matches B to the same set of women.*

Proof. B equals $S_{t_i} \setminus S_{t_{i-1}}$ for some i. Applying Lemma 1 to S_{t_i} and to $S_{t_{i-1}}$ proves the Lemma.

Lemma 3. *Consider a woman w_n who is matched by $\mu_{\mathcal{M}}$ and let $B = (l, r]$ denote her block. Let x denote the number of men from a better block that are ranked by w_n between a man of B and m_n:*

$$x = |\{i \leq l \mid \exists j > l, \; m_j \succ_{w_n} m_i \succ_{w_n} m_n\}|.$$

Then in w_n's preference list, the difference of ranks between w_n's worst and best stable partners is at most $x + r - l - 1$.

Proof. Since $\mu_{\mathcal{M}}$ is woman-pessimal by Theorem 1, m_n is the last stable husband in w_n's preference list. Let m_j denote her best stable husband.

In w_n's preference list, the interval from m_j to m_n contains men from her own block, plus possibly some additional men. Such a man m_i comes from outside her block $(l, r]$ and she prefers him to m_n: since r is a prefix separator, we must have $i \leq l$. Thus x counts the number of men who do not belong to her block but who in her preference list are ranked between m_j and m_n.

On the other hand, the number of men who belong to her block and who in her preference list are ranked between m_j and m_n (inclusive) is at most $r - l$.

Together, the difference of ranks between w_n's worst and best stable partners is at most $x + (r - l) - 1$. See Fig. 1 for an illustration.

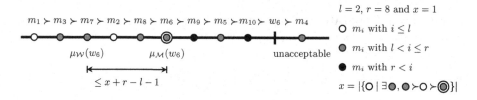

Fig. 1. Preference list of w_n, with $n = 6$. The block of w_n is defined by a left separator at $l = 2$ and a right separator at $r = 8$. Colors white, gray and black corresponds to blocks, and are defined in the legend. All stable partners of w_n must be gray. Men in black are all ranked after $m_n = \mu_{\mathcal{M}}(w_n)$. The difference in rank between w_n's worst and best partner is at most the number of gray men (here $r - l = 6$), minus 1, plus the number of white men ranked after a gray man and before m_n (here $x = 1$).

2.2 Conditioning on the Man Optimal Stable Matching When Preferences Are Random

We study the case where each person draws her preference list from an arbitrary distribution. The preference lists are random variables, that are independent but not necessarily identically distributed.

Intuitively, we use the *principle of deferred decision* and construct preference lists in an online manner. By Theorem 1 the man-optimal stable matching $\mu_{\mathcal{M}}$ is computed by Algorithm 1, and the remaining randomness can be used for a stochastic analysis of each person's stable partners. To be more formal, we define a random variable $\mathcal{H}$, and inspection of Algorithm 1 shows that $\mathcal{H}$ contains enough information on each person's preferences to run Algorithm 1 deterministically.

Definition 5. *Let $\mathcal{H} = (\mu_{\mathcal{M}}, (\sigma_m)_{m \in \mathcal{M}}, (\pi_w)_{w \in \mathcal{W}})$ denote the random variable consisting of (1) the man-optimal stable matching $\mu_{\mathcal{M}}$, (2) each man's ranking of the women he prefers to his partner in $\mu_{\mathcal{M}}$, and (3) each woman's ranking of the men who prefer her to their partner in $\mu_{\mathcal{M}}$.*

2.3 Analyzing the Number x of Men from Other Blocks

Lemma 4. *Recall the sequence $(u_k)_{k \geq 1}$ defined in the statement of Theorem 1:*

$$\forall k \geq 1, \quad u_k = \max_{w,i} \left\{ \frac{\mathbb{P}[m_{i+k} \succ_w m_i]}{\mathbb{P}[m_i \succ_w m_{i+k}]} \,\middle|\, w \text{ finds both } m_i \text{ and } m_{i+k} \text{ acceptable} \right\}$$

Let w be a woman. Given a subset of her acceptable men and a ranking of that subset $a_1 \succ_w \cdots \succ_w a_p$, we condition on the event that in w's preference list, $a_1 \succ_w \cdots \succ_w a_p$ holds. Let $m_i = a_1$ be w's favorite man in that subset. Let J_i be a random variable, equal to the highest $j \geq i$ such that woman w prefers m_j to m_i. Formally, $J_i = \max\{j \geq i \mid m_j \succeq_w m_i\}$. Then, for all $k \geq 1$, we have

$$\mathbb{P}[J_i < i + k \mid J_i < i + k + 1] \geq \exp(-u_k), \quad \text{and} \quad \mathbb{P}[J_i < i + k] \geq \exp(-\textstyle\sum_{\ell \geq k} u_\ell).$$

Proof. J_i is determined by w's preference list. We construct w's preference list using the following algorithm: initially we know her ranking σ_A of the subset $A = \{a_1, a_2, \ldots, a_p\}$ of acceptable men, and $m_i = a_1$ is her favorite among those. For each j from N to i in decreasing order, we insert m_j into the ranking according to the distribution of w's preference list, stopping as soon as some m_j is ranked before m_i (or when $j = i$ is that does not happen). Then the step $j \geq i$ at which this algorithm stops equals J_i.

To analyze the algorithm, observe that at each step $j = N, N-1, \ldots$, we already know w's ranking of the subset $S = \{m_{j+1}, \ldots, m_N\} \cup \{a_1, \ldots, a_p\} \cup$ {men who are not acceptable to w}. If m_j is already in S, w prefers m_i to m_j, thus the algorithm continues and $J_i < j$. Otherwise the algorithm inserts m_j into the existing ranking: by definition of regular distributions (Definition 1), the probability that m_j beats m_i given the ranking constructed so far is at most the unconditional probability $\mathbb{P}[m_j \succ_w m_i]$.

$$\mathbb{P}[J_i < j \mid w\text{'s partial ranking at step } j] \geq 1 - \mathbb{P}[m_j \succ_w m_i].$$

By definition of u_{j-i}, we have $1 - \mathbb{P}[m_j \succ_w m_i] = \left(1 + \frac{\mathbb{P}[m_j \succ_w m_i]}{\mathbb{P}[m_i \succ_w m_j]}\right)^{-1} \geq (1 + u_{j-i})^{-1} \geq \exp(-u_{j-i})$.

Summing over all rankings σ_S of S that are compatible with σ_A and with $J_i \leq j$,

$$\mathbb{P}[J_i < j \mid J_i \leq j] = \sum_{\substack{\sigma_S \text{ compatible with} \\ J_i \leq j \text{ and with } \sigma_A}} \mathbb{P}[\sigma_S \mid \sigma_A] \cdot \mathbb{P}[J_i < j \mid \sigma_S]$$

$$\geq \sum_{\sigma_S} \mathbb{P}[\sigma_S \mid \sigma_A] \cdot \exp(-u_{j-i}) = \exp(-u_{j-i}).$$

Finally, $\mathbb{P}[J_i < j] = \prod_{\ell=j}^{N} \mathbb{P}[J_i < \ell \mid J_i \leq \ell] \geq \prod_{k \geq j-i} \exp(-u_k)$.

Recall from Lemma 3 that $r - l - 1 + x$ is an upper bound on the difference of rank of woman w_n's worst and best stable husbands. We first bound the expected value of the random variable x defined in Lemma 3.

Lemma 5. *Given a woman w_n, define the random variable x as in Lemma 3: conditioning on $\mathcal{H}$, $x = |\{i \leq l \mid \exists j > l, \ m_j \succ_{w_n} m_i \succ_{w_n} m_n\}|$ is the number of men in a better block, who can be ranked between w_n's worst and best stable husbands. Then $\mathbb{E}[x] \leq \sum_{k \geq 1} k u_k$.*

Proof. Start by conditioning on $\mathcal{H}$, and let $m_n = a_1 \succ_w a_2 \succ_w \cdots \succ_w a_p$ be w_n's ranking of men who prefer her to their partner in $\mu_{\mathcal{M}}$. We draw the preference lists of each woman w_i with $i < n$, and use Algorithm 2 to compute the value of l.

For each $i \leq l$, we proceed as follows. If $m_n \succ_{w_n} m_i$, then m_i cannot be ranked between w_n's worst and best stable partners. Otherwise, we are in a situation where $m_i \succ_{w_n} a_1 \succ w_n \cdots \succ_{w_n} a_p$. Using notations from Lemma 4, w prefers m_i to all m_j with $j > l$ if and only if $J_i < l + 1$. By Lemma 4 this occurs with probability at least $\exp(-\sum_{k \geq l+1-i} u_k)$. Thus

$$\mathbb{P}[\exists j > l, \ m_j \succ_{w_n} m_i \succ_{w_n} m_n \mid \mathcal{H}, l] \leq 1 - \exp\left(-\sum_{k \geq l+1-i} u_k\right) \leq \sum_{k \geq l+1-i} u_k.$$

Summing this probability for all $i \leq l$, we obtain $\mathbb{E}[x \mid \mathcal{H}, l] \leq \sum_{i \leq l} \sum_{k \geq l+1-i} u_k \leq \sum_{k \geq 1} k u_k$.

2.4   Analyzing the Block Size

Lemma 6. *Consider w_n who is matched by μ_M. Then Algorithm 2 outputs the block containing w_n.*

Algorithm 2. Computing a block

Initialization:
>   Compute the man optimal stable matching $\mu_{\mathcal{M}}$.
>   Relabel women so that w_i denotes the wife of m_i in $\mu_{\mathcal{M}}$
>   Pick a woman w_n who is married in $\mu_{\mathcal{M}}$.

Left prefix separator: initialize $l \leftarrow n - 1$
>   **while** there exists $i \leq l$ and $j > l$ such that $m_j \succ_{w_i} m_i$:
> $l \leftarrow \min\{i \leq l \mid \exists j > l, \ m_j \succ_{w_i} m_i\} - 1$.

Right prefix separator: initialize $r \leftarrow n$.
>   **while** there exists $j > r$ and $i \leq r$ such that $m_j \succ_{w_i} m_i$:
> $r \leftarrow \max\{j > r \mid \exists i \leq r, \ m_j \succ_{w_i} m_i\}$.

Output: $(l, r]$.

Proof. Algorithm 2 is understood most easily by following its execution on Fig. 2. Algorithm 2 applies a right-to-left greedy method to find the largest prefix separator l which is $\leq n - 1$. By definition of prefix separators, a witness that some t is not a prefix separator is a pair (m_j, w_i) where $j > t \geq i$ and woman w_i prefers man m_j to her partner: $m_j >_{w_i} m_i$. Then the same pair also certifies that no $t' = t, t-1, t-2, \ldots, i$ can be a prefix separator either, so the algorithm jumps to $i - 1$ and looks for a witness again. When there is no witness, a prefix separator has been found, thus l is the largest prefix separator $\leq n - 1$. Similarly, Algorithm 2 computes the smallest prefix separator r which is $\geq n$. Thus, by definition of blocks, $(l, r]$ is the block containing w_n.

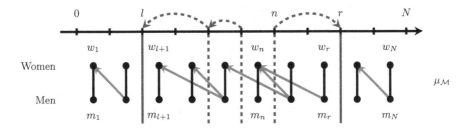

Fig. 2. Computing the block containing woman w_n. The vertical black edges correspond to the men-optimal stable matching $\mu_{\mathcal{M}}$. There is a light gray arc (m_j, w_i) if $j > i$ and woman w_i prefers man m_j to her partner: $m_j \succ_{w_i} m_i$. The prefix separators correspond to the solid red vertical lines which do not intersect any gray arc. Algorithm 2 applies a right-to-left greedy method to find the largest prefix separator l which is $\leq n - 1$, jumping from dashed red line to dashed red line, and a similar left-to-right greedy method again to find the smallest prefix separator r which is $\geq n$. This determines the block $(l, r]$ containing n. (Color figure online)

Definition 6. *Let X be the random variable defined as follows. Let $(\Delta_t)_{t \geq 0}$ denote a sequence of i.i.d.r.v.'s taking non-negative integer values with the following distribution:*

$$\forall \delta > 0, \quad \mathbb{P}[\Delta_t < \delta] = \exp\left(-\sum_{k \geq \delta} k u_k\right)$$

Then $X = \Delta_0 + \Delta_1 + \cdots + \Delta_{T-1}$, where T is the first $t \geq 0$ such that $\Delta_t = 0$.

The proofs of the following Lemmas can be found in [13].

Lemma 7. *Given a woman w_n, let $(l, r]$ denote the block containing n. Conditioning on $\mathcal{H}$, l and r are integer random variable, such that $r - n$ and $n - 1 - l$ are stochastically dominated by X.*

Lemma 8. *We have $\mathbb{E}[X] \leq \exp(\sum_{k \geq 1} k u_k) \sum_{k \geq 1} k^2 u_k$.*

2.5 Putting Everything Together

Proof (Proof of Theorem 1). Without loss of generality, we may assume that $N = M \leq W$ and that each man is matched in the man-optimal stable matching $\mu_{\mathcal{M}}$: to see that, for each man m we add a "virtual" woman w as his least favorite acceptable partner, such that m is the only acceptable partner of w. A man is single in the original instance if and only if he is matched to a "virtual" woman in the new instance.

We start our analysis by conditioning on the random variable $\mathcal{H}$ (see Definition 5). Algorithm 1 then computes μ_M, which matches each woman to her worst stable partner. Up to *relabeling* the women, we may also assume that for all $i \leq N$ we have $w_i := \mu_{\mathcal{M}}(m_i)$.

Let w_n be a woman who is married in $\mu_{\mathcal{M}}$. From there, we use Lemma 3 to bound the difference of rank between her worst and best stable partner by $x + r - l - 1 = x + (r - n) + (n - l - 1)$. We bound the expected value of x using Lemma 5, and the expected values of both $r - n$ and $n - l - 1$ using Lemmas 7 and 8.

3   Unique Stable Partner: Proof of Theorem 2

Theorem 2. *Assume that each woman independently draws her preference list from a regular distribution. Let u_k be an upper bound on the odds that man m_{i+k} is ranked before man m_i:*

$$\forall k \geq 1, \quad u_k = \max_{w,i} \left\{ \frac{\mathbb{P}[m_{i+k} \succ_w m_i]}{\mathbb{P}[m_i \succ_w m_{i+k}]} \; \middle| \; w \text{ finds both } m_i \text{ and } m_{i+k} \text{ acceptable} \right\}$$

Further assume that all preferences are complete, that $u_k = \exp(-\Omega(k))$, and that men have uniformly random preferences. Then, in expectation the fraction of persons who have multiple stable partners converges to 0.

The proof first continues the analysis of blocks started in Sect. 2.4. When $u_k = \exp(-\Omega(k))$, it can be tightened with a mathematical analysis to prove (Corollary 1) that with high probability, no block size exceeds $O(\log n)$, and that in addition, in her preference list no woman switches the relative ordering of two men m_i and $m_{i+\Omega(\log n)}$. The rest of the proof assumes that those properties hold. The only remaining source of randomness comes from the preference lists of men.

The intuition is that it is hard for man m_i to have another stable partner from his block. Because of the random uniform assumption on m_i's preference list, between w_i and the next person from his block, his list is likely to have some woman w_j with $j \gg i$. Woman w_j likes m_i better than her own partner, because of the no-switching property, and m_i likes her better than his putative second stable partner, so they form a blocking pair preventing m_i's second stable partner. Transforming that intuition into a proof requires care because of the need to condition on several events.

Definition 7. *Let $C = O(1)$ be a constant to be defined later. Let $\mathcal{K}$ denote the event that every block has size at most $C \ln N$, and every woman prefers man m_i to man m_{i+k}, whenever $k \geq C \ln N$.*

The proofs of the following Lemmas can be found in [13].

Lemma 9. *Assume that women have preferences drawn from regular distributions such that $u_k = \exp(-\Omega(k))$. Then, the size of each man's block is a random variable with an exponential tail:*

$$\forall i, \quad \mathbb{P}[block\ containing\ m_i\ has\ size\ \geq k] = \exp(-\Omega(k)).$$

Corollary 1. *One can choose $C = \mathcal{O}(1)$ such that the probability of event $\mathcal{K}$ is $\geq 1 - 1/N^2$.*

Lemma 10. *Fix $i \in [1, N]$. Conditioning on $\mathcal{H}$ and on $\mathcal{K}$, the probability that woman w_i has more than one stable husband is at most $3C \ln N/(N + C \ln N - i)$.*

Proof (Proof of Theorem 2). As in the previous proof, in our analysis we condition on event $\mathcal{H}$ (see Definition 5), i.e. on (1) the man-optimal stable matching $\mu_{\mathcal{M}}$, (2) each man's ranking of the women he prefers to his partner in $\mu_{\mathcal{M}}$, and (3) each woman's ranking of the men who prefer her to their partner in $\mu_{\mathcal{M}}$. As before, a person who is not matched in $\mu_{\mathcal{M}}$ remains single in all stable matchings, hence, without loss of generality, we assume that $M = W = N$, and that $w_i = \mu_{\mathcal{M}}(m_i)$ for all $1 \leq i \leq N$.

Let Z denote the number of women with several stable partners. We show that in expectation $Z = \mathcal{O}(\ln^2 N)$, hence the fraction of persons with multiple stable partners converges to 0. We separate the analysis of Z according to whether event $\mathcal{K}$ holds. When $\mathcal{K}$ does not hold, we bound that number by N, so by Corollary 1: $\mathbb{E}[Z] \leq (1/N^2) \times N + (1 - 1/N^2) \times \mathbb{E}(Z|\mathcal{K})$.

Conditioning on $\mathcal{H}$ and switching summations, we write:

$$\mathbb{E}(Z|\mathcal{K}) = \sum_{\mathcal{H}} \mathbb{P}[\mathcal{H}] \cdot \mathbb{E}(Z|\mathcal{K}, \mathcal{H}) = \sum_i \sum_{\mathcal{H}} \mathbb{P}[\mathcal{H}] \cdot \mathbb{P}[w_i \text{ has several stable husbands} \mid \mathcal{K}, \mathcal{H}]$$

By Lemma 10, we can write: $\mathbb{P}[w_i \text{ has several stable husbands} \mid \mathcal{K}, \mathcal{H}] \leq 3C \ln N/(N + C \ln N - i)$. Hence the expected number of women who have several stable partners is at most $1/N$ plus

$$\sum_{i=1}^{N} \frac{3C \ln N}{N + C \ln N - i} = \sum_{i=0}^{N-1} \frac{3C \ln N}{i + C \ln N}$$

$$\leq 3C \ln N \int_{C \log N - 1}^{C \log N - 1 + N} \frac{dt}{t}$$

$$= 3C \ln N \ln \left(\frac{C \log N - 1 + N}{C \log N - 1} \right)$$

When N is large enough, we can simplify this bound to $3C \ln^2 N$.

Acknowledgements. This work was partially funded by the grant ANR-19-CE48-0016 from the French National Research Agency (ANR).

References

1. Abdulkadiroğlu, A., Pathak, P.A., Roth, A.E.: The New York city high school match. Am. Econ. Rev. **95**(2), 364–367 (2005)
2. Abdulkadiroğlu, A., Pathak, P.A., Roth, A.E., Sönmez, T.: The Boston Public School match. Am. Econ. Rev. **95**(2), 368–371 (2005)
3. Ashlagi, I., Braverman, M., Thomas, C., Zhao, G.: Tiered random matching markets: rank is proportional to popularity. In: 12th Innovations in Theoretical Computer Science Conference (ITCS). Schloss Dagstuhl-Leibniz-Zentrum für Informatik (2021)
4. Ashlagi, I., Kanoria, Y., Leshno, J.D.: Unbalanced random matching markets: the stark effect of competition. J. Polit. Econ. **125**(1), 69–98 (2017)
5. Azevedo, E.M., Leshno, J.D.: A supply and demand framework for two-sided matching markets. J. Polit. Econ. **124**(5), 1235–1268 (2016)
6. Banerjee, A., Duflo, E., Ghatak, M., Lafortune, J.: Marry for what? Caste and mate selection in modern India. Am. Econ. J. Microecon. **5**(2), 33–72 (2013)
7. Biró, P., Hassidim, A., Romm, A., Shorrer, R.I., Sóvágó, S.: Need versus merit: the large core of college admissions markets. arXiv preprint arXiv:2010.08631 (2020)
8. Correa, J., et al.: School choice in Chile. In: Proceedings of the 2019 ACM Conference on Economics and Computation, pp. 325–343 (2019)
9. Demange, G., Gale, D., Sotomayor, M.: A further note on the stable matching problem. Discret. Appl. Math. **16**(3), 217–222 (1987)
10. Dubins, L.E., Freedman, D.A.: Machiavelli and the Gale-Shapley algorithm. Am. Math. Mon. **88**(7), 485–494 (1981)
11. Gale, D., Shapley, L.S.: College admissions and the stability of marriage. Am. Math. Mon. **69**(1), 9–15 (1962)
12. Gale, D., Sotomayor, M.: Some remarks on the stable matching problem. Discret. Appl. Math. **11**(3), 223–232 (1985)
13. Gimbert, H., Mathieu, C., Mauras, S.: Two-sided matching markets with strongly correlated preferences. arXiv preprint arXiv:1904.03890 (2019)
14. Gusfield, D.: Three fast algorithms for four problems in stable marriage. SIAM J. Comput. **16**(1), 111–128 (1987)
15. Gusfield, D., Irving, R.W.: The Stable Marriage Problem: Structure and Algorithms. MIT Press, Cambridge (1989)
16. Hitsch, G.J., Hortaçsu, A., Ariely, D.: Matching and sorting in online dating. Am. Econ. Rev. **100**(1), 130–63 (2010)
17. Immorlica, N., Mahdian, M.: Incentives in large random two-sided markets. ACM Trans. Econ. Comput. **3**(3), 14 (2015)
18. Kanoria, Y., Min, S., Qian, P.: In which matching markets does the short side enjoy an advantage? In: Proceedings of the 2021 ACM-SIAM Symposium on Discrete Algorithms (SODA), pp. 1374–1386. SIAM (2021)
19. Knuth, D.E., Motwani, R., Pittel, B.: Stable husbands. In: Proceedings of the First Annual ACM-SIAM Symposium on Discrete Algorithms, pp. 397–404 (1990)
20. Kojima, F., Pathak, P.A.: Incentives and stability in large two-sided matching markets. Am. Econ. Rev. **99**(3), 608–27 (2009)
21. Lee, S.: Incentive compatibility of large centralized matching markets. Rev. Econ. Stud. **84**(1), 444–463 (2016)
22. Lennon, C., Pittel, B.: On the likely number of solutions for the stable marriage problem. Comb. Probab. Comput. **18**(3), 371–421 (2009)

23. Pathak, P.A., Sönmez, T.: Leveling the playing field: sincere and sophisticated players in the Boston mechanism. Am. Econ. Rev. **98**(4), 1636–52 (2008)
24. Pittel, B.: The average number of stable matchings. SIAM J. Discret. Math. **2**(4), 530–549 (1989)
25. Pittel, B.: On likely solutions of a stable marriage problem. Ann. Appl. Probab. **2**, 358–401 (1992)
26. Pittel, B., Shepp, L., Veklerov, E.: On the number of fixed pairs in a random instance of the stable marriage problem. SIAM J. Discret. Math. **21**(4), 947–958 (2007)
27. Rheingans-Yoo, R., Street, J.: Large random matching markets with localized preference structures can exhibit large cores. Technical report, Mimeo (2020)
28. Roth, A.E.: The economics of matching: stability and incentives. Math. Oper. Res. **7**(4), 617–628 (1982)
29. Roth, A.E., Peranson, E.: The redesign of the matching market for American physicians: some engineering aspects of economic design. Am. Econ. Rev. **89**(4), 748–780 (1999)

Communicating Finite State Machines and an Extensible Toolchain for Multiparty Session Types

Nobuko Yoshida(✉) ⓘ, Fangyi Zhou ⓘ, and Francisco Ferreira ⓘ

Imperial College London, London, UK
{n.yoshida,fangyi.zhou15,f.ferreira-ruiz}@imperial.ac.uk

Abstract. Multiparty session types (MPST) provide a typing discipline for message passing concurrency, ensuring deadlock freedom for distributed processes. This paper first summarises the relationship between MPST and communicating finite state machines (CFSMs), which offers not only theoretical justifications of MPST but also a guidance to implement MPST in practice. As one of the applications, we present νSCR (NuSCR), an extensible toolchain for MPST-based multiparty protocols. The toolchain can convert multiparty protocols in the SCRIBBLE protocol description language into global types in the MPST theory; global types are projected into local types, and local types are converted to their corresponding CFSMs. The toolchain also generates APIs from CFSMs that implement endpoints in the protocol. Our design allows for language-independent code generation, and opens possibilities to generate APIs in various programming languages. We design our toolchain with modularity and extensibility in mind, so that extensions of core MPST can be easily integrated within our framework. As a case study, we show the implementation of the nested protocol extension in νSCR, to showcase our extensibility.

Keywords: Session Types · Communicating Finite State Machines · Distributed programming · Scribble · Protocols

1 Introduction

In the modern era of distributed and concurrent programming, how to achieve *safety* with minimal effort (i.e. lightweight formal methods) becomes a hot area of research. *Session types* [19] provide a typing discipline for message passing concurrency, by assigning *session types* to communication channels, in terms of a sequence of actions over a channel. Session types, initially only able to describe communications between *two* ends of a channel, are later extended to *multiparty* [20,21], giving rise to the *multiparty session types (MPST)* theory. The MPST typing discipline guarantees that a set of well-typed communicating processes are free from deadlocks or communication mismatches.

© Springer Nature Switzerland AG 2021
E. Bampis and A. Pagourtzis (Eds.): FCT 2021, LNCS 12867, pp. 18–35, 2021.
https://doi.org/10.1007/978-3-030-86593-1_2

1.1   Communicating Finite State Machines and Session Types

Motivation: Why CFSMs? *Communicating Automata* [2], also known as *Communicating Finite State Machines* (CFSMs), are a classical model for protocol specification and verification. Before being used in many industrial contexts, CFSMs have been a pioneer theoretical formalism, in which distributed safety properties could be formalised and studied.

Establishing a formal connection between CFSMs and session types allows the use of CFSMs to build theoretically well-founded tools for MPST. The first work that utilised CFSMs in practice is Demangeon et al. [11], a toolchain for monitoring multiparty communications at runtime for large scientific cyber-infrastructures developed by the Ocean Observatories Initiative [41].

From the theoretical side, the CFSM framework offers canonical justifications for session types to answer open questions which have been asked since [20]. The **1st** question is about *expressiveness*: to which class of CFSMs do session types correspond? The **2nd** question concerns the *semantic correspondence* between session types and CFSMs: how do the safety properties that session types guarantee relate to those of CFSMs? The **3rd** question is about *efficiency*: why do session types provide efficient algorithms for type-checking or verifying distributed programs, while general CFSMs are undecidable? CFSMs can be also seen as generalised endpoint specifications, therefore an excellent target for a common ground for comparing protocol specification languages.

To answer the three questions above, we need to identify a *sound and complete* subset of CFSMs that corresponds to MPST behaviour, which we explain below.

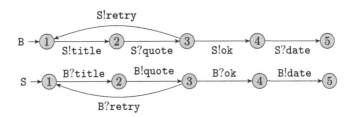

Fig. 1. Two dual communicating automata: the buyer and the seller

Binary Session Types as CFSMs. The subclass that fully characterises binary session types [19] was actually proposed by Gouda, Manning and Yu [16] in a pure automata context (independently from the discovery of session types [45]). Consider a simple business protocol between a Buyer and a Seller. From the Buyer's viewpoint, the Buyer sends the title of a book, then the Seller answers with a quote. If the Buyer is satisfied by the quote, then they send their address and the Seller sends back the delivery date; otherwise they retry the same conversation. This can be specified by the two machines of the Buyer and the Seller in Fig. 1. We can observe that these CFSMs satisfy three conditions: First, the communications are *deterministic*: messages that are part of the same choice, **ok** and `retry` here, are distinct. Secondly, there is no mixed state (each state has either only sending actions, or only receiving actions). Third, these two machines have *compatible* traces (i.e. *dual*): the Seller machine can be defined by exchanging sending and receiving actions of the Buyer machine. Breaking one of these conditions allows deadlock situations and breaking one of the first two conditions makes the compatibility checking undecidable [16].

Essentially, the same characterisation is given in binary session types [19]. Consider the following session type of the Buyer.

$$\mu t. \oplus title. \& quote. \oplus \{ok : \oplus addrs. \& date.\, \textbf{end} \quad retry : \textbf{t}\} \tag{1}$$

The session type above describes the communication pattern using several constructs. The operator $\oplus title$ denotes an output of the title, whereas $\& quote$ denotes an input of a quote. The output choice features the two options *ok* and *retry* and . denotes sequencing. **end** represents the termination of the session, and μt is recursion. The simplicity and tractability of binary sessions come from the notion of *duality* in interactions [15], which corresponds to compatibility of CFSMs. In (1), not only the Buyer's behaviour, but also the whole conversation structure is already represented in this single type: the interaction pattern of the Seller is fully given as the dual of the type in (1) (exchanging input $\oplus$ and output $\&$). When composing two parties, we only have to check they have mutually dual types, and the resulting communication is guaranteed to be deadlock-free.

Multiparty Session Types and CFSMs. The notion of duality is no longer effective in *multiparty* communication, where the whole conversation cannot be reconstructed from only the behaviour of a single machine. Instead of directly trying to decide whether the communication of a system satisfies safety (which is undecidable in the general case), we devise a compatible, decidable condition of a set of machines, which forces them to *collaborate* together. We define a complete characterisation of global type behaviours into CFSMs: a set of CFSMs satisfy some *compatible conditions*, if and only if the CFSMs can mimic the expected behaviour of a given global type. A good global type means the global type can only generate *safe* CFSMs by *endpoint projection*, which satisfies realisability.

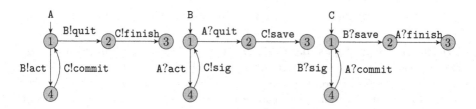

Fig. 2. CFSMs for the commit protocol

We give a simple example to illustrate the proposal. The Commit protocol in Fig. 2 involves three machines: Alice, Bob and Carol. Alice orders Bob to **act** or **quit**. If **act** is sent, Bob sends a **signal** to Carol, and Alice sends a **commitment** to Carol and continues. Otherwise Bob informs Carol to save the data, and Alice gives the final notification to Carol to finish the protocol.

Deniélou and Yoshida [12] present a decidable notion of *multiparty compatibility* as a generalisation of duality of binary sessions, for a given set of (more than two) CFSMs. The idea is that any single machine can see the rest of the machines as a single machine, up to *unobservable actions* (like a τ-transition in CCS). Therefore, we check the *duality* between each automaton and the rest, up to internal communications (*1-bounded executions* in the terminology of CFSMs) that the other machines will independently perform. For example, in Fig. 2, to check the compatibility of trace AB!quit·AC!finish in Alice, we observe the dual trace AB?quit · AC?finish from Bob and Carol, executing the internal communication between Bob and Carol: BC!save · BC?save. If this extended duality is valid for all the machines from any 1-bounded reachable state, then they satisfy multiparty compatibility and can characterise a well-formed global type.

Our motivation to study this general compatibility comes from the need for using global types to develop tools for choreographic distributed testing in web service software [44] and distributed monitoring for large cyber-infrastructures [41], where local specifications are often updated independently and one needs to refine the original global specification according to the local updates.

The 1-boundedness and multiparty compatibility conditions are extended to the k-bounded condition in [28] (called *k-multiparty compatibility*). Another flexible form of safe and more asynchronous CFSMs (which do not rely on duality or buffer bounds) is studied in [7,14] (called *asynchronous subtyping*). Unfortunately, the asynchronous subtyping relation is undecidable, even if limited to only two machines; currently its decidable sound algorithms are restricted to either binary session types [3] or finite MPSTs [6].

Practically, a direct analysis based on CFSMs is computationally expensive, even if the shapes of CFSMs are limited. For building a toolchain for practical programming languages, we take the *safe-by-construction* approaches—we start from specifying a global type, and project it to endpoint types or CFSMs, for code generation into various programming languages, and/or other purposes. Interestingly, multiparty compatibility also helps enlarge the well-formedness condition of global types [23]. See Sect. 5.

1.2 νSCR: An Extensible Toolchain for Multiparty Session Types

We present a new toolchain for multiparty protocols, νSCR (NUSCR), for handling protocols written in the SCRIBBLE language. The implementation of MPST has three main aspects: **(1)** a language for specifying global interactions—specifically, the SCRIBBLE protocol description language [18]; **(2)** a tool to manipulate specifications and generate implementable APIs (SCRIBBLE [44] is already a mature industrial-strength tool able to generate APIs in multiple programming languages); and **(3)** a theory backing the safety guarantees, such as Featherweight SCRIBBLE [35], which minds the gap between the practical SCRIBBLE protocol description language, and the theoretical MPST specifications [21].

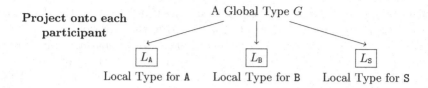

Fig. 3. Top-down methodology

The aim of this implementation is to be *lightweight* and *extensible*. Whilst the SCRIBBLE language describes more expressive protocols than the original MPST theory [21], νSCR handler a *core*, well-defined subset of SCRIBBLE protocols that have a corresponding MPST global type, following the formalisation of [35]. We do so in anticipation that further extensions of the MPST theory can be easily implemented in νSCR, and a researcher can smoothly integrate their own MPST design/theory in νSCR. For this purpose, we use a modular design that does not only enable future extensions, but also makes them easy to implement.

The rest of the paper is structured as follows: Sect. 2 introduces multiparty session types, the theoretical foundation of our tool; Sect. 3 introduces the νSCR toolchain; Sect. 4 presents a case study of extending νSCR with *nested protocols* [10]; and Sect. 5 summarises related work and concludes this paper. νSCR is publicly available at https://github.com/nuscr/nuscr/ under the GPLv3 license.

2   Multiparty Session Types (MPST)

In this section, we introduce the theoretical foundation of our tool—Multiparty Session Types (MPST) [21], a typing discipline for concurrent processes.

The main design philosophy of MPST follows a top-down approach (see Fig. 3): a *global type* describes a global view of a communication protocol between a number of participants. Each participant has their own perspective of the protocol, prescribed by their *local type*, which are obtained via an operation called *projection*. A local type for a participant can be used for code generation or type-checking to ensure that the participating process follows the local type. If all participating processes follow their corresponding local types, obtained via projection from a global type, these processes are free from communication mismatches or deadlocks, guaranteed by the MPST typing discipline.

Global and Local Types. We show the syntax of global types and local types in Fig. 4. The global type $p \rightarrow q \{l_i(S_i).G_i\}_{i \in I}$ is a message from p to q, where $p \neq q$ and $I \neq \emptyset$. The message carries a *label* l_i and payload type S_i, selected from a non-empty index set I, and the protocol continues as G_i. We write $p \rightarrow q : l(S).G$ when $|I| = 1$. **end** denotes a type that is *terminated*. The local type $p\&\{l_i(S_i).L_i\}_{i \in I}$ (resp. $p\oplus\{l_i(S_i).L_i\}_{i \in I}$) denotes an *external choice* (resp. *internal choice*), where participant carrying this local type will *receive* (resp.

$S ::= \texttt{int} \mid \texttt{bool} \mid \ldots$	Base Types	$L ::=$	Local Types
$G ::=$	Global Types	$\mid\ \texttt{p\&}\{l_i(S_i).L_i\}_{i \in I}$	External Choice
$\mid\ \texttt{p} \rightarrow \texttt{q}\,\{l_i(S_i).G_i\}_{i \in I}$	Message	$\mid\ \texttt{p} \oplus \{l_i(S_i).L_i\}_{i \in I}$	Internal Choice
$\mid\ \mu\texttt{t}.G$	Recursion	$\mid\ \mu\texttt{t}.L$	Recursion
$\mid\ \texttt{t} \mid \texttt{end}$	Type Var., End	$\mid\ \texttt{t} \mid \texttt{end}$	Type Var., End

Fig. 4. Syntax of multiparty session types, in the style of [48]

send) a message from (resp. to) the participant $\texttt{p}$, among the index set I. Recursive types are realised by $\mu\texttt{t}.G$ (resp. $\mu\texttt{t}.L$) and $\texttt{t}$, by taking a equi-recursive view (However, we require types to be contractive, e.g. $\mu\texttt{t}.\texttt{t}$ is not allowed).

We can obtain local types by *projecting* a global type upon a participant. Projection is defined as a *partial* function, since not all global types are implementable—these types might be unable to be implemented in a type-safe way. We say a global type is *well-formed*, if the projection of the global type upon all participants are *defined*. Well-formed global types can be implemented by a collection of concurrent processes, each implementing their projected local type. Well-typed processes will enjoy the benefit of the MPST typing discipline, are free from deadlocks or communication mismatches. Curious readers may refer to [48] for more details.

From Local Types to Communicating Finite State Machines. A local type describes the behaviour of a specific role in a given global type, which can be represented by a *communicating finite state machine*[1] *(CFSM)* [2]. As shown by Deniélou and Yoshida [12] and Neykova and Yoshida [35], there is an algorithm to construct a CFSM that is trace-equivalent to the local type.

***Relation to* Scribble.** SCRIBBLE [44,49] is a toolchain for implementing multiparty protocols. In particular, the syntax of the SCRIBBLE protocol description language correlates closely to the theory of MPST. Neykova and Yoshida [35] give a formal description of the SCRIBBLE protocol description language, known as Featherweight SCRIBBLE, and establish a correspondence between global protocols in Featherweight SCRIBBLE and global types in the MPST theory (Sect. 4).

3 νSCR: An Extensible Implementation of Multiparty Session Types in OCAML

In this section, we describe the structure of νSCR and highlight the correspondence to the multiparty session type theory. νSCR is written in OCAML in around 8000 lines of code, implementing the core part of the SCRIBBLE language, with various extensions to the original MPST. νSCR also has a web interface (https://nuscr.dev/), so that users can perform quick prototyping in browsers, saving the need for installation (see Fig. 5 for a screenshot).

---

[1] Also known as *endpoint finite state machine (EFSM)* [22].

vScr live

Global protocol

```
global protocol Adder(role C, role S)
{ choice at C
 { add(int) from C to S;
   add(int) from C to S;
   sum(int) from S to C;
   do Adder(C, S); }
 or
 { bye() from C to S;
   bye() from S to C; } }
```

examples/annot/Adder.scr   ˅   Analyse

Local types

- C@Adder [Project] [FSM]
- S@Adder [Project] [FSM]

Projected on to C@Adder :
```
rec __Adder_C_S {
  choice at C [
    add(int) to S;
    add(int) to S;
    sum(int) from S;
    continue __Adder_C_S;
  } or {
    bye() to S;
    bye() from S;
    end
  }
}
```

Fig. 5. A screenshot of the νSCR web interface, showing an `Adder` protocol

Overview. νSCR is designed to be extensible, so that researchers working on MPST theories can find it easy to implement their extensions upon the code base of νSCR. Inspired by HASKELL, we use *language pragmas* to control language extensions, so that users do not need to download different versions of the software for different language extensions. Currently, two major extensions are implemented, namely nested protocols [10,13] and refinement types [51]. Protocols in the SCRIBBLE description language are accepted by νSCR, and then converted into an MPST global type.

From a global type, νSCR is able to project upon a specified participant to obtain their local type, and subsequently obtain the corresponding CFSM. Moreover, νSCR is able to generate code for implementing the participant in various programming languages, from their local type or CFSM. νSCR can be used either as a standalone command line application, or as an OCAML library for manipulating multiparty protocols.

Code Layout. The codebase of νSCR can be briefly split into 4 components: `syntax`, `mpst`, `codegen` and `utils`. We introduce the components in detail.

Syntax. The `syntax` component handles the syntax of the SCRIBBLE protocol description language, the core part of which is shown in Fig. 6. We use OCAM-LLEX and MENHIR to generate the lexer and parser respectively. A SCRIBBLE

Protocol Declarations $P ::=$ **global protocol** p (**role** $r_1, \cdots,$ **role** r_n)$\{G\}$

Protocol Constructs $G ::=$	$1(S)$ **from** r_1 **to** $r_2; G'$	Single Message
	$\mid$ **choice at** r $\{G_1\}$ **or** $\cdots$ **or** $\{G_n\}$	Branches
	$\mid$ **rec** X $\{G'\}$ $\mid$ **continue** X	Recursion / Var.
	$\mid$ **end** (omitted in practice)	Termination
	$\mid$ **do** $p(r_1, \cdots, r_n)$	Protocol Call
Base Types $S ::=$	**int** $\mid$ **bool** $\mid \cdots$	

Fig. 6. Syntax of core SCRIBBLE language

```
1  global protocol Adder(role C, role S)
2  { choice at C
3    { add(int) from C to S;
4      add(int) from C to S;
5      sum(int) from S to C;
6      do Adder(C, S); }
7    or
8    { bye() from C to S;
9      bye() from S to C; } }
```

$$G_{\mathtt{Adder}} = \mu t.\mathsf{C} \rightarrow \mathsf{S} \left\{ \begin{array}{l} add(\texttt{int}). \\ \mathsf{C} \rightarrow \mathsf{S} : add(\texttt{int}). \\ \mathsf{S} \rightarrow \mathsf{C} : sum(\texttt{int}). \\ t; \\ bye(). \\ \mathsf{S} \rightarrow \mathsf{C} : bye(). \\ \text{end} \end{array} \right\}$$

(a) Adder Protocol in SCRIBBLE

(b) Global Type of Adder Protocol

Fig. 7. Adder protocol and its corresponding global type

module consists of multiple protocol declarations P. In the syntax, protocol names are represented by p, role names by r, label names by 1, and recursion variable names by X. The four kinds of names range over string identifiers. They are separated in distinct name spaces in νSCR, and appropriately distinguished.

As a running example, we show a simple SCRIBBLE protocol describing an Adder protocol in Fig. 7a, where a Client is able to make various requests to **add** two **int**s, before they decide to finish the protocol with a **bye** message.

Multiparty Session Types. We show the key pipeline of handling multiparty session types in Fig. 8, implemented in the **mpst** component. An input file is parsed into a SCRIBBLE *module*, by the **syntaxtree** component described in the previous paragraph. The protocols are then converted into a *global type* (defined in Gtype module), which describes an overall protocol between multiple roles. A global type is *projected* into a *local type* (defined in Ltype module), given a specified role, which describes the local communication behaviour. We construct a corresponding *communicating finite state machine (CFSM)* [2] (defined in Efsm module) for the local type, and it can be used for API generation.

To obtain a global type, we extract it from the syntax tree of the SCRIBBLE protocol file. During this extraction process, we perform syntactic checks on the protocol, e.g. validating whether role names, recursion variables, and protocol names have been defined before they are used. We show the global type of the Adder protocol in Fig. 7b.

Fig. 8. Workflow of νSCR

```
1   global protocol NonDirected (role A, role B, role C)
2   { choice at A // A sends to either B or to C in this choice
3       { Foo() from A to B; // either send to B
4         Bar() from A to C; }
5     or { Bar() from A to C; // or send to C
6         Foo() from A to B; } }
```

Fig. 9. Non-directed choice in SCRIBBLE

It is important to note that, syntactically correct protocols may fall out of the expressiveness of the original MPST theory, e.g. the protocol shown in Fig. 9. The role A makes a choice of sending Foo to B first, or sending Bar to C first, which has no corresponding construct in the syntax (Fig. 4). Whilst some protocols fall out of the scope of the core MPST theory, an extension to the core theory may accept such protocols with non-directed choices.

The projection from global types upon participants is implemented in the Ltype module, and the projected local types can be converted into their corresponding communicating automata, using the technique described in [12]. We use the graph library OCAMLGRAPH [8], to represent the CFSM as a directed graph. We show the local type for Client in Fig. 10a, and its corresponding CFSM in Fig. 10b. Both local types and communicating automata can be used for code generation purposes, which will be introduced in the next paragraph.

Code Generation. The codegen component generates APIs for implementing distributed processes using the MPST theory. Following the MPST design methodology, processes should follow the projected local type from the prescribed global type. By the means of code generation, the processes implemented using generated APIs will be *correct by construction.*

Currently, νSCR supports code generation in OCAML, GO (with the nested protocol extension [13]) and F⋆ (with the refinement type extension [51]). Moreover, νSCR can export the CFSM as a GRAPHVIZ DOT file, and code generation backends can be implemented separately from νSCR. This approach has been used to support code generation in RUST [9], SCALA and TYPESCRIPT [31].

To generate code in OCAML, we use a CFSM-based generation technique, as proposed in [22]; however, we do not follow the class-based APIs, i.e. states in the CFSMs are classes, and state transitions are methods on classes in [22]. While it would be possible to implement a similar object oriented approach in OCAML, it does not fit well in the functional programming paradigm. νSCR uses a *callback-based* approach [51] for API generation, and generates functions for transitions and maintains the state *internally* in a finite state machine runner.

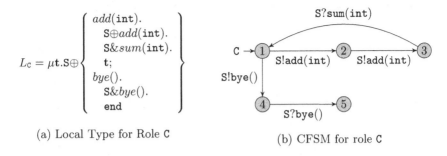

(a) Local Type for Role C

(b) CFSM for role C

Fig. 10. Local type and CFSM for role C in the Adder protocol

```
1   module type Callbacks = sig
2     type t (* An abstract type for user-maintained state *)
3
4     (* Sending callbacks return the state and a labelled value to send *)
5     val state1Send : t → t * [`bye of unit | `add of int]
6     val state2Send : t → t * [`add of int]
7
8     (* Receiving callbacks take received value as arguments,
9      * and return the state *)
10    val state3Receivesum : t → int → t
11    val state4Receivebye : t → unit → t
12  end
```

Fig. 11. Generated module type in OCAML for role C

API Style. The generated API separates the program logic and communication aspects of the endpoint program, in contrast to existing approaches of code generation [22]. We generate type signatures of *callback functions*, corresponding to state transitions in the CFSM, for handling the program logic. The signatures are collected in the form of a module type, named `Callbacks`. We show the generated module signature in Fig. 11, for implement the Client in the Adder protocol (Fig. 10). Since we use a graph representation for CFSMs, the generation process is done by iterating through the edges of the graph.

For a complete endpoint, We generate an OCAML functor taking a module of type `Callbacks` to an implementation module. The module exposes a runner, which executes the CFSM when provided connections to other communicating roles. The runner handles the communication with other roles, so that the callback module does not need to involve any sending and receiving primitives.

Optional Monadic APIs. To enable asynchronous execution, we optionally generate code compatible with monadic communication primitives. This allows users to implement the endpoint program with popular asynchronous execution libraries in OCAML, such as LWT [40].

```
1  (*# NestedProtocols #*)          1  global protocol ForkJoin
2  nested protocol Fork             2    (role M, role W)
3    (role M; new role W)           3  { choice at M
4  { choice at M                    4    { Task() from M to W;
5    { Task() from M to W;          5      M calls Fork(M);
6      M calls Fork(M);             6      Result() from W to M; }
7      Result() from W to M; }      7    or
8    or                             8    { 3lugleTask() ſɩoɯ M ʟʋ W,
9    { End() from M to W; } }       9      Result() from W to M; } }
```

Fig. 12. A nested fork join protocol in SCRIBBLE [13, Fig. 7.3]

Utilities. The `utils` component contain miscellaneous modules fulfilling various utility functions. A few notable modules in this component, relevant to future extensions of νSCR, include:

- `Names` module defines separated namespaces for all kinds of names occurring in global and local types, e.g. payload type names, payload label names, recursion variable names, etc.
- `Err` module defines all kinds of errors that occur throughout all components.
- `Pragma` module defines language pragmas, controlling the enabled extensions.

4   Extending νSCR

The modular design of νSCR allows extensions of the MPST theory to be implemented easily. The language pragmas, implemented as a special comment at the beginning of an input file, control which extensions are enabled when handling the protocols. So far, two major extensions have been added to νSCR: *nested protocols* implemented by Echarren Serrano [13], and *refinement types* implemented by Zhou et al. [51]; and additional extensions are being implemented: *choreography automata* [1] by Neil Sayers, *parallel types* by Francisco Ferreira, and *crash handling* by Adam D. Barwell.

We use the *nested protocol* extension by Echarren Serrano [13] as a case study to demonstrate how an extension can be implemented in νSCR. Nested protocols [10] allow dynamic creation of participants and sub-sessions in a protocol, extending the expressiveness of global types. In Fig. 12, we show a fork join protocol, described in SCRIBBLE with the nested protocol extension.

Creating a New Pragma. The first line of Fig. 12 enables the nested protocols extension using the *pragma* `NestedProtocols` (wrapped in (*# #*)). New pragmas are added in the `Pragma` module in the `utils` component, including a new constructor for the new pragma, and functions to get and set whether the extension is toggled. An implementer may also check for conflicting pragmas when processing all pragmas, so that incompatible extensions are not enabled at the same time. In order to preserve the behaviour when the extension is not enabled, it is essential that subsequent implementations of the extension should query whether the extension is enabled before proceeding.

Extending the Syntax. The extension allows *nested* protocols to be defined using the keyword `nested`, with the possibility to dynamically create new participating roles. In addition, a new construct `calls` is introduced to create a sub-session that follows a nested protocol, where new roles may be created to participate in the sub-session.

To implement these new syntactic constructs, an implementer should extend the `syntaxtree` component. To begin with, the concrete syntax tree (in the `Syntax` module) is to be extended with constructors for the new syntax, e.g. the new `calls` constructs in protocol body. Additional lexing or parsing rules should be added accordingly in the corresponding module.

Extending the MPST Theory. The crucial part is to implement the theory extension in the `mpst` component, where global and local types are defined. Within the component, global (resp. local) types are defined using the OCaml type `Gtype.t` (resp. `Ltype.t`). We add new constructors for new global types (`CallG` for protocol calls) and new local types (`InviteCreateL` for inviting and creating dynamic roles, and `AcceptL` for accepting invitations). Projection can be extended accordingly in the `Ltype` module, which we will not explain in detail.

However, extending the global and local types does not complete the extension—the implementer needs to connect the concrete syntax of Scribble global protocols to the abstract syntax of MPST global types. The extraction is defined at `Gtype.of_protocol`, where a global type is obtained from a global protocol. When processing the new syntactic constructs added by the extension, the implementer should remember to call `Pragma.nested_protocol_enabled` (which will return `true` when the pragma is set) to avoid interference with the core MPST, i.e. when the extension is not enabled.

Extending the Code Generation. Section 3 describes OCaml code generation from CFSMs. However, constructing CFSMs for nested protocols is an open problem. Hence, for this extension the Go code generation is instead based on *local types*. νScr code generation backends in the `codegen` component is free to choose any representation in the `mpst` component, so implementers may pick whichever representation that suits best their code generation approach.

The generated Go APIs also use callbacks, and the message passing primitives in Go with *channels* and concurrent execution with *goroutines* fit the setup of session types very well. The code generator creates type definitions for different channels used in the communication, message exchanges, and callbacks. New participants in the protocol can be spawned when needed using goroutines.

5   Related and Future Work

Non-Scribble-Based MPST Implementations. Scalas and Yoshida [43] (accompanying artefact) implement a toolkit for analysing synchronous multiparty protocols. The underlying theory for this tool is the *generalised* multiparty session types, where the type system is parameterised on a safety property.

The toolkit uses a model checker (mCRL2 [46]) to decide whether the desired safety property holds. A shortcoming of this approach is that (1) the verification power is bound by the model checker—for example, mCRL2 allows to verify only finite-controlled local session types (no parallel compositions under recursion) and cannot verify channel passing; and (2) the approach is not scalable to asynchronous communication with unbounded buffers (as safety becomes undecidable). Several prototype tools that analyse safety of general forms of local types (or CFSMs) are developed in the context of multiparty CFSMs [27,28] and binary CFSMs [3]. The tool in [27] enables a *bottom-up approach*, which builds a global type from a set of safe local types. These approaches are, in general, high in complexity (requiring a global analysis to a set of CFSMs), and difficult to integrate with real programming languages because of the need to extract local types from source languages. For example, Ng and Yoshida [38] develop a tool (based on [27]) to build a global graph from local session types extracted from GO source codes, in order to check deadlock-freedom. Only a subset of GO syntax is supported [50].

Our top-down approach is based on the original, less general multiparty session type theory, yet we implement an extensible toolchain with possibilities to generate OCAML code for execution. Imai et al. [24] implement multiparty session types in OCAML with protocol *combinators*, whereas our approach takes inputs from SCRIBBLE protocols. Their tool uses features such as variant and object types in OCAML to encode external and internal choices in the local types, and supports session delegation. Our callback-based approach does not support delegation, but also does not require sophisticated type system features.

For more advanced applications of MPST, global types with motion primitives of Cyber Physical Systems [29,30] provide a collision freedom guarantee for concurrent robotics applications. Castro-Perez and Yoshida [6] use global types to uniformly predict communication costs of parallel algorithms and distributed protocols implemented in different languages.

Scribble-*Based MPST Implementations and Extensions*. The SCRIBBLE toolchain provides a language-agnostic description language for multiparty protocols, targeting a variety of programming languages: JAVA [22], SCALA [42], GO [4], TYPESCRIPT [31], PURESCRIPT [25], RUST [9,26], F♯ [32], F⋆ [51], ERLANG [33], PYTHON [11,34], MPI-C [36,37], C [39], etc.

The SCRIBBLE toolchain describes multiparty protocols, including some that are not expressible in the MPST theory, e.g. the choice constructs in SCRIBBLE name a role making an internal choice, whereas the MPST global type has form $p \rightarrow q\{l_i(S_i).G_i\}$, where two roles are named. The protocol in Fig. 9 is expressible in SCRIBBLE, although not in the core MPST theory we implement.

The SCRIBBLE toolchain implements a number of extensions of MPST, e.g. explicit connections [23], interruptible protocols [11]. νSCR implements the core MPST theory by our design choice, so that extensions of the MPST theory can be easily implemented, with the usual syntactical projections. Whilst the SCRIBBLE toolchain uses model checking and other validation techniques to verify the safety of the multiparty protocol to enlarge well-formedness, the same technique might not be applicable when extending the original MPST [21]. νSCR keeps an underlying core syntax and its validation faithful to the literature, so that other users can easily integrate their own MPST theory. We demonstrate our extensibility via the case study with nested protocols.

Besides the SCRIBBLE toolchain itself, Voinea et al. [47] provide a tool, STMUNGO, to translate a SCRIBBLE multiparty protocol to a typestate specification in JAVA. The typestate specification can be checked via MUNGO, a static typechecker for typestates in JAVA. Developers can use the generated typestate APIs to implement the multiparty protocol safely. Harvey et al. [17] use the SCRIBBLE toolchain with explicit connections [23] to develop a tool for affine multiparty session types with adaptations.

Future Work. Recently, Castro-Perez et al. [5] propose ZOOID, a domain specific language for certified multiparty communication, embedded in COQ and implemented atop their mechanisation framework of asynchronous MPST. For future work, we would like to produce a certified version of SCRIBBLE—CERTISCR, extending the ZOOID framework, for the core of νSCR. We would like to mechanise the extraction process from SCRIBBLE to global types in [5], and the CFSM construction process from local types in COQ, completing the picture of fully mechanised toolchain from SCRIBBLE protocols to CFSMs. The COQ code can then be extracted into OCAML to produce a formally verified νSCR core.

Acknowledgements. We thank Simon Castellan for the initial collaboration on νSCR project. The work is supported by EPSRC EP/T006544/1, EP/K011715/1, EP/K034413/1, EP/L00058X/1, EP/N027833/1, EP/N028201/1, EP/T006544/1, EP/T014709/1 and EP/V000462/1, and NCSS/EPSRC VeTSS.

References

1. Barbanera, F., Lanese, I., Tuosto, E.: Choreography automata. In: Bliudze, S., Bocchi, L. (eds.) COORDINATION 2020. LNCS, vol. 12134, pp. 86–106. Springer, Cham (2020). https://doi.org/10.1007/978-3-030-50029-0_6
2. Brand, D., Zafiropulo, P.: On communicating finite-state machines. J. ACM **30**(2), 323–342 (1983). https://doi.org/10.1145/322374.322380
3. Bravetti, M., Carbone, M., Lange, J., Yoshida, N., Zavattaro, G.: A sound algorithm for asynchronous session subtyping and its implementation. Log. Methods Comput. Sci. **17**(1), March 2021. https://lmcs.episciences.org/7238
4. Castro, D., Hu, R., Jongmans, S.S., Ng, N., Yoshida, N.: Distributed programming using role-parametric session types in go: statically-typed endpoint APIs for dynamically-instantiated communication structures. Proc. ACM Program. Lang. 3 (POPL), January 2019. https://doi.org/10.1145/3290342
5. Castro-Perez, D., Ferreira, F., Gheri, L., Yoshida, N.: Zooid: a DSL for certified multiparty computation: from mechanised metatheory to certified multiparty processes. In: Proceedings of the 42nd ACM SIGPLAN International Conference on Programming Language Design and Implementation, PLDI 2021, New York, NY, USA, pp. 237–251. Association for Computing Machinery (2021). https://doi.org/10.1145/3453483.3454041

6. Castro-Perez, D., Yoshida, N.: CAMP: cost-aware multiparty session protocols. Proc. ACM Program. Lang. 4 (OOPSLA), November 2020. https://doi.org/10.1145/3428223

7. Chen, T.C., Dezani-Ciancaglini, M., Scalas, A., Yoshida, N.: On the preciseness of subtyping in session types. Log. Methods Comput. Sci. **13**(2), June 2017. https://lmcs.episciences.org/3752

8. Conchon, S., Filliâtre, J.C., Signoles, J.: OCamlgraph: An OCaml Graph Library (2017). http://ocamlgraph.lri.fr/index.en.html. Accessed 21 May 2021

9. Cutner, Z., Yoshida, N.: Safe session-based asynchronous coordination in rust. In: Damiani, F., Dardha, O. (eds.) COORDINATION 2021. LNCS, vol. 12717, pp. 80–89. Springer, Cham (2021). https://doi.org/10.1007/978-3-030-78142-2_5

10. Demangeon, R., Honda, K.: Nested protocols in session types. In: Koutny, M., Ulidowski, I. (eds.) CONCUR 2012. LNCS, vol. 7454, pp. 272–286. Springer, Heidelberg (2012). https://doi.org/10.1007/978-3-642-32940-1_20

11. Demangeon, R., Honda, K., Hu, R., Neykova, R., Yoshida, N.: Practical interruptible conversations: distributed dynamic verification with multiparty session types and Python. Formal Methods Syst. Des. **46**(3), 197–225 (2014). https://doi.org/10.1007/s10703-014-0218-8

12. Deniélou, P.-M., Yoshida, N.: Multiparty compatibility in communicating automata: characterisation and synthesis of global session types. In: Fomin, F.V., Freivalds, R., Kwiatkowska, M., Peleg, D. (eds.) ICALP 2013. LNCS, vol. 7966, pp. 174–186. Springer, Heidelberg (2013). https://doi.org/10.1007/978-3-642-39212-2_18

13. Echarren Serrano, B.: Nested multiparty session programming in go. Master's thesis, Imperial College London (2020). https://becharrens.files.wordpress.com/2020/07/final_report.pdf

14. Ghilezan, S., Pantović, J., Prokić, I., Scalas, A., Yoshida, N.: Precise subtyping for asynchronous multiparty sessions. Proc. ACM Program. Lang. 5 (POPL), January 2021. https://doi.org/10.1145/3434297

15. Girard, J.Y.: Linear logic. Theor. Comput. Sci. **50**(1), 1–101 (1987). https://www.sciencedirect.com/science/article/pii/0304397587900454

16. Gouda, M.G., Manning, E.G., Yu, Y.T.: On the progress of communication between two machines. In: Maekawa, M., Belady, L.A. (eds.) IBM 1980. LNCS, vol. 143, pp. 369–389. Springer, Heidelberg (1982). https://doi.org/10.1007/3-540-11604-4_62

17. Harvey, P., Fowler, S., Dardha, O., Gay, S.J.: Multiparty session types for safe runtime adaptation in an actor language. In: Møller, A., Sridharan, M. (eds.) 35th European Conference on Object-Oriented Programming (ECOOP 2021). Leibniz International Proceedings in Informatics (LIPIcs), vol. 194, pp. 10:1–10:30. Schloss Dagstuhl - Leibniz-Zentrum für Informatik, Dagstuhl, Germany (2021). https://drops.dagstuhl.de/opus/volltexte/2021/14053

18. Honda, K., Mukhamedov, A., Brown, G., Chen, T.-C., Yoshida, N.: Scribbling interactions with a formal foundation. In: Natarajan, R., Ojo, A. (eds.) ICDCIT 2011. LNCS, vol. 6536, pp. 55–75. Springer, Heidelberg (2011). https://doi.org/10.1007/978-3-642-19056-8_4

19. Honda, K., Vasconcelos, V.T., Kubo, M.: Language primitives and type discipline for structured communication-based programming. In: Hankin, C. (ed.) ESOP 1998. LNCS, vol. 1381, pp. 122–138. Springer, Heidelberg (1998). https://doi.org/10.1007/BFb0053567

20. Honda, K., Yoshida, N., Carbone, M.: Multiparty asynchronous session types. In: Proceedings of the 35th Annual ACM SIGPLAN-SIGACT Symposium on Principles of Programming Languages, POPL 2008, New York, NY, USA, pp. 273–284. ACM (2008). http://doi.acm.org/10.1145/1328438.1328472

21. Honda, K., Yoshida, N., Carbone, M.: Multiparty asynchronous session types. J. ACM **63**, 1–67 (2016)

22. Hu, R., Yoshida, N.: Hybrid session verification through endpoint API generation. In: Stevens, P., Wasowski, A. (eds.) FASE 2016. LNCS, vol. 9633, pp. 401–418. Springer, Heidelberg (2016). https://doi.org/10.1007/978-3-662-49665-7_24

23. Hu, R., Yoshida, N.: Explicit connection actions in multiparty session types. In: Huisman, M., Rubin, J. (eds.) FASE 2017. LNCS, vol. 10202, pp. 116–133. Springer, Heidelberg (2017). https://doi.org/10.1007/978-3-662-54494-5_7

24. Imai, K., Neykova, R., Yoshida, N., Yuen, S.: Multiparty session programming with global protocol combinators. In: Hirschfeld, R., Pape, T. (eds.) 34th European Conference on Object-Oriented Programming (ECOOP 2020), pp. 9:1–9:30. Leibniz International Proceedings in Informatics (LIPIcs), Schloss Dagstuhl-Leibniz-Zentrum für Informatik, Dagstuhl, Germany (2020). https://drops.dagstuhl.de/opus/volltexte/2020/13166

25. King, J., Ng, N., Yoshida, N.: Multiparty session type-safe web development with static linearity. In: Martins, F., Orchard, D. (eds.) Proceedings Programming Language Approaches to Concurrency- and Communication-cEntric Software, Prague, Czech Republic, 7th April 2019. Electronic Proceedings in Theoretical Computer Science, vol. 291, pp. 35–46. Open Publishing Association (2019)

26. Lagaillardie, N., Neykova, R., Yoshida, N.: Implementing multiparty session types in rust. In: Bliudze, S., Bocchi, L. (eds.) COORDINATION 2020. LNCS, vol. 12134, pp. 127–136. Springer, Cham (2020). https://doi.org/10.1007/978-3-030-50029-0_8

27. Lange, J., Tuosto, E., Yoshida, N.: From communicating machines to graphical choreographies. In: Proceedings of the 42nd Annual ACM SIGPLAN-SIGACT Symposium on Principles of Programming Languages, POPL 2015, New York, NY, USA, pp. 221–232. Association for Computing Machinery (2015). https://doi.org/10.1145/2676726.2676964

28. Lange, J., Yoshida, N.: Verifying asynchronous interactions via communicating session automata. In: Dillig, I., Tasiran, S. (eds.) CAV 2019. LNCS, vol. 11561, pp. 97–117. Springer, Cham (2019). https://doi.org/10.1007/978-3-030-25540-4_6

29. Majumdar, R., Pirron, M., Yoshida, N., Zufferey, D.: Motion session types for robotic interactions (brave new idea paper). In: Donaldson, A.F. (ed.) 33rd European Conference on Object-Oriented Programming (ECOOP 2019). Leibniz International Proceedings in Informatics (LIPIcs), vol. 134, pp. 28:1–28:27. Schloss Dagstuhl-Leibniz-Zentrum fuer Informatik, Dagstuhl, Germany (2019). http://drops.dagstuhl.de/opus/volltexte/2019/10820

30. Majumdar, R., Yoshida, N., Zufferey, D.: Multiparty motion coordination: from choreographies to robotics programs. Proc. ACM Program. Lang. 4 (OOPSLA), November 2020. https://doi.org/10.1145/3428202

31. Miu, A., Ferreira, F., Yoshida, N., Zhou, F.: Communication-safe web programming in TypeScript with routed multiparty session types. In: Proceedings of the 30th ACM SIGPLAN International Conference on Compiler Construction, CC 2021, New York, NY, USA, pp. 94–106. Association for Computing Machinery (2021). https://doi.org/10.1145/3446804.3446854

32. Neykova, R., Hu, R., Yoshida, N., Abdeljallal, F.: A session type provider: compile-time API generation of distributed protocols with refinements in F#. In: Proceedings of the 27th International Conference on Compiler Construction, CC 2018, New York, NY, USA, pp. 128–138. ACM (2018). http://doi.acm.org/10.1145/3178372.3179495

33. Neykova, R., Yoshida, N.: Let it recover: multiparty protocol-induced recovery. In: Proceedings of the 26th International Conference on Compiler Construction, CC 2017, New York, NY, USA, pp. 98–108. Association for Computing Machinery (2017). https://doi.org/10.1145/3033019.3033031

34. Neykova, R., Yoshida, N.: Multiparty session actors. Log. Methods Comput. Sci. **13**(1), March 2017. https://lmcs.episciences.org/3227

35. Neykova, R., Yoshida, N.: Featherweight scribble. In: Boreale, M., Corradini, F., Loreti, M., Pugliese, R. (eds.) Models, Languages, and Tools for Concurrent and Distributed Programming. LNCS, vol. 11665, pp. 236–259. Springer, Cham (2019). https://doi.org/10.1007/978-3-030-21485-2_14

36. Ng, N., de Figueiredo Coutinho, J.G., Yoshida, N.: Protocols by default. In: Franke, B. (ed.) CC 2015. LNCS, vol. 9031, pp. 212–232. Springer, Heidelberg (2015). https://doi.org/10.1007/978-3-662-46663-6_11

37. Ng, N., Yoshida, N.: Pabble: parameterised Scribble. SOCA **9**(3), 269–284 (2014). https://doi.org/10.1007/s11761-014-0172-8

38. Ng, N., Yoshida, N.: Static deadlock detection for concurrent go by global session graph synthesis. In: Proceedings of the 25th International Conference on Compiler Construction, CC 2016, New York, NY, USA, pp. 174–184. Association for Computing Machinery (2016). https://doi.org/10.1145/2892208.2892232

39. Ng, N., Yoshida, N., Honda, K.: Multiparty session C: safe parallel programming with message optimisation. In: Furia, C.A., Nanz, S. (eds.) TOOLS 2012. LNCS, vol. 7304, pp. 202–218. Springer, Heidelberg (2012). https://doi.org/10.1007/978-3-642-30561-0_15

40. Ocsigen: Lwt Manual (2021). https://ocsigen.org/lwt/latest/manual/manual. Accessed 21 May 2021

41. OOI: Ocean Observatories Initiative (2020). http://www.oceanobservatories.org/

42. Scalas, A., Dardha, O., Hu, R., Yoshida, N.: A linear decomposition of multiparty sessions for safe distributed programming. In: Müller, P. (ed.) 31st European Conference on Object-Oriented Programming (ECOOP 2017). Leibniz International Proceedings in Informatics (LIPIcs), vol. 74, pp. 24:1–24:31. Schloss Dagstuhl-Leibniz-Zentrum fuer Informatik, Dagstuhl, Germany (2017). http://drops.dagstuhl.de/opus/volltexte/2017/7263

43. Scalas, A., Yoshida, N.: Less is more: multiparty session types revisited. Proc. ACM Program. Lang. 3 (POPL), January 2019. https://doi.org/10.1145/3290343

44. Scribble Authors: Scribble: Describing Multi Party Protocols (2015). http://www.scribble.org/. Accessed 21 May 2021

45. Takeuchi, K., Honda, K., Kubo, M.: An interaction-based language and its typing system. In: Halatsis, C., Maritsas, D., Philokyprou, G., Theodoridis, S. (eds.) PARLE 1994. LNCS, vol. 817, pp. 398–413. Springer, Heidelberg (1994). https://doi.org/10.1007/3-540-58184-7_118

46. Technische Universiteit Eindhoven: mCRL2 (2018). https://www.mcrl2.org/web/user_manual/index.html

47. Voinea, A.L., Dardha, O., Gay, S.J.: Typechecking Java protocols with [St]Mungo. In: Gotsman, A., Sokolova, A. (eds.) FORTE 2020. LNCS, vol. 12136, pp. 208–224. Springer, Cham (2020). https://doi.org/10.1007/978-3-030-50086-3_12

48. Yoshida, N., Gheri, L.: A very gentle introduction to multiparty session types. In: Hung, D.V., D'Souza, M. (eds.) ICDCIT 2020. LNCS, vol. 11969, pp. 73–93. Springer, Cham (2020). https://doi.org/10.1007/978-3-030-36987-3_5

49. Yoshida, N., Hu, R., Neykova, R., Ng, N.: The Scribble protocol language. In: Abadi, M., Lluch Lafuente, A. (eds.) TGC 2013. LNCS, vol. 8358, pp. 22–41. Springer, Cham (2014). https://doi.org/10.1007/978-3-319-05119-2_3

50. Yuan, T., Li, G., Lu, J., Liu, C., Li, L., Xue, J.: GoBench: a benchmark suite of real-world go concurrency bugs. In: 2021 IEEE/ACM International Symposium on Code Generation and Optimization (CGO), pp. 187–199 (2021)

51. Zhou, F., Ferreira, F., Hu, R., Neykova, R., Yoshida, N.: Statically verified refinements for multiparty protocols. Proc. ACM Program. Lang. 4 (OOPSLA), November 2020. https://doi.org/10.1145/3428216

Contributed Papers

First-Order Logic and Its Infinitary Quantifier Extensions over Countable Words

Bharat Adsul[1], Saptarshi Sarkar[1], and A. V. Sreejith[2(✉)]

[1] IIT Bombay, Powai, Mumbai 400076, Maharashtra, India
{adsul,sapta}@cse.iitb.ac.in
[2] IIT Goa, Farmagudi, Ponda 403401, Goa, India
sreejithav@iitgoa.ac.in

Abstract. We contribute to the refined understanding of the language-logic-algebra interplay in the context of *first-order* properties of countable words. We establish decidable algebraic characterizations of one variable fragment of FO as well as boolean closure of existential fragment of FO via a strengthening of Simon's theorem about piecewise testable languages. We propose a new extension of FO which admits infinitary quantifiers to reason about the inherent infinitary properties of countable words. We provide a very natural and hierarchical block-product based characterization of the new extension. We also explicate its role in view of other natural and classical logical systems such as WMSO and FO[cut] - an extension of FO where quantification over Dedekind-cuts is allowed. We also rule out the possibility of a finite-basis for a block-product based characterization of these logical systems. Finally, we report simple but novel algebraic characterizations of one variable fragments of the hierarchies of the new proposed extension of FO.

Keywords: Countable words · First-order logic · Monoids

1 Introduction

Over finite words, we have a foundational language-logic-algebra connection (see [11,18]) which equates regular-expressions, MSO-logic, and (recognition by) finite monoids/automata. In fact, one can effectively associate, to a regular language, its finite *syntactic monoid*. This canonical algebraic structure carries a rich amount of information about the corresponding language. Its role is highlighted by the classical Schützenberger-McNaughton-Papert theorem (see, for instance, [12]) which shows that *aperiodicity* property of the syntactic monoid coincides with describability using star-free expressions as well as definability in First-Order (FO) logic. So, we arrive at a refined understanding of the language-logic-algebra connection to an important subclass of regular languages: it equates star-free regular expressions, FO-logic, and aperiodic finite monoids.

© Springer Nature Switzerland AG 2021
E. Bampis and A. Pagourtzis (Eds.): FCT 2021, LNCS 12867, pp. 39–52, 2021.
https://doi.org/10.1007/978-3-030-86593-1_3

A variety of algebraic tools have been developed and crucially used to obtain deeper insights. Some of these tools [12,15,17] are: ordered monoids, the so-called Green's relations, wreath/block products and related principles etc. Let us mention Simon's celebrated theorem [14] - which equates piecewise-testable languages, Boolean closure of the existential fragment of FO-logic and J-trivial[1] finite monoids. It is important to note that this is an effective characterization, that is, it provides a decidable characterization of the logical fragment. There have been several results of this kind (see the survey [7]). Another particularly interesting set of results is in the spirit of the fundamental Krohn-Rhodes theorem. These results establish a block-product based decompositional characterization of a logical fragment and have many important applications [15]. The prominent examples are a characterization of FO-logic (resp. FO^2, the two-variable fragment) in terms of strongly (resp. weakly) iterated block-products of copics of the unique 2-element aperiodic monoid.

One of the motivations for this work is to establish similar results in the theory of regular languages of countable words. We use the overarching algebraic framework developed in the seminal work [5] to reason about languages of countable words. This framework extends the language-logic-algebra interplay to the setting of countable words. It develops fundamental algebraic structures such as *finite* ⊛-monoids and ⊛-algebras and equates MSO-definability with recognizability by these algebraic structures. A detailed study of a variety of sub-logics of MSO over countable words is carried out in [6]. This study also extends classical Green's relations to ⊛-algebras and makes heavy use of it. Of particular interest to us are the results about algebraic equational characterizations of FO, FO[cut] – an *extension* of FO that allows quantification over Dedekind cuts and WMSO – an *extension* of FO that allows quantification over finite sets. A decidable algebraic characterization of FO^2 over countable words is also presented in [10]. Another recent development [1] is the seamless integration of block products into the countable setting. The work introduces the block product operation of the relevant algebraic structures and establishes an appealing block product principle. Further, it naturally extends the above-mentioned block product characterizations of FO and FO^2 to countable words.

In this work, we begin our explorations into the *small* fragments of FO over countable words, guided by the choice of results in [7]. We arrive at the language-logic-algebra connection for FO^1 – the one variable fragment of FO. Coupled with earlier results about FO^2 and $FO = FO^3$ (see [8]), this completes our algebraic understanding of FO fragments defined by the number of permissible variables. We next extend Simon's theorem on piecewise testable languages to countable words and provide a natural algebraic characterization of the Boolean closure of the existential-fragment of FO. Fortunately or unfortunately, depending on the point of view, this landscape of small fragments of FO over countable words parallels very closely the same landscape over finite words. This can be attributed to the limited expressive power of FO over countable words. For instance, Bès and Carton [4] showed that the seemingly natural 'finiteness'

---

[1] Here J is one of the fundamental Green's equivalence relations.

property (that the set of all positions is a finite set) of countable words can not be expressed in FO!

One of the main contributions of this work is the introduction of new *infinitary* quantifiers to FO. The works [3,9] also extend FO over arbitrary structures by *cardinality/finitary-counting* quantifiers and studies decidable theories thereof. An extension of FO over finite and ω-words by *modulus-counting quantifiers* is algebraically characterized in [16]. The main purpose of our new quantifiers is to naturally allow expression of infinitary features which are inherent in the countable setting and study the resulting definable formal languages in the algebraic framework of [5]. An example formula using such an infinitary quantifier is: $\exists^{\infty_1}x : a(x) \wedge \neg\exists^{\infty_1}x : b(x)$. In its natural semantics, this formula with one variable asserts that there are infinitely many a-labelled positions and only finitely many b-labelled positions. We propose an extension of FO called FO[∞] that supports *first-order* infinitary quantifiers of the form $\exists^{\infty_k}x$ to talk about existence of higher-level infinitely (more accurately, Infinitary rank k) many witnesses x. We organize FO[∞] in a natural hierarchy based on the maximum allowed infinitary-level of the quantifiers.

We now summarize the key technical results of this paper. We establish a hierarchical block product based characterization of FO[∞]. Towards this, we identify an appropriate simple family of ⊛-algebras and show that this family (in fact, its initial fragments) serve as a basis in our hierarchical block product based characterization. We establish that FO[∞] properties can be expressed simultaneously in FO[cut] as well as WMSO. We also show that the language-logic-algebra connection for FO^1 admits novel generalizations to the one variable fragments of the new extension of FO. We finally present 'no finite block product basis' theorems for our FO extensions, FO[cut], and the class FO[cut] ∩ WMSO. This is in contrast with [1] where the unique 2-element ⊛-algebra is a basis for a block-product based characterization of FO.

The rest of the paper is organized as follows. Section 2 recalls basic notions about countable words and summarizes the necessary algebraic background from the framework [5]. Section 3 deals with the small fragments of FO: FO^1 and the Boolean closure of the existential fragment of FO. Section 4 contains the extensions FO[∞] and results relevant to it. Section 5 is concerned with 'no finite block product basis' theorems. See [2] for complete details and proofs.

2   Preliminaries

In this section we briefly recall the algebraic framework developed in [5].

Countable Words. A *countable linear ordering* (or simply ordering) $\alpha = (X, <)$ is a non-empty countable set X equipped with a total order: X is the *domain* of α. An ordering $\beta = (Y, <)$ is called a *subordering* of α if $Y \subseteq X$ and the order on Y is induced from that of X. We denote by $\omega, \omega^*, \delta, \eta$ the orderings $(\mathbb{N}, <), (-\mathbb{N}, <), (\mathbb{Z}, <), (\mathbb{Q}, <)$ respectively. A *Dedekind cut* (or simply a cut) is a left-closed subset $Y \subseteq X$ of α. Given disjoint linear orderings $(\beta_i)_{i \in \alpha}$ indexed with a linear ordering α, their *generalized sum* $\sum_{i \in \alpha} \beta_i$ is the linear ordering over the union of the domains of the β_i's, with the order defined by $x < y$ if either $x \in \beta_i$ and $y \in \beta_j$ with $i < j$, or $x, y \in \beta_i$ for some i, and $x < y$ in β_i. The book [13] contains a detailed study of linear orderings.

An *alphabet* Σ is a finite set of symbols called *letters*. Given a linear ordering α, a *countable word* (henceforth called word) over Σ of domain α is a mapping $w : \alpha \to \Sigma$. The domain of a word is denoted $dom(w)$. For a subset $I \subseteq dom(w)$, $w|_I$ denotes the *subword* obtained by restricting w to the domain I. If I is an *interval* ($\forall x, y \in I$, $x < z < y \to z \in I$) then $w|_I$ is called a *factor* of w. The set of all words is denoted $\Sigma^{\circledast}$ and the set of all non-empty (resp. finite) words $\Sigma^{\oplus}$ (resp. Σ^*). A *language* (of countable words) is a subset of $\Sigma^{\circledast}$. The *generalized concatenation* of the words $(w_i)_{i \in \alpha}$ indexed by a linear ordering α is $\prod_{i \in \alpha} w_i$ and denotes the word w of domain $\sum_{i \in \alpha} \beta_i$ where β_i are disjoint and such that $w|_{\beta_i}$ is isomorphic to w_i for all $i \in \alpha$.

The *empty word* ε, is the only word of empty domain. The ω-*power* of a word u is defined as $u^{\omega} ::= \prod_{i \in \omega} u$. The ω^*-*power* of a word u, denoted by u^{ω^*}, is $\prod_{i \in \omega^*} u$. The *perfect shuffle* for a non-empty finite set of letters $A \subseteq \Sigma$ (denoted by A^{η}) is a word of domain $(\mathbb{Q}, <)$ in which only letters from A occur and, all non-empty and non-singleton intervals contain at least one occurrence of each letter in A. This word is unique up to isomorphism (see [13]). We can extend the notion of perfect shuffle to a finite set of words $W = \{w_1, \ldots, w_k\}$. We define W^{η} to be $\prod_{i \in \mathbb{Q}} w_{f(i)}$ where $f : (\mathbb{Q}, <) \to \{1, 2, \ldots, k\}$ is the unique perfect shuffle over the set of letters $\{1, 2, \ldots, k\}$.

The Algebra. A $\circledast$-*monoid* $\mathbf{M} = (M, \pi)$ is a set M equipped with an operation π, called the *product*, from $M^{\circledast}$ to M, that satisfies $\pi(a) = a$ for all $a \in M$, and the *generalized associativity* property: for every words u_i over M with i ranging over a countable linear ordering α, $\pi\left(\prod_{i \in \alpha} u_i\right) = \pi\left(\prod_{i \in \alpha} \pi(u_i)\right)$. We reserve the notation id for the *identity element* $\mathrm{id} = \pi(\varepsilon)$; it is called the neutral element in [5]. An example of a $\circledast$-monoid is the free $\circledast$-monoid $(\Sigma^{\circledast}, \varepsilon, \prod)$ over the alphabet Σ with the product being the generalized concatenation. Now we discuss some natural algebraic notions. A *morphism* from a $\circledast$-monoid (M, π) to a $\circledast$-monoid (M', π') is a map $h : M \to M'$ such that, for every $w \in M^{\circledast}$, $h(\pi(w)) = \pi'(\bar{h}(w))$ where $\bar{h}$ is the pointwise extension of h to words. We skip the notions sub-$\circledast$-*monoid* and direct products since they are as expected. We say $\mathbf{M} = (M, \pi)$ *divides* $\mathbf{M}' = (M', \pi')$ if there exists a sub $\circledast$-monoid $\mathbf{M}'' = (M'', \pi'')$ of $\mathbf{M}'$ and a surjective morphism from $\mathbf{M}''$ to $\mathbf{M}$.

A $\circledast$-monoid $\mathbf{M} = (M, \pi)$ is said to be finite if M is so. Note that, even for a finite $\circledast$-monoid, the product operation π has an infinitary description. It turns out that π can be captured using finitely presentable *derived operations*. Corresponding to a $\circledast$-monoid (M, π) there is an induced $\circledast$-*algebra* $\mathbf{M} = (M, \mathrm{id}, \cdot, \tau, \tau^*, \kappa)$ where the operations are defined as following: for all $a, b \in M$, $a \cdot b = \pi(ab)$, $a^{\tau} = \pi(a^{\omega})$, $a^{\tau^*} = \pi(a^{\omega^*})$ and for all $\emptyset \neq E \subseteq M$, $E^{\kappa} = \pi(E^{\eta})$. For a singleton set $\{m\}$, we write $m^{\kappa} = \{m\}^{\kappa}$. These derived operators satisfy certain natural axioms; see [5] for details. It has been established in [5] that an *arbitrary* finite $\circledast$-algebra $\mathbf{M} = (M, \mathrm{id}, \cdot, \tau, \tau^*, \kappa)$ satisfying these natural axioms is induced by a unique ⊛-monoid $\mathbf{M} = (M, \pi)$. It is rather straightforward to define the notions of morphisms, subalgebras, direct-products as well as division for ⊛-algebras.

It follows from the definition of a ⊛-algebra $\mathbf{M} = (M, \mathrm{id}, \cdot, \tau, \tau^*, \kappa)$ that $(M, \mathrm{id}, \cdot)$ is a *monoid*, that is the operation $\cdot$ is associative with identity id. Note that, for all $m \in M$, $m \cdot \mathrm{id} = \mathrm{id} \cdot m = m$ and for all $\emptyset \neq E \subseteq M$, $E^\kappa = (E \cup \{\mathrm{id}\})^\kappa$. Further, $\mathrm{id}^\tau = \mathrm{id}^{\tau^*} = \mathrm{id}^\kappa = \mathrm{id}$. As a result, in our definitions of ⊛-algebras later in the paper, we restrict the descriptions of derived operators to $M \setminus \{\mathrm{id}\}$. An *idempotent* is an element e where $e \cdot e = e$. In a finite monoid, every element m has a unique positive power m^k which is an idempotent; we denote this idempotent by $m^!$ and refer to it as the *idempotent power*.

An *evaluation tree* over a word $u \in M^⊛ \setminus \{\varepsilon\}$ is a tree $\mathcal{T} = (T, h)$ such that every branch/path of $\mathcal{T}$ is of finite length and where every vertex in T is a factor of u, the root is u and $h : T \to M$ is a map such that:

- A leaf is a singleton letter $a \in M$ such that $h(a) = a$.
- Internal nodes have either two or ω or ω^* or $\mathbb{Q}$ many children.
- If w has children v_1 and v_2, then $w = v_1 v_2$ and $h(w) = h(v_1) \cdot h(v_2)$.
- If w has ω many children $\langle v_1, v_2, \dots \rangle$, then there is an idempotent e such that $e = h(v_i)$ for all $i \geq 1$, and $w = \prod_{i \in \omega} v_i$ and $h(w) = e^\tau$.
- If w has ω^* many children $\langle \dots, v_{-2}, v_{-1} \rangle$, then there is an idempotent f such that $f = h(v_i)$ for all $i \leq -1$, and $w = \prod_{i \in \omega^*} v_i$ and $h(w) = f^{\tau^*}$.
- If w has $\mathbb{Q}$ many children $\langle v_i \rangle_{i \in \mathbb{Q}}$, then $w = \prod_{i \in \mathbb{Q}} v_i$ where for the perfect shuffle f over an $E = \{a_1, \dots, a_k\} \subseteq M$, $h(v_i) = a_{f(i)}$, and $h(w) = E^\kappa$.

The *value* of $\mathcal{T}$ is defined to be $h(u)$. It was shown in [5, Proposition 3.8 and 3.9] that every word u has an evaluation tree and the values of two evaluation trees of u are equal and they are equal to $\pi(u)$. Therefore, a ⊛-algebra defines the generalized associativity product $\pi : M^⊛ \to M$. The correspondence between finite ⊛-monoids and ⊛-algebras permits interchangeability; we exploit it implicitly.

A morphism from the *free ⊛-monoid* $\Sigma^⊛$ to $\mathbf{M}$ is described (determined) by a map $h' : \Sigma \to M$; we simply write $h' : \Sigma \to \mathbf{M}$. With h' also denoting its pointwise extension $h' : \Sigma^⊛ \to M^⊛$, given a word $u \in \Sigma^⊛$, we can use the evaluation tree over the word $h'(u) \in M^⊛$ to obtain $\pi(h'(u)) \in M$. By further abuse of notation, $h' : \Sigma^⊛ \to M$ also denotes the morphism which sends u to $\pi(h'(u))$. We say that L is recognized by $\mathbf{M}$ if there exists a map/morphism $h' : \Sigma^⊛ \to \mathbf{M}$ such that $L = h'^{-1}(h'(L))$. The fundamental result of [5] states that *regular languages* (MSO definable languages) are exactly those recognized by finite ⊛-monoids (equivalently ⊛-algebras). It is important to note that, every regular language L is associated a finite (canonical/minimal) syntactic ⊛-monoid (and a corresponding syntactic ⊛-algebra) which divides every ⊛-monoid (resp. ⊛-algebra) that recognizes L.

Example 1. The $\circledast$-monoid $U_1 = (\{id, 0\}, \pi)$ and its induced $\circledast$-algebra are shown on the left and right respectively.

$$\pi(u) = \begin{cases} id & \text{if } u \in \{id\}^{\circledast} \\ 0 & \text{otherwise} \end{cases}$$

$\cdot$	id	0	τ	τ^*
id	id	0	id	id
0	0	0	0	0

$$S^{\kappa} = \begin{cases} id & \text{if } S = \{id\} \\ 0 & \text{otherwise} \end{cases}$$

Let $\Sigma = \{a, b\}$ and L be the set of words which contain an occurrence of letter a. It is easy to see that the map $h : \Sigma \to U_1$ sending $h(a) = 0, h(b) = id$ recognizes L as $L = h^{-1}(0)$. In fact, U_1 is the syntactic $\circledast$-monoid of L.

Example 2. Consider the $\circledast$-algebra $Gap = (\{id, [\,], (\,], [\,), (\,), g\}, id, \cdot, \tau, \tau^*, \kappa)$. We let $\Sigma = \{a\}$ and define the map $h : \Sigma \to Gap$ as $h(a) = [\,]$. The resulting morphism maps a word u to $h(u) = g$ iff the word u admits a gap; that is a cut with no maximum and its complement has no minimum. Other words are mapped to their correct 'ends-type': for instance, $h(u) = [\,)$ iff $dom(u)$ has a minimum and no maximum. For a word $v = a^{\omega}a^{\omega^*}$, the pointwise extension $v' = h(v) = [\,]^{\omega}[\,]^{\omega^*}$. An example evaluation tree $\mathcal{T}$ for v' consists of root with two children. The left (resp. right) child has ω (resp. ω^*) many children $[\,]$ and has value $[\,]^{\tau}$ (resp. $[\,]^{\tau^*}$). As a result, the value of $\mathcal{T}$ is $[\,]^{\tau} \cdot [\,]^{\tau^*} = [\,) \cdot (\,] = g$.

$\cdot$	$[\,]$	$[\,)$	$(\,]$	$(\,)$	g	τ	τ^*
$[\,]$	$[\,]$	$[\,)$	$[\,]$	$[\,)$	g	$[\,)$	$(\,]$
$[\,)$	$[\,]$	$[\,)$	g	g	g	$[\,)$	$(\,)$
$(\,]$	$(\,]$	$(\,)$	$(\,]$	$(\,)$	g	$(\,)$	$(\,]$
$(\,)$	$(\,]$	$(\,)$	g	g	g	g	g
g	g	g	g	g	g	g	g

$$S^{\kappa} = \begin{cases} id & \text{if } S = \{id\} \\ g & \text{otherwise} \end{cases}$$

We can characterize $\circledast$-monoids using equational *identities*. For example, **M** is a *commutative* $\circledast$-monoid if and only if **M** satisfies the equation $x \cdot y = y \cdot x$. This means that the equation holds for any assignment of elements in the **M** to the variables x and y. We say **M** is aperiodic if it satisfies the profinite identity $x = x \cdot x^{!}$. Like in the case of monoids, the set of $\circledast$-monoids satisfying a set of equations are closed under subsemigroup, division and *direct product* [6].

The *block product* of $\circledast$-monoids **M** and **N**, is denoted by **M$\square$N** and is the semidirect product of **M** and $\mathbf{K} = \mathbf{N}^{M \times M}$ with respect to the *canonical* left and right 'action' of **M** on **K**. The details are given in [1]. The *block product principle* characterizes languages defined by block product of $\circledast$-monoids. Towards this, fix a map $h : \Sigma \to \mathbf{M}\square\mathbf{N}$ such that $h(a) = (m_a, f_a)$ where $m_a \in M$ and $f_a : M \times M \to N$. The map $h_1 : \Sigma \to \mathbf{M}$ setting $h_1(a) = m_a$ defines a morphism $h_1 : \Sigma^{\circledast} \to M$. We define the *transducer* $\sigma : \Sigma^{\circledast} \to (M \times \Sigma \times M)^{\circledast}$ as follows: let $u \in \Sigma^{\circledast}$ with domain α. The word $u' = \sigma(u)$ has domain α and for a position $x \in \alpha$, $u'(x) = (h_1(u_{<x}), u(x), h_1(u_{>x}))$. Here $u_{<x}$ (resp. $u_{>x}$) is the subword of u on positions strictly less (resp. greater) than x.

Proposition 1 (Block Product Principle [1]). *Let $L \subseteq \Sigma^{\circledast}$ be recognized by $h : \Sigma \to \mathbf{M}\square\mathbf{N}$. Then L is a boolean combination of languages of the form*

L_1 and $\sigma^{-1}(L_2)$ where L_1 and L_2 are recognized by $\mathbf{M}$ and $\mathbf{N}$ respectively and $\sigma : \Sigma^{\circledast} \to (M \times \Sigma \times M)^{\circledast}$ is the aforementioned state-based transducer.

3 Small Fragments of FO

In this section, we focus on two particularly small fragments of first-order logic interpreted over countable words. First-order logic uses variables $x, y, z, \ldots$ which are interpreted as positions in the domain of a word. The syntax of *first-order logic* (FO) is: $x < y \mid a(x) \mid \phi_1 \wedge \phi_2 \mid \phi_1 \vee \phi_2 \mid \neg\phi \mid \exists x \; \phi$, for all $a \in \Sigma$.

We skip the natural semantics. A language L of countable words is said to be FO-definable if there exists an FO-sentence ϕ such $L = \{u \in \Sigma^{\circledast} \mid u \models \phi\}$.

Recall that the classical Schützenberger-McNaughton-Papert theorem characterizes FO-definabilty of a regular language of finite words in terms of aperiodicity of its finite syntactic monoid. The survey [7] presents similar decidable characterizations of several interesting small fragments of FO-logic such as FO^1, FO^2, $B(\exists^*)$ – boolean closure of the existential first-order logic. It is known [8] that, over finite *as well as countable words*, $\mathrm{FO} = \mathrm{FO}^3$. As mentioned in the introduction, over countable words, we already have decidable algebraic characterizations of FO^3 from [6] and FO^2 from [10]. Here we identify decidable algebraic characterizations, over countable words, for FO^1 and $B(\exists^*)$.

3.1 FO with Single Variable

The fragment FO^1 has access to only one variable. We recall that over finite words a regular language is FO^1-definable iff its syntactic monoid is commutative and idempotent. We henceforth focus our attention to FO^1 on countable words.

Clearly, FO^1 can recognize all words with a particular letter. With a single variable the logic cannot talk about order of letters or count the number of occurrence of a letter. This gives an intuition that the syntactic $\circledast$-monoid of a language definable in FO^1 is commutative and idempotent.

We say that a $\circledast$-algebra $\mathbf{M} = (M, \mathrm{id}, \cdot, \boldsymbol{\tau}, \boldsymbol{\tau}^*, \boldsymbol{\kappa})$ is *shuffle-trivial* if it satisfies the equational identity: $\{x_1, \ldots, x_p\}^{\kappa} = x_1 \cdot x_2 \cdot \ldots \cdot x_p$. Note that shuffle-triviality implies commutativity: $x \cdot y = \{x, y\}^{\kappa} = \{y, x\}^{\kappa} = y \cdot x$. Moreover, every element of $\mathbf{M}$ is a *shuffle-idempotent*: for all $m \in M, m^{\kappa} = m$. It is a consequence of the axioms of a $\circledast$-algebra that a shuffle-idempotent is an idempotent.

Theorem 1. Let $L \subseteq \Sigma^{\circledast}$ be a regular language. The following are equivalent.

1. L is recognized by some finite shuffle-trivial $\circledast$-algebra.
2. L is a boolean combination of languages of the form $B^{\circledast}$ where $B \subseteq \Sigma$.
3. L is definable in FO^1.
4. L is recognized by direct product of U_1s.
5. The syntactic $\circledast$-algebra of L is shuffle-trivial.

3.2   Boolean Closure of Existential FO

Let us first recall the characterization of $B(\exists^*)$ - the boolean closure of existential FO over finite words. This is precisely the content of the theorem due to Simon [14]. The usual presentation of Simon's theorem refers to *piecewise testable languages* which are easily seen to be equivalent to $B(\exists^*)$-definable languages. *Simon's theorem* states that a regular language of finite words is $B(\exists^*)$-definable iff its syntactic monoid is J-trivial. We refer to [12] for a detailed study of Green's relations and its use in the proof of Simon's theorem.

The original proof of Simon's theorem uses the congruence $\sim_n$, parametrized by $n \in \mathbb{N}$, on finite words Σ^*: for $u, v \in \Sigma^*$, $u \sim_n v$ if u and v have the same set of subwords of length less than or equal to n. Note that $\sim_n$ has finite index.

We fix $n \in \mathbb{N}$ and work with $\sim_n$ defined on countable words $\Sigma^{\circledast}$: for $u, v \in \Sigma^{\circledast}$, $u \sim_n v$ if u and v have the same set of subwords of length less than or equal to n. It is immediate that $\sim_n$ is an equivalence relation on $\Sigma^{\circledast}$ of finite index. We let $S_n = \Sigma^{\circledast} / \sim_n$ denote the finite set of $\sim_n$-equivalence classes. For a word w, $[w]_n$ denotes the $\sim_n$-equivalence class which contains w.

Lemma 1. *There is a natural well-defined product operation* $\pi : S_n^{\circledast} \to S_n$ *as follows:* $\pi\left(\prod_{i \in \alpha}[w_i]_n\right) = \left[\prod_{i \in \alpha} w_i\right]_n$. *This operation* π *satisfies the generalized associativity property. As a result,* $\mathbf{S_n} = (S_n, \mathrm{id} = [\varepsilon]_n, \pi)$ *is a* $\circledast$*-monoid.*

Note that the lemma implies that $h_n : \Sigma^{\circledast} \to \mathbf{S_n}$ mapping w to $[w]_n$ is a morphism of $\circledast$-*monoids*.

We say that a $\circledast$-algebra is *shuffle-power-trivial* if it satisfies the (profinite) identity: $\{x_1, \dots, x_p\}^\kappa = (x_1 \cdot x_2 \cdot \dots \cdot x_p)^!$. Note that, every idempotent of such a $\circledast$-algebra is a shuffle-idempotent: $x^! = x$ implies $x^\kappa = x$. Further, it can be shown that, in this case, the underlying monoid is J-trivial.

Theorem 2. *Let* $L \subseteq \Sigma^{\circledast}$ *be a regular language. The following are equivalent.*

1. *L is recognized by a finite shuffle-power-trivial $\circledast$-algebra.*
2. *L is recognized by the quotient morphism $h_n : \Sigma^{\circledast} \to \mathbf{S_n}$ for some n.*
3. *L is definable in $B(\exists^*)$.*
4. *The syntactic $\circledast$-algebra of L is shuffle-power-trivial.*

4   First Order Logic with infinitary quantifiers

Our results in the previous section resemble very closely the corresponding results over finite words. This can be attributed to the limited capability of the operators τ, τ^* and κ in the $\circledast$-monoids we witnessed. As mentioned in the Introduction, FO cannot define the language of infinite number of a's. An existential quantifier is a threshold counting quantifier - it says there exists at least one position satisfying a property. Using multiple such quantifiers, FO can count up to any finite constant but not more. Over countable words, it is natural to ask for stronger threshold quantifiers. We introduce natural infinitary versions of the existential quantifier which precisely serve this purpose.

We define $\mathcal{I}_0$ to be the set of all non-empty finite orderings. For any number $n \in \mathbb{N}$, we define the set $\mathcal{I}_n$ to be the set of all orderings of the form $\sum_{i \in \mathbb{Z}} \alpha_i$ where $\alpha_i \in \mathcal{I}_{n-1} \cup \{\varepsilon\}$ and is closed under finite sum. We define the Infinitary rank (or simply *rank*) of a linear ordering α (denoted by $\infty\text{-}rank(\alpha)$) as the least n (if it exists) where $\alpha \in \mathcal{I}_n$. If there is no such n we say that the rank is infinite. For example, $\infty\text{-}rank(\omega) = \infty\text{-}rank(\omega + \omega) = \infty\text{-}rank(\omega^* + \omega) = 1$, $\infty\text{-}rank(\omega^2) = \infty\text{-}rank(\omega^2 + \omega^*) = 2$, and the rank of $\eta = (\mathbb{Q}, <)$ is infinite.

We introduce the logic FO$[\infty]$ extending FO with *infinitary quantifiers* : $\exists^{\infty_0} x\, \varphi \mid \exists^{\infty_1} x\, \varphi \mid \ldots \mid \exists^{\infty_n} x\, \varphi \mid \ldots$ for all $n \in \mathbb{N}$.

Note that all the variables are first order and they are interpreted as positions, that is, elements of the underlying linear ordering. The semantics of the infinitary quantifier $\exists^{\infty_n} x$ for an $n \geq 0$ is: for a word w and an assignment s, we say $w, s \models \exists^{\infty_n} x\, \varphi$ if there exists a subordering $X \subseteq dom(w)$ such that $\infty\text{-}rank(X) = n$ and $w, s[x = i] \models \varphi$ for all $i \in X$. Hence, $\exists^{\infty_0} x\, \varphi$ is equivalent to $\exists x\, \varphi$ since both formulas are true if and only if there is at least one position x which satisfies ϕ. The logic FO$[(\infty_j)_{j \leq n}]$ denote the fragment containing only the infinitary quantifiers $\exists^{\infty_j} x$ for all $j \leq n$. Clearly the following relationship is maintained among the logics:

$$\text{FO} = \text{FO}[(\infty_j)_{j \leq 0}] \subseteq \text{FO}[(\infty_j)_{j \leq 1}] \subseteq \text{FO}[(\infty_j)_{j \leq 2}] \subseteq \ldots$$

We also denote by FO$^1[(\infty_j)_{j \leq n}]$ the corresponding one variable fragment of FO$[(\infty_j)_{j \leq n}]$.

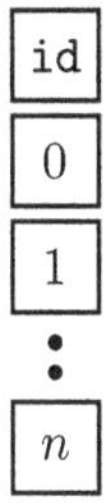

Example 3. The formula $\exists^{\infty_1} x\, a(x)$ denotes the set of all countable words with infinitely many positions labelled a. Since FO cannot express this, it shows FO $\subsetneq$ FO$[(\infty_j)_{j \leq 1}]$.

Fig. 1. Δ_n-chain

For $n \geq 0$, we define ⊛-algebra Δ_n-chain as: $(\{\text{id}, 0, 1, \ldots, n\}, \text{id}, \cdot, \tau, \tau^*, \kappa)$ where for all $0 \leq i \leq j \leq n$, $i \cdot j = j \cdot i = \max(i, j) = j$ and for all $0 \leq k < n$, $k^\tau = k^{\tau^*} = k + 1$ and $n^\tau = n^{\tau^*} = n$. That is, $k^\tau = k^{\tau^*} = \min(k + 1, n)$. Moreover, $\text{id}^\kappa = \text{id}$ and $S^\kappa = n$ for any S where $S \backslash \{\text{id}\} \neq \emptyset$. In short, the descriptions of the derived operators restricted to $\Delta_n \backslash \{\text{id}\}$ are

$$(i, j) \mapsto \max(i, j), \quad i \xmapsto{\tau} \min(i + 1, n), \quad i \xmapsto{\tau^*} \min(i + 1, n), \quad S \xmapsto{\kappa} n$$

Note that Δ_n is both commutative and idempotent. It is also the syntactic ⊛-algebra for the language defined by $\exists^{\infty_n} x\, a(x)$. Further, Δ_0 is isomorphic to U_1 from Example 1. See (Fig. 1).

4.1 FO$[\infty]$ with single variable

In this section we characterize languages definable in the one variable fragment FO$^1[(\infty_j)_{j \leq n}]$ as those which can be recognized by the direct product of Δ_n.

Theorem 3. *Languages recognized by direct product of Δ_n are exactly those definable in* $\mathrm{FO}^1[(\infty_j)_{j \leq n}]$.

Proof. We first show that languages recognized by Δ_n are definable in $\mathrm{FO}^1[(\infty_j)_{j \leq n}]$. Let $h \colon \Sigma^{\circledast} \to \Delta_n$ be a morphism. It suffices to show that for any element $m \in \Delta_n$, $h^{-1}(m)$ is definable in $\mathrm{FO}^1[(\infty_j)_{j \leq n}]$. In the rest of the discussion we adopt the convention that $\mathrm{id} < 0$. Let $\uparrow m$ denote the set $\{m' \mid m' \geq m\}$. Note that for an $m < n$, $h^{-1}(m) = h^{-1}(\uparrow m) \setminus h^{-1}(\uparrow(m+1))$ and $h^{-1}(n) = h^{-1}(\uparrow n)$. Therefore, it is sufficient to show that $h^{-1}(\uparrow m)$ is definable in $\mathrm{FO}^1[(\infty_j)_{j \leq n}]$. For each $m \in \Delta_n$, we define the language $L(m)$ as $\{w \mid$ there exists a letter a in w such that $h(a) = j \neq \mathrm{id}$ and either $j \geq m$ or there is a set of positions α labelled a such that $\infty\text{-}rank(\alpha) = j'$ and $j + j' \geq m\}$ The following $\mathrm{FO}^1[(\infty_j)_{j \leq n}]$ sentence defines the language $L(m)$.

$$\bigvee_{a \in \Sigma,\, h(a) \geq m} \exists x\, a(x) \quad \vee \quad \bigvee_{a \in \Sigma,\, 0 \leq h(a) < m} \exists^{\infty_{m-h(a)}} x\, a(x)$$

We show that $L(m) = h^{-1}(\uparrow m)$ by induction on m. The base case holds since $\uparrow \mathrm{id} = \Delta_n$, $h^{-1}(\uparrow \mathrm{id}) = \Sigma^{\circledast}$ and $L(\mathrm{id}) = \Sigma^{\circledast}$. To prove the induction hypothesis assume the claim holds for all $j < m$. Consider a word w. By a second induction on the height of an evaluation tree (T, h) for w we show for all words $v \in T$, $v \in h^{-1}(\uparrow m)$ if and only if $v \in L(m)$. In each of the following cases we assume that the children of the node (if they exist) satisfy the second induction hypothesis.

1. Case v is a letter: The hypothesis clearly holds
2. Case v is a concatenation of two words v_1 and v_2: There are two cases to consider - $\{v_1, v_2\} \cap h^{-1}(\uparrow m) \neq \emptyset$ or not. In the first case, let for an $i \in \{1, 2\}$ we have $h(v_i) \geq m$ and $v_i \in L(m)$. Clearly $h(v) = h(v_1 v_2) \geq m$ and $v \in L(m)$. For the second case, let us assume $h(v_1) = i$ and $h(v_2) = j$ such that $i \leq j < m$ and both $v_1, v_2 \notin L(m)$. From the definition of Δ_n, it follows that $h(v) = h(v_1 v_2) = j$. Let the a-labelled suborderings in v_1 and v_2 be α_1 and α_2 respectively where $\infty\text{-}rank(\alpha_1) \leq \infty\text{-}rank(\alpha_2) = j'$. It follows from the definition that $\infty\text{-}rank(\alpha_1 + \alpha_2) = j'$ and therefore $v \notin L(m)$.
3. Case v is an ω-sequence of words $\langle v_1, v_2, \ldots, \rangle$ such that $h(v_i) = k$, for all i, and k is an idempotent (in Δ_n all elements are idempotents): Firstly, if $k \geq m$ and $v_i \in L(m)$ then clearly $h(v) \geq m$ and $v \in L(m)$. The non-trivial case is $k = m - 1$. From the second induction hypothesis $v_i \notin L(m)$ for all i. From the definition of Δ_n, $h(v) = k^{\tau} = m$. We need to show that $v \in L(m)$. By first induction hypothesis, each v_i has a letter a_i and an a_i-labelled set of positions α_i such that $h(a_i) + \infty\text{-}rank(\alpha_i) = k$. Since $|\Sigma|$ is finite, ω-many of these a_is are the same letter, say a. Hence the a-labelled set of positions $\alpha = \sum_{i : a_i = a} \alpha_i$ in v satisfies $\infty\text{-}rank(\alpha) = \infty\text{-}rank(\alpha_i) + 1$. As a consequence, $h(a) + \infty\text{-}rank(\alpha) = k + 1$ or in other words $v \in L(m)$.
4. Case v is an ω^*-sequence: This case is symmetric to the above case.
5. Case v is a perfect shuffle, $h(v) = S^{\kappa}$: It is easy to see that the induction hypothesis holds if $S = \{\mathrm{id}\}$. So, assume $S \cap \{\mathrm{id}\} \neq \emptyset$. Hence $h(v) = n$.

Since, there are $\mathbb{Q}$-many children u where $h(u) \neq \mathtt{id}$, there is a letter a such that a-labelled set of positions in v has infinite rank or $v \in L(n)$.

The other direction of the proof follows from the fact that a one variable quantifier free formula is essentially a disjunction of letter predicates and therefore the boolean combination of formulas can be recognized by direct products of Δ_k.

4.2 The General FO[∞] logic

In this section, we consider the full logic $\mathrm{FO}[(\infty_j)_{j \leq n}]$ and observe that they define exactly those languages recognized by block products of Δ_n.

Theorem 4. *The languages defined by* $\mathrm{FO}[(\infty_j)_{j \leq n}]$ *are exactly those recognized by finite block products of* Δ_n. *Moreover, the languages defined by* $\mathrm{FO}[\infty]$ *are exactly those recognized by finite block products of* $\{\Delta_n \mid n \in \mathbb{N}\}$.

Proof. We first show that languages recognizable by finite block products of Δ_n are definable in $\mathrm{FO}[(\infty_j)_{j \leq n}]$. The proof is via induction on the number of Δ_n in an iterated block product. The base case follows from Theorem 3.

For the inductive step, consider a morphism $h \colon \Sigma^{\circledast} \to M \square \Delta_n$. Let $h_1 \colon \Sigma^{\circledast} \to M$ be the induced morphism to M, and let σ be the associated transducer. By the block product principle (see Proposition 1), any language recognized by h is a boolean combination of languages $L_1 \subseteq \Sigma^{\circledast}$ recognized by M and $\sigma^{-1}(L_2)$ where $L_2 \subseteq (M \times \Sigma \times M)^{\circledast}$ is recognized by Δ_n. By induction hypothesis, L_1 is $\mathrm{FO}[(\infty_j)_{j \leq n}]$ definable. By the base case L_2 is $\mathrm{FO}[(\infty_j)_{j \leq n}]$ definable but over the alphabet $M \times \Sigma \times M$. To complete the proof, one needs to show for any word $w \in \Sigma^{\circledast}$ and assignment s, and for any $\mathrm{FO}[(\infty_j)_{j \leq n}]$ formula φ over the alphabet $M \times \Sigma \times M$, there exists a $\mathrm{FO}[(\infty_j)_{j \leq n}]$ formula $\hat{\varphi}$ over the alphabet Σ such that $w, s \models \hat{\varphi}$ if and only if $\sigma(w), s \models \varphi$. For instance, suppose $\varphi = \exists^{\infty^i} x \, (m_1, c, m_2)(x)$, and inductively ϕ_{m_1} (resp. ϕ_{m_2}) are $\mathrm{FO}[(\infty_j)_{j \leq n}]$ formula characterizing words over $\Sigma^{\circledast}$ that are mapped by h_1 to m_1 (resp. m_2). Then $\hat{\varphi}$ is $\exists^{\infty^i} x \, (\phi_{m_1}|_{<x} \wedge c(x) \wedge \phi_{m_2}|_{>x})$, where $\phi_{m_1}|_{<x}$ is the formula ϕ_{m_1} with all its variables relativised to less than the variable x. This way, one proves that $\sigma^{-1}(L_2)$ is $\mathrm{FO}[(\infty_j)_{j \leq n}]$ definable. This completes the proof of this direction.

The other direction of the proof is a standard generalization of the proof of equivalence of FO and the block product closure of Δ_0 given in [1, Theorem 2]. The block product principle allows us to "simulate" infinitary quantifiers using block products of Δ_n and vice-versa. We can then inductively recognize languages defined by formulas using iterated block products.

We claim that both first order logic with cuts (FO[cut]) and weak monadic second order logic (WMSO) can define the languages definable in $\mathrm{FO}[\infty]$.

Theorem 5. $\mathrm{FO}[\infty] \subseteq \mathrm{FO}[\mathrm{cut}] \cap \mathrm{WMSO}.$[2]

---

[2] Henceforth, by a slight abuse of notation, $\mathrm{FO}[\infty]$, $\mathrm{FO}[\mathrm{cut}]$, WMSO also denote the language-classes defined by the corresponding logics.

5 No Finite Basis Theorems

The main goal of this section is to prove that $FO[\infty], FO[cut]$ and $FO[cut] \cap$ WMSO over countable words do not admit a block product based characterization which uses only a *finite* set of ⊛-monoids. This is in stark contrast with the result in [1] which shows that a language of countable words is FO-definable iff it is recognized by a strong iteration of block product of copies of Λ_0 (alternately called U_1). This is abbreviated by saying that FO has a block-product based characterization using a basis which contains the single ⊛-monoid Δ_0. Notice that, it follows from the results in the previous section that $FO[\infty]$ admits a block product based characterization using the *natural infinite basis* $\{\Delta_n\}_{n \in \mathbb{N}}$.

Fix a finite ⊛-algebra $\mathbf{M} = (M, \mathrm{id}, \cdot, \tau, \tau^*, \kappa)$. For every $n \in \mathbb{N}$, we define the operation $\gamma_n : M \to M$ which maps x to x^{γ_n}. The inductive definition of γ_n is as follows (recall idempotent power): $x^{\gamma_0} = x^!$ and $x^{\gamma_n} = ((x^{\gamma_{n-1}})^\tau (x^{\gamma_{n-1}})^{\tau^*})^!$.

Lemma 2. *For each $m \in M$, there exists n such that $\forall n' \geq n, m^{\gamma_n} = m^{\gamma_{n'}}$.*

We now define the *gap-nesting-length* of $\mathbf{M}$ (in notation, $\mathrm{gnlen}(\mathbf{M})$) to be the smallest n such that for all $m \in M$, $m^{\gamma_n} = m^{\gamma_{n+1}}$. It follows from the previous lemma that a finite ⊛-algebra has a finite gap-nesting-length. It is a simple computation that, for each k, $\mathrm{gnlen}(\Delta_k) = k$. The following main technical lemma is the key to our no-finite-basis theorems.

Lemma 3. *For finite aperiodic[3] ⊛-algebras $\mathbf{M}$ and $\mathbf{N}$,*

1. *We have, $\mathrm{gnlen}(\mathbf{M}\square\mathbf{N}) \leq \max(\mathrm{gnlen}(\mathbf{M}), \mathrm{gnlen}(\mathbf{N}))$.*
2. *If $\mathbf{M}$ divides $\mathbf{N}$ then $\mathrm{gnlen}(\mathbf{M}) \leq \mathrm{gnlen}(\mathbf{N})$.*

Corollary 1. $FO[(\infty_j)_{j \leq n}] \subsetneq FO[(\infty_j)_{j \leq n+1}]$.

Proof. By Theorem 4, the syntactic ⊛-algebra $\mathbf{M}$ of any $FO[(\infty_j)_{j \leq n}]$-definable language divides a block product of copies of Δ_n. By Lemma 3 and the fact that $\mathrm{gnlen}(\Delta_n) = n$, $\mathrm{gnlen}(\mathbf{M}) \leq n$. Note that, Δ_{n+1} is the syntactic ⊛-algebra for the language L defined by the $FO[(\infty_j)_{j \leq n+1}]$ formula $\exists^{\infty_{n+1}} x \, a(x)$. As $\mathrm{gnlen}(\Delta_{n+1}) = n + 1$, it follows that L cannot be defined in $FO[(\infty_j)_{j \leq n}]$.

Theorem 6. *There is no finite basis for a block product based characterization for any of these logical systems $FO[\infty], FO[cut], FO[cut] \cap$ WMSO.*

Proof. Fix one of the logics $\mathcal{L}$ mentioned in the statement of the theorem. It follows from Theorem 5 and the algebraic characterization of $FO[cut]$ from [6] that the syntactic ⊛-algebras of $\mathcal{L}$-definable languages are aperiodic. Now suppose, for contradiction, that $\mathcal{L}$ admits a finite basis B of aperiodic ⊛-algebras for its block product based characterization. Since B is finite, there exists $n \in \mathbb{N}$ such that for all ⊛-algebras $\mathbf{M}$ in B, $\mathrm{gnlen}(\mathbf{M}) \leq n$. It follows by Lemma 3 that the syntactic ⊛-algebra $\mathbf{N}$ of *every* $\mathcal{L}$-definable language has the property $\mathrm{gnlen}(\mathbf{N}) \leq n$.

---

[3] This simply means that the underlying monoid of a ⊛-algebra is aperiodic.

Now consider the language L defined by the FO$[\infty]$ sentence $\phi = \exists^{\infty_{n+1}} x \, a(x)$. By Theorem 5, L is $\mathcal{L}$-definable. Hence, the gap-nesting-length of the syntactic $\circledast$-algebra K of L is less than or equal to n. However, K is simply Δ_{n+1} and gnlen(Δ_{m+1}) $= n + 1$. This leads to a contradiction.

6 Conclusion

Over countable words, we have obtained decidable characterizations of the one variable fragment FO^1 and the Boolean closure of the existential-fragment of FO. More importantly, we have enriched FO with new infinitary quantifiers and established hierarchical block-product based characterization of the resulting extension FO$[\infty]$. We also show that FO$[\infty]$ properties can be expressed simultaneously in FO[cut] as well as WMSO. We do not know if the converse also holds. If true, it will provide a syntactic means to describe the semantic 'class' FO[cut] $\cap$ WMSO. We have also shown that these natural logical systems can not have a block-product based characterization using a finite basis.

References

1. Adsul, B., Sarkar, S., Sreejith, A.V.: Block products for algebras over countable words and applications to logic. In: 34th Annual ACM/IEEE Symposium on Logic in Computer Science, LICS 2019, pp. 1–13. IEEE (2019)
2. Adsul, B., Sarkar, S., Sreejith, A.V.: First-order logic and its infinitary quantifier extensions over countable words. CoRR abs/2107.01468 (2021). https://arxiv.org/abs/2107.01468
3. Baudisch, A., Seese, D., Tuschik, H.P., Weese, M.: Decidability and Generalized Quantifiers. Akademie Verlag, Berlin (1980)
4. Bès, A., Carton, O.: Algebraic characterization of FO for scattered linear orderings. In: Computer Science Logic, 20th Annual Conference of the EACSL, CSL 2011. LIPIcs, vol. 12, pp. 67–81. Schloss Dagstuhl - Leibniz-Zentrum für Informatik (2011)
5. Carton, O., Colcombet, T., Puppis, G.: An algebraic approach to MSO-definability on countable linear orderings. J. Symb. Log. 83(3), 1147–1189 (2018)
6. Colcombet, T., Sreejith, A.V.: Limited set quantifiers over countable linear orderings. In: Halldórsson, M.M., Iwama, K., Kobayashi, N., Speckmann, B. (eds.) ICALP 2015. LNCS, vol. 9135, pp. 146–158. Springer, Heidelberg (2015). https://doi.org/10.1007/978-3-662-47666-6_12
7. Diekert, V., Gastin, P., Kufleitner, M.: A survey on small fragments of first-order logic over finite words. Int. J. Found. Comput. Sci. 19(3), 513–548 (2008)
8. Gabbay, D.M., Hodkinson, I., Reynolds, M.: Temporal Logic: Mathematical Foundations and Computational Aspects, vol. 1. Oxford University Press, Oxford (1994)
9. Gradel, E., Otto, M., Rosen, E.: Two-variable logic with counting is decidable. In: Proceedings of Twelfth Annual IEEE Symposium on Logic in Computer Science, pp. 306–317 (1997)
10. Manuel, A., Sreejith, A.V.: Two-variable logic over countable linear orderings. In: 41st International Symposium on Mathematical Foundations of Computer Science, MFCS 2016, pp. 66:1–66:13 (2016)

11. Pin, J.E.: Handbook of Formal Languages, Vol. 1. chap. Syntactic Semigroups, pp. 679–746. Springer-Verlag, Heidelberg (1997)
12. Pin, J.É.: Mathematical foundations of automata theory (2020)
13. Rosenstein, J.G.: Linear Orderings. Academic Press, New York (1981)
14. Simon, I.: Piecewise testable events. In: Brakhage, H. (ed.) GI-Fachtagung 1975. LNCS, vol. 33, pp. 214–222. Springer, Heidelberg (1975). https://doi.org/10.1007/3-540-07407-4_23
15. Straubing, H.: Finite automata, formal logic, and circuit complexity. Birkhauser Verlag, Basel, Switzerland (1994)
16. Straubing, H., Thérien, D., Thomas, W.: Regular languages defined with generalized quantifiers. In: Lepistö, T., Salomaa, A. (eds.) ICALP 1988. LNCS, vol. 317, pp. 561–575. Springer, Heidelberg (1988). https://doi.org/10.1007/3-540-19488-6_142
17. Straubing, H., Weil, P.: Varieties. CoRR abs/1502.03951 (2015). http://arxiv.org/abs/1502.03951
18. Thomas, W.: Handbook of Formal Languages, vol. 3. chap. Languages, Automata, and Logic, pp. 389–455. Springer-Verlag Inc., New York (1997). http://dl.acm.org/citation.cfm?id=267871.267878

From Symmetry to Asymmetry: Generalizing TSP Approximations by Parametrization

Lukas Behrendt[1], Katrin Casel[1(✉)], Tobias Friedrich[1],
J. A. Gregor Lagodzinski[1], Alexander Löser[1], and Marcus Wilhelm[2]

[1] Hasso Plattner Institute, University of Potsdam, Potsdam, Germany
{lukas.behrendt,alexander.loeser}@student.hpi.de,
{katrin.casel,tobias.friedrich,gregor.lagodzinski}@hpi.de
[2] Karlsruhe Institute of Technology, Karlsruhe, Germany
marcus.wilhelm@kit.edu

Abstract. We generalize the tree doubling and Christofides algorithm to parameterized approximations for ATSP. The parameters we consider for the respective generalizations are upper bounded by the number of *asymmetric distances*, which yields algorithms to efficiently compute good approximations also for moderately asymmetric TSP instances. As generalization of the Christofides algorithm, we derive a parameterized 2.5-approximation, where the parameter is the size of a vertex cover for the subgraph induced by the asymmetric distances. Our generalization of the tree doubling algorithm gives a parameterized 3-approximation, where the parameter is the minimum number of asymmetric distances in a minimum spanning arborescence. Further, we combine these with a notion of symmetry relaxation which allows to trade approximation guarantee for runtime. Since the two parameters we consider are theoretically incomparable, we present experimental results which show that generalized tree doubling frequently outperforms generalized Christofides with respect to parameter size.

Keywords: Parameterized approximation · Stability of approximation · TSP vs. ATSP

1 Introduction

The ubiquitous traveling salesman problem asks for a shortest round trip through a given set of cities. Its relation to the Hamiltonian cycle problem does not only imply NP-hardness, but also implies that efficient approximation is impossible for unrestricted instances, which is why distances are usually assumed to satisfy the triangle inequality. This restriction to *metric* instances is one of the most extensively studied problems in combinatorial optimization, yet its approximability prevails as an active research area. Despite the breakthrough by Svensson

© Springer Nature Switzerland AG 2021
E. Bampis and A. Pagourtzis (Eds.): FCT 2021, LNCS 12867, pp. 53–66, 2021.
https://doi.org/10.1007/978-3-030-86593-1_4

et al. [32], particularly the difference between symmetric and asymmetric distances remains rather poorly understood. In this paper we employ the tools of parameterized complexity as a new approach to explicitly study the effects of asymmetry on the approximability of the metric traveling salesman problem.

1.1 Motivation

Symmetric distance, meaning that traveling from A to B has the same cost as traveling from B to A, is certainly the most common assumption to the metric traveling salesman problem. In fact, it is so common that the name (metric) traveling salesman problem (TSP) is usually associated with this symmetric version, while the more general case is explicitly referred to as *asymmetric* (ATSP).

It appears as if symmetry plays a vital role in view of approximations. For TSP it was known for over 40 years that a $\frac{3}{2}$-approximation is possible with the famous algorithm of Christofides [10] (or Christofides-Serdyukov, see [4]). Recently, Karlin et al. [22] showed a randomized approximation with an expected ratio of $\frac{3}{2} - \varepsilon$ for a small constant $\varepsilon > 0$. For ATSP, Svensson et al. [32] answered the longstanding open question for the existence of a constant factor approximation in the affirmative. Although the current state of the art was very recently established by Traub and Vygen [33] with the ratio of $22 + \varepsilon$, this still leaves a significant gap between the positive results for TSP and ATSP, whereas the currently known lower bounds by Karpinski et al. [23] of $\frac{123}{122}$ for TSP and $\frac{75}{74}$ for ATSP do not indicate such a vast difference. This raises the question of how symmetry truly affects approximability.

The assumption of symmetry does not seem very natural. A study by Martínez Mori and Samaranayake [28] shows that road networks exhibit asymmetry even when only the lengths of the shortest paths are considered. Phenomena like road blocks, one-way streets and rush hour can result in unbounded violations of symmetry while the triangle inequality remains satisfied. In comparison, restricting to distances that satisfy the triangle inequality is a reasonable assumption in all scenarios where visiting cities more than once is acceptable. Finding a shortest tour that visits each city *at least* once translates to metric TSP by taking the shortest path metric, also called *metric closure.*

The *asymmetry factor* is the maximum ratio between the length of the shortest paths from A to B and B to A over all cities A, B. The investigation in [28] revealed that most asymmetries are insignificantly small. With these few but existing significant asymmetries in mind, we consider spending exponential time with respect to some measure of the degree of asymmetry. Our basic objective is to salvage the approximability of TSP for ATSP by allowing this increase in runtime. Formally, we give parameterized approximations (see e.g., [26]), which means a guaranteed performance ratio and a runtime of the form $poly(n)f(k)$, where f is an arbitrary function, n is the size of the instance and k is a measure for asymmetry. This approach aims to offer efficiency for instances of low asymmetry and to improve our understanding of the challenges asymmetric distances pose to the design of approximation algorithms.

1.2   Our Results

We derive parameterized approximations based on the Christofides and the tree doubling algorithm with respective suitable parameters. Both parameters under study are bounded by the number of *asymmetric distances*, i.e., pairs of vertices (u, v) for which traveling from u to v is cheaper than traveling from v to u. Further, we combine these parameters with the asymmetry factor Δ [28] in the sense that they treat distances with asymmetry factor $\Delta \leq \beta$ for some $\beta \geq 1$ as symmetric, which shrinks both parameters to consider only the more severe β-*asymmetries* (distance from v to u is at least β times the distance from u to v). In particular, we derive parameterized approximations with

- ratio $\frac{7}{4} + \frac{3}{4}\beta$ for parameter $k = $ size of a vertex cover for the subgraph induced by the β-asymmetric distances (*generalized Christofides*);
- ratio $2 + \beta$ for parameter $z = $ minimum number of β-asymmetric distances in a minimum spanning arborescence (*generalized tree doubling*).

For $\beta = 1$, we prove the ratio of 2.5 to be tight for generalized Christofides. The lack of such a tightness result and further observations lead us to conjecture that generalized tree doubling is actually a 2-approximation for $\beta = 1$. Since the two parameters k and z are theoretically incomparable, we conduct experiments which show that generalized tree doubling frequently outperforms generalized Christofides with respect to parameter size.

The paper is organized as follows. In Sect. 3 we generalize the Christofides algorithm. Our main result, the more elaborate generalized tree doubling algorithm, is presented in Sect. 4. In Sect. 5 we give the combination with the asymmetry factor and Sect. 6 describes our experimental results. For the full version of this extended abstract, see [3].

1.3   Related Work

Conceptually, our approach can be seen as a study of *stability* with respect to asymmetry in the framework of *stability of approximation* by Böckenhauer et al. [6]. Probably the most extensively studied stability measure for (A)TSP is the β-triangle inequality, also called *parameterized* triangle inequality, which refers to the requirement $c(u, v) \leq \beta(c(u, w) + c(w, v))$ for all $u, v, w \in V$ with $u \neq v \neq w$. For ATSP with β-triangle inequality, the $\frac{1}{2(1-\beta)}$-approximation derived by Kowalik and Mucha [25] for $\beta \in (\frac{1}{2}, 1)$ improves upon a series of previous results [5,9,34] and is also known to be tight with respect to the cycle cover relaxation as lower bound. For TSP, the survey of Klasing and Mömke [24] gives a summary of the known results with β-triangle inequality.

Martínez Mori and Samaranayake [28] showed that the Christofides algorithm is $\frac{3}{2}$-stable with respect to the asymmetry factor, meaning that it can be used to compute a $\frac{3}{2}\Delta$-approximation for instances with asymmetry factor at most Δ.

So far, there are only a few parameterized approximations for (variations of) TSP. Marx et al. [27] consider ATSP on a restricted graph class called k-nearly-embeddable. They derive approximations where the ratio and the runtime depend on structural parameters of the given instance. A true parameterized approximation for a TSP type problem is given by Böckenhauer et al. in [7] for deadline TSP, a generalization of TSP where some cities have to be reached by the tour within a given deadline. They give a 2.5-approximation that requires exponential time only with respect to the number of cities with deadline.

Another interesting approach to invest moderate exponential time is given by Bonnet et al. in [8]. They derive a routine that allows to compute for any $r \leq n$ a $\log r$ approximation for ATSP that requires time $\mathcal{O}^*(2^{\frac{n}{r}})$.

2 Preliminaries

Throughout the paper, instances of ATSP are always simple complete directed graphs denoted by $G = (V, A, c)$ with non-negative cost function c on A. For $u, v \in V$, (u, v) denotes the *arc* from u to v and $c(u, v)$ denotes its cost. For an arc $(u, v) \in A$ we call $(v, u) \in A$ the *opposite* arc. To refer to the connections between vertices without regarding any directedness, for an arc $(u, v) \in A$ and its opposite arc, we call $\{u, v\}$ an *arc-pair* or simply *link*. Links can be thought of like edges in an undirected graph. If the cost function c satisfies the triangle inequality, i.e., $c(u, v) \leq c(u, w) + c(w, v)$ for all $u, v, w \in V$, we call G *metric*. If the graph is not clear from context, we use $V[G]$ and $A[G]$ to denote the vertices and arcs of G, respectively.

In a not necessarily complete graph G', a *trail* is a sequence of vertices where each vertex is equal to or has an arc to its successor. A *path* is a trail containing no vertex twice. *Circuit* and *cycle* denote a trail and a path where the last vertex has an arc to the first vertex, respectively. We denote a trail by $v_1, \ldots, v_n$ and a circuit by $(v_1, \ldots, v_n)$. A *tour* of G' is a cycle that visits each vertex of G'.

If G is metric, every trail can be turned into a path visiting the same vertices via a *metric shortcut* without increasing the cost, where metric shortcut means removing multiple occurrences of each vertex. All tours in G are valid ATSP solutions, and we use $c^*(G)$ to denote the cost of an optimal solution for G.

For $G = (V, A, c)$ we denote the vertex-induced subgraph of $V' \subseteq V$ by $G[V']$, the arc-induced subgraph of $A' \subseteq A$ by $G[A']$ and also the link-induced subgraph of a set of links E by $G[E]$. Slightly abusing notation, $G[V']$ and $G[A']$ then also inherit the weights of G. Further, for a subgraph G' of G, we use $c(G')$ to denote the sum of all arc costs in G'. We observe:

Lemma 1. *Let G be a metric graph and $V' \subseteq V$. Then, $G[V']$ is metric as well.*

Lemma 2. *Let G be a metric graph and $V' \subseteq V$. Then, $c^*(G[V']) \leq c^*(G)$.*

We also use one other transformation we call *minor*. Here, G' is a *minor* of G if there is a series of *contractions* which, starting from G, result in G'. A contraction of (u, v) replaces u and v with a single vertex uv and sets $c(w, uv) = \min\{c(w, u), c(w, v)\}$ and $c(uv, w) = \min\{c(u, w), c(v, w)\}$ for all $w \in V \setminus \{u, v\}$.

3 Generalized Christofides Algorithm

The Christofides algorithm [10] is a polynomial approximation for TSP with performance ratio $\frac{3}{2}$. On instance G it first computes a minimum spanning tree T

for G and then adds a minimum cost perfect matching M on the vertices V' of odd degree in T. The resulting subgraph is connected and each vertex has an even degree, so it is possible to compute an Eulerian cycle for it, which is a circuit of cost $c(T) + c(M)$ that visits all vertices. Metric shortcuts turn this circuit into a tour. Since taking every second edge in an optimal tour for $G[V']$ gives a perfect matching for the vertices of odd degree, the edges in M have a cost of at most $\frac{1}{2}c^*(G[V']) \leq \frac{1}{2}c^*(G)$. Together with the bound of $c^*(G)$ on the cost of T, this proofs the approximation ratio of $\frac{3}{2}$.

Regarding ATSP, the most dire problem of this approach is that combining T and M to an Eulerian circuit is impossible if some arcs point in the wrong direction, and it is unclear how to restrict T and M accordingly while keeping the relation of their cost to the optimum value. Due to this conceptual problem, we use a reduction to a TSP instance for which the Christofides algorithm can be applied. Observe that such a reduction cannot simply be designed by brute-force guessing the correct set of asymmetries in an optimal solution; fixing a subset of arcs to be in a solution cannot be modeled as an undirected instance. The design of our algorithm is instead based on a simple structural insight that allows the use of the Christofides algorithm on a symmetric subgraph.

We first explain an easier variant of the algorithm. The idea is to divide the graph into an asymmetric and a symmetric subgraph. For $G = (V, A, c)$ we define the set of *asymmetric links* by $E_a = \{\{u, v\} \mid u, v \in V, c(u, v) \neq c(v, u)\}$ and the set of *asymmetric* and *symmetric vertices* by $V_a = \{v \in V \mid \{u, v\} \in E_a \text{ for some } u \in V\}$, and $V_s = V \setminus V_a$, respectively.

We define the asymmetric subgraph by $G[V_a \cup \{v\}]$, where v is an arbitrary vertex in V_s, and the symmetric one by $G[V_s]$. Note that tours through both subgraphs can be merged at the overlap in v and turned into a tour of the whole graph with metric shortcuts. Combining in this way an exact solution for $G[V_a \cup \{v\}]$ and a $\frac{3}{2}$-approximate solution for $G[V_s]$, computed by the Christofides algorithm, overall yields a parameterized $\frac{5}{2}$-approximation with parameter $|V_a|$.

To improve this, consider a vertex cover VC of $G[E_a]$. The complement of VC forms an independent set in $G[E_a]$, implying that G contains no asymmetric links between vertices in $V_s \cup (V_a \setminus VC)$. This can be exploited to consider the smaller structural parameter z, the size of a vertex cover in $G[E_a]$. The improved algorithm uses a vertex cover VC in $G[E_a]$, selects a vertex $v \in V_s$ and considers $G[VC \cup \{v\}]$ as the asymmetric and $G[V \setminus VC]$ as the symmetric subgraph.

Using a simple $\mathcal{O}(m + 2^z z^2)$ algorithm (e.g. branching on the k^2-kernel as discussed in the introduction of [11] for "Bar Fight Prevention") for the minimal vertex cover for $G[E_a]$, and the dynamic programming algorithm by Held and Karp [17] for ATSP on $G[VC \cup \{v\}]$ in $\mathcal{O}(2^z z^2)$ yields the following.

Theorem 1. *Metric ATSP can be $\frac{5}{2}$-approximated in $\mathcal{O}(n^3 + 2^z z^2)$ where z is the size of a minimum vertex cover of the subgraph induced by all asymmetric links.*

Instead of exact algorithms for the vertex cover for $G[E_a]$ and the solution on $G[VC \cup \{v\}]$, we can also use approximations. A 2-approximation for vertex cover and the $\frac{2}{3} \log n$-approximation of Feige and Singh [14] on $G[VC \cup \{v\}]$ yields the following interesting result.

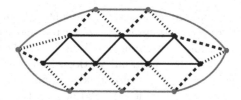

Fig. 1. G_k for $k = 7$: Black and gray links are symmetric with cost 2, dotted links are symmetric with cost 1. Dashed links are asymmetric, with cost 1 from gray to black vertex and cost 2 from black to gray vertex.

Corollary 1. *Metric ATSP can be* $(\frac{2}{3}\log x + \frac{3}{2})$-*approximated in polynomial time, where* $x = \min(2z + 1, |V_a|)$, V_a *is the set of asymmetric vertices and* z *is the size of a minimum vertex cover for the subgraph induced by all asymmetric links.*

This improves upon the approximation ratio of $\frac{2}{3}\log n$ if $\frac{x}{n} < 2^{-\frac{9}{4}}$, meaning that $G[VC \cup \{v\}]$ only contains a sufficiently small fraction of the vertices. We note that the result of Asadapour et al. [2] gives a polynomial $(8\log(z)/\log\log(z) + \frac{3}{2})$-approximation, which is asymptotically stronger but less suitable for the instances with small values of z we are interested in.

Further, note that one can also use any approximation for TSP (not just the Christofides algorithm) for the symmetric subgraph and obtain an $(\alpha + 1)$-approximation for ATSP from any α-approximation for TSP.

It remains to see if this approach can be improved. Aiming for a smaller parameter seems difficult as this would not split off a symmetric subgraph. Regarding a possible improvement of the ratio, one might hope to salvage the ratio of $\frac{3}{2}$ for TSP, obtained by the Christofides algorithm, for ATSP. However, such an improvement requires a different algorithmic strategy as the ratio in Theorem 1 is asymptotically tight, which can be shown as follows.

We define a family of graphs G_k for $k \in \mathbb{N}$, $k > 2$ such that the approximation ratio converges to 2.5 for increasing k, Fig. 1 describes G_7. The black zig-zag pattern is the textbook example for the tightness of the Christofides algorithm. The idea is that the gray vertices build the minimum vertex cover such that the black zig-zag pattern becomes the symmetric instance. The gray cycle is then the asymmetric subgraph and solving it exactly yields a tour of cost $2k$. Together with the approximation on the symmetric subgraph, which converges to $3k$, this results in a tour of length $5k$. As the optimal tour takes the dotted and dashed links in the cheaper direction and has cost $2k$, we deduce that 2.5 is asymptotically tight for Theorem 1.

4   Generalized Tree Doubling Algorithm

One other widely known approximation for TSP is the tree doubling algorithm. It computes a minimum spanning tree (MST) and doubles every edge in it to

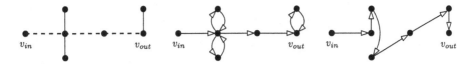

Fig. 2. Exemplary construction for a suitable path χ_i. Left: spanning tree with P_i dashed; Middle: trail through partially doubled edges; Right: resulting path.

ensure the existence of an Eulerian circuit. Since the circuit uses every MST edge exactly twice, it is twice as expensive as the tree, which itself is at most as expensive as the optimum tour. Thus, transforming the circuit with metric shortcuts gives a 2-approximation. To adapt this approach to ATSP we use a *minimum spanning arborescence* (MSA) as the directed variant of an MST. Tree doubling then runs into trouble when the cost of an opposite arc is arbitrarily higher than the direction contained in the MSA. These arcs are the core of the problem and hence our basis to generalize the tree doubling algorithm.

Formally, we call $(u,v) \in A$ a *one-way arc* in $G = (V, A, c)$ if $c(u,v) < c(v,u)$. In a nutshell, our algorithm removes all one-way arcs from an MSA, computes a tour for each resulting connected component by an altered tree doubling routine and uses exponential time in the number of removed one-way arcs to connect these subtours to a solution for the whole graph. For a best runtime, we hence want to keep the number of one-way arcs in the starting MSA as small as possible. For our parametrization, we formally define k to be the minimum number of one-way arcs in an MSA for G.

At first glance, it might seem that finding an MSA with k one-way arcs is a difficult task. However this can be accomplished by searching for an MSA with the altered weight function c' defined by $c'(e) = |V|c(e) + 1$ if e is a one-way arc, and $c'(e) = |V|c(e)$, otherwise. Trying every possible root vertex with this altered weight function, and the Chu–Liu/Edmonds algorithm [13] with Fibonacci heaps [15] to compute the MSA, yields the following result.

Lemma 3. *Let G be a metric ATSP instance, then an MSA of G with a minimum number of one-way arcs can be computed in $\mathcal{O}(n^3)$.*

With this best MSA, we can describe our generalized tree doubling algorithm. Let T be the MSA for G computed with Lemma 3, and let $T_1, \ldots, T_{k+1}$ be the connected components in the graph created by deleting all k one-way arcs from T. We construct a graph M by contracting each set of vertices $V[T_i]$ to one vertex v_i^M with our notion of contraction to a minor. This results in $V[M] = \{v_1^M, \ldots, v_{k+1}^M\}$ and for all $v_i^M, v_j^M \in V[M]$ with $i \neq j$, $c(v_i^M, v_j^M) = \min \left(\{ c(t_i, t_j) \mid t_i \in V[T_i], t_j \in V[T_j] \} \right)$.

Lemma 4. *Let G be a metric ATSP instance and M be minor of G, then $c^*(M) \leq c^*(G)$.*

Since M only contains $k + 1$ vertices, we brute-force an optimal tour τ' for M. It remains to extend τ' to a tour of G. Consider a vertex v_i^M in M (which corresponds to the component T_i) and assume w.l.o.g. that in τ' it is preceded by v_{i-1}^M

and precedes v_{i+1}^M. Further, let $(v_{out}^{T_{i-1}}, v_{in}^{T_i})$ and $(v_{out}^{T_i}, v_{in}^{T_{i+1}})$ be the cheapest arc between T_{i-1} and T_i, and T_i and T_{i+1}, respectively. The goal is to find a path χ_i that starts in $v_{in}^{T_i}$, ends in $v_{out}^{T_i}$, and spans all vertices in T_i (formally a solution to s-t-path TSP for $G[T_i]$ with $s = v_{in}^{T_i}$ and $t = v_{out}^{T_i}$). Replacing v^{T_i} in τ' by χ_i for each i turns τ' into a tour. However, the cost of χ_i has to be bounded.

Such a path χ_i through T_i can be found by adapting the tree doubling algorithm. We treat T_i as undirected and double all its edges that are not on the shortest path P_i from $v_{in}^{T_i}$ to $v_{out}^{T_i}$. The resulting graph contains an Eulerian trail from $v_{in}^{T_i}$ to $v_{out}^{T_i}$, which is turned into a path by metric shortcuts ensuring that $v_{in}^{T_i}$ and $v_{out}^{T_i}$ remain start and end node, see Fig. 2 for an example. Observe that we cannot use any of the better approximations for s-t-path TSP, such as [19], since the subgraph induced by the vertices in T_i is not necessarily completely symmetric. Further, even if this was possible, the only information we can use to compare the tour through T_i with the optimum for the whole graph G are the arcs from the MSA, which in the worst case always results in a ratio of 2.

For the cost of χ_i, note that it contains for each arc (u, v) in T_i at most both (u, v) and (v, u). Since there are no one-way arcs in T_i, any opposite arc is at most as expensive as the original arc in T_i. Consequently, the cost of χ_i is at most twice the cost of the arcs in T_i and the sum of all χ_i is at most $2c^*(G)$. In combination with the cost of at most $c^*(G)$ for τ', this yields:

Theorem 2. *Metric ATSP can be 3-approximated in $\mathcal{O}(2^k \cdot k^2 + n^3)$, where k is the minimum number of one-way arcs in a minimum spanning arborescence.*

Contrary to the approach in Sect. 3, we cannot plug in some approximation to find a good tour τ' for M to derive something like Corollary 1. Note that the minor M is not necessarily metric since contractions do not preserve the triangle inequality. Still, one might ask if M, as minor of a metric graph, has useful structural properties. However, the following result discourages such ideas.

Lemma 5. *Let G be a complete, directed graph with cost function c. Then, there exists a complete, metric graph $\hat{G}$ of which G is a minor.*

Computing a tour for M is related to the *generalized traveling salesman problem* (GTSP) which can be tracked back to publications of Henry-Labordère and Saksena [18, 31]. Given a partition of the cities into r sets, GTSP asks for a minimum cost tour containing (at least) one vertex from each of the r sets. Unfortunately, there are no known efficient ways to solve or approximate GTSP. However, we observe that using an optimal GTSP tour for the vertex sets corresponding to $T_1, \ldots, T_{k+1}$ instead of the tour through M still yields a 3-approximation. In fact, this remains true even if we fix one arbitrary city for each set, which yields a graph M' that is just an induced subgraph and hence metric. For this simplified approach, the ratio 3 is indeed asymptotically tight.

Aside from the fact that we did not find a tight example for Theorem 2, seeing that the choice of any arbitrary vertex still yields a 3-approximation causes us to conjecture that our more sophisticated generalization of the tree doubling algorithm is in fact a 2-approximation. Proving such a ratio however requires an exploitable connection between the cost for the paths χ_i and the cost of τ'.

5   Trading Approximation Quality for Runtime

In real life, we expect instances with many small asymmetries which have little impact but lead to relatively large parameter values. Therefore, ignoring asymmetric links where both directions have similar cost and trading some approximation quality for running time yields an intriguing perspective. As a formal way to describe moderate asymmetry, we use the asymmetry factor of Martínez Mori and Samaranayake [28] as introduced in Sect. 1.3. Since Δ is commonly used for the maximum degree, and we want to describe variable restrictions of the asymmetry factor, we use β instead. For $\beta \geq 1$ we call a link $\{u, v\}$ or arc (u, v) β-symmetric if $\frac{1}{\beta} \leq \frac{c(u,v)}{c(v,u)} \leq \beta$, otherwise it is called β-asymmetric. We show that our algorithms support a quality-runtime trade-off with respect to β.

5.1   Relaxed Generalized Christofides Algorithm

For a given β we modify the algorithm presented in Sect. 3 by treating every β-symmetric link as symmetric. This results in parametrization by the vertex cover of the subgraph induced by all β-asymmetric links. We denote this parameter by z_β. Since the β-symmetric subgraph is not completely symmetric, the Christofides algorithm cannot be directly used. Martínez Mori and Samaranayake [28] showed that it is $\frac{3}{2}$-stable by replacing every link with an undirected edge and assigning it the cost of the more expensive direction. Combined with the arguments used for Theorem 1, this gives a parameterized $(\frac{3}{2}\beta + 1)$-approximation for parameter z_β. This can be improved by turning the β-symmetric subgraph symmetric by assigning the cost of the cheaper direction. Although this may not yield a metric graph, it suffices that the original graph is metric to prove that the Christofides algorithm yields a good solution.

Theorem 3. *For any $\beta \geq 1$, metric ATSP can be $(\frac{3}{4}\beta + \frac{7}{4})$-approximated in $\mathcal{O}(n^3 + 2^{z_\beta} z_\beta^2)$ where z_β is the size of a minimum vertex cover of the subgraph induced by all β-asymmetric links.*

5.2   Relaxed Generalized Tree Doubling Algorithm

For the generalized tree doubling algorithm, we define a β-one-way arc as a one-way arc that is β-asymmetric. We denote by k_β the minimum number of β-one-way arcs in an MSA. Note that the strategy in Lemma 3 can also be used to find an MSA with k_β β-one-way arcs. The generalization of Theorem 2 is straightforward, instead of deleting all one-way arcs we only delete β-one-way arcs. This results in fewer components and a smaller graph M_β. The drawback is a change to the cost analysis: so far, we considered the component trees T_i to be symmetric. Now for every $(u, v) \in A[T_i]$, the opposite arc (v, u) can be up to β times as expensive. The adjusted tree doubling algorithm for the path through T_i uses every arc in T_i and its opposite arc at most once, which in total costs at most $(1 + \beta)c^*(G)$. Combined with the cost of at most $c^*(G)$ for an optimum tour through M_β, this yields:

Theorem 4. *For any $\beta \geq 1$, metric ATSP can be $(2 + \beta)$-approximated in $\mathcal{O}(2^{k_\beta} k_\beta^2 + n^3)$ where k_β is the minimum number of β-one-way arcs in an MSA.*

6 Experimental Results

To test the practical viability of our algorithms, we implemented them in their relaxed form (see Sect. 5) to also observe their behavior when certain asymmetries are ignored. We evaluated on the asymmetric graphs from the TSPLIB collection [29], the standard benchmark for TSP solvers, and on a set of specific ATSP instances extracted from road networks by Rodríguez and Ruiz [30].

6.1 Implementation Details

Our implementation (available on GitHub[1]) is written in *Python 3*, except for the vertex cover solver, which is written in *Java*. We used the Python library *NetworkX* [16] for graph manipulation, the C++ library *Lemon* [12] for computing MSAs, and *Concorde* [1] for solving TSP exactly. Since *Concorde* is a TSP solver, we transformed the ATSP instances into TSP instances with the transformation presented by Jonker and Volgenant [20,21].

We note that the runtime of our implementations is incomparable to state of the art ATSP solvers. Among others, the reason is Python's inherently low performance and the inefficiency of solving ATSP with Concorde. However, this is of no importance for our evaluation of approximation ratio, parameter size, and the proof of concept.

6.2 Experiments

In the TSPLIB there are 19 asymmetric instances ranging from 17 up to 443 vertices. As some of the instances are not metric, we computed the metric closure of each graph. The instances' names contain the number of vertices (e.g., *ftv33*) and similar names indicate similar properties. For example, instances starting with *rbg* have relatively high symmetry and a high number of zero-cost arcs. Contrasting that, the instances with prefix *ftv* contain little symmetry, but most asymmetric links are only moderately asymmetric. Most instances are rather small, with only 6 of the instances having more than 70 vertices. We ignored the instance *br17* as its metric closure is completely symmetric.

For each TSPLIB instance we executed each algorithm five times with different values for β and recorded the value of the parameter as well as the approximation ratio. Starting with $\beta = 1$ (which corresponds to 100% of the asymmetric links), we raised the value of β each step, reducing the number of asymmetric links treated as asymmetric to a quarter of the previous experiment. Some instances include many zero-cost arcs, so there is no value of β ignoring those. We considered zero-cost arcs to have a small positive cost (set to 0.1) when calculating the asymmetry factor, thus treating links with a small additive error

---
[1] https://github.com/Blaidd-Drwg/atsp-approximation.

Table 1. Experimental results on TSPLIB instances with percentage of asymmetric links that were treated as asymmetric shown in the column header. Each cell contains parameter value and approximation factor, separated by a slash (trivial parameter value 0 omitted in 0% column). Superiority in the sense of smaller kernel or better approximation ratio is highlighted with bold font.

	Generalized Christofides algorithm					Generalized tree doubling algorithm				
	100%	25%	6.25%	1.56%	0%	100%	25%	6.25%	1.56%	0%
ft53	**53/1.00**	29/1.54	13/1.70	6/1.69	**1.72**	**45/1.08**	**25/1.36**	**6/1.42**	**1/1.57**	1.97
ft70	69/1.02	34/1.24	12/1.26	7/1.41	**1.24**	**64/1.02**	**27/1.13**	**4/1.20**	**2/1.21**	1.28
ftv170	155/1.17	123/1.38	**97/1.57**	**64/1.85**	2.37	**108/1.14**	**107/1.14**	103/1.21	**75/1.46**	**1.81**
ftv33	29/**1.12**	19/1.45	11/**1.43**	5/1.56	**1.33**	19/1.34	16/**1.34**	11/1.44	**2/1.23**	1.50
ftv35	32/**1.07**	21/1.51	12/1.55	6/1.49	**1.38**	**23/1.15**	**17/1.23**	11/**1.47**	**2/1.28**	1.58
ftv38	33/**1.13**	23/1.38	12/**1.43**	7/1.47	**1.39**	**23/1.24**	**18/1.33**	12/1.54	**3/1.30**	1.62
ftv44	40/**1.09**	32/1.38	19/1.46	10/1.56	**1.54**	**32/1.24**	**25/1.41**	**18/1.41**	**7/1.50**	1.79
ftv47	44/**1.05**	32/1.47	19/1.66	13/1.65	1.66	**35/1.09**	**30/1.16**	**19/1.34**	**9/1.38**	**1.58**
ftv55	49/**1.13**	38/1.44	**23/1.57**	15/1.65	1.84	**37/1.20**	**32/1.26**	25/1.34	**12/1.58**	2.00
ftv64	57/**1.11**	46/1.46	**30/1.66**	18/1.73	1.72	**50/1.10**	**43/1.15**	31/1.29	**14/1.71**	1.45
ftv70	63/**1.11**	50/1.43	**32/1.64**	20/1.72	1.96	**53/1.26**	**47/1.14**	33/1.21	**16/1.57**	1.51
kro124p	99/1.11	86/1.30	65/1.36	40/1.41	**1.24**	**81/1.06**	**70/1.13**	**57/1.20**	**34/1.28**	1.37
p43	15/1.01	6/1.01	0/1.01	0/1.01	1.01	**0/1.01**	**0/1.01**	0/1.01	0/1.01	1.01
rbg323	**148/1.02**	59/**1.17**	43/**1.19**	18/1.30	1.34	235/1.09	22/1.27	6/1.27	0/1.30	**1.30**
rbg358	**108/1.01**	47/**1.13**	27/**1.15**	22/**1.14**	1.18	232/1.03	**39/1.14**	18/1.19	13/1.20	1.22
rbg403	**125/1.01**	41/**1.12**	11/1.26	11/1.26	1.17	**113/1.05**	**30/1.14**	0/1.24	0/1.24	1.24
rbg443	**138/1.00**	43/**1.14**	12/1.24	12/1.24	1.15	**127/1.04**	**32/1.17**	0/1.24	0/1.24	1.24
ry48p	47/1.20	37/1.40	23/1.46	11/1.47	1.16	**28/1.10**	**22/1.14**	11/1.24	5/1.29	1.21

as symmetric in case of these otherwise undauntedly asymmetric one-way arcs of cost 0. Note that we did not alter the instance, but only used these additive errors for relaxation decisions. Finally, β was set to ∞, such that the graph is treated as completely symmetric. This results in the non-generalized versions of the tree doubling and Christofides algorithm. The results are shown in Table 1.

The second dataset contains 450 ATSP instances based on travel distances between random points sampled across different regions and cities in Spain. The graphs in this dataset have between 50 and 500 vertices. On average 98.8% of the links are asymmetric (std. dev. 1.08%) and no graph contains arcs of cost zero. Most links are however only slightly asymmetric: denoting by *asymmetry factor* the relative difference between the cost of a links more expensive arc and its opposite arc, the mean asymmetry factor is 3.55% on average over all graphs (std. dev. 0.040%). The median asymmetry factor is 1.32% (std. dev. 1.56%) on average. There are however also links with large asymmetry factor. The highest asymmetry factor is 15.0 (std. dev. 58.8) on average. Overall this makes the graphs in the second dataset very relevant to the algorithms we present. Unfortunately, due to computational constraints and the size of the dataset and the graphs therein, we could only determine the values of the parameters and not the cost of all optimal tours and the obtained approximation ratio. Figure 3 presents the relative value of z and k for different values of β.

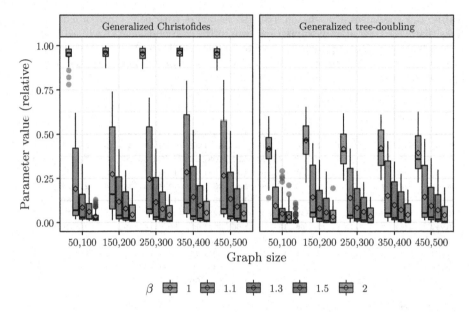

Fig. 3. Parameter values relative to graph size for generalized Christofides and tree doubling algorithms for different values of β. Each box spans the second and third quartile of the data and whiskers extend for 1.5 inter-quartile-ranges. The median is marked as a line, the mean as a rhombus and outliers as disks.

6.3   Evaluation

First, we note that most graphs in the TSPLIB contain very little symmetry. This leads to large parameter values for $\beta = 1$, i.e., only some graphs with more than 10% symmetry have parameter values below 50% of the graph size. Still, we observe that the approximation factor is always far below the upper bound, never exceeding even 2.0. Also, we see that interpolating β to reduce the number of relevant asymmetric links produces a valuable trade-off between approximation quality and parameter value. Comparing both algorithms, we observe that on the majority of instances and values for β the generalized tree doubling algorithm produces smaller parameter values.

This can also be observed on the instances of the second dataset, which we consider to be more representative of realistic inputs. We want to highlight that the parameters are significantly smaller than the size of the input graphs even for small values of β. E.g., for the generalized tree doubling algorithm with $\beta = 1.1$ the median relative parameter value over all instances is 0.045. It also seems that the relative size of the parameters is stable for different input sizes.

These results underline the practicality of our approach, especially with regards to the parameter values obtained by choosing a suitable β.

References

1. Applegate, D., Bixby, R., Cook, W., Chvátal, V.: On the solution of traveling salesman problems. Doc. Math. **111**, 645–656 (1998)
2. Asadpour, A., Goemans, M.X., Madry, A., Gharan, S.O., Saberi, A.: An O(log n/log log n)-approximation algorithm for the asymmetric traveling salesman problem. Oper. Res. **65**(4), 1043–1061 (2017). https://doi.org/10.1287/opre.2017.1603
3. Behrendt, L., Casel, K., Friedrich, T., Lagodzinski, J.A.G., Löser, A., Wilhelm, M.: From symmetry to asymmetry: generalizing TSP approximations by parametrization. CoRR abs/1911.02453 (2019). http://arxiv.org/abs/1911.02453
4. van Bevern, R., Slugina, V.A.: A historical note on the 3/2-approximation algorithm for the metric traveling salesman problem. Hist. Math. **53**, 118–127 (2020). https://www.sciencedirect.com/science/article/pii/S0315086020300240
5. Bläser, M., Manthey, B., Sgall, J.: An improved approximation algorithm for the asymmetric TSP with strengthened triangle inequality. J. Discret. Algorithms **4**(4), 623–632 (2006). https://doi.org/10.1016/j.jda.2005.07.004
6. Böckenhauer, H., Hromkovic, J., Klasing, R., Seibert, S., Unger, W.: Towards the notion of stability of approximation for hard optimization tasks and the traveling salesman problem. Theor. Comput. Sci. **285**(1), 3–24 (2002). https://doi.org/10.1016/S0304-3975(01)00287-0
7. Böckenhauer, H., Hromkovic, J., Kneis, J., Kupke, J.: The parameterized approximability of TSP with deadlines. Theor. Comput. Sci. **41**(3), 431–444 (2007). https://doi.org/10.1007/s00224-007-1347-x
8. Bonnet, É., Lampis, M., Paschos, V.T.: Time-approximation trade-offs for inapproximable problems. J. Comput. Syst. Sci. **92**, 171–180 (2018). https://doi.org/10.1016/j.jcss.2017.09.009
9. Chandran, L.S., Ram, L.S.: Approximations for ATSP with parametrized triangle inequality. In: Alt, H., Ferreira, A. (eds.) STACS 2002. LNCS, vol. 2285, pp. 227–237. Springer, Heidelberg (2002). https://doi.org/10.1007/3-540-45841-7_18
10. Christofides, N.: Worst-case analysis of a new heuristic for the travelling salesman problem. Technical report 388, Graduate School of Industrial Administration, Carnegie Mellon University (1976)
11. Cygan, M., et al.: Parameterized Algorithms. Springer, Cham (2015). https://doi.org/10.1007/978-3-319-21275-3
12. Dezső, B., Jüttner, A., Kovács, P.: LEMON - an open source C++ graph template library. Electron. Notes Theor. Comput. Sci. **264**(5), 23–45 (2011). https://doi.org/10.1016/j.entcs.2011.06.003
13. Edmonds, J.: Optimum branchings. J. Res. Natl. Bur. Stan. Sect. B Math. Math. Phys. **71B**(4), 233 (1967)
14. Feige, U., Singh, M.: Improved approximation ratios for traveling salesperson tours and paths in directed graphs. In: Charikar, M., Jansen, K., Reingold, O., Rolim, J.D.P. (eds.) APPROX/RANDOM -2007. LNCS, vol. 4627, pp. 104–118. Springer, Heidelberg (2007). https://doi.org/10.1007/978-3-540-74208-1_8
15. Gabow, H.N., Galil, Z., Spencer, T.H., Tarjan, R.E.: Efficient algorithms for finding minimum spanning trees in undirected and directed graphs. Combinatorica **6**(2), 109–122 (1986). https://doi.org/10.1007/BF02579168
16. Hagberg, A., Schult, D., Swart, P.: Exploring network structure, dynamics, and function using NetworkX. In: Proceedings of the 7th Python in Science Conference, pp. 11–15 (2008). http://conference.scipy.org/proceedings/SciPy2008/paper_2/

17. Held, M., Karp, R.M.: A dynamic programming approach to sequencing problems. J. Soc. Ind. Appl. Math. **10**(1), 196–210 (1962)

18. Henry-Labordère, A.L.: The record balancing problem: a dynamic programming solution of a generalized traveling salesman problem. RAIRO **B-2**, 43–49 (1969)

19. Hoogeveen, J.A.: Analysis of Christofides' heuristic: some paths are more difficult than cycles. Oper. Res. Lett. **10**(5), 291–295 (1991). https://doi.org/10.1016/0167-6377(91)90016-I

20. Jonker, R., Volgenant, T.: Transforming asymmetric into symmetric traveling salesman problems. Oper. Res. Lett. **2**(4), 161–163 (1983)

21. Jonker, R., Volgenant, T.: Transforming asymmetric into symmetric traveling salesman problems: erratum. Oper. Res. Lett. **5**(4), 215–216 (1986)

22. Karlin, A., Klein, N., Gharan, S.O.: A (slightly) improved approximation algorithm for metric TSP. In: Khuller, S., Williams, V.V. (eds.) Proceedings of the STOC 2021, pp. 32–45. ACM (2021). https://doi.org/10.1145/3406325.3451009

23. Karpinski, M., Lampis, M., Schmied, R.: New inapproximability bounds for TSP. J. Comput. Syst. Sci. **81**(8), 1665–1677 (2015). https://doi.org/10.1016/j.jcss.2015.06.003

24. Klasing, R., Mömke, T.: A modern view on stability of approximation. In: Böckenhauer, H.-J., Komm, D., Unger, W. (eds.) Adventures Between Lower Bounds and Higher Altitudes. LNCS, vol. 11011, pp. 393–408. Springer, Cham (2018). https://doi.org/10.1007/978-3-319-98355-4_22

25. Kowalik, Ł, Mucha, M.: Two approximation algorithms for ATSP with strengthened triangle inequality. In: Dehne, F., Gavrilova, M., Sack, J.-R., Tóth, C.D. (eds.) WADS 2009. LNCS, vol. 5664, pp. 471–482. Springer, Heidelberg (2009). https://doi.org/10.1007/978-3-642-03367-4_41

26. Marx, D.: Parameterized complexity and approximation algorithms. Comput. J. **51**(1), 60–78 (2008). https://doi.org/10.1093/comjnl/bxm048

27. Marx, D., Salmasi, A., Sidiropoulos, A.: Constant-factor approximations for asymmetric TSP on nearly-embeddable graphs. In: Proceedings of APPROX/RANDOM 2016, pp. 16:1–16:54. Schloss Dagstuhl - Leibniz-Zentrum für Informatik (2016). https://doi.org/10.4230/LIPIcs.APPROX-RANDOM.2016.16

28. Mori, J.C.M., Samaranayake, S.: Bounded asymmetry in road networks. Sci. Rep. **9**(11951), 1–9 (2019). https://doi.org/10.1038/s41598-019-48463-z

29. Reinelt, G.: TSPLIB–a traveling salesman problem library. INFORMS J. Comput. **3**(4), 376–384 (1991). https://doi.org/10.1287/ijoc.3.4.376

30. Rodríguez, A., Ruiz, R.: The effect of the asymmetry of road transportation networks on the traveling salesman problem. Comput. Oper. Res. **39**(7), 1566–1576 (2012). https://doi.org/10.1016/j.cor.2011.09.005

31. Saksena, J.P.: Mathematical model of scheduling clients through welfare agencies. Comput. Oper. Res. J. **8**, 185–200 (1970)

32. Svensson, O., Tarnawski, J., Végh, L.A.: A constant-factor approximation algorithm for the asymmetric traveling salesman problem. J. ACM **67**(6), 37:1-37:53 (2020). https://doi.org/10.1145/3424306

33. Traub, V., Vygen, J.: An improved approximation algorithm for ATSP. In: Proceedings of STOC 2020, pp. 1–13. ACM (2020). https://doi.org/10.1145/3357713.3384233

34. Zhang, T., Li, W., Li, J.: An improved approximation algorithm for the ATSP with parameterized triangle inequality. J. Algorithms **64**(2–3), 74–78 (2009). https://doi.org/10.1016/j.jalgor.2008.10.002

A Poly-log Competitive Posted-Price Algorithm for Online Metrical Matching on a Spider

Max Bender[1]([⊠]), Jacob Gilbert[2]([⊠]), and Kirk Pruhs[1]([⊠])[iD]

[1] Computer Science Deparment, University of Pittsburgh,
Pittsburgh, PA 15213, USA
{mcb121,kirk}@pitt.edu
[2] Computer Science Department, University of Maryland, College Park, USA
jgilber8@umd.edu

Abstract. Motivated by demand-responsive parking pricing systems we consider posted-price algorithms for the online metrical matching problem. Our main result is a polylog competitive posted-price algorithm in the case that the metric space is a spider.

1 Introduction

1.1 Motivation and Problem Statement

SFpark is San Francisco's system for managing the availability of on-street parking [2,3,29]. The goal of SFpark is to reduce the time and fuel wasted by drivers searching for an open parking spot. The system monitors parking usages using sensors embedded in the pavement and distributes this information in real-time to drivers via SFpark.org and phone apps. SFpark periodically adjusts parking meter pricing to manage demand, to lower prices in underutilized areas, and to raise prices in overutilized areas. Prices can range from a minimum of 25 cents to a maximum of 7 dollars per hour during normal hours, with a 18 dollars per hour cap for special events such as baseball games or street fairs. Several other cities in the world have similar demand-responsive parking pricing systems, for example Calgary has had the ParkPlus system since 2008 [1].

One natural simple model of the problem of centrally assigning drivers to parking spots to minimize time and fuel usage is the online metrical matching problem. The setting for online metrical matching consists of a collection of k servers (the parking spots) located at various locations within a metric space. The algorithm then sees an online sequence of requests over time that arrive at various locations in the metric space (the drivers arriving to look for a parking spot). In response to a request, the online algorithm must match the request

Supported in part by NSF grants CCF-1421508 and CCF-1535755, and an IBM Faculty Award.

E. Bampis and A. Pagourtzis (Eds.): FCT 2021, LNCS 12867, pp. 67–84, 2021.
https://doi.org/10.1007/978-3-030-86593-1_5

(car) to some server (parking spot) that has not been previously matched. Conceptually we interpret this matching as the request (car) moving to the location of the matched server (parking spot). The objective goal is to minimize the aggregate distance traveled by the requests (cars).

In order to be implementable within the context of SFpark, algorithms for online metric matching must be posted-price algorithms. In this setting, posted-price means that before each request arrives, the algorithm sets a price on each unused server (parking spot) without knowing the location where the next request will arrive. We assume each request is a selfish agent who moves to the available server (parking spot) that minimizes the sum of the price of that server (parking spot) and the distance to that server (parking spot). The objective remains to minimize the aggregate distance traveled by the requests (cars). Thus conceptually, the objective of the parking pricing agency is minimizing social cost, not maximizing revenue.

Research into posted-price algorithms for online metrical matching was initiated in [14] as part of a broader study of the use of posted-price algorithms to minimize social cost in online optimization problems. As a posted-price algorithm is a valid online algorithm, one can not expect to obtain a better competitive ratio for posted-price algorithms than what is achievable by general online algorithms. So this research line has primarily focused on problems where the optimal competitive ratio achievable by an online algorithm is (perhaps approximately) known and seeks to determine whether a comparable competitive ratio can be (again perhaps approximately) achieved by a posted-price algorithm. The higher level goal is to determine the increase in social cost that is necessitated by the restriction that the algorithm has to use posted prices to incentivize selfish agents, instead of being able to mandate agent behavior.

Before stating our results on posted-price algorithms for online metric matching we want to lay the groundwork by reviewing past work on posted-price algorithm design techniques, past work on general online algorithms for online metric matching, and past work on posted-price algorithms for online metric matching.

1.2 Past Work

The most obvious algorithmic design approach for posted-price problems is to directly design a pricing algorithm from scratch, as is done for metrical task systems in [15]. Two less direct algorithmic design paradigms have emerged in the literature. The first algorithmic design paradigm is what we will call *mimicry*. A posted-price algorithm A *mimics* an online algorithm B if the probability that B will take a particular action is equal the probability that a self-interested agent will choose this same action when the prices of actions are set using A. For example, [18] shows how to set prices to mimic the $O(1)$-competitive algorithm Slow-Fit from [8,9] for the problem of minimizing makespan on related machines. For some problems it is not possible to mimic known competitive algorithms using posted prices. For such problems, another algorithmic design paradigm is what we will call *monotonization*. In the monotonization algorithm design approach, one first seeks to characterize the online algorithms that can be mimicked, and then design such a mimicable online algorithm (the reason for using this terminology is that in all the examples in the literature, this characterization involves some sort of monotonicity property). For example, monotonization is used in [15] to obtain an $O(k)$-competitive posted-price algorithm for the k-server problem on a line metric, in [16] to obtain an $O(k)$-competitive posted-price algorithm for the k-server problem on a tree metric, and in [21] to obtain an $O(1)$-competitive posted-price algorithm for minimizing maximum flow time on related machines. In all of these examples, the competitive ratio achievable by the posted-price algorithm is comparable to the best competitive ratio achievable by a general online algorithm, thus showing that there is minimal increase in social cost necessitated by the use of posted-prices.

Let us now turn to known results for general algorithms for online metric matching in a general metric. There is a deterministic online algorithm that is $(2k - 1)$-competitive, and no deterministic online algorithm can achieve a better competitive ratio in a star metric [22,23]. Using HST's (Hierarchically Separated Trees) [11,12,17,25] obtained an $O(\log^3 k)$-competitive randomized algorithm. The algorithm in [25] can be viewed as combining two online randomized metric matching algorithms:

HST Algorithm: An $O(\log k)$-competitive algorithm for $O(\log k)$-HST's.

Uniform Metric Space Algorithm: The natural $O(\log k)$-competitive for a uniform metric space.

Later [10] obtained an $O(\log^2 k)$-competitive randomized algorithm by replacing the HST algorithm used by [25] by a $O(\log k)$-competitive algorithm for 2-HST's.

Better competitive ratios for online metric matching can be achieved on some natural metric spaces. Let us first consider a line metric. An $O(k^{.59})$-competitive deterministic algorithm was given in [4]. Subsequently several different $O(\log k)$-competitive randomized algorithms are given in [20]. These algorithms leverage special properties of HST's constructed from a line metric. [20] also showed that the natural Harmonic algorithm is $O(\log \Delta)$-competitive, where Δ is the ratio of the distance between the furthest two servers and the distance between the closest two servers. And by applying a standard doubling approach, [20] showed how to convert this Harmonic algorithm into an $O(\log k)$-competitive algorithm. An $O(\log^2 k)$-competitive deterministic online algorithm, called RM, was given in [26], and this was later improved to $O(\log k)$ in [28]. An $\Omega(\log k)$ lower bound on the competitive ratio for certain types of natural algorithms is given in [5,24]. A 9.001 lower bound on the competitive ratio of deterministic algorithm is given in [19]. This was improved to an $\Omega(\sqrt{\log k})$ lower bound on the competitive ratio for randomized algorithms in [27].

[26] also showed that for every metric space the competitive ratio of the RM algorithm is at most $O(\log^2 k)$ times the optimal competitive ratio achievable by a deterministic algorithm on that metric space.

We now turn to posted-price algorithms for online metric matching. [14] shows that the online algorithm Harmonic is mimicable on a line metric, and thus, using the results from [20], obtain an $O(\log \Delta)$-competitive randomized posted-price algorithm for a line metric.

[13] considered posted-price algorithms for tree metrics. [13] adopts the monotization approach, and identifies a monotonicity property that characterizes mimicable algorithms for online metrical matching in tree metrics. This monotization property is that, as a request arrival location moves closer to a server location, the probability that the request uses that server can not decrease. While this monotization property might seem innocuous at first, standard algorithmic approaches are seemingly hopelessly nonmonotone. For example, there is no hope to mimic any of the online algorithms that are based on HST's as HST's by their very nature lose too much information about the structure of a tree metric. [13] developed a type of hierarchical tree, which they call a grove, that is a refinement of an HST that retains more information about the topology of the original metric space, and showed how to approximate a tree metric by a grove in a similar way to which one can approximate a tree metric by an HST. One way to think of an HST is as a unit star where each of the leaves can be recursively thought of as a scaled-down HST. In this vein, a grove is unit tree where each of the nodes can be recursively thought of as a scaled-down grove. (Unit here means the distance of every edge is 1) One can then develop algorithms for a grove, as one does for a HST, that is, design one algorithm for the grove/HST and another algorithm for the base metric space. However, for groves the base metric is a unit tree (instead of a uniform metric space as it is for an HST). [13] gave a monotone grove algorithm that, if combined with certain types of "low-hop" monotone algorithms for a unit tree, yields a poly-log competitive monotone algorithm for a general tree metric (unfortunately its not quite a black box reduction). An algorithm for a unit tree is low-hop if the number of servers that have to move a positive distance to a parking spot is at most

$$
L\left((1+\epsilon)H + \frac{\ln k}{\epsilon}\right)
$$

where H is the diameter of the unit tree, $\frac{1}{\epsilon}$ is poly-log bounded, and L is the optimal/minimum number of servers that have to move a positive distance to a parking spot. The multiplicative $(1 + \epsilon)$ term has to be so small because the grove algorithm is going to apply the unit tree algorithm recursively to a poly-log depth. [13] then developed a monotone low-hop algorithm for a unit tree for the online metric search problem, which is a special case of the online metric matching problem in that there is a promise that there is an optimal matching with only one nonzero length edge (most lower bounds for online metric matching are of this special type). This low-hop algorithm is based on the classic multiplicative weights algorithm in the setting of online learning from experts [6]. Conceptually there is one expert for each leaf of the tree, and this expert always recommends that the car/request travel toward this leaf. Putting this all together, the main result of [13] is a $O(\log^6 \Delta \log^2 n)$-competitive posted-price algorithm for online metric searching (not matching) on a general tree metric.

1.3   Our Contribution

Our main result is a $O(\log^5 \Delta \; \log^2 n)$-competitive posted-price algorithm for online metric matching on a spider metric. A spider is a rooted tree T in which the root ρ (sometimes called the head) is the only node of degree greater than 2. We use d to denote the degree of the root and use the term leg to refer to the leaf to root paths in the spider. We achieve our main result by designing a monotone low-hop algorithm Spider-Match for a unit spider, and then applying the techniques developed in [13] for groves.

As in [13], the starting point in the design of our low-hop algorithm Spider-Match is the classic multiplicative weights algorithm in the setting of online learning from experts [6]. However, because we consider online metric matching, instead of online metric search as in [13], we have to surmount technical difficulties that did not arise in [13]. In Sect. 2 we give an overview of some of the key algorithmic ideas behind the design of Spider-Match, and an explanation of some of these technical difficulties. We strongly recommend that the reader read this section before launching into the subsequent technical sections.

In Sect. 3 we give some preparatory notation and terminology. In Sect. 4 we give a formal definition of the Spider-Match algorithm. In Sect. 5 we give the analysis of Spider-Match. Finally in the appendix we show how these results can be combined with the results from [13] to obtain a $O(\log^5 \Delta \log^2 n)$-competitive posted-price algorithm for online metric matching on a spider metric (This is more or less the same as the grove analysis in [13], except for a slight refinement that cuts off a factor of $\log \Delta$).

2   Intuitive Overview on a Simple Instance

Assume in our unit spider T that each leg is of infinite length, that the root ρ contains no servers, and that each node other than ρ contains a single server. Furthermore, suppose the request sequence $r_1, r_2, \ldots$ is such that

- The first $m < d$ requests, $r_1, \ldots, r_m$ all arrived at the root ρ, and
- if a request r_t, $t > m$, arrives on a leg ℓ then it must arrive at the vertex on leg ℓ that is closest to the root ρ among those vertices where a request has not yet arrived.

To model this within the setting of online learning from experts [6], we assume that there are $\binom{d}{m}$ experts, one for each of the possible collection R of m distinct legs. Initially, the expert R recommends the first m requests move down the legs in R (one request per leg) to the first available server. Subsequently each expert R recommends requests that arrive on a leg $\ell \in R$ move down to the next available server on leg ℓ, and recommends that requests that arrive on a leg $\ell \notin R$ use the server at the location where the request arrived (which exists by assumption). So each expert incurs a cost of 1 hop when a request arrives on one of its recommended legs, and a cost of 0 otherwise. (Recall that in the definition of a low-hop algorithm, all moves of nonzero distance cost one hop).

The classic multiplicative weights algorithm maintains a probability distribution π over experts R, which in turn induces a probability $\pi(s)$ that server s is available if an expert is picked according to probability distribution π. If one could design an online algorithm that also maintained the invariant that each server s is available with probability $\pi(s)$ then one could conclude that the expected number of hops used by this algorithm is at most:

$$m\left((1+\epsilon)H + \frac{\ln(d)}{\epsilon} + 1\right),$$

where H is the minimum total cost of all the experts and is thus low-hop, using the standard analysis of the multiplicative weights algorithm [6].

To maintain the invariant that each server is available with probability $\pi(s)$ we need to design a probability distribution q^r for each possible location r where the next request might arrive, where $q^r(\ell)$ is the probability that our algorithm will move a request arriving at location r to the next available server on leg ℓ. We need that probability distributions q^r to satisfy two properties:

– The probability that a server s is available will be $\pi(s)$, so the experts distribution is matched.
– These probabilities are monotone. So as r moves closer to a server s the probability that the algorithm will use server s can not decrease.

Again, at first impression the monotonicity requirement can seem innocuous, but it is actually very limiting. Our eventual design of such probability distributions q^r was by significant trial and error; we do not have a principled explanation why these are in some sense the "right" distributions.

An even bigger technical challenge arises when we relax the requirement that the m requests at the root arrive at the start, and instead may arrive anywhere in the sequence. Then the algorithm's estimate m of the number of root requests, and thus the number $\binom{d}{m}$ of experts, can increase when a request arrives at the root. This request then yields in a new probability distribution $\tilde{\pi}$ over $\binom{d}{m+1}$ experts that replaces the prior probability distribution π over $\binom{d}{m}$ experts. First, we need to show how to adapt the standard analysis of the multiplicative weights algorithm to handle this. Then we need to design a probability distribution $\tilde{q}$ where $\tilde{q}(\ell)$ gives the probability that this request at the root will move to the first available server on leg ℓ. We need the $\tilde{q}$ to satisfy the following two properties:

– If the next request arrives at the root then the probability that a server s is available will be $\tilde{\pi}(s)$, so the new experts distribution is matched.
– The probability distributions q^r and $\tilde{q}$ are monotone. So as r moves closer to a server s the probability that the algorithm will use server s can not decrease.

As the probability distributions q^r and $\tilde{q}$ are designed to match different experts distributions, it is challenging to attain monotonicity. Again our design of $\tilde{q}$ derived from significant trial and error.

This is illustrative of a general phenomenon, it is difficult to maintain monotonicity for any sort of algorithmic design that has the form:

If a property P is true about a new request then take action A, else take action B.

when some request locations will make property P true and some requests locations will make property P false. In this case, the actions A and B have to be carefully coordinated with each other, and with the locations where property P is true, to maintain monotonicity.

Roughly speaking, on a general instance, our algorithm Spider-Match works as follows. If there is an available server between the current request and the root then the request moves to the first available server on the path to the root. Intuitively, these are "easy" requests that we account for without having to invoke multiplicative weights. Otherwise if there is a hole between the current request and the root, then the request is matched to the available server on leg ℓ that is closest to the root with a probability $q(\ell)$. Roughly speaking, a hole is a server location that does not contain an available server, but that could have contained an available server if the prior random events internal to the algorithm had been different. Otherwise, the request is matched to the available server on leg ℓ that is closest to the root with a probability $\tilde{q}(\ell)$.

3   Notation and Terminology

Definition 1. – A spider metric is a rooted tree metric $(T = ((V, \rho), E), x)$, an acyclic connected graph consisting of vertices V and edges E with a particular vertex ρ marked as the root, along with a distance metric $x : V \times V \to \mathbb{R}$ satisfying
   i) $x(u, v) = 0$ if and only if $u = v$,
   ii) $x(u, v) = x(v, u)$, and
   iii) $x(u, v) \leq x(u, w) + x(w, v)$
   for all $u, v, w \in V$.
- The degree d of a tree is the degree of the root ρ.
- A spider metric is a rooted tree metric $(T = ((V, \rho), E), x)$ where ρ is the single vertex of degree greater than 2.
- A server s is a leaf-server if there are no other servers in the subtree rooted at s.
- Let $L(T) = \{\mu_1, ..., \mu_d\}$ denote the collection of leaf-servers.
- For $\ell \in [d]$, define $T_\ell \subseteq V$ as the set of servers on the path from the root ρ to μ_ℓ, inclusive. T_ℓ is referred to as the ℓth leg of T.
- A server s on T_ℓ becomes a *hole* when either
   a) a request r_t arrives on T_λ where $\lambda \neq \ell$ and r_t is matched to s,
   b) or a request r_t arrives on the path from ρ to s (inclusive of ρ, non-inclusive of s) and matches to s.
- A hole s is *filled* and loses its status as a hole when a request r_t arrives such that

a) s is on the path from r_t to ρ, inclusive,

b) and there is no available server on the path from r_t to s.

- The number of holes at time t is denoted by m_t.
- We define T_ℓ to be alive if there is still an available server or a hole in T_ℓ and dead otherwise.
- Let $\mathcal{A}^t = \{\ell \in [d] \mid T_\ell \text{ is alive just before the arrival of } r_t\}$.
- If $r_t \neq \rho$, let $\ell_t = \ell$ such that $r_t \subset T_\ell$. Otherwise, if $r_t = \rho$, let $\ell_t = 0$.
- Let χ_t denote the sum of available servers and holes on the path from r_t to the root ρ, inclusive.

4   The Spider-Match Algorithm Design

We formally define the `Spider-Match` algorithm on a given rooted spider T with root ρ. The probability distributions $q_\sigma^t(\ell)$ and $\tilde{q}_\sigma^t(\ell)$, used in the algorithm description, are defined after the algorithm (as this seems more natural).

Definition 2 (Spider-Match Algorithm). `Spider-Match` maintains an internal setting σ, initialized to $\emptyset$, representing which legs of the spider `Spider-Match` currently has holes on. The algorithm operates in two phases, starting with the core phase and possibly transitioning to the epilogue. During the **core phase**, at the arrival of r_t:

1. If $\chi_t > 1$, match r_t to the closest available server on the path from r_t to the root, inclusive (as we shall see, in this case there must be at least one such available server).
2. If $\chi_t = 1$:
   (a) If there is an available server on the path from r_t to ρ, match r_t to that server.
   (b) Otherwise, if r_t does not kill ℓ_t, then r_t is matched to the first available server in T_ℓ with probability $q_\sigma^t(\ell)$. In this case, $\sigma \leftarrow \sigma \backslash \{\ell_t\} \cup \{\ell\}$.
   (c) Otherwise, r_t is matched to the first available server in T_ℓ with probability $\tilde{q}_\sigma^t(\ell)$. In this case, $\sigma \leftarrow \sigma \backslash \{\ell_t\} \cup \{\ell\}$.
   In this case, we say that r_t was *collocated* if the server s that contributed to χ_t was collocated with r_t; otherwise, r_t is *non-collocated*.
3. If $\chi_t = 0$, r_t is matched to the first available server in T_ℓ with probability $\tilde{q}_\sigma^t(\ell)$. In this case, $\sigma \leftarrow \sigma \cup \{\ell\}$. If $m_{t+1} = |\mathcal{A}^{t+1}|$, the epilogue immediately begins.

In the **epilogue phase**, at the arrival of r_t:

1. If there is an available server on the path from r_t to ρ, match r_t to the closest available server on the path from r_t to the root, inclusive.
2. Else if $r_t = \rho$ or T_{ℓ_t} is dead, match r_t to the first available server on T_ℓ where ℓ is chosen from $\mathcal{A}^t$ uniformly at random.
3. Else match r_t to the first available server in T_{ℓ_t}.

In order to define q and $\tilde{q}$, we let

$$n_\ell^t = |\{r_i \mid i \le t,\ r_i \in T_\ell,\ \text{and}\ \chi_i = 1\}|,$$
$$w_\ell^t = (1 - \epsilon)^{n_\ell^t}\ \text{for}\ \ell \in [d],$$
$$w_R^t = \prod_{\ell \in R} w_\ell^t = (1 - \epsilon)^{\sum_{\ell \in R} n_\ell^t}\ \text{for}\ R \in 2^{[d]},$$
$$W^t(\mathcal{X}) = \sum_{R \in \mathcal{X}} w_R^t\ \text{for}\ \mathcal{X} \in 2^{2^{[d]}}.$$

This lets us now define q and $\tilde{q}$ as follows:

$$q_\sigma^t(\ell) = \begin{cases} 0 & \text{if}\ \ell \in (\sigma \backslash \{\ell_t\}) \cup ([d] \backslash \mathcal{A}_t) \\[2mm] \dfrac{\epsilon w_\ell^t \sum_{T \in \binom{\mathcal{A}^t \backslash \{\ell_t, \ell\}}{m_t - 1}} \frac{w_T^t}{m_t - |\sigma \cap T|}}{(1 - \epsilon) w_{\ell_t}^t W^t\left(\binom{\mathcal{A}^t \backslash \{\ell_t\}}{m_t - 1}\right) + W^t\left(\binom{\mathcal{A}^t \backslash \{\ell_t\}}{m_t}\right)} & \text{if}\ \ell \in \mathcal{A}^t \backslash \sigma \\[4mm] 1 - \sum_{\lambda \ne \ell_t} q_\sigma^t(\lambda) & \text{if}\ \ell = \ell_t \end{cases}$$

$$\tilde{q}_\sigma^t(\ell) = \begin{cases} 0 & \text{if}\ \ell \in \sigma \cup ([d] \backslash \mathcal{A}^t) \\[2mm] \dfrac{w_\ell^t \sum_{T \in \binom{\mathcal{A}^t \backslash \{\ell\}}{m_t}} \frac{w_T^t}{m_t + 1 - |\sigma \cap T|}}{W^t\left(\binom{\mathcal{A}^t}{m_t + 1}\right)} & \text{if}\ \ell \in \mathcal{A}^t \backslash \sigma \end{cases}$$

5   Analysis of the Spider-Match Algorithm

In this section, we first show in Lemma 2 that `Spider-Match` is well defined. We then show in Lemma 3 that `Spider-Match` follows the multiplicative updates method experts' distribution to maintain its holes. In Lemma 4 we show that `Spider-Match` is monotone. Lastly, we show the main theorem 1 of this section, where we bound the number of times `Spider-Match` pays to handle requests.

Lemma 1. By construction of `Spider-Match`, it follows that until the epilogue there is at most one hole per subtree T_ℓ.

Proof. This follows from the fact that `Spider-Match` either matches a request to an available server and thus does not make a new hole, or creates a new hole on a leg without a hole since by the definitions of q and $\tilde{q}$, there is 0 probability of matching to a leg in $\sigma \backslash \{\ell_t\}$. Note that there will always be a leg without a hole for a request to be to matched to in the core phase, else we move to the epilogue phase.

Lemma 2. `Spider-Match` is well defined.

Proof. We will show that

a) $\sum_{\ell \in [d]} q_R^t(\ell) = 1$,
b) $\sum_{\ell \in [d]} \tilde{q}_R^t(\ell) = 1$

for any $R \in 2^{[d]}$.

For a), note that $q_R^t(\ell_t) = 1 - \sum_{\lambda \neq \ell_t} q_R^t(\lambda)$, so it suffices to show that $q_R^t(\ell) \leq 1$ for all ℓ. Since all terms in the denominator are positive, $\epsilon < 1$, and $\frac{1}{m_t - |R \cap T|} \leq 1$, it follows that $q_R^t(\ell) \leq 1$ for all ℓ.

For b), since $\tilde{q}_R^t(\ell) = 0$ for $\ell \in R \cup ([d] \setminus \mathcal{A}^t)$, it remains to show that $\sum_{\ell \in \mathcal{A}^t \setminus R} \tilde{q}_R^t(\ell) = 1$.

$$\sum_{\ell \in \mathcal{A}^t \setminus R} \tilde{q}_R^t(\ell) = \frac{\sum_{\ell \in \mathcal{A}^t \setminus R} w_\ell^t \sum_{T \in \binom{\mathcal{A}^t \setminus \{\ell\}}{m_t}} \frac{w_T^t}{m_t + 1 - |R \cap T|}}{W^t\left(\binom{\mathcal{A}^t}{m_t + 1}\right)} \qquad \text{by Def. 2}$$

$$= \frac{\sum_{\ell \in \mathcal{A}^t \setminus R} \sum_{T \in \binom{\mathcal{A}^t \setminus \{\ell\}}{m_t}} \frac{w_{T \cup \{\ell\}}^t}{m_t + 1 - |R \cap (T \cup \{\ell\})|}}{W^t\left(\binom{\mathcal{A}^t}{m_t + 1}\right)} \qquad \begin{array}{c}\text{since weights are} \\ \text{multiplicative and } \ell \notin R\end{array}$$

$$= \frac{\sum_{T \in \binom{\mathcal{A}^t}{m_t + 1}} \frac{|T \setminus R|}{m_t + 1 - |R \cap T|} w_T^t}{W^t\left(\binom{\mathcal{A}^t}{m_t + 1}\right)}$$

$$= \frac{\sum_{T \in \binom{\mathcal{A}^t}{m_t + 1}} w_T^t}{W^t\left(\binom{\mathcal{A}^t}{m_t + 1}\right)} \qquad \begin{array}{c}\text{as } |T \setminus R| = |T| - |T \cap R| \\ = m_t + 1 - |T \cap R| \text{ since} \\ T \in \binom{\mathcal{A}_t}{m_t + 1}\end{array}$$

$$= \frac{W^t\left(\binom{\mathcal{A}^t}{m_t + 1}\right)}{W^t\left(\binom{\mathcal{A}^t}{m_t + 1}\right)} = 1 \qquad \text{by Def. 2}$$

Definition 3 (p_T^t **and** π_T^t). We denote p_T^t as the probability that the internal parameter σ of the algorithm is $T \in \binom{\mathcal{A}^t}{m_t}$ just before the arrival of r_t. We define

$$\pi_T^t = \frac{w_T^t}{W^t\left(\binom{[d]}{m_t}\right)}$$

Lemma 3. For any $R \in \binom{[d]}{m_t}$, we have $p_R^t \geq \pi_R^t$.

Proof. To conserve space, this proof has been moved to the appendix.

We next show that this algorithm is monotone and hence induces a pricing scheme as shown in [13].

Lemma 4. Spider-Match is monotone.

Proof. Let $r_t \rightarrow_{\text{Spider-Match}} s$ denote the event that r_t is matched by Spider-Match to s. We must show that $\Pr[r_t \rightarrow_{\text{Spider-Match}} s \mid r_t = u] \leq \Pr[r_t \rightarrow_{\text{Spider-Match}} s \mid r_t = v]$ for all $u, v \in V$ and $s \in S$ where v is on the path from u to s. Note that u and v can belong to separate subtrees; and if $s = v$, then $\Pr[r_t \rightarrow_{\texttt{Spider-Match}} s \mid r_t = v] = 1$, so we can assume that s is not collocated with v. The claim is also trivial for the case where $u = v$. Letting $\chi_t^v = \chi_t \mid r_t = v$, we break the proof into the following cases:

a. $\chi_t^u = 0$ and
   i) $\chi_t^v = 0$
   ii) $\chi_t^v = 1$
   iii) $\chi_t^v > 1$
b. $\chi_t^u > 1$ and
   i) $\chi_t^v = 0$
   ii) $\chi_t^v = 1$
   iii) $\chi_t^v > 1$
c. $\chi_t^u = 1$ and
   i) $\chi_t^v > 1$
   ii) $\chi_t^v = 1$
   iii) $\chi_t^v = 0$

Throughout this proof, we will define ℓ_v such that v is on T_{ℓ_v}.

For a.i), we have that $\Pr[r_t \rightarrow_{\texttt{Spider-Match}} s \mid r_t = u] = \Pr[r_t \rightarrow_{\texttt{Spider-Match}} s \mid r_t = v]$, so the claim is trivially true.

For a.ii), note that if $\ell_v \in \sigma$ then $\Pr[r_t \rightarrow_{\texttt{Spider-Match}} s \mid r_t = u] = 0$. However, if $\ell_v \notin \sigma$, then there is an available server in between u and v, so $\Pr[r_t \rightarrow_{\texttt{Spider-Match}} s \mid r_t = u] = 0$.

For a.iii), note that there must be some available server between u and v, as $\chi_t^v > 1$; thus we have that $\Pr[r_t \rightarrow_{\texttt{Spider-Match}} s \mid r_t = u] = 0$.

For b.i) and b.ii), note that there must be some available server between u and v, as $\chi_t^u > 1 \geq \chi_t^v$; thus we have that $\Pr[r_t \rightarrow_{\texttt{Spider-Match}} s \mid r_t = u] = 0$.

For b.iii), if u and v are on T_ℓ and T_λ respectively, where $\ell \neq \lambda$, then there are available servers between u and v and thus $\Pr[r_t \rightarrow_{\texttt{Spider-Match}} s \mid r_t = u] = 0$. If u and v are on the same T_ℓ and if s is the closest available server to u on the path from u to ρ, then $\Pr[r_t \rightarrow_{\texttt{Spider-Match}} s \mid r_t = u] = \Pr[r_t \rightarrow_{\texttt{Spider-Match}} s \mid r_t = v] = 1$. Otherwise, $\Pr[r_t \rightarrow_{\texttt{Spider-Match}} s \mid r_t = u] = 0$.

For c.i), note that there must be some available server between u and v, as $\chi_t^v > 1$; thus we have that $\Pr[r_t \rightarrow_{\texttt{Spider-Match}} s \mid r_t = u] = 0$.

For c.ii), first if $\ell_u = \ell_v$ then $\Pr[r_t \rightarrow_{\texttt{Spider-Match}} s \mid r_t = u] = \Pr[r_t \rightarrow_{\texttt{Spider-Match}} s \mid r_t = v]$. Otherwise, note that if $\ell_v \in \sigma$ then $\Pr[r_t \rightarrow_{\texttt{Spider-Match}} s \mid r_t = u] = 0$. However, if $\ell_v \notin \sigma$, then there is an available server in between u and v, so $\Pr[r_t \rightarrow_{\texttt{Spider-Match}} s \mid r_t = u] = 0$.

For c.iii), first if $\ell_u \notin \sigma$ then $\Pr[r_t \rightarrow_{\texttt{Spider-Match}} s \mid r_t = u] = 0$. Otherwise, if $\ell_t = \ell_u$ would kill ℓ_u, then $\Pr[r_t \rightarrow_{\texttt{Spider-Match}} s \mid r_t = u] = \Pr[r_t \rightarrow_{\texttt{Spider-Match}} s \mid r_t = v]$. Otherwise, we must show that $q_\sigma^t(\ell) \leq \tilde{q}_\sigma^t(\ell)$. Since $\epsilon \leq \frac{1}{3}$ we note

$$q_\sigma^t(\ell) = \frac{\epsilon w_\ell^t \sum_{T \in \binom{A^t \setminus \{\ell_t, \ell\}}{m_t - 1}} \frac{w_T^t}{m_t - |\sigma \cap T|}}{(1 - \epsilon) w_{\ell_t}^t W^t\left(\binom{A^t \setminus \{\ell_t\}}{m_t - 1}\right) + W^t\left(\binom{A^t \setminus \{\ell_t\}}{m_t}\right)} \leq \frac{\epsilon}{1 - \epsilon} \frac{w_\ell^t \sum_{T \in \binom{A^t \setminus \{\ell_t, \ell\}}{m_t - 1}} \frac{w_T^t}{m_t - |\sigma \cap T|}}{w_{\ell_t}^t W^t\left(\binom{A^t \setminus \{\ell_t\}}{m_t - 1}\right) + W^t\left(\binom{A^t \setminus \{\ell_t\}}{m_t}\right)}$$

$$= \frac{\epsilon}{1 - \epsilon} \frac{w_\ell^t \sum_{T \in \binom{A^t \setminus \{\ell_t, \ell\}}{m_t - 1}} \frac{w_T^t}{m_t - |\sigma \cap T|}}{W^t\left(\binom{A^t}{m_t}\right)} \leq \frac{1}{2} \frac{w_\ell^t \sum_{T \in \binom{A^t \setminus \{\ell_t, \ell\}}{m_t - 1}} \frac{w_T^t}{m_t - |\sigma \cap T|}}{W^t\left(\binom{A^t}{m_t}\right)}. \tag{1}$$

Hence, it suffices to prove that $\dfrac{w_\ell^t \sum_{T \in \binom{A^t \setminus \{\ell_t, \ell\}}{m_t - 1}} \frac{w_T^t}{m_t - |\sigma \cap T|}}{W^t\left(\binom{A^t}{m_t}\right)} \leq 2\tilde{q}_\sigma^t(\ell)$. By cross multiplication, this becomes:

$$W^t\left(\binom{A^t}{m_t + 1}\right) \sum_{T \in \binom{A^t \setminus \{\ell_t, \ell\}}{m_t - 1}} \frac{w_T^t}{m_t - |\sigma \cap T|} \leq 2 W^t\left(\binom{A^t}{m_t}\right) \sum_{T \in \binom{A^t \setminus \{\ell\}}{m_t}} \frac{w_T^t}{m_t + 1 - |\sigma \cap T|} \tag{2}$$

Letting $\mathcal{P} = A^t \setminus \{\ell_t, \ell\}$, we note

$$W^t\left(\binom{A^t}{m_t + 1}\right) = w_{\{\ell_t, \ell\}}^t W^t\left(\binom{\mathcal{P}}{m_t - 1}\right) + (w_{\ell_t}^t + w_\ell^t) W^t\left(\binom{\mathcal{P}}{m_t}\right) + W^t\left(\binom{\mathcal{P}}{m_t + 1}\right), \tag{3}$$

$$W^t\left(\binom{A^t}{m_t}\right) = w_{\{\ell_t, \ell\}}^t W^t\left(\binom{\mathcal{P}}{m_t - 2}\right) + (w_{\ell_t}^t + w_\ell^t) W^t\left(\binom{\mathcal{P}}{m_t - 1}\right) + W^t\left(\binom{\mathcal{P}}{m_t}\right), \tag{4}$$

and $\displaystyle\sum_{T \in \binom{A^t \setminus \{\ell\}}{m_t}} \frac{w_T^t}{m_t + 1 - |\sigma \cap T|} = w_{\ell_t}^t \sum_{T \in \binom{\mathcal{P}}{m_t - 1}} \frac{w_T^t}{m_t - |\sigma \cap T|} + \sum_{T \in \binom{\mathcal{P}}{m_t}} \frac{w_T^t}{m_t + 1 - |\sigma \cap T|} \tag{5}$

We will break down the left-hand side of Eq. 2 into the following disjoint inequalities, where the right-hand sides of the following inequalities are all disjoint sums from the right-hand side of Eq. 2. Hence to show Eq. 2 it suffices to show the following three inequalities:

1) $\displaystyle W^t\left(\binom{\mathcal{P}}{m_t - 1}\right) \sum_{T \in \binom{\mathcal{P}}{m_t - 1}} \frac{w_T^t}{m_t - |\sigma \cap T|}$

$\displaystyle \leq 2 W^t\left(\binom{\mathcal{P}}{m_t - 2}\right) \sum_{T \in \binom{\mathcal{P}}{m_t}} \frac{w_T^t}{m_t + 1 - |\sigma \cap T|} + 2 W^t\left(\binom{\mathcal{P}}{m_t - 1}\right) \sum_{T \in \binom{\mathcal{P}}{m_t - 1}} \frac{w_T^t}{m_t - |\sigma \cap T|}$

2) $\displaystyle W^t\left(\binom{\mathcal{P}}{m_t}\right) \sum_{T \in \binom{\mathcal{P}}{m_t - 1}} \frac{w_T^t}{m_t - |\sigma \cap T|} \leq 2 W^t\left(\binom{\mathcal{P}}{m_t - 1}\right) \sum_{T \in \binom{\mathcal{P}}{m_t}} \frac{w_T^t}{m_t + 1 - |\sigma \cap T|}$

3) $\displaystyle W^t\left(\binom{\mathcal{P}}{m_t + 1}\right) \sum_{T \in \binom{\mathcal{P}}{m_t - 1}} \frac{w_T^t}{m_t - |\sigma \cap T|} \leq 2 W^t\left(\binom{\mathcal{P}}{m_t}\right) \sum_{T \in \binom{\mathcal{P}}{m_t}} \frac{w_T^t}{m_t + 1 - |\sigma \cap T|}$

Note that the first one is trivially true. For 2 and 3, note that we are effectively taking a summation on both sides over weighted multisets. In order to show the result, we will try to match terms on the left hand side to terms at least as big on the right hand side.

Call a function $f : A \times B \to C \times D$ *useful* if f satisfies:

1. f is an injection
2. if $f(a, b) = (c, d)$, then $c = a\backslash\{\mu\}$ and $d = b \cup \{\mu\}$ for some $\mu \in a\backslash b$.

Suppose $f : \binom{P}{m_t} \times \binom{P}{m_t-1} \to \binom{P}{m_t-1} \times \binom{P}{m_t}$ is useful. Let $g : \binom{P}{m_t} \times \binom{P}{m_t-1} \to \binom{P}{1}$ be defined such that $f(a, b) = (a\backslash g(a,b),\ b \cup g(a,b))$. Then

$$W^t\left(\binom{P}{m_t}\right) \sum_{T \in \binom{P}{m_t-1}} \frac{w_T^t}{m_t - |\sigma \cap T|} = \sum_{R \in \binom{P}{m_t}} \sum_{T \in \binom{P}{m_t-1}} \frac{w_R^t w_T^t}{m_t - |\sigma \cap T|}$$

$$\leq 2 \sum_{R \in \binom{P}{m_t}} \sum_{T \in \binom{P}{m_t-1}} \frac{w_R^t w_T^t}{m_t + 1 - |\sigma \cap T|}$$

$$= 2 \sum_{R \in \binom{P}{m_t}} \sum_{T \in \binom{P}{m_t-1}} \frac{w_{R\backslash g(R,T)}^t w_{T \cup g(R,T)}^t}{m_t + 1 - |\sigma \cap T|}$$

$$\leq 2 \sum_{R \in \binom{P}{m_t}} \sum_{T \in \binom{P}{m_t-1}} \frac{w_{R\backslash g(R,T)}^t w_{T \cup g(R,T)}^t}{m_t + 1 - |\sigma \cap (T \cup g(R,T))|}$$

$$\leq 2 \sum_{R \in \binom{P}{m_t-1}} \sum_{T \in \binom{P}{m_t}} \frac{w_R^t w_T^t}{m_t + 1 - |\sigma \cap T|}$$

$$= 2W^t\left(\binom{P}{m_t-1}\right) \sum_{T \in \binom{P}{m_t}} \frac{w_T^t}{m_t + 1 - |\sigma \cap T|}$$

Similarly, suppose $f : \binom{P}{m_t+1} \times \binom{P}{m_t-1} \to \binom{P}{m_t} \times \binom{P}{m_t}$ is useful. Let $g : \binom{P}{m_t+1} \times \binom{P}{m_t-1} \to \binom{P}{1}$ be defined such that $f(a, b) = (a\backslash g(a,b),\ b \cup g(a,b))$. Then

$$W^t\left(\binom{P}{m_t+1}\right) \sum_{T \in \binom{P}{m_t-1}} \frac{w_T^t}{m_t - |\sigma \cap T|} = \sum_{R \in \binom{P}{m_t+1}} \sum_{T \in \binom{P}{m_t-1}} \frac{w_R^t w_T^t}{m_t - |\sigma \cap T|}$$

$$\leq 2 \sum_{R \in \binom{P}{m_t+1}} \sum_{T \in \binom{P}{m_t-1}} \frac{w_R^t w_T^t}{m_t + 1 - |\sigma \cap T|}$$

$$= 2 \sum_{R \in \binom{P}{m_t+1}} \sum_{T \in \binom{P}{m_t-1}} \frac{w_{R\backslash g(R,T)}^t w_{T \cup g(R,T)}^t}{m_t + 1 - |\sigma \cap T|}$$

$$\leq 2 \sum_{R \in \binom{P}{m_t+1}} \sum_{T \in \binom{P}{m_t-1}} \frac{w_{R\backslash g(R,T)}^t w_{T \cup g(R,T)}^t}{m_t + 1 - |\sigma \cap (T \cup g(R,T))|}$$

$$\leq 2 \sum_{R \in \binom{P}{m_t}} \sum_{T \in \binom{P}{m_t}} \frac{w_R^t w_T^t}{m_t + 1 - |\sigma \cap T|}$$

$$\leq 2W^t\left(\binom{P}{m_t}\right) \sum_{T \in \binom{P}{m_t}} \frac{w_T^t}{m_t + 1 - |\sigma \cap T|}$$

Hence if these useful functions exist, then the proof is complete.

We now turn to showing the existence of these functions.

1. A useful function $f : \binom{P}{m_t} \times \binom{P}{m_t-1} \to \binom{P}{m_t-1} \times \binom{P}{m_t}$ exists.

2. A useful function $f : \binom{P}{m_t+1} \times \binom{P}{m_t-1} \to \binom{P}{m_t} \times \binom{P}{m_t}$ exists.

For tho first statement, consider a bipartite graph $G = ((X,Y), E)$ where $X = \binom{P}{m_t} \times \binom{P}{m_t-1}$, $Y = \binom{P}{m_t-1} \times \binom{P}{m_t}$, and $\{(R_1, T_1), (R_2, T_2)\} \in E$ if $R_2 = R_1 \backslash \{\ell\}$ and $T_2 = T_1 \cup \{\ell\}$ for some $\ell \in R_1 \backslash T_1$. Note that by this choice of edges, the graph is divided into disjoint unions where for each maximal connected subgraph G_Q there exists some multi set Q so that for each element (R, T) of the subgraph G_Q we have that $Q = R \uplus T$. Furthermore, this subgraph is $m_t - |\Delta^Q|$-regular, where $\Delta^Q = \{x \in Q \mid n_x^Q > 1\}$, so by Hall's theorem there exists a perfect matching. Since there exists a perfect matching on each maximal connected subgraph, there is a perfect matching between the two sets. The function induced by this matching is useful by construction.

For the second statement, consider a bipartite graph $G = (X, Y), E$ where $X = \binom{P}{m_t+1} \times \binom{P}{m_t-1}$, $Y = \binom{P}{m_t} \times \binom{P}{m_t}$, and $\{(R_1, T_1), (R_2, T_2)\} \in E$ if $R_2 = R_1 \backslash \{\ell\}$ and $T_2 = T_1 \cup \{\ell\}$ for some $\ell \in R_1 \backslash T_1$. Note that by this choice of edges, the graph is divided into disjoint unions where for each maximal connected subgraph $G_Q = ((X_Q, Y_Q), E_Q)$ there exists some multi set Q so that for each element (R, T) of the subgraph G_Q we have that $Q = R \uplus T$. Furthermore, this subgraph is $(m_t + 1 - |\Delta^Q|, m_t - |\Delta^Q|)$-regular, where $\Delta^Q = \{x \in Q \mid n_x^Q > 1\}$, so by Hall's theorem there exists a matching saturating X_Q. Thus, there is a matching saturating X. The function induced by this matching is useful by construction.

Lastly, we show the main theorem of this section, where we bound the number of times we pay a positive cost as a function of the height H of our tree and the degree d of the root. For the analysis of our algorithm, we will consider a metric search problem which we think will help the reader understand the analysis of Theorem 1. In this metric search problem, the algorithm sees a set of available parking spots located within a metric space. Over time, two types of events can occur: a car can enter the space, or a parking spot can be decommissioned. The algorithm's job is to always keep cars matched to available parking spaces, where each space can be filled by just one car. When a car arrives, it must be moved to an available parking space; when a parking space is decommissioned, if a car had been parked at it, then that car must be moved to a different available parking space. Clearly, any request sequence can be simulated by equating the servers to parking spaces and requests to cars. Alternatively, if a request arrives at a server for which no other request has arrived at, this can equivalently be thought of as decommissioning that parking spot. As such, given r, we will create a sequence η of events for the parking problem described above according to the following deterministic function:

Definition 4 (The η sequence). Given r_t, define η_t by:

1. If $\chi_t = 0$: η_t is the event that a new car enters the space at the location of r_t.
2. If $\chi_t \geq 1$:
   (a) if r_t is collocated with an available server or if r_t is collocated with a parked car which has already moved: η_t is the event that the parking spot at r_t is decommissioned.
   (b) else: η_t is the event that a new car enters the space at the location of r_t.

Let L_t denote the number of cars in the system after the η_t event. Then, any optimal matching solution must contain at least L_t positive cost matchings to handle the requests $\{r_1, ..., r_t\}$. Furthermore, it follows from the definition of m_t that $m_t \leq L_t$.

Theorem 1.

$$\mathbf{E}\left[\sum_{t=1}^{n} \mathbf{1}^r(t)\right] \leq L_t \left((1+\epsilon)H + \frac{\ln d}{\epsilon} + 1\right)$$

where $\mathbf{1}^r(t)$ is an indicator random variable that is 1 if **Spider-Match** *pays positive cost to match r_t and 0 otherwise.*

Proof. Our proof relies on the following claim:

Claim. Let r_τ denote the first request such that $\chi_t = 0$ and let κ denote the number of requests r_t such that $\chi_t = 0$. First, let r' be the request sequence r after removing any request for which $\chi_t = 0$. Now consider the alternate request sequence r'' defined such that $r''_t = r_t$ for $t < \tau$, $r''_t = \rho$ for $\tau \leq t < \tau + \kappa$, and $r''_{\tau+\kappa+t} = r'_{\tau+t}$ for $t \geq 0$. Then $\mathbf{E}\left[\sum_{t=1}^{n} \mathbf{1}^r(t)\right] \leq \mathbf{E}\left[\sum_{t=1}^{n} \mathbf{1}^{r''}(t)\right]$.

First, we show that with this claim the result follows: because of the claim, we can assume $r = r''$, so that for $t \notin \{\tau, ..., \tau + \kappa - 1\}$ it follows that $\chi_t \neq 0$. We define cost vectors

$$c_R^t = \begin{cases} 1 & \ell_t \in R \text{ or } R \cap \mathcal{D}^t \neq \emptyset \\ 0 & \text{otherwise.} \end{cases}$$

where $\mathcal{D}^t$ is the set of dead legs at time t.
     Define δ_R^t such that

$$p_R^t = \pi_R^t + \delta_R^t \tag{6}$$

for all $R \in \binom{[d]}{m_t}$. By Lemma 3, it follows that $\delta_R^t \geq 0$ for all R. Then we have that

$$\sum_{R \in \binom{A^t}{m_t}} p_R^t = \sum_{R \in \binom{A^t}{m_t}} \pi_R^t + \sum_{R \in \binom{A^t}{m_t}} \delta_R^t, \tag{7}$$

where $\sum_{R \in \binom{A^t}{m_t}} p_R^t = 1$ by construction.

Then, if $\chi_t = 1$ and η_t is a decommissioning of a parking spot, we have

$$
\mathbf{E}\left[\mathbf{1}^r(t)\right] = \sum_{\substack{R \in \binom{A^t}{m_t} \\ \text{s.t. } \ell_t \in R}} p_R^t
$$

$$
= \sum_{\substack{R \in \binom{A^t}{m_t} \\ \text{s.t. } \ell_t \in R}} \delta_R^t + \sum_{\substack{R \in \binom{A^t}{m_t} \\ \text{s.t. } \ell_t \in R}} \pi_R^t \qquad \text{definition of } \delta_R^t
$$

$$
\leq \sum_{R \in \binom{A^t}{m_t}} \delta_R^t + \sum_{\substack{R \in \binom{A^t}{m_t} \\ \text{s.t. } \ell_t \in R}} \pi_R^t
$$

$$
= \sum_{R \cap \mathcal{D}^t \neq \emptyset} \pi_R^t + \sum_{\substack{R \in \binom{A^t}{m_t} \\ \text{s.t. } \ell_t \in R}} \pi_R^t \qquad \text{definition of } \delta_R^t
$$

$$
= \vec{c}^t \cdot \vec{\pi}^t \qquad \text{definition of } \vec{c}^t.
$$

Let $\mathcal{T} = \{t \mid \eta_t \text{ is a decommissioning}\}$.

The Multiplicative Weights guarantee from [7] gives us that

$$
\forall R \in \binom{[d]}{m_t}, \ \sum_{t \in \mathcal{T}} \vec{c}^t \cdot \vec{\pi}^t \leq (1 + \epsilon) \sum_{t \in \mathcal{T}} c_R^t + \frac{\ln \binom{[d]}{m_t}}{\epsilon}
$$

Let us choose R to be the last m_t legs that die; the result follows.

Lastly, to see that the claim holds, let $\tilde{\tau}$ denote the last time t such that $\chi_t = 0$. Then for $t \notin \{\tau, ..., \tilde{\tau}\}$, it follows that $\mathbf{E}\left[\mathbf{1}^r(t)\right] = \mathbf{E}\left[\mathbf{1}^{r''}(t)\right]$. Next, for any $t \in \{\tau, ..., \tilde{\tau}\}$, let $r_{t'}$ denote the request corresponding to r_t''. Then the number of holes at time t under r'' is at least the number of holes at time t' under r by construction of r'', thus $\mathbf{E}\left[\mathbf{1}^r(t)\right] \geq \mathbf{E}\left[\mathbf{1}^{r''}(t)\right]$. The claim follows, completing the proof.

6   Conclusion

The obvious immediate open question is whether a poly-log competitive posted-price algorithm exists for online metric matching on a general tree metric. The most immediate problem that one runs into when trying to apply the approach of [13], and that we apply here, is that it is not at all clear what the "right" choice of experts is. Each of the natural choices has fundamental issues that seem challenging to overcome.

Acknowledgements. We thank Anupam Gupta, Aditya Krishnan, and Alireza Samadian for extensive helpful discussions.

References

1. Calgary ParkPlus Homepage. https://www.calgaryparking.com/parkplus
2. SFpark Homepage. http://sfpark.org/
3. SFpark Wikipedia page. https://en.wikipedia.org/wiki/SFpark
4. Antoniadis, A., Barcelo, N., Nugent, M., Pruhs, K., Scquizzato, M.: A $o(n)$-competitive deterministic algorithm for online matching on a line. In: Workshop on Approximation and Online Algorithms, pp. 11–22 (2014)
5. Antoniadis, A., Fischer, C., Tönnis, A.: A collection of lower bounds for online matching on the line. In: Bender, M.A., Farach-Colton, M., Mosteiro, M.A. (eds.) LATIN 2018. LNCS, vol. 10807, pp. 52–65. Springer, Cham (2018). https://doi.org/10.1007/978-3-319-77404-6_5
6. Arora, S., Hazan, E., Kale, S.: The multiplicative weights update method: a meta-algorithm and applications. Theory Comput. **8**, 121–164 (2012)
7. Arora, S., Hazan, E., Kale, S.: The multiplicative weights update method: a meta-algorithm and applications. Theory Comput. **8**(6), 121–164 (2012). https://doi.org/10.4086/toc.2012.v008a006. http://www.theoryofcomputing.org/articles/v008a006
8. Aspnes, J., Azar, Y., Fiat, A., Plotkin, S., Waarts, O.: On-line routing of virtual circuits with applications to load balancing and machine scheduling. J. ACM **44**(3), 486–504 (1997)
9. Azar, Y., Kalyanasundaram, B., Plotkin, S.A., Pruhs, K., Waarts, O.: On-line load balancing of temporary tasks. J. Algorithms **22**(1), 93–110 (1997)
10. Bansal, N., Buchbinder, N., Gupta, A., Naor, J.: A randomized $O(\log^2 k)$-competitive algorithm for metric bipartite matching. Algorithmica **68**(2), 390–403 (2014)
11. Bartal, Y.: Probabilistic approximation of metric spaces and its algorithmic applications. In: Symposium on Foundations of Computer Science, pp. 184–193 (1996)
12. Bartal, Y.: On approximating arbitrary metrics by tree metrics. In: ACM Symposium on Theory of Computing, pp. 161–168 (1998)
13. Bender, M., Gilbert, J., Krishnan, A., Pruhs, K.: Competitively pricing parking in a tree. In: Chen, X., Gravin, N., Hoefer, M., Mehta, R. (eds.) WINE 2020. LNCS, vol. 12495, pp. 220–233. Springer, Cham (2020). https://doi.org/10.1007/978-3-030-64946-3_16
14. Cohen, I.R., Eden, A., Fiat, A., Jez, L.: Pricing online decisions: beyond auctions. CoRR abs/1504.01093 (2015). http://arxiv.org/abs/1504.01093
15. Cohen, I.R., Eden, A., Fiat, A., Jez, L.: Pricing online decisions: beyond auctions. In: ACM-SIAM Symposium on Discrete Algorithms, pp. 73–91 (2015)
16. Cohen, I.R., Eden, A., Fiat, A., Jez, L.: Dynamic pricing of servers on trees. In: Approximation, Randomization, and Combinatorial Optimization. Algorithms and Techniques. LIPIcs, vol. 145, pp. 10:1–10:22 (2019)
17. Fakcharoenphol, J., Rao, S., Talwar, K.: A tight bound on approximating arbitrary metrics by tree metrics. J. Comput. Syst. Sci. **69**(3), 485–497 (2004)
18. Feldman, M., Fiat, A., Roytman, A.: Makespan minimization via posted prices. In: ACM Conference on Economics and Computation, pp. 405–422 (2017)
19. Fuchs, B., Hochstättler, W., Kern, W.: Online matching on a line. Theoret. Comput. Sci. **332**(1–3), 251–264 (2005)
20. Gupta, A., Lewi, K.: The online metric matching problem for doubling metrics. In: Czumaj, A., Mehlhorn, K., Pitts, A., Wattenhofer, R. (eds.) ICALP 2012. LNCS, vol. 7391, pp. 424–435. Springer, Heidelberg (2012). https://doi.org/10.1007/978-3-642-31594-7_36

21. Im, S., Moseley, B., Pruhs, K., Stein, C.: Minimizing maximum flow time on related machines via dynamic posted pricing. In: European Symposium on Algorithms, pp. 51:1–51:10 (2017)
22. Kalyanasundaram, B., Pruhs, K.: Online weighted matching. J. Algorithms **14**(3), 478–488 (1993)
23. Khuller, S., Mitchell, S.G., Vazirani, V.V.: On-line algorithms for weighted bipartite matching and stable marriages. Theoret. Comput. Sci. **127**(2), 255–267 (1994)
24. Koutsoupias, E., Nanavati, A.: The online matching problem on a line. In: Solis-Oba, R., Jansen, K. (eds.) WAOA 2003. LNCS, vol. 2909, pp. 179–191. Springer, Heidelberg (2004). https://doi.org/10.1007/978-3-540-24592-6_14
25. Meyerson, A., Nanavati, A., Poplawski, L.J.: Randomized online algorithms for minimum metric bipartite matching. In: ACM-SIAM Symposium on Discrete Algorithms, pp. 954–959 (2006)
26. Nayyar, K., Raghvendra, S.: An input sensitive online algorithm for the metric bipartite matching problem. In: Symposium on Foundations of Computer Science, pp. 505–515 (2017)
27. Peserico, E., Scquizzato, M.: Matching on the line admits no $o(\sqrt{\log n})$-competitive algorithm. CoRR abs/2012.15593 (2020). https://arxiv.org/abs/2012.15593
28. Raghvendra, S.: Optimal analysis of an online algorithm for the bipartite matching problem on a line. In: Symposium on Computational Geometry. LIPIcs, vol. 99, pp. 67:1–67:14 (2018)
29. Shoup, D., Pierce, G.: SFpark: pricing parking by demand (2013). https://www.accessmagazine.org/fall-2013/sfpark-pricing-parking-demand/

Computational Complexity of Covering Disconnected Multigraphs

Jan Bok[1]([✉]) [iD], Jiří Fiala[2] [iD], Nikola Jedličková[2] [iD], Jan Kratochvíl[2] [iD], and Michaela Seifrtová[2] [iD]

[1] Computer Science Institute, Faculty of Mathematics and Physics, Charles University, Prague, Czech Republic
bok@iuuk.mff.cuni.cz
[2] Department of Applied Mathematics, Faculty of Mathematics and Physics, Charles University, Prague, Czech Republic
{fiala,jedlickova,honza,mikina}@kam.mff.cuni.cz

Abstract. The notion of graph covers is a discretization of covering spaces introduced and deeply studied in topology. In discrete mathematics and theoretical computer science, they have attained a lot of attention from both the structural and complexity perspectives. Nonetheless, disconnected graphs were usually omitted from the considerations with the explanation that it is sufficient to understand coverings of the connected components of the target graph by components of the source one. However, different (but equivalent) versions of the definition of covers of connected graphs generalize to nonequivalent definitions of disconnected graphs. The aim of this paper is to summarize this issue and to compare three different approaches to covers of disconnected graphs: 1) locally bijective homomorphisms, 2) globally surjective locally bijective homomorphisms (which we call *surjective covers*), and 3) locally bijective homomorphisms which cover every vertex the same number of times (which we call *equitable covers*). The standpoint of our comparison is the complexity of deciding if an input graph covers a fixed target graph. We show that both surjective and equitable covers satisfy what certainly is a natural and welcome property: covering a disconnected graph is polynomial time decidable if such it is for every connected component of the graph, and it is NP-complete if it is NP-complete for at least one of its components. Despite of this, we argue that the third variant, equitable covers, is the right one, when considering covers of colored (multi)graphs. Moreover, the complexity of surjective and equitable covers differ from the fixed parameter complexity point of view. We conclude the paper by a complete characterization of the complexity of covering 2-vertex colored multigraphs with semi-edges. We present the results in the utmost generality and strength. In accord with the current trends we consider (multi)graphs with semi-edges, and, on the other hand, we aim at proving the NP-completeness results for simple input graphs.

© Springer Nature Switzerland AG 2021
E. Bampis and A. Pagourtzis (Eds.): FCT 2021, LNCS 12867, pp. 85–99, 2021.
https://doi.org/10.1007/978-3-030-86593-1_6

1   Introduction

The notion of graph covering is motivated by the notion of covering of topological spaces. It has found numerous applications in graph theory, in construction of highly symmetric graphs of requested further properties (cf. [3,9]), but also in models of local computation [2,5,6]. The application in computer science led Abello, Fellows, and StIlwell [1] to pose the problem of characterizing those (multi)graphs for which one can decide in polynomial time if they are covered by an input graph. They have noticed that because of the motivation, it is natural to consider multigraphs, i.e., the situation when multiple edges and loops are allowed. Kratochvíl, Proskurowski and Telle [8] showed that in order to fully characterize the complexity of covering simple graphs, it is necessary but also sufficient to characterize the complexity of covering colored mixed multigraphs of minimum degree at least three. Informally, semi-edges have, compared to the usual edges in graph theory, only one endpoint. In modern topological graph theory it has now become standard to consider graphs with semi-edges since these occur naturally in algebraic graph reductions. Bok et al. initiated the study of the computational complexity of covering graphs with semi-edges in [4].

In all the literature devoted to the computational aspects of graph covers, only covers of connected graphs have been considered so far. The authors of [7] justify this by claiming in Fact 2.b that "For a disconnected graph H, the H-cover problem is polynomially solvable (NP-complete) if and only if the H_i-cover problem is polynomially solvable (NP-complete) for every (for some) connected component H_i of H". Though this seems to be a plausible and desirable property, a closer look shows that the validity of this statement depends on the exact definition of covers for disconnected graphs. Namely in the case of multigraphs with semi-edges when the existence of a covering projection does not follow from the existence of a degree-obedient vertex mapping anymore.

The purpose of this paper is to have a closer look at covers of disconnected graphs in three points of view: the definition, complexity results, and the role of disconnected subgraphs in colored multigraphs. In Sect. 3 we first discuss what are the possible definitions of covers of disconnected graphs – locally bijective homomorphisms are a natural generalization from the algebraic graph theory standpoint, globally surjective locally bijective homomorphisms (which we call *surjective covers*) seem to have been understood by the topological graph theory community as the generalization from the standpoint of topological motivation, and a novel and more restrictive definition of *equitable covers*, in which every vertex of the target graph is required to be covered by the same number of vertices of the source one. The goal of the paper is to convince the reader that the most appropriate definition is the last one. In Sect. 4 we inspect the three possible definitions under the magnifying glass of computational complexity. The main result is that the above mentioned Fact 2.b is true for surjective covers, and remains true also for the newly proposed definition of equitable covers of disconnected graphs. The NP-hardness part of the statement is proven for instances when the input graphs are required to be simple. Lastly, in Sect. 5

we review the concept of covers of colored graphs and show that in this context the notion of equitable covers is indeed the most natural one. We justify our approach by providing a characterization of polynomial/NP-complete instances of the H-COVER problem for colored mixed multigraphs with semi-edges. It is worth noting that in Sect. 2 we also introduce a new notion of *being stronger*, a relation between (multi)graphs that generalizes the covering relation and which we utilize in the NP-hardness reductions in Sect. 4. We believe that this notion is interesting on its own and that its further study would deepen the understanding of graph covers.

2   Covers of Connected Graphs

In this section we formally define what we call *graphs*, we review the notion of a covering projection for connected graphs and we introduce a quasi-ordering of connected graphs defined by the existence of their simple covers. In the rest of the paper we drop the prefix *multi-* when we speak about (multi-)graphs – from now on, *graphs* are allowed to have multiple edges, loops, and/or semi-edges (we talk about *simple graphs* if none of these are allowed).

A very elegant description of ordinary edges, loops and semi-edges through the concept of *darts* is used in more algebraic-based papers on covers. The following formal definition is inspired by the one given in [11].

Definition 1. *A* graph *is a triple* (D, V, Λ), *where D is a set of* darts, *and V and Λ are each a partition of D into disjoint sets. Moreover, all sets in Λ have size one or two.*

Vertices *are here the sets of darts forming the partition V. The set of* links Λ *splits into three disjoint sets* $\Lambda = E \cup L \cup S$, *where E represents the* edges, *i.e., those links of Λ that intersect two distinct vertices from V, L are the* loops, *i.e., those 2-element sets of Λ that are subsets of some set from V, and S are the* semi-edges, *i.e., the 1-element sets from Λ.*

The usual terminology that a vertex $v \in V$ is *incident* with an edge $e \in E$ or that distinct vertices u and v are *adjacent* can be expressed as $v \cap e \neq \emptyset$ and as $\exists e \in E : u \cap e \neq \emptyset \wedge v \cap e \neq \emptyset$, respectively.

A graph is usually defined as an ordered triple (V, Λ, ι), for $\Lambda = E \cup L \cup S$, where ι is the *incidence mapping* $\iota : \Lambda \longrightarrow V \cup \binom{V}{2}$ such that $\iota(e) \in V$ for all $e \in L \cup S$ and $\iota(e) \in \binom{V}{2}$ for all $s \in E$. We use both approaches in this paper and employ advantages of each of them in different situations. See an illustrative example in Fig. 1.

The *degree* of a vertex $v \in V$ is $\deg(v) = |v|$. The fact that a loop contributes 2 to the degree of its vertex may seem not automatic at first sight, but becomes natural when graph embeddings on surfaces are considered.

The *multiedge* between u and v is an inclusion-wise maximal subset of links that connect u and v, i.e. $\{e \in E : e \cap v \neq \emptyset \wedge v \cap e \neq \emptyset\}$ and the cardinality of this set is the *multiplicity* of the (multi)edge uv. In the same way we define the multiplicity of a loop or of a semi-edge.

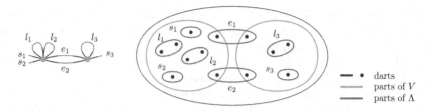

Fig. 1. An example of a graph drawn in a usual graph-theoretical way (left) and using the alternative dart-based definition (right). The figure appeared first in [4].

In case of simple graphs we use also the traditional notation for an edge as $e = uv$ and we write $G = (V, E)$.

A graph $H = (D', V', \Lambda')$ is a *subgraph* of a graph $G = (D, V, \Lambda)$ if their sets of darts satisfy $D' \subseteq D$ and their partitions fulfill $V' = V|_{D'}$ and $\Lambda' = \Lambda|_{D'}$.

A *path* in graph G is a sequence of distinct darts such that consecutive darts either constitute an edge or a vertex of degree 2. A path is *closed* if the first pair as well as the last pair constitute edges. In such a case we say that it connects the vertex containing the first dart to the vertex of the last one. If the first pair and the last pair are vertices then the path is *open*. In all other cases (including a sequence of length 1) the path is *half-way*.

By a *component* of a graph we mean an inclusion-wise maximal induced subgraph such that every two of its vertices are connected by a subgraph isomorphic to a path. We say that a graph is *connected* if it has only a single component.

It shall be useful for our purposes to specifically denote one-vertex and two-vertex graphs. Let us denote by $F(b, c)$ the one-vertex graph with b semi-edges and c loops and by $W(k, m, \ell, p, q)$ the two-vertex graph with k semi-edges and m loops at one vertex, p loops and q semi-edges at the other one, and $\ell > 0$ multiple edges connecting the two vertices (these edges are referred to as *bars*). In other words, $W(k, m, \ell, p, q)$ is obtained from the disjoint union of $F(k, m)$ and $F(q, p)$ by connecting their vertices by ℓ parallel edges. Note that the graph in Fig. 1 is in fact $W(2, 2, 2, 1, 1)$.

2.1   Covers of Connected Graphs

Though there is no ambiguity in the definition of graph covers of connected graphs, the standard definition used e.g. in [8] or [10] becomes rather technical especially when semi-edges are allowed. The following simple-to-state yet equivalent definition was introduced in [4].

Definition 2. *We say that a graph $G = (D_G, V_G, \Lambda_G)$ covers a connected graph $H = (D_H, V_H, \Lambda_H)$ (denoted as $G \longrightarrow H$) if there exists a map $f : D_G \to D_H$ such that:*

- *The map f is surjective.*
- *For every $u \in V_G$, there is a $u' \in V_H$ such that the restriction of f onto u is a bijection between u and u'.*
- *For every $e \in \Lambda_G$, there is an $e' \in \Lambda_H$ such that $f(e) = e'$.*

We write $G \longrightarrow H$ to express that G covers H when both G and H are connected graphs. This compact and succinct definition emphasizes the usefulness of the dart definition of graphs in contrast with the lengthy and technical definition of covers in the standard way which is recalled in the following proposition.

Proposition 3. *A graph G covers a graph H if and only if G allows a pair of mappings $f_V : V(G) \longrightarrow V(H)$ and $f_\Lambda : \Lambda(G) \longrightarrow \Lambda(H)$ such that*

1. *$f_\Lambda(e) \in L(H)$ for every $e \in L(G)$ and $f_\Lambda(e) \in S(H)$ for every $e \in S(G)$,*
2. *$\iota(f_\Lambda(e)) = f_V(\iota(e))$ for every $e \in L(G) \cup S(G)$,*
3. *for every link $e \in \Lambda(G)$ such that $f_\Lambda(e) \in S(H) \cup L(H)$ and $\iota(e) = \{u, v\}$, we have $\iota(f_\Lambda(e)) = f_V(u) = f_V(v)$,*
4. *for every link $e \in \Lambda(G)$ such that $f_\Lambda(e) \in E(H)$ and $\iota(e) = \{u, v\}$ (note that it must be $f_V(u) \neq f_V(v)$), we have $\iota(f_\Lambda(e)) = \{f_V(u), f_V(v)\}$,*
5. *for every loop $e \in L(H)$, $f^{-1}(e)$ is a disjoint union of loops and cycles spanning all vertices $u \in V(G)$ such that $f_V(u) = \iota(e)$,*
6. *for every semi-edge $e \in S(H)$, $f^{-1}(e)$ is a disjoint union of edges and semi-edges spanning all vertices $u \in V(G)$ such that $f_V(u) = \iota(e)$, and*
7. *for every edge $e \in E(H)$, $f^{-1}(e)$ is a disjoint union of edges (i.e., a matching) spanning all vertices $u \in V(G)$ such that $f_V(u) \in \iota(e)$.*

Finally, H-COVER is the associated decision problem having a graph G on input and asking if G covers H.

2.2 A Special Relation Regarding Covers

Graph covering is a transitive relation among connected graphs. Thus when $A \longrightarrow B$ for graphs A and B, every graph G that covers A also covers B. Surprisingly, the conclusion may hold true also in cases when A does not cover B, if we only consider simple graphs G. To describe this phenomenon, we introduce the following definition, which will prove useful in several reductions later on.

Definition 4. *Given connected graphs A, B, we say that A is stronger than B, and write $A \triangleright B$, if every simple graph that covers A also covers B.*

The smallest nontrivial example of such a pair of graphs are two one-vertex graphs: $F(2, 0)$ with a pair of semi-edges and $F(0, 1)$, one vertex with a loop. While $F(0, 1)$ is covered by any cycle, only cycles of even length cover $F(2, 0)$. So $F(2, 0) \triangleright F(0, 1)$.

Observe that whenever A is simple, then $(A \triangleright B)$ if and only if $(A \longrightarrow B)$.

One might also notice that $\triangleright$ defines a quasi-order on connected graphs. Many pairs of graphs are left incomparable with respect to this relation, even those covering a common target graph. On the other hand, the equivalence classes of pair-wise comparable graphs may be nontrivial, and the graphs within one class might have different numbers of vertices. For example, $W(0, 0, 2, 0, 0)$ and $F(2, 0)$ form an equivalence class of $\triangleright$, as for both of these graphs, the class of simple graphs covering them is exactly the class of even cycles. We believe the relation of being stronger is a concept interesting on its own. In particular, the following questions remain open and seem relevant.

Problem 5. Do there exist two graphs A and B such that A has no semi-edges, $A \not\longmapsto B$ and yet $A \triangleright B$?

Problem 6. Do there exist two $\triangleright$-equivalent graphs such that none of them covers the other one?

3   What is a Cover of a Disconnected Graph?

From now on, we assume that we are given two (possibly disconnected) graphs G and H and we are interested in determining whether G covers H. In particular in this section we discus what it means that G covers H. We assume that G has p components of connectivity, $G_1, G_2, \ldots, G_p$, and H has q components, $H_1, H_2, \ldots, H_q$. It is reasonable to request that a covering projection must map each component of G onto some component of H, and this restricted mapping must be a covering. The questions we are raising are:

1. Should the covering projection be globally surjective, i.e., must the preimage of every vertex of H be nonempty?
2. Should the preimages of the vertices of H be of the same size?

Both these questions are the first ones at hand when trying to generalize graph covers to disconnected graphs, since the answer is "yes" in the case of connected graphs (and it is customary to call a projection that covers every vertex k times a *k-fold cover*).

Definition 7. *Let G and H be graphs and let us have a mapping $f \colon G \longrightarrow H$.*

- *We say that f is a* locally bijective homomorphism *of G to H if for each component G_i of G, the restricted mapping $F|_{G_i} \colon G_i \longrightarrow H$ is a covering projection of G_i onto some component of H. We write $G \longrightarrow_{lb} H$ if such a mapping exists.*
- *We say that f is a* surjective covering projection *of G to H if for each component G_i of G, the restricted mapping $F|_{G_i} \colon G_i \longrightarrow H$ is a covering projection of G_i onto some component of H, and f is surjective. We write $G \longrightarrow_{sur} H$ if such a mapping exists.*
- *We say that f is an* equitable covering projection *of G to H if for each component G_i of G, the restricted mapping $F|_{G_i} \colon G_i \longrightarrow H$ is a covering projection of G_i onto some component of H, and for every two vertices $u, v \in V(H)$, $|f^{-1}(u)| = |f^{-1}(v)|$. We write $G \longrightarrow_{equit} H$ if such a mapping exists.*

A useful tool both for describing and discussing the variants, as well as for algorithmic considerations, is introduced in the following definition.

Definition 8. *Given graphs G and H with components of connectivity $G_1, G_2,$ $\ldots, G_p$, and $H_1, H_2, \ldots, H_q$, respectively, the* covering pattern *of the pair G, H is the weighted bipartite graph* $\mathrm{Cov}(G, H) = (\{g_1, g_2, \ldots, g_p, h_1, h_2, \ldots, h_q\}, \{g_i h_j : G_i \longrightarrow H_j\})$ *with edge weights* $r_{ij} = r(g_i h_j) = \frac{|V(G_i)|}{|V(H_j)|}$.

The following observation follows directly from the definitions, but will be useful in the computational complexity considerations.

Observation 9. *Let G and H be graphs. Then the following holds.*

- *We have $G \longrightarrow_{lb} H$ if and only if the degree of every vertex $g_i, i = 1, 2, \ldots, p$ in $\mathrm{Cov}(G, H)$ is greater than zero.*
- *We have $G \longrightarrow_{sur} H$ if and only if the degree of every vertex $g_i, i = 1, 2, \ldots, p$ in $\mathrm{Cov}(G, H)$ is greater than zero and $\mathrm{Cov}(G, H)$ has a matching of size q.*
- *We have $G \longrightarrow_{equit} H$ if and only if $\mathrm{Cov}(G, H)$ has a spanning subgraph $\mathrm{Map}(G, H)$ such that every vertex $g_i, i = 1, 2, \ldots, p$ has degree 1 in $\mathrm{Map}(G, H)$ and for every vertex h_j of $\mathrm{Cov}(G, H)$, $\sum_{i : g_i h_j \in E(\mathrm{Map}(G,H))} r_{ij} = k$, where $k = \frac{|V(G)|}{|V(H)|}$.*

4   Complexity Results

We feel the world will be on the right track if H-COVER is polynomial time solvable whenever H_i-COVER is polynomial time solvable for every component H_i of H, while H-COVER is NP-complete whenever H_i-COVER is NP-complete for some component H_i of H. To strengthen the results, we allow arbitrary input graphs (i.e., with multiple edges, loops and/or semi-edges) when considering polynomial time algorithms, while we restrict the inputs to simple graphs when we aim at NP-hardness results. In some cases we are able to prove results also from the Fixed Parameter Tractability standpoint.

The following lemma is simple, but useful. Note that though we are mostly interested in the time complexity of deciding $G \longrightarrow H$ for a fixed graph H and input graph G, this lemma assumes both the source and the target graphs to be part of the input. The size of the input is measured by the number of edges.

Lemma 10. *Let $\varphi(A, B)$ be the best running time of an algorithm deciding if $A \longrightarrow B$ for connected graphs A and B, and let $\varphi(n, B)$ be the worst case of $\varphi(A, B)$ over all connected graphs of size n. Then for given input graphs G and H with components of connectivity $G_1, G_2, \ldots, G_p$, and $H_1, H_2, \ldots, H_q$, respectively, the covering pattern $\mathrm{Cov}(G, H)$ can be constructed in time $O(pq \cdot \max_{j=1}^{q} \varphi(n, H_j)) = O(n^2 \cdot \max_{j=1}^{q} \varphi(n, H_j))$, where n is the input size, i.e., the sum of the numbers of edges in G and H.* □

Corollary 11. *Constructing the covering pattern of input graphs G and H is in the complexity class XP when parameterized by the maximum size of a component of the target graph H, provided the H_j-COVER problem is polynomial time solvable for every component H_j of H.*

Corollary 12. *The covering pattern of input graphs G and H can be constructed in polynomial time provided all components of H have bounded size and the H_j-COVER problem is solvable in polynomial time for every component H_j of H.* □

In the following subsections, we discuss and compare the computational complexity of deciding the existence of locally bijective homomorphisms, surjective covers and equitable covers. The corresponding decision problems are denoted by LBHOM, SURJECTIVECOVER, and EQUITABLECOVER. If the target graph is fixed to be H, we write H-LBHOM (and analogously for the other variants).

4.1 Locally Bijective Homomorphisms

Though the notion of locally bijective homomorphisms is seemingly the most straightforward generalization of the fact that in a graph covering projection to a connected graph "the closed neighborhood of every vertex of the source graph is mapped bijectively to the closed neighborhood of its image", we show in this subsection that it does not behave as we would like to see it from the computational complexity perspective. Proposition 14 shows that there are infinitely many graphs H with only two components each such that H-LBHOM is polynomial time solvable, while H_i-LBHOM is NP-complete for one component H_i of H. The polynomial part of the desired properties is, however, fulfilled, even in some cases when both graphs are part of the input:

Theorem 13. *If H_i-COVER is polynomial time solvable for every component H_i of H, then*

  i) the H-LBHOM problem is polynomial time solvable,
 ii) the LBHOM problem is in XP when parameterized by the maximum size of a component of the target graph H,
iii) the LBHOM problem is solvable in polynomial time, provided the components of H have bounded size.

However, it is not true that H-LBHOM is NP-complete whenever H_j-COVER is NP-complete for some component H_j of H. Infinitely many examples can be constructed by means of the following proposition. These examples provide another argument for our opinion that the notion of locally bijective homomorphism is not the right generalization of graph covering to disconnected graphs.

In the following propositon, $+$ represents the operation of disjoint union of two graphs.

Proposition 14. *Let $H = H_1 + H_2$ for connected graphs H_1 and H_2 such that $H_1 \triangleright H_2$. Then H-LBHOM for simple input graphs is polynomially reducible to H_2-LBHOM for simple input graphs. In particular, if H_2-LBHOM is polynomial time decidable, then so is H-LBHOM as well.*

4.2 Surjective Covers

The notion of surjective covers is favored by topologists since it captures the fact that every vertex (point) of the target graph (space) is covered [Nedela, private communication 2020]. We are happy to report that this notion behaves as we would like to see from the point of view of computational complexity.

Theorem 15. *If H_i-COVER is polynomial time solvable for every component H_i of H, then*

- *i) the H-SURJECTIVECOVER problem is polynomial time solvable,*
- *ii) the SURJECTIVECOVER problem is in XP when parameterized by the maximum size of a component of the target graph H, and*
- *iii) the SURJECTIVECOVER problem is solvable in polynomial time if the components of H have bounded size.*

For surjective covers, the NP-hardness of the problem of deciding if there is a covering of one component of H propagates to NP-hardness of deciding if there is a surjective covering of entire H, even when our attention is restricted to simple input graphs.

Theorem 16. *The H-SURJECTIVECOVER problem is NP-complete for simple input graphs if H_i-COVER is NP-complete for simple input graphs for at least one component H_i of H.*

Proof. Without loss of generality suppose that H_1-COVER is NP-complete for simple input graphs. Let G_1 be a simple graph for which $G_1 \longrightarrow H_1$ is to be tested. We show that there exists a polynomial time reduction from H_1-COVER to H-SURJECTIVECOVER. For every $j = 2, \ldots, q$, fix a simple connected graph G_j that covers H_j such that $G_j \longrightarrow H_1$ if and only if $H_j \rhd H_1$ (in other words, G_j is a witness which does not cover H_1 when H_j is not stronger than H_1). Note that the size of each G_j, $j = 2, \ldots, q$, is a constant which does not depend on the size of the input graph G_1. Note also, that since H is a fixed graph, we do not check algorithmically whether $H_j \rhd H_1$ when picking G_j. We are only proving the existence of a reduction, and for this we may assume the relation $H_j \rhd H_1$ to be given by a table.

Let G be the disjoint union of G_j, $j = 1, \ldots, q$. We claim that $G \longrightarrow_{sur} H$ if and only if $G_1 \longrightarrow H_1$. The "if" part is clear. We map G_j onto H_j for every $j = 1, 2, \ldots, q$ by the covering projections that are assumed to exist. Their union is a surjective covering projection of G to H.

For the "only if" direction, suppose that $f \colon V(G) \longrightarrow V(H)$ is a surjective covering projection. Since f must be globally surjective and G and H have the same number of components, namely q, different components of G are mapped onto different components of H by f. Define $\widetilde{f} \in Sym(q)$ by setting $\widetilde{f}(i) = j$ if and only if f maps G_i onto H_j. Then $\widetilde{f}$ is a permutation of $\{1, 2, \ldots, q\}$. Consider the cycle containing 1. Let it be $(i_1 = 1, i_2, i_3, \ldots, i_t)$, which means that $G_{i_j} \longrightarrow H_{i_{j+1}}$ for $j = 1, 2, \ldots, t-1$, and $G_{i_t} \longrightarrow H_{i_1}$. By reverse induction on j, from $j = t$ down to $j = 2$, we prove that $H_j \vartriangleright H_1$. Indeed, for $j = t$, $G_{i_t} \longrightarrow H_1$ means that H_{i_t} is stronger than H_1, since we would have set G_{i_t} as a witness that does not cover H_1 if it were not. For the inductive step, assume that $H_{i_{j+1}} \vartriangleright H_1$ and consider G_{i_j}. Now G_{i_j} covers $H_{i_{j+1}}$ since $\widetilde{f}(i_j) = i_{j+1}$. Because G_{i_j} is a simple graph and $H_{i_{j+1}}$ is stronger than H_1, this implies that $G_{i_j} \longrightarrow H_1$. But then H_{i_j} must itself be stronger than H_1, otherwise we would have set G_{i_j} as a witness that does not cover H_1. We conclude that $H_{i_2} \vartriangleright H_1$, and hence $G_1 \longrightarrow H_1$ follows from the fact that the simple graph G_1 covers H_{i_2}.

4.3 Equitable Covers

Equitable covers also behave nicely from the computational complexity point of view in the crucial aspects:

Theorem 17. *The H-EQUITABLECOVER problem is polynomial time solvable if H_i-COVER is polynomial time solvable for every component H_i of H.*

Proof. First construct the covering pattern $\mathrm{Cov}(G, H)$. Since H is a fixed graph, this can be done in time polynomial in the size of the input, i.e., G, as it follows from Corollary 12.

Using dynamic programming, fill in a table $M(s, k_1, k_2, \ldots, k_q)$, $s = 0, 1, \ldots, p$, $k_j = 0, 1, \ldots, k = \frac{|V(G)|}{|V(H)|}$ for $j = 1, 2, \ldots, q$, with values true and false. Its meaning is that $M(s, k_1, k_2, \ldots, k_q) = \text{true}$ if and only if $G_1 \cup G_2 \cup \ldots \cup G_s$ allows a locally bijective homomorphism f to H such that for every j and every $u \in V(H_j)$, $|f^{-1}(u)| = k_j$. The table is initialized by setting $M(0, k_1, \ldots, k_q)$ to true, if $k_1 = k_2 = \ldots = k_q = 0$ and to false otherwise.

In the inductive step assume that all values for some s are filled in correctly, and move on to $s + 1$. For every edge $g_{s+1} h_j$ of $\mathrm{Cov}(G, H)$ and every q-tuple $k_1, k_2, \ldots, k_q$ such that $M(s, k_1, k_2, \ldots, k_q) = \text{true}$, set $M(s + 1, k_1, k_2, \ldots, k_j + r_{s+1,j}, \ldots, k_q) = \text{true}$, provided $k_j + r_{s+1,j} \leq k$. Clearly, the loop invariant is fulfilled, and hence G is a k-fold (equitable) cover of H if and only if $M(p, k, k, \ldots, k)$ is set to true.

The table M has $(p + 1) \cdot (k + 1)^q = O(n^{q+1})$ entries and the inductive step changes $O((k + 1)^q \cdot q)$ values. So processing the table can be concluded in $O((k + 1)^q (1 + pq)) = O(n^{q+1})$ steps.

If both G and H are part of the input, we do not know how to avoid q in the exponent.

Proposition 18. *The EQUITABLECOVER problem is in XP when parameterized by the number q of components of H plus the maximum size of a component of the target graph H, provided H_i-COVER is polynomial time solvable for every component H_i of H.*

Problem 19. Is the EQUITABLECOVER problem in XP when parameterized by the maximum size of a component of the target graph H, provided each H_i-COVER is polynomial time solvable for every component H_i of H?

The NP-hardness theorem holds true as well:

Theorem 20. *The H-EQUITABLECOVER problem is NP-complete for simple input graphs if H_i-COVER is NP-complete for simple input graphs for at least one component H_i of H.*

5   Covering Colored Two-Vertex Graphs

We now introduce the last generalization and consider coverings of graphs which come with links and vertices equipped with additional information, which we simply refer to as a color. The requirement is that the covering projection respects the colors, both on the vertices and on the links. This generalization is not purposeless as it may seem. It is shown in [8] that to fully characterize the complexity of H-COVER for simple graphs H, it is necessary and suffices to understand the complexity of H-COVER for colored mixed multigraphs of valency greater than 2. The requirement on the minimum degree of H gives hope that the characterization can be more easily described. We will first describe the concept of covers of colored graphs with semi-edges in detail in Subsect. 5.1, where we also give our final argument in favor of equitable covers. Then we extend the characterization of the computational complexity of covering colored 2-vertex graphs without semi-edges presented in [8] to general graphs in Subsect. 5.2.

5.1   Covers of Colored Graphs

Definition 21. *We say that a graph G is* colored, *if it is equipped with a function $c : D \cup V \to \mathbb{N}$. Furthermore, a colored graph covers a colored graph H if G covers H via a mapping f and this mapping respects the colors, i.e., $c_G = c_H \circ f$ on D and every $u \in V_G$ satisfies $c_G(u) = c_H(f(u))$.*

Note that one may assume without loss of generality that all vertices are of the same color, since we can add the color of a vertex as a shade to the colors of its darts. However, for the reductions described below, it is convenient to keep the intermediate step of coloring vertices as well.

The final argument that equitable covers are the most proper generalization to disconnected graphs is given by the following observation. (Note that color induced subgraphs of a connected graph may be disconnected).

Observation 22. *Let a colored graph H be connected and let $f : G \longrightarrow H$ be a color preserving mapping that respects the links (i.e., for every link $e \in \Lambda_G$, there is a link $e' \in \Lambda_H$ such that $f(e) = e'$). Then f is a covering projection if and only if $f : G_{i,j} \longrightarrow H_{i,j}$ is an equitable covering projection for every two (not necessarily distinct) colors i, j, where $G_{i,j}, H_{i,j}$ denote the subgraphs of G and H induced by the links e such that $c(e) = \{i, j\}$.*   □

Kratochvíl et al. [8] proved that the existence of a covering between two (simple) graphs can be reduced to the existence of a covering between two colored graphs of minimum degree three. Their concept of *colored directed multigraph* is equivalent to our concept of colored graphs (without semi-edges), namely:

Fig. 2. Reduction of a graph to a colored graph of minimum degree 3. Distinct colors represent distinct integers.

- The vertex color encoding the collection of trees (without semi-edges) stemming from a vertex is encoded as the vertex color in the exactly same way.
- The link color encoding a subgraph isomorphic to colored induced path between two vertices of degree at least three is encoded as the pair of colors of the edge or a loop that is used for the replacement of the path.
- When the path coloring is symmetric, we use the same color twice for the darts of the replaced arc which could be viewed as an undirected edge of the construction of [8].
- On the other hand, when the coloring is not symmetric and the replaced arc hence needed to be directed in [8], we use a pair of distinct colors on the two darts, which naturally represents the direction.

When semi-edges are allowed we must take into account one more possibility. The color used on the two darts representing a symmetric colored path of even number of vertices may be used also to represent a half-way path with the identical color pattern ended by a semi-edge. A formal description follows:

By a *pattern* P we mean a finite sequence of positive integers $(p_1, \ldots, p_k)$. A pattern is symmetric if $p_i = p_{k+1-i}$, and the reverse pattern is $\overline{P} = (p_k, \ldots, p_1)$.

The pattern of a closed path $d_1, \ldots, d_{2k}$ in a colored graph G is the sequence of colors $c(u_0), c(d_1), c(d_2), c(u_1), c(d_3), c(d_4), c(u_2), \ldots, c(d_{2k}), c(u_k)$, where $\forall i \in \{1, \ldots, k\}$ the vertex u_i contains the dart d_{2i}, and in addition u_0 contains the dart d_1. Analogously we define patterns of open and half-way paths—the sequence of dart colors is augmented by the vertex colors.

Now, a half-way path of pattern P that starts in a vertex of degree 3 and ends by a semi-edge will be replaced by a single dart whose color is identical to those used for the two darts used for the replacement of closed paths whose pattern is the concatenation $P\overline{P}$, see Fig. 2.

5.2 Two-Vertex Graphs

We say that a colored graph G is *regular* if all its vertices have the same color and for every $i \in \mathbb{N}$ all vertices are incident with the same number of darts of color i. Kratochvíl et al. [8] completely characterized the computational complexity of the H-COVER problem on colored graphs on at most two vertices without semi-edges. Their result implies the following:

Proposition 23. *Let H be a connected colored graph on at most two vertices without semi-edges. The H-COVER problem is polynomially solvable if:*

1. *The graph H contains only one vertex or*
2. *H is not regular or*
3. *(a) for every color $i \in \mathbb{N}$, the H_i-EQUITABLECOVER problem is solvable in polynomial time, where H_i is the colored subgraph of H induced by the links colored by i, and*
   (b) for every pair of colors $i, j \in \mathbb{N}$, the $H_{i,j}$-EQUITABLECOVER problem is solvable in polynomial time, where $H_{i,j}$ is the colored subgraph of H induced by the links $l \in \Lambda$ such that $c(l) = \{i, j\}$.

Otherwise, the H-COVER problem is NP-complete.

Informally, the NP-completeness persists if and only if H has two vertices which have the same degree in every color, and the NP-completeness appears on a monochromatic subgraph (either undirected or directed). Such a subgraph must contain both vertices and be connected. We extend this characterization to include semi-edges as well:

Theorem 24. *Let H be a colored graph on at most two vertices. The H-EQUITABLECOVER problem is polynomially solvable if:*

1. *The graph H contains only one vertex and for every i, H_i-COVER is solvable in polynomial time, where H_i is the subgraph of H induced by the loops and semi-edges colored by i or*
2. *H is not regular and for every i and each vertex $u \in V_H$, the H_i^u-COVER problem is solvable in polynomial time, where H_i^u is the colored subgraph of H induced by the loops and semi-edges incident with u colored by i or*
3. *H is regular on two vertices and*
   (a) for every color $i \in \mathbb{N}$, the H_i-EQUITABLECOVER problem is solvable in polynomial time, where H_i is the colored subgraph of H induced by the links colored by i, and
   (b) for every pair of colors $i, j \in \mathbb{N}$, the $H_{i,j}$-EQUITABLECOVER problem is solvable in polynomial time, where $H_{i,j}$ is the subgraph of H induced by the links $l \in \Lambda$ such that $c(l) = \{i, j\}$.

Otherwise, the H-EQUITABLECOVER problem is NP-complete.

Observe that the cases when every color induces in H a subgraph without semi-edges are covered by Proposition 23.

6   Conclusion

The main goal of this paper was to point out that the generalization of the notion of graph covers of connected graphs to disconnected ones is not obvious. We have presented three variants, depending of the requirement if the projection should or need not be globally surjective, and if all vertices should be covered the same number of times. We argue that the most restrictive variant, which we call equitable covers, is the most appropriate one, namely from the point of view of covers of colored graphs.

We have compared the computational complexity aspects of these variants and show that two of them, surjective and equitable covers, possess the naturally desired property that H-COVER is polynomially solvable if covering each component of H is polynomially solvable, and NP-complete if covering at least one component of H is NP-complete. However, we identified some open question from the point of view of fixed parameter tractability.

In the last section we review the extension of graph covers to covers of colored graphs, recall that colors can be encoded by non-coverable patterns in simple graphs, and discuss this issue in detail for the case when semi-edges are allowed. With this new feature we conclude the complete characterization of the computational complexity of covering 2-vertex colored graphs, initiated (and proved for graphs without semi-edges) 24 years ago in [8].

Acknowledgments. – Jan Bok and Nikola Jedličková: Supported by research grant GAČR 20-15576S of the Czech Science Foundation and by SVV–2020–260578. The authors were also partially supported by GAUK 1580119.

– Jiří Fiala and Jan Kratochvíl: Supported by research grant GAČR 20-15576S of the Czech Science Foundation.

– Michaela Seifrtová: Supported by research grant GAČR 19-17314J of the Czech Science Foundation.

References

1. Abello, J., Fellows, M.R., Stillwell, J.C.: On the complexity and combinatorics of covering finite complexes. Aust. J. Combin. **4**, 103–112 (1991)
2. Angluin, D.: Local and global properties in networks of processors. In: Proceedings of the 12th ACM Symposium on Theory of Computing, pp. 82–93 (1980)
3. Biggs, N.: Algebraic Graph Theory. Cambridge University Press, Cambridge (1974)
4. Bok, J., Fiala, J., Hliněný, P., Jedličková, N., Kratochvíl, J.: Computational complexity of covering two-vertex multigraphs with semi-edges. CoRR abs/2103.15214 (2021). https://arxiv.org/abs/2103.15214. To appear in proceedings of MFCS 2021
5. Chalopin, J., Métivier, Y., Zielonka, W.: Local computations in graphs: the case of cellular edge local computations. Fund. Inform. **74**(1), 85–114 (2006)
6. Chaplick, S., Fiala, J., van 't Hof, P., Paulusma, D., Tesař, M.: Locally constrained homomorphisms on graphs of bounded treewidth and bounded degree. In: Gąsieniec, L., Wolter, F. (eds.) FCT 2013. LNCS, vol. 8070, pp. 121–132. Springer, Heidelberg (2013). https://doi.org/10.1007/978-3-642-40164-0_14
7. Kratochvíl, J., Proskurowski, A., Telle, J.A.: Complexity of graph covering problems. In: Mayr, E.W., Schmidt, G., Tinhofer, G. (eds.) WG 1994. LNCS, vol. 903, pp. 93–105. Springer, Heidelberg (1995). https://doi.org/10.1007/3-540-59071-4_40
8. Kratochvíl, J., Proskurowski, A., Telle, J.A.: Covering directed multigraphs I. Colored directed multigraphs. In: Möhring, R.H. (ed.) WG. LNCS, vol. 1335, pp. 242–257. Springer, Heidelberg (1997)

9. Malnič, A., Nedela, R., Škoviera, M.: Lifting graph automorphisms by voltage assignments. Eur. J. Comb. **21**(7), 927–947 (2000)
10. Matoušek, J., Nešetřil, J.: Invitation to Discrete Mathematics. Oxford University Press, Oxford (1998)
11. Mednykh, A.D., Nedela, R.: Harmonic Morphisms of Graphs: Part I: Graph Coverings, 1st edn. Vydavatelstvo Univerzity Mateja Bela v Banskej Bystrici (2015)

The Complexity of Bicriteria Tree-Depth

Piotr Borowiecki[1,2], Dariusz Dereniowski[1(✉)], and Dorota Osula[1]

[1] Faculty of Electronics, Telecommunications and Informatics,
Gdańsk University of Technology, Gdańsk, Poland
deren@eti.pg.edu.pl
[2] Institute of Control and Computation Engineering, University of Zielona Góra,
Zielona Góra, Poland

Abstract. The tree-depth problem can be seen as finding an elimination tree of minimum height for a given input graph G. We introduce a bicriteria generalization in which additionally the *width* of the elimination tree needs to be bounded by some input integer b. We are interested in the case when G is the line graph of a tree, proving that the problem is $\mathcal{NP}$-hard and obtaining a polynomial-time additive $2b$-approximation algorithm. This particular class of graphs received significant attention, mainly due to potential applications. These include purely combinatorial applications like searching in tree-like partial orders (which generalizes binary search in sorted data), or practical ones in parallel processing.

Keywords: Elimination tree · Graph ranking · Parallel assembly · Tree-depth

1 Introduction

The problem of computing tree-depth has a long history in the realm of parallel computations. It was considered for the first time under the name of *minimum height elimination trees* where it played an important role in speeding-up parallel factorization of (sparse) matrices [17]. Then, the problem re-appeared under the name of *vertex ranking* [2]. More applications have been brought up, including parallel assembly of multi-part products, where tree-like structures have been mostly considered. Another application in this realm includes parallel query processing in relational databases [19]. It turns out that in order to design an efficient parallel schedule for performing the query, a vertex ranking of a line graph of a spanning tree of G is computed. Later on, the same problem has been introduced under different names in a number of other applications: *LIFO-search* [11], searching in tree-like partial orders [1], which generalizes the classical binary search in sorted arrays, *ordered coloring*, graph ranking [12], and more recently,

Work partially supported under Ministry of Science and Higher Education (Poland) subsidy for Gdańsk University of Technology. Moreover, D. Dereniowski and D. Osula have been partially supported by National Science Centre (Poland) grant number 2018/31/B/ST6/00820.

© Springer Nature Switzerland AG 2021
E. Bampis and A. Pagourtzis (Eds.): FCT 2021, LNCS 12867, pp. 100–113, 2021.
https://doi.org/10.1007/978-3-030-86593-1_7

tree-depth [21]. Through the connection to searching partial orders, it is worth pointing out that tree-depth computation (i.e., a search strategy for the partial order) can be used to automated finding of software bugs [1]. In this application, again, the line graphs of trees and their tree-depth are of interest.

Related Work. The tree-depth problem is $\mathcal{NP}$-complete for arbitrary line graphs [15]. The smallest superclass of trees for which the problem is $\mathcal{NP}$-complete are chordal graphs [9]. For the trees themselves, the problem can be solved in linear-time [23]. The problem turned out to be much more challenging and interesting for line graphs of trees. A number of papers have been published, see e.g. [25–27], that gradually reduced the complexity from $O(n^4 \log^3 n)$ [1] and $O(n^3 \log n)$ [24] to the final linear-time algorithms [16,20], where n is the order of the input tree. Motivated e.g. by applications, the edge-weighted case has been introduced [6]: it is strongly $\mathcal{NP}$-hard even for restricted classes of trees [4–6]. For algorithmic results on weighted paths see e.g. [4,14]. For weighted trees, there is a number of works improving on possible approximation ratio achievable in polynomial-time [4–6], with the best one of $O(\sqrt{\log n})$ [8], and it remains as a challenging open question whether a constant-factor approximation is feasible. There is also a number of searching models that generalize searching in (rooted) tree-like partial orders to more general classes of partial orders see e.g. [3,7,22]. These general graph-theoretic models find applications e.g. in machine learning [10]. The tree-depth generalization that we consider for line graphs of trees has been introduced for trees [28] under the name of *vertex ranking with capacity*: there exists an $O^*(2.5875^n)$-time optimal algorithm for general graphs, and $f(n)$-approximate solution to vertex ranking can be transformed to an $(f(n)+1)$-approximate solution to vertex ranking with capacity. Moreover, for trees, this problem admits a polynomial time absolute $O(\log b)$-approximation [28].

Problem Statement and Our Results. In this section we introduce a generalization of the concept of the classical elimination tree by considering elimination forests with level functions explicitly defined on their vertex sets. In this context a *rooted forest* is meant as a disjoint union of rooted trees. We start with a notion of the classical elimination tree that we call here a free elimination tree. For the sake of correctness we point out that all graphs $G = (V, E)$ considered in this paper are finite, simple and undirected, with vertex set V and edge set E.

Definition 1. *A* free elimination tree *for a connected graph G is a rooted tree T defined recursively as follows:*

1. *let $V(T) = V(G)$ and let an arbitrary vertex $r \in V(T)$ be the root of T,*
2. *if $|V(G)| = 1$, then let $E(T) = \emptyset$. Otherwise, let $E(T) = \bigcup_{i=1}^{k} \left(E(T_i) \cup \{e_i\} \right)$ where k is the number of connected components of $G - r$, and T_i stands for an elimination tree for the i-th connected component of $G - r$ with the root $r(T_i)$ joined by the edge e_i with the root r of T, i.e., $e_i = \{r(T_i), r\}$.*

Definition 2. *Let G be a graph with k connected components. A* free elimination forest *for G is the disjoint union of k free elimination trees, each of which is determined for distinct connected component of G.*

Given two vertices u and v of a rooted forest F, we say that v is *an ancestor* of u if v belongs to the path with the end-vertices in u and the root of the connected component of F that contains u. If v is an ancestor of u and $\{v, u\} \in E(F)$, then v is the *parent* of u while u is a *child* of v.

Definition 3. *Let F be a free elimination forest for a graph G, and let f : $V(F) \to \mathbb{Z}^1$ be a level function, i.e. a function such that $f(u) < f(v)$ whenever v is an ancestor of u. A free elimination forest F with a level function f is called an* elimination forest *for G and it is denoted by F_f.*

For an elimination forest F_f its *height* $h(F_f)$ is defined as $\max\{f(v) \mid v \in V(F_f)\}$. Clearly, the maximum can be attained only for the roots of the connected components of F_f. Now, for every $i \in \{1, \ldots, h(F_f)\}$ we define the *i-th level* $L_i(F_f)$ of F_f as the set of vertices v in $V(F_f)$ for which $f(v) = i$ (notice that we allow empty levels). In a natural way the *width* $w(F_f)$ of elimination forest F_f is defined as $\max_i |L_i(F_f)|$, where $i \in \{1, \ldots, h(F_f)\}$.

We point out that the above definitions do not impose the placement of the roots of all k connected components of F_f at the highest level. In fact, each root can be placed at an arbitrary level. Moreover, it is also not required that adjacent vertices of an elimination forest occupy consecutive levels. Also note that though the definition of the classical elimination tree does not explicitly give any level function, one of the possible functions can be, and usually is, implicitly deduced by assuming that each level is formed by a single recursive step in Definition 1.

Definition 4. *Let G be a graph and let b be a positive integer. The* bounded-width tree-depth *$btd(G, b)$ of a graph G is the minimum k for which there exists an elimination forest F_f for a graph G such that $h(F_f) = k$ and $w(F_f) \leq b$.*

Note that for every G and $b > 0$ there always exists some elimination forest of width bounded by b. In what follows we omit subscript f whenever level function f is clear from the context. We can now formulate our main problem.

BOUNDED-WIDTH TREE-DEPTH (BTD)

Input: A graph G, positive integers k and b.
Question: Does $btd(G, b) \leq k$ hold?

The above seemingly small differences in the classical and our definitions play an important role both in our $\mathcal{NP}$-completeness reduction and algorithm. They significantly affect the bounded-width tree-depth problem complexity thus making it different than that of the classical tree-depth. The classical tree-depth problem can be solved in linear time for line graphs of trees [16, 20]. The generalization we consider turns out to be $\mathcal{NP}$-complete.

Theorem 1. BTD *problem is $\mathcal{NP}$-complete for line graphs of trees.*

The proofs given in Sect. 3 reveal that bounded-width tree-depth behaves differently than the original tree-depth problem strongly depending on the connectivity of the graph. Specifically, for tree-depth, only connected graphs are of interest since tree-depth of a non-connected graph is just the maximum tree-depth taken over its connected components. In the bounded-width tree-depth problem, the connected components interact. En route of proving Theorem 1, we first obtain hardness of the problem for line graphs of forests and then we extend it to get the $\mathcal{NP}$-completeness of BTD for line graphs of trees.

On the positive side, we develop an approximation algorithm for line graphs of trees. Here, the fact that minimum height elimination tree (or equivalently an optimal tree-depth) can be found efficiently [16, 20] turns out to be very useful—we start with such a tree (with some additional preprocessing) and squeeze it down so that its width becomes as required. The squeezing-down is done in a top-down fashion, by considering the highest level that contains more than b vertices, and moving the excess vertices downwards. The algorithm moves down those vertices that are the roots of the subtrees of the smallest height. This leads to polynomial-time approximation algorithm with an additive error of $2b$.

Theorem 2. *There exists a polynomial-time additive $2b$-approximation algorithm for* BTD *problem for line graphs of trees.*

We note that our algorithm is different from the one for trees in [28]. Finding balanced vertex separators for trees is easy an hence in [28] it is guaranteed that after an initial recursive search one gets a forest with components of roughly similar size. Then, a solution can be constructed by putting the vertices of each component on pairwise distinct levels. In contrast, for line graphs of trees balanced separators do not exist in general. Thus we initialize our algorithm with a minimum height elimination tree, for otherwise using involved bottom-up dynamic programming as in [16, 20] seems unavoidable. Another difference is that it is unknown if there exists an optimal polynomial-time algorithm for trees.

2 Preliminaries

For the sake of clarity and to avoid involved notation and argument, necessary when carrying the proofs directly on line graphs, we use $\mathcal{E}_G$ to denote an elimination forest for the line graph $L(G)$ of a graph G. Consequently, since by the definitions of line graph and elimination forest, there is a natural one-to-one correspondence between the edges of a graph G and the vertices of its line graph and hence the vertices of an elimination forest $\mathcal{E}_G$, for *each edge* of G we shortly say that it *corresponds* to the appropriate vertex v of $\mathcal{E}_G$ and that it *belongs* to the level of $\mathcal{E}_G$ that contains v. Also, with a small abuse of notation the above one-to-one correspondence allows the use of any level function ℓ, determined for an elimination forest $\mathcal{E}_G$, as if it was defined on $E(G)$. Thus for every edge e in $E(G)$ we say that $\ell(e) = p$ if for the corresp. vertex v of $\mathcal{E}_G$ it holds $v \in L_p(\mathcal{E}_G)$.

Concerning the properties of level functions in the above-mentioned common context of a graph G and elimination forest $\mathcal{E}_G$, we note that if two distinct edges e_1, e_2 are adjacent in G, then they cannot belong to the same level of $\mathcal{E}_G$, i.e., $\ell(e_1) \neq \ell(e_2)$. Similarly, it is not hard to see that in the recursive elimination process performed on the vertices of $L(G)$ (according to Definition 1) all vertices eliminated at the same recursive step belong to distinct components of the processed graph, and for each of them the value of a level function is greater than for the vertices eliminated in further steps. In other words, for distinct edges e_1, e_2 of G such that $\ell(e_1) = \ell(e_2)$ every path with end-edges e_1, e_2 contains an edge e' such that $\ell(e') > \ell(e_1)$. Note that considering ℓ as a function defined on $E(G)$ and satisfying both of the above-mentioned properties, ℓ can be equivalently seen as an edge ranking of a graph G (see e.g. [13]).

We use the same common context to define the visibility of a level from a vertex in G. Namely, for a vertex v in G we say that the p-th level is *visible* in G *from* v if there *exists* an edge $e = \{u_1, u_2\}$ with $\ell(e) = p$ and G contains a path with end-vertices v, u, where $u \in \{u_1, u_2\}$ and for each edge e' of the path $\ell(e') \leq p$. The set of levels visible in G from v is denoted by $\mathrm{vis}(G, v)$. When determining the levels admissible for a given edge in a graph G (or for the corresponding vertex in elimination forest $\mathcal{E}_G$) we need to consider and forbid all levels visible from both end-vertices of that edge, i.e., the level p is *admissible* for $e = \{u_1, u_2\}$ if neither $p \in \mathrm{vis}(G - e, u_1)$ nor $p \in \mathrm{vis}(G - e, u_2)$.

3 $\mathcal{NP}$-completeness of BTD

Technically, we prove $\mathcal{NP}$-completeness of BTD performing a polynomial-time reduction from the Minimum hitting set (MHS) problem.

Minimum Hitting Set (MHS)
Input: A set $A = \{a_1, \ldots, a_n\}$, subsets $A_1, \ldots, A_m$ of A, an integer $t \geq 0$.
Question: Is there an $A' \subseteq A$ such that $|A'| \leq t$ and $A' \cap A_j \neq \emptyset$, $j \in \{1, \ldots, m\}$?

Construction. On the basis of the input to the MHS problem, we construct an appropriate forest F consisting of the tree T, called the *main* component, and some number of *additional* connected components created on the basis of 'template' trees $\overline{T}_d$ defined later on. By S_n we denote an n-vertex star, and for a vertex v in G we use $S(v)$ to denote its subgraph induced by v and its neighbors of degree 1. An important parameter in our construction is an integer M given at the end of this section. First, we focus on the structure of the main component T obtained by the identification of distinguished vertices r_i of the trees $T(a_i)$ constructed for the corresponding $a_i \in A$, $i \in \{1, \ldots, n\}$. The common vertex r, resulting from the identification, becomes the root of T. Next, we add $M + 3(m-1) + 4$ edges incident to r (hence $S(r)$ is a star $S_{M+3(m-1)+5}$). As the building blocks of $T(a_i)$ we need graphs G_α with $\alpha \geq M + 1$ and graphs $G(a_i)$ (for their structure, see Fig. 1). The graph G_α has one distinguished vertex called a *connector*, denoted by w_1, while $G(a_i)$ has two *connectors* r_i and v_i. The stars in the 'loops' in Fig. 1 have their central vertex marked in black. From a slightly informal perspective, we describe $T(a_i)$ as 'star-shaped' structure formed with $m + 1$ graphs $T_0, \ldots, T_m$, where $T_0 = G(a_i)$, $T_j = G_{\varphi(j)}$ with

$$\varphi(j) = M + 3(j - 1) + 1$$

for every $j \in \{1, \ldots, m\}$. More formally, $T(a_i)$ is formed by the identification of the connector v_i of T_0 with the connector w_1, done for each T_j, $j \in \{1, \ldots, m\}$.

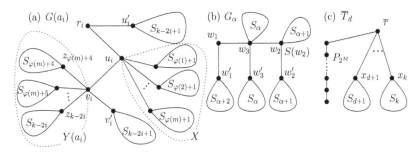

Fig. 1. The graphs used in Construction: (a) $G(a_i)$, (b) G_α, (c) $\overline{T}_d$, where $M \le d < \alpha$.

To complete the construction of $T(a_i)$, for every A_j containing a_i we take a copy of an 2^M-vertex path P_{2^M} and identify one of its end-vertices with a leaf of $S(w_2)$ of the corresponding tree T_j in $T(a_i)$ (P_{2^M} is now *attached* at T_j). At this point, we remark on the structure of T. We note that $G(a_i)$ depends on a_i which determines the degrees of u_i', v_i' and v_i, and hence the number of vertices in $\{z_{\varphi(m)+4}, \ldots, z_{k-2i}\}$ (see Fig. 1(a)). Also note that for a given $a_i \in A$ the trees $T_1, \ldots, T_m$ in $T(a_i)$ are pairwise different and their structure depends on j. On the other hand $T_1, \ldots, T_m$ are independent of a_i. For distinct elements $a_{i_1}, a_{i_2} \in A$ the structure of $T(a_{i_1})$ and $T(a_{i_2})$ differs, which follows from dissimilarity of $G(a_{i_1})$ and $G(a_{i_2})$ and varying 'attachment patterns' of paths P_{2^M}. To finish the construction of the forest F we use the trees $\overline{T}_d$ with $d \ge M$ (see Fig. 1(c); the path P_{2^M} is said to be *attached* at $\overline{T}_d$). Namely, the forest F is composed of a single copy of T, $2n - |A_j| + 1$ copies of $\overline{T}_{\varphi(j)-1}$ for $j = 1$, and $n - |A_j| + 1$ copies of $\overline{T}_{\varphi(j)-1}$ for each $j \in \{2, \ldots, m\}$, as well as $n - 1$ copies of $\overline{T}_{\varphi(j)}$ and n copies of $\overline{T}_{\varphi(j)+1}$ with both templates taken for each $j \in \{1, \ldots, m\}$. In what follows, referring to a path P_{2^M} we always mean one of the paths P_{2^M} used in this construction. Define $\overline{m}$:

$$\overline{m} = |E(F) \setminus \bigcup E(P_{2^M})|, \text{ where the sum runs over all paths } P_{2^M}. \quad (1)$$

We remark that $\overline{m}$ depends on k because the construction of F depends on k. For the BTD problem, we set the input parameters b and k to be:

$$k = M + 1 + 3m + 2n + t, \quad b = n2^M + \overline{m} \quad (2)$$

where M is chosen as a minimum integer satisfying $M \ge 2\lceil \log_2 \overline{m} \rceil + 1$. This can be rewritten as $2^M > 4\overline{m}^2$ (we use this form in the proof). We remark that the values of the parameters can be set to be polynomial in m, n and t. It follows from the construction that $\overline{m}$ is polynomially bounded in k, that is, $\overline{m} \le c_1 k^c$ for some constants c_1 and c. Take M and k so that $M \ge 2c \log_2 k + 2 \log_2 c_1 + 3$. This can be done in view of (2). This fixes the values of $\overline{m}$ and b. Since $2c \log_2 k + 2 \log_2 c_1 + 3 \ge 2\lceil \log_2 \overline{m} \rceil + 1$, it holds $2^M > 4\overline{m}^2$. This bound on M implies $k = O(m + n + t)$ and thus $\overline{m}$, 2^M and b are polynomial in m, n and t.

The Idea of the Proof. First of all, we note that the vast majority of edges in the forest F belongs to the additional components, which due to their structure

fit into precisely planned levels of an elimination forest $\mathcal{E}_F$, thus leaving exactly
the right amount of capacity on those levels where the main component gadgets come into play. The positioning of elimination subtrees corresponding to
appropriate paths P_{2M} is determined in Lemma 4; also see Fig. 2 for a sketch of
elimination subtrees fitting into appropriate levels. Though most of the capacity
consumed by the main component can be attributed to the paths P_{2M} (attached
at leaves of respective instances of the gadget G_α) the role of just a few edges
of $G(a_i)$ and G_α cannot be overestimated. As we will see, the assignment of
the edge $\{u_i, v_i\}$ to the level $k - 2i + 2$ is equivalent to including the element
a_i in the solution A'. Due to the sensitivity of the gadget G_α to what levels
are visible from its connector w_1 'outside' G_α (see Lemmas 1 and 2) we get a
coupling between the level of $\{u_i, v_i\}$ and the highest level that can be occupied
by the so called 'root edge' of the path P_{2M} attached at G_α, when particular
instance $G_{\varphi(j)}$ of the gadget in $T(a_i)$ corresponds to the set A_j containing a_i.
More specifically, if $\ell(\{u_i, v_i\}) = k - 2i + 2$, then the root edge of such a paths
P_{2M} can be moved to $\varphi(j)$ from $\varphi(j) - 1$ allowing all of its other edges to be
moved one level up, thus gaining the increase of the free space at the level that
has been previously occupied by roughly half of its edges.

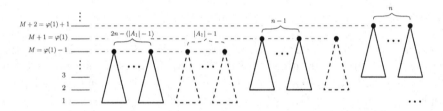

Fig. 2. Positioning of elimination subtrees corresp. to P_{2M}'s in the main and additional
components (dashed and solid lines, resp.) of the forest F. A snapshot for $j = 1$.

The bound t on the size of solution A' is met by attaching appropriate number
of edges pending at the root r of the main component T. The number of such
edges depends on t and it is calculated in such a way that in at most t of n
subgraphs $T(a_i)$ the edge $\{r, u_i\}$ will be allowed at level not exceeding $k - 2n$. In
Lemma 3 we show that either $\{r, u_i\}$ or $\{u_i, v_i\}$ must occupy the level $k - 2i + 2$
and hence there will be at most t subgraphs $T(a_i)$ with the edge $\{u_i, v_i\}$ assigned
to the level $k - 2i + 2$ and triggering the above-mentioned process of lifting.

Proof of $\mathcal{NP}$-Completeness. We always treat the gadgets as if they were subgraphs of the forest F, e.g., when analyzing admissibility of levels for particular
edges of G_α, we use vis(F', w_1) to refer to the levels visible from w_1 in F', where
F' is a graph induced by $V(F) \backslash V(G_\alpha - w_1)$.

Lemma 1. *If $\alpha + 2$ or $\alpha + 3$ belongs to vis(F', w_1), then for every elimination
tree $\mathcal{E}_{G_\alpha}$ it holds $h(\mathcal{E}_{G_\alpha}) > \alpha + 3$. If neither $\alpha + 2$ nor $\alpha + 3$ belongs to vis(F', w_1)
and $\alpha + 1 \in$ vis(F', w_1), then in every elimination tree $\mathcal{E}_{G_\alpha}$ of height $\alpha + 3$ the*

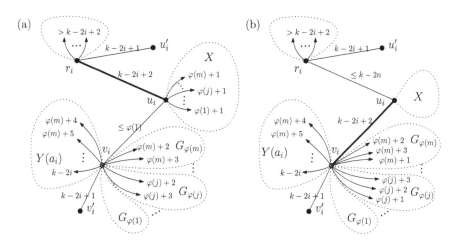

Fig. 3. The two major cases of visibility (the arrows point the levels visible from v_i).

levels admissible for $S(w_2)$ are in $\{1, \ldots, \alpha\}$, and the only levels visible from w_1 in G_α are $\alpha + 2$ and $\alpha + 3$.

Lemma 2. *If neither $\alpha + 1$, $\alpha + 2$ nor $\alpha + 3$ belongs to $\mathrm{vis}(F', w_1)$, then there exists an elimination tree $\mathcal{E}_{G_\alpha}$ of height $\alpha + 3$ with all edges of $S(w_2)$ at levels in $\{2, \ldots, \alpha + 1\}$ and such that only the levels $\alpha + 1, \alpha + 2$ and $\alpha + 3$ are visible from w_1 in G_α. Moreover, there is no elimination tree $\mathcal{E}_{G_\alpha}$ with $h(\mathcal{E}_{G_\alpha}) < \alpha + 3$.*

The next lemma describes the mutual interaction between distinguished edges of $G(a_i)$, which manifests as a 'switching' property of the gadget and allows 'lifting' of appropriate elimination subtrees preserving the bound b on the level size.

Lemma 3. *If $\mathcal{E}_T$ is an elimination tree of height k, then for each $i \in \{1, \ldots, n\}$ either $\ell(\{r, u_i\}) = k - 2i + 2$ or $\ell(\{u_i, v_i\}) = k - 2i + 2$, and each level $p \in \{k - 2i + 1, \ldots, k\}$ is visible from r in the subgraph induced by the vertices of $T(a_1), \ldots, T(a_i)$.*

Consequently, if $\mathcal{E}_T$ is an elimination tree of height k, then each level in $\{k - 2n + 1, \ldots, k\}$ is visible in T from r and if $\ell(\{u_i, v_i\}) = k - 2i + 2$, then $\ell(\{r, u_i\}) \le k - 2n$, $i \in \{1, \ldots, n\}$. Note that if $\ell(\{r, u_i\}) = k - 2i + 2$, then due to the visibility of particular levels, the edge $\{u_i, v_i\}$ must be assigned to a level p satisfying $p < M$, which is crucial in our reduction. In fact, pushing $\{u_i, v_i\}$ to a low level triggers the above-mentioned 'switching' by making certain levels in the subgraph X visible from the vertex v_i (see Fig. 3). In particular, for each $j \in \{1, \ldots, m\}$ the level $\varphi(j) + 1$ becomes visible in X from the vertex $w_1 \in V(G_{\varphi(j)})$ which by Lemma 1 results in impossibility of 'lifting' $\mathcal{E}_{P_{2M}}$ for P_{2M} is attached at $G_{\varphi(j)}$.

The *root edge* of the path P_{2M}, denoted by $r(P_{2M})$, is the edge at the highest level of $\mathcal{E}_{P_{2M}}$. An important aspect of lifting the elimination subtrees is that of estimating the highest possible levels at which the root edges of P_{2M}'s can be placed in $\mathcal{E}_F$. In what follows R is the set of all root edges, and R_m, R_a are the subsets of those in the main and additional components, respectively.

Lemma 4. *Let F be the forest corresponding to an instance of the* MHS *problem and let $\mathcal{E}_F$ be b-bounded. For each $i \in \{1, \ldots, n\}$ and $j \in \{1, \ldots, m\}$ it holds:*

(a) *If P_{2M} is attached at T_j of $T(a_i)$, then $\ell(r(P_{2M})) \in \{\varphi(j) - 1, \varphi(j)\}$.*
(b) *If P_{2M} is attached at $\overline{T}_d$ with $d = \varphi(j) - 1$, then $r(P_{2M}) \in L_d(\mathcal{E}_F)$.*
(c) *$|L_{\varphi(j)-1}(\mathcal{E}_F) \cap R| \leq \eta$, where $\eta = 2n$ if $j = 1$, and $\eta = n$ if $j \in \{2, \ldots, m\}$.*

Theorem 3. *The* BTD *problem is $\mathcal{NP}$-complete for line graphs of forests.*

Proof (sketch). Let A, t and $A_1, \ldots, A_m$ form an instance of the MHS problem, and let F and b be as in Construction (recall $k = M + 3m + t + 2n + 1$).

($\Rightarrow$) We are going to argue that if there exists a b-bounded elimination forest $\mathcal{E}_F$ such that $h(\mathcal{E}_F) \leq k$, then there exists a solution A' to the MHS problem such that $|A'| \leq t$. A solution to the MHS problem is defined as follows:

$$a_i \in A' \text{ if and only if } \ell(\{u_i, v_i\}) = k - 2i + 2, \tag{3}$$

for each $i \in \{1, \ldots, n\}$. First we prove that $|A'| \leq t$. Let H be a connected component of the graph obtained from F by the removal of all edges e for which in $\mathcal{E}_F$ it holds $\ell(e) > k - 2n$ and such that the root r of T belongs to H. Clearly, $h(\mathcal{E}_H) \leq k - 2n$. On the contrary, suppose that $|A'| > t$. We know that if $\ell(u_i, v_i) = k - 2i + 2$ (i.e. when by (3) $a_i \in A'$), then $\{r, u_i\} \in E(H)$. Moreover, $|E(S(r))| = k - 2n - t$ and hence $E(S(r)) \subseteq E(H)$. Thus $d_H(r) \geq |E(S(r))| + |A'| = k - 2n - t + |A'| > k - 2n$, which in turn implies $h(\mathcal{E}_H) > k - 2n$, a contradiction. Now, we prove a 'hitting property', i.e. $A' \cap A_j \neq \emptyset$ for each A_j with $j \in \{1, \ldots, m\}$. By Lemma 4(b), $\eta - |A_j| + 1$ elements in R_a belong to $L_{\varphi(j)-1}(\mathcal{E}_F)$. If $|A_j|$ elements in R_m were additionally contained in $L_{\varphi(j)-1}(\mathcal{E}_F)$, then $|L_{\varphi(j)-1}(\mathcal{E}_F) \cap R| > \eta$, which would contradict Lemma 4(c). Therefore at least one element in R_m, say the one related to P_{2M} attached at T_j of $T(a_{i^*})$, must be assigned to a level different from $\varphi(j) - 1$ which by Lemma 4(a) is exactly the level $\varphi(j)$. It holds (we omit details) that if $\ell(\{r, u_i\}) = k - 2i + 2$, then in every $\mathcal{E}_{T[u_i]}$ of height $k - 2i + 1$ for each edge e of $S(w_2)$ in T_j we have $\ell(e) \in \{1, \ldots, \varphi(j)\}$. Thus $\ell(\{r, u_{i^*}\}) \neq k - 2i^* + 2$ and hence by Lemma 3 we have $\ell(\{u_{i^*}, v_{i^*}\}) = k - 2i^* + 2$ which by (3) results in $a_{i^*} \in A'$.

($\Leftarrow$) Now, we show that if there exists a solution A' to the MHS problem such that $|A'| \leq t$, then there exists a b-bounded elimination forest $\mathcal{E}_F$ with $h(\mathcal{E}_F) \leq k$. Given A', we have to define on $E(F)$ a level function ℓ with values at most k. Starting with the edges $\{u_i, v_i\}$ and $\{r, u_i\}$ for all $i \in \{1, \ldots, n\}$ we set $\ell(\{u_i, v_i\}) = k - 2i + 2$ if $a_i \in S'$, and $\ell(\{r, u_i\}) = k - 2i + 2$ otherwise. For the remaining edges, the values of ℓ follow from omitted proofs of Lemmas 1–4, properties of our gadgets, and the analysis of P_{2M}'s attached at subgraphs T_j of the main component T. We focus on the last aspect. By assumption, for each $j \in \{1, \ldots, m\}$ there exists $a_{i^*} \in A' \cap A_j$ and hence we set $\ell(\{u_{i^*}, v_{i^*}\}) = k - 2i^* + 2$. Omitting details, we observe that there exists an elimination tree $\mathcal{E}_{T[v_i]}$ of height

$k - 2i + 1$ such that the levels in $\{2, \ldots, \varphi(j) + 1\}$ are admissible for the edges of $S(w_2)$ in each T_j of $T(a_{i*})$. Let $\ell(e) = \varphi(j) + 1$, where e is an edge of $S(w_2)$ sharing an end-vertex with the P_{2^M} attached at T_j. This allows $r(P_{2^M})$ at level $\varphi(j)$ (with other edges of P_{2^M} assigned to the levels $\varphi(j) - M + 1, \ldots, \varphi(j) - 1$). For the remaining $|A_j| - 1$ paths P_{2^M}, their root edges we assign to the level $\varphi(j) - 1$. Considering the elements in R_a, independently, for each $j \in \{1, \ldots, m\}$ we get $\eta - |A_j| - 1$, $n - 1$ and n root edges (of the paths P_{2^M} attached at components $\overline{T}_d$ with $d \in \{\varphi(j) - 1, \varphi(j), \varphi(j) + 1\}$) that by Lemma 4(b) are already assigned to respective levels d. Thus summing the elements in R_m and R_a for each level $p \in \{\varphi(j) - 1, \varphi(j), \varphi(j) + 1\}$ we obtain $|L_p(\mathcal{E}_F) \cap R| \leq \eta$. The proof is completed by showing that each level contains at most b elements. Namely, each level p contains 2^s vertices corresponding to the edges of P_{2^M} rooted at level $p + s$ for each $s \in \{0, \ldots, M - 1\}$. Therefore, if W denotes the set of all edges of F that do not belong to the paths P_{2^M}, we get

$$|L_p(\mathcal{E}_F)| \leq \eta \sum_{s=0}^{M-1} 2^s + |W \cap L_p(\mathcal{E}_F)| \leq n2^M + \overline{m} \overset{(2)}{=} b. \qquad \square$$

Transition from a forest to a tree is done by spanning all connected components of a forest F into specific tree T. Then the reduction from BTD for forests is used as the crux of the proof of Theorem 1.

4   The Approximation Algorithm

For an elimination tree $\mathcal{E}_T$ and its vertex v, we denote by $\mathcal{E}_T[v]$ the subtree of $\mathcal{E}_T$ induced by v and all its descendants in $\mathcal{E}_T$. By *lowering* a vertex v we mean a two-phase operation of moving all vertices in $\mathcal{E}_T[v]$ one level down in $\mathcal{E}_T$ (i.e., decreasing $\ell(u)$ by 1 for each $u \in V(\mathcal{E}_T[v])$) and if for the resulting function there is a vertex $u \in V(\mathcal{E}_T[v])$ with $\ell(u) \leq 0$, then incrementing $\ell(w)$ for each $w \in V(\mathcal{E}_T)$. (The former 'normalization' is to ensure that levels are positive integers.) Though always feasible, it may produce an elimination tree with height larger than the initial one. We say that an elimination tree $\mathcal{E}_T$ is *compact* if every subtree of $\mathcal{E}_T$ occupies a set of consecutive levels in $\mathcal{E}_T$. This can be easily ensured in linear time. Namely, for each $\{v, u\} \in E(\mathcal{E}_T)$ such that $\ell(v) < \ell(u) - 1$, increment the level of v and all its descendants. Given an elimination tree $\mathcal{E}_T$, we distinguish a specific type of the root-leaf paths. Namely, by a *trunk* we mean a path $(v_1, \ldots, v_k)$ from the root v_k to an arbitrary leaf v_1 at level 1 of $\mathcal{E}_T$ (note that only leaves at level 1 are admissible). We use trunks to define branches at the vertices v_i with $i \in \{2, \ldots, k\}$ of the trunk $(v_1, \ldots, v_k)$, where for particular vertex v_i a *branch* is understood as a subtree $\mathcal{E}_T[v]$ with v being a child of v_i such that $v \neq v_{i-1}$. Since elimination trees we consider are binary trees, a branch with respect to a given trunk is uniquely determined. For any elimination tree $\mathcal{E}_T$, the b highest levels are called its *prefix*. A subtree of $\mathcal{E}_T$ is *thin* if in each level it has at most one vertex. A level i is *full* if $|L_i(\mathcal{E}_T)| \geq b$.

Preprocessing Steps. In this section we introduce operations of stretching and sorting of elimination trees that constitute two major steps of the preprocessing phase of our algorithm. We start with an operation of *stretching* a subtree $\mathcal{E}_T[v]$. If subtree is thin or its height is at least b, then stretching leaves the subtree unchanged. Otherwise, we note later on, that it is enough to consider the case of a vertex v at level l with two children u_1 and u_2 at level $l-1$ such that $\mathcal{E}_T[u_1]$ and $\mathcal{E}_T[u_2]$ are thin and compact. Let l_i denote the lowest level occupied by a vertex of $\mathcal{E}_T[u_i]$, $i \in \{1,2\}$, so the levels occupied by the vertices of $\mathcal{E}_T[u_i]$ are $l_i, \ldots, l-1$. Then, stretching $\mathcal{E}_T[v]$ is realized by lowering $l-l_1$ times the vertex u_2 so that it is placed at the level $l_1 - 1$. Consequently, $\mathcal{E}_T[v]$ becomes thin and occupies at most $2l - l_1 - l_2 + 1$ levels with the root v at level l, the vertices of $\mathcal{E}_T[u_1]$ remaining at levels in $l_1, \ldots, l-1$ and the vertices of $\mathcal{E}_T[u_2]$ moved to the levels in $l_1 - l + l_2, \ldots, l_1 - 1$. (For simplicity, we referred here to the levels of all vertices according to the level function of the initial elimination tree, i.e., before the 'normalization' applied after each lowering of a vertex.) For an elimination tree $\mathcal{E}_T$, a *stretched* elimination tree $\mathcal{E}_T'$ is obtained by stretching each subtree $\mathcal{E}_T[v]$ of $\mathcal{E}_T$, where the roots v are selected in postorder fashion, i.e., for each v we stretch the subtrees rooted at the children of v before stretching $\mathcal{E}_T[v]$. Clearly, the transition from $\mathcal{E}_T$ to $\mathcal{E}_T'$ can be computed in linear time.

Observation 1. *If $\mathcal{E}_T'$ is a stretched elimination tree obtained from a compact elimination tree $\mathcal{E}_T$, then $h(\mathcal{E}_T') \leq h(\mathcal{E}_T) + 2b$ and $\mathcal{E}_T'$ is compact. Moreover, if for a vertex v, $|V(\mathcal{E}_T[v])| \geq b$, then $\mathcal{E}_T'[v]$ has height at least b.*

Suppose that $\mathcal{E}_T$ is an elimination tree with a vertex z_1 and its child z_2. A *switch* of z_1 and z_2 is the operation on $\mathcal{E}_T$ as shown in Fig. 4 (intuitively, the switch results in exchanging the roles of the subtrees $\mathcal{E}_T[z_3]$ and $\mathcal{E}_T[z_4]$ while the aim of switching is to obtain a type of ordering of the subtrees). We note that this concept has been used on a wider class of non-binary elimination trees under the names of tree rotations or reorderings (see, e.g. [18]). Here we define switching with no connection to whether z_1 and z_2 belong to a given trunk or not. Later on, we use this operation with respect to the location of particular trunks. If switch is performed on an elimination tree for some graph G, then the resulting tree is also an elimination tree for G. We say that $\mathcal{E}_T$ is *sorted along* a trunk $(v_1, \ldots, v_k)$ (recall that v_k is the root) if the height of the branch at v_i is at most

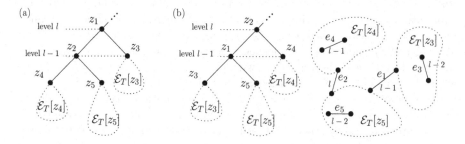

Fig. 4. A switch of z_1 and z_2: (a) $\mathcal{E}_T$; (b) $\mathcal{E}_T$ after the switch; (c) the levels after the switch shown on the tree T, where each z_i in $\mathcal{E}_T$ corresponds to the edge e_i in T;

that of the branch at v_{i+1}, where $i \in \{2, \ldots, k-1\}$. We say that $\mathcal{E}_T$ is *sorted* if for each vertex v of $\mathcal{E}_T$, the subtree $\mathcal{E}_T[v]$ is sorted along each of its trunks. It is important to note that the notion of a trunk pertains to any elimination tree and hence can be equally applied not only to $\mathcal{E}_T$ but also to any of its subtrees.

Lemma 5. *For each tree T, there exists a sorted $\mathcal{E}_T$ of minimum height and it can be computed in polynomial time.*

Formulation of the Algorithm. Informally, the algorithm starts with a minimum height elimination tree that is stretched and sorted. In each iteration of the main loop the following takes place. If the current elimination tree is b-bounded, then the algorithm finishes. Otherwise, in the highest level that has more than b vertices, the lowest subtree rooted at a vertex v is located. Then, v is lowered.

Algorithm BTD: bounded-width tree-depth approximation for $L(T)$

---

1   Let $\mathcal{E}_T$ be a minimum height compact elimination tree for $L(T)$
2   Obtain $\mathcal{E}_T^0$ by first making $\mathcal{E}_T$ stretched and then sorted, and set $t \leftarrow 0$
3   **while** $\mathcal{E}_T^t$ is not b-*bounded* **do**
4   Find the highest level l_t such that $|L_{l_t}(\mathcal{E}_T^t)| > b$
5      $v \leftarrow \arg\min_u h(\mathcal{E}_T^t[u])$, where u iterates over all vertices in $L_{l_t}(\mathcal{E}_T^t)$
6   Obtain $\mathcal{E}_T^{t+1}$ by lowering v in $\mathcal{E}_T^t$ and increment t
7   **return** $\mathcal{E}_T^t$

Analysis (Sketch). The complexity follows from [16, 20].

Since lowering any vertex gives a valid elimination tree, the final $\mathcal{E}_T^\tau$ is a b-bounded elimination tree. An informal outline of its height analysis is as follows. We introduce the following structural property of an elimination tree. Let $l_1 < \cdots < l_d$ be the full levels of an elimination tree $\mathcal{E}_T$ except for the highest $b-1$ levels. We say that $\mathcal{E}_T$ is *structured* if either all levels with at least b vertices belong to the prefix or $l_d = h(\mathcal{E}_T) - b + 1$ and $l_i = l_{i-1} + 1$ for each $i \in \{2, \ldots, d\}$. In other words, in a structured elimination tree the full levels that are not in the prefix form a consecutive segment that reaches the lowest level of the prefix. It turns out that $\mathcal{E}_T^0$ obtained in line 2 of Algorithm BTD is structured.

The two following invariants along the iterations of the loop are maintained, when transforming $\mathcal{E}_T^{t-1}$ to $\mathcal{E}_T^t$. If some level $h(\mathcal{E}_T^t) - l$ becomes full, then the level $h(\mathcal{E}_T^t) - l + 1$ should be also full at this point, and both levels should stay full in the future iterations. This essentially means that the trees $\mathcal{E}_T^t$, $t > 0$, remain structured. Therefore, if for the final elimination tree $\mathcal{E}_T^\tau$, $h(\mathcal{E}_T^\tau) > h(\mathcal{E}_T^0)$, then at most b levels are not full: the highest $b-1$ ones and level 1. Thus, $h(\mathcal{E}_T^\tau) \leq b + \frac{m}{b}$, where m is the number of edges in T. This ensures the required approximation bound, since in the case $h(\mathcal{E}_T^\tau) = h(\mathcal{E}_T^0)$ by Observation 1 we know that this height is additively at most b from $btd(L(T), b)$. This proves Theorem 2.

References

1. Ben-Asher, Y., Farchi, E., Newman, I.: Optimal search in trees. SIAM J. Comput. **28**(6), 2090–2102 (1999)
2. Bodlaender, H.L., et al.: Rankings of graphs. SIAM J. Discrete Math. **11**(1), 168–181 (1998)
3. Carmo, R., Donadelli, J., Kohayakawa, Y., Laber, E.S.: Searching in random partially ordered sets. Theor. Comput. Sci. **321**(1), 41–57 (2004)
4. Cicalese, F., Jacobs, T., Laber, E.S., Valentim, C.D.: The binary identification problem for weighted trees. Theor. Comput. Sci. **459**, 100–112 (2012)
5. Cicalese, F., Keszegh, B., Lidický, B., Pálvölgyi, D., Valla, T.: On the tree search problem with non-uniform costs. Theor. Comput. Sci. **647**, 22–32 (2016)
6. Dereniowski, D.: Edge ranking of weighted trees. Discret. Appl. Math. **154**(8), 1198–1209 (2006)
7. Dereniowski, D.: Edge ranking and searching in partial orders. Discret. Appl. Math. **156**(13), 2493–2500 (2008)
8. Dereniowski, D., Kosowski, A., Uznański, P., Zou, M.: Approximation strategies for generalized binary search in weighted trees. In: ICALP 2017, pp. 84:1–84:14 (2017)
9. Dereniowski, D., Nadolski, A.: Vertex rankings of chordal graphs and weighted trees. Inf. Process. Lett. **98**(3), 96–100 (2006)
10. Emamjomeh-Zadeh, E., Kempe, D.: A general framework for robust interactive learning. In: NIPS 2017, pp. 7082–7091 (2017)
11. Giannopoulou, A.C., Hunter, P., Thilikos, D.M.: LIFO-search: a min-max theorem and a searching game for cycle-rank and tree-depth. Discret. Appl. Math. **160**(15), 2089–2097 (2012)
12. Iyer, A.V., Ratliff, H.D., Vijayan, G.: Optimal node ranking of trees. Inf. Process. Lett. **28**(5), 225–229 (1988)
13. Iyer, A.V., Ratliff, H.D., Vijayan, G.: On an edge ranking problem of trees and graphs. Discret. Appl. Math. **30**(1), 43–52 (1991). https://doi.org/10.1016/0166-218X(91)90012-L
14. Laber, E.S., Milidiú, R.L., Pessoa, A.A.: On binary searching with nonuniform costs. SIAM J. Comput. **31**(4), 1022–1047 (2002)
15. Lam, T.W., Yue, F.L.: Edge ranking of graphs is hard. Discret. Appl. Math. **85**(1), 71–86 (1998)
16. Lam, T.W., Yue, F.L.: Optimal edge ranking of trees in linear time. Algorithmica **30**(1), 12–33 (2001)
17. Liu, J.W.: The role of elimination trees in sparse factorization. SIAM J. Matrix Anal. Appl. **11**(1), 134–172 (1990)
18. Liu, J.: Equivalent sparse matrix reorderings by elimination tree rotations. SIAM J. Sci. Stat. Comput. **9**(3), 424–444 (1988)
19. Makino, K., Uno, Y., Ibaraki, T.: On minimum edge ranking spanning trees. J. Algorithms **38**(2), 411–437 (2001)
20. Mozes, S., Onak, K., Weimann, O.: Finding an optimal tree searching strategy in linear time. In: SODA 2008, pp. 1096–1105 (2008)
21. Nešetřil, J., de Mendez, P.O.: Tree-depth, subgraph coloring and homomorphism bounds. Eur. J. Comb. **27**(6), 1022–1041 (2006)
22. Onak, K., Parys, P.: Generalization of binary search: searching in trees and forest-like partial orders. In: FOCS 2006, pp. 379–388 (2006)

23. Schäffer, A.A.: Optimal node ranking of trees in linear time. Inf. Process. Lett. **33**(2), 91–96 (1989)
24. de la Torre, P., Greenlaw, R., Schäffer, A.A.: Optimal edge ranking of trees in polynomial time. Algorithmica **13**(6), 592–618 (1995)
25. Zhou, X., Kashem, M.A., Nishizeki, T.: Generalized edge-rankings of trees (extended abstract). In: d'Amore, F., Franciosa, P.G., Marchetti-Spaccamela, A. (eds.) WG 1996. LNCS, vol. 1197, pp. 390–404. Springer, Heidelberg (1997). https://doi.org/10.1007/3-540-62559-3_31
26. Zhou, X., Nishizeki, T.: An efficient algorithm for edge-ranking trees. In: van Leeuwen, J. (ed.) ESA 1994. LNCS, vol. 855, pp. 118–129. Springer, Heidelberg (1994). https://doi.org/10.1007/BFb0049402
27. Zhou, X., Nishizeki, T.: Finding optimal edge-rankings of trees. In: SODA 1995, pp. 122–131
28. Zwaan, R.: Vertex ranking with capacity. In: van Leeuwen, J., Muscholl, A., Peleg, D., Pokorný, J., Rumpe, B. (eds.) SOFSEM 2010. LNCS, vol. 5901, pp. 767–778. Springer, Heidelberg (2010). https://doi.org/10.1007/978-3-642-11266-9_64

TS-Reconfiguration of Dominating Sets in Circle and Circular-Arc Graphs

Nicolas Bousquet and Alice Joffard[(✉)]

LIRIS, CNRS, Université Claude Bernard, Lyon, France
{nicolas.bousquet,alice.joffard}@liris.cnrs.fr

Abstract. We study the dominating set reconfiguration problem with the token sliding rule. Let G be a graph $G = (V, E)$ and two dominating sets D_s and D_t of G. The goal is to decide if there exists a sequence $S = \langle D_1 := D_s, \ldots, D_\ell := D_t \rangle$ of dominating sets of G such that for any two consecutive dominating sets D_r and D_{r+1} with $r < t$, $D_{r+1} = (D_r \setminus u) \cup \{v\}$, where $uv \in E$.

In a recent paper, Bonamy et al. [3] studied this problem and raised the following questions: what is the complexity of this problem on circular-arc graphs? On circle graphs? In this paper, we answer both questions by proving that the problem is polynomial on circular-arc graphs and PSPACE-complete on circle graphs.

Keywords: Reconfiguration · Dominating sets · Token sliding · Circle graphs · Circular-arc graphs

1 Introduction

Reconfiguration problems consist, given an instance of a problem, in determining if (and in how many steps) we can transform one of its solutions into another one via a sequence of elementary operations keeping a solution along this sequence. The sequence is called a *reconfiguration sequence*.

Let Π be a problem and $\mathcal{I}$ be an instance of Π. Another way to describe a reconfiguration problem is to define the *reconfiguration graph* $\mathcal{R}_\mathcal{I}$, whose vertices are the solutions of $\mathcal{I}$ and in which two solutions are adjacent if and only if we can transform the first into the second in one elementary step. In this paper, we focus on the REACHABILITY problem that, given two solutions I, J of $\mathcal{I}$, returns true if and only if there exists a reconfiguration sequence from I to J. Other works have focused on different problems such as the connectivity of the reconfiguration graph or its diameter, see e.g. [4,7]. Reconfiguration problems arise in various fields such as combinatorial games, motion of robots, random sampling, or enumeration. It has been intensively studied for various rules and problems such as satisfiability constraints [7], graph coloring [1,6], vertex covers and independent sets [10,11,13] or matchings [2]. The reader is referred to the

This work was supported by ANR project GrR (ANR-18-CE40-0032).

E. Bampis and A. Pagourtzis (Eds.): FCT 2021, LNCS 12867, pp. 114–134, 2021.
https://doi.org/10.1007/978-3-030-86593-1_8

surveys [14, 16] for a more complete overview on reconfiguration problems. In this work, we focus on dominating set reconfiguration. Throughout the paper, all the graphs are finite and simple.

Let $G = (V, E)$ be a graph. A *dominating set* of G is a subset of vertices X such that, for every $v \in V$, either $v \in X$ or v has a neighbor in X. A dominating set can be seen as a subset of tokens placed on vertices that dominates the graph. Three types of elementary operations, called *reconfiguration rules*, have been studied for the reconfiguration of dominating sets.

- The *token addition-removal* rule (TAR) where each operation consists in either removing a token from a vertex, or adding a token on any vertex.
- The *token jumping* rule (TJ) where an operation consists in moving a token from a vertex to any vertex of the graph.
- The *token sliding* rule (TS) where an operation consists in sliding a token from a vertex to an adjacent vertex.

In this paper, we focus on the reconfiguration of dominating sets with the token sliding rule. Note that we allow (as well as in the other papers on the topic, see [3]) the dominating sets to be multisets. In other words, several tokens can be put on the same vertex. Bonamy et al. observed in [3] that this choice can modify the reconfiguration graph and the set of dominating sets that can be reached from the initial one. More formally, we consider the following problem:

DOMINATING SET RECONFIGURATION UNDER TOKEN SLIDING (DSR$_{TS}$)
Input: A graph G, two dominating sets D_s and D_t of G.
Output: Does there exist a dominating set reconfiguration sequence from D_s to D_t under the token sliding rule ?

Dominating Set Reconfiguration under Token Sliding. The dominating set reconfiguration problem has been widely studied with the token addition-removal rule. Most of the earlier works focused on the conditions that ensure that the reconfiguration graph is connected in function of several graph parameters, see e.g. [5, 8, 15]. From a complexity point of view, Haddadan et al. [9], proved that the reachability problem is PSPACE-complete under the addition-removal rule, even when restricted to split or bipartite graphs. They also provide linear time algorithms in trees and interval graphs.

More recently, Bonamy et al. [3] studied the token sliding rule. They proved that DSR$_{TS}$ is PSPACE-complete, even restricted to split, bipartite or bounded treewidth graphs. They also provide polynomial time algorithms for cographs and dually chordal graphs (which contain interval graphs). In their paper, they raise the following question: is it possible to generalize the polynomial time algorithm for interval graphs to circular-arc graphs ?

They also ask if there exists a class of graphs for which the maximum dominating set problem is NP-complete but its TS-reconfiguration counterpart is polynomial. They propose the class of circle graphs as a candidate.

Our Contribution. In this paper, we answer the questions raised in [3]. First, we prove the following:

Theorem 1. DSR$_{TS}$ *is polynomial in circular-arc graphs.*

The idea of the proof is that if we fix a vertex of the dominating set then we can unfold the rest of the graph to get an interval graph. We can then use the algorithm of Bonamy et al. on interval graphs to determine if we can slide the fixed vertex to a better position. Our second main result is the following:

Theorem 2. DSR$_{TS}$ *is PSPACE-complete in circle graphs.*

This is answering a second question of [3]. The proof is inspired from the proof that DOMINATING SET IN CIRCLE GRAPHS is NP-complete [12] but has to be adapted in the reconfiguration framework by adding gadgets to have control on the possible dominating sets of optimal size plus one.

The proofs of the statements marked with $\star$ are not given in this proceeding.

2   Preliminaries

Let $G = (V, E)$ be a graph. Given a vertex $v \in V$, $N(v)$ denotes the *open neighborhood* of v, i.e. the set $\{y \in V : vy \in E\}$.

A *multiset* is defined as a set but an element can appear several times. The number of times it appears is its *multiplicity*. The multiplicity of an element that does not appear in the multiset is 0. Let A and B be multisets. The *union* of A and B, denoted by $A \cup B$, is the multiset containing only elements of A or B, and in which the multiplicity of each element is the sum of their multiplicities in A and B. The *difference* $A \setminus B$ is the multiset containing only elements of A, and in which the multiplicity of each element is the difference between its multiplicity in A and its multiplicity in B (if the result is negative then the element is not in $A \setminus B$). By abuse of language, all along this paper, we refer to multisets as sets.

Under the token sliding rule, a *move* $v_i \rightsquigarrow v_j$, from a set S_r to a set S_{r+1}, denotes the token sliding operation along the edge $v_i v_j$ from v_i to v_j, i.e. $S_{r+1} = (S_r \cup \{v_j\}) \setminus \{v_i\}$ with v_j is in S_r. We say that a set S *is before* a set S' in a reconfiguration sequence S if S contains a subsequence starting with S and ending with S'.

3   A Polynomial Time Algorithm for Circular-Arc Graphs

An *interval graph* $G = (V, E)$ is an intersection graph of intervals of the real line. In other words, the set of vertices is a set of real intervals and two vertices are adjacent if their corresponding intervals intersect. A *circular-arc graph* $G = (V, E)$ is an intersection graph of arcs of a circle. In other words, every vertex is associated an arc A and there is an edge between two vertices if their corresponding arcs intersect. By abuse of notation, we refer to the vertices by their image arc. circular-arc graphs strictly contain interval graphs. Bonamy et al. proved the following result in [3] that we will use as a black-box:

Theorem 3 (Bonamy et al. [3]). *Let G be a connected interval graph, and D_s, D_t be two dominating sets of G of the same size. There exists a TS-reconfiguration sequence from D_s to D_t.*

One can naturally wonder if Theorem 3 can be extended to circular-arc graphs. The answer is negative since, for every k, the cycle C_{3k} is a circular-arc graph that has only three dominating sets of size exactly k (the ones containing vertices i mod 3 for $i \in \{0, 1, 2\}$), which are pairwise non adjacent for the TS-rule.

However, we prove that we can decide in polynomial time if we can transform one dominating set into another. The remaining of this section is devoted to prove Theorem 1.

Let $G = (V, E)$ be a circular-arc graph and D_s, D_t be dominating sets of G of size k. Assume first that there exists an arc $v \in V$ that is the whole circle. So $\{v\}$ is a dominating set of G and for any dominating sets D_s and D_t, we can move a token from D_s to v, then move every other token of D_s to a vertex of D_t (in at most two steps passing through v), and finally move the token on v to the last vertex of D_t. Since a token stays on v, we keep a dominating set. So if such an arc exists, there exists a reconfiguration sequence from D_s to D_t.

From now on we assume that no arc contains the whole circle (and that no vertex dominates the graph). For any arc $v \in V$, the *left extremity* of v, denoted by $\ell(v)$, is the first extremity of v we meet when we follow the circle clockwise, starting from a point outside of v. The other extremity of v is called the *right extremity* and is denoted by $r(v)$.

We start by showing a simple lemma that is used in the proof of Theorem 1.

Lemma 1. *(⋆) Let $G = (V, E)$ be a graph, and $u, v \in V$, where $N[u] \subseteq N[v]$. If S is a TS-reconfiguration sequence in G, and S' is obtained by replacing any occurrence of u by v in the dominating sets of S, then S' also is a TS-reconfiguration sequence.*

For the proof of Theorem 1, we also need the following auxiliary graph G_u (see Fig. 1 for an illustration). Let u be a vertex of G such that no arc strictly contains u. For any $v \neq u$ not contained in u, we create an arc v' that is the closure of $v \setminus u$[1]. Since u is maximal by inclusion, v' is an arc. Let G'_u be the circular-arc graph containing all the arcs v' defined above plus u. Note that the set of edges of G'_u might be smaller than E. That being said, any dominating set D of G containing u can be adapted into a dominating set D' of G'_u containing u and the image v' of any vertex $v \in D$. We now construct G_u from G'_u. First remove the vertex u. Note that after this deletion, no arc intersects the open interval $(\ell(u), r(u))$ so the resulting graph is an interval graph. We can unfold it in such a way that the first vertex starts at position $\ell(u)$ and the last vertex ends at position $r(u)$ (see Fig. 1). We add two new vertices, u' and u'', that correspond to each extremity of u. One has interval $(-\infty, \ell(u))$ and the other

---

[1] v' is the part of v that is not included in u. Note that the fact that v' is the closure of that arc ensures that u and v' intersect.

has interval $(r(u), +\infty)$. Since no arc but u' (resp. u'') intersects $(-\infty, \ell(u)]$ (resp. $[r(u), +\infty))$, we can create $(n + 2)$ new vertices only adjacent to u' (resp. u''). These $2n + 4$ vertices are called the *leaves* of G_u.

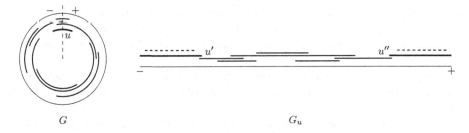

G G_u

Fig. 1. The interval graph G_u obtained from the circular-arc graph G. The thick interval in G correponds to u. The thick intervals in G_u correspond to u' and u'', and the intervals above are the added leaves.

Given a set D of G that contains u, we define a set D_u of G_u such that u', u'' are in D_u and, for every $v \neq u$ in $V(G)$, v' is in D_u if and only if $v \in D$.

Lemma 2. *(⋆) If D is a dominating set of G that contains u then D_u is a dominating set of G_u.*

Note that D_u has size $|D| + 1$.

Lemma 3. *(⋆) If D is a dominating set of G that contains u, then the following hold:*

(i) *All the dominating sets of G_u of size $|D| + 1$ contain u' and u''.*
(ii) *For every dominating set X of G_u of size $|D| + 1$, $(X \cap V) \cup \{u\}$ is a dominating set of G of size at most $|D|$.*
(iii) *Every reconfiguration sequence in G_u between two dominating sets D_s, D_t of G_u of size at most $|D| + 1$ that does not contain any leaf can be adapted into a reconfiguration sequence in G from $(D_s \setminus \{u', u''\}) \cup \{u\}$ to $(D_t \setminus \{u', u''\}) \cup \{u\}$.*

Using Lemmas 2, 3 and Theorem 3, we can prove the following:

Corollary 1. *Let G be a circular-arc graph, and D_s, D_t be two dominating sets of G, with $D_s \cap D_t \neq \emptyset$. There exists a TS-reconfiguration sequence from D_s to D_t.*

We now have all the ingredients to prove Theorem 1.

Proof (of Theorem 1). Let $G = (V, E)$ be a circular-arc graph, and let D_s and D_t be two dominating sets of G. Free to slide tokens, we can assume that the intervals of D_s and D_t are maximal by inclusion. By Lemma 1, we can also assume that all the vertices of the dominating sets we consider are maximal by inclusion. Moreover, by Corollary 1, we can assume that $D_s \cap D_t = \emptyset$. By abuse of notation, we say that in G, an arc v is the *first arc on the left* (resp. *on the right*) of another arc u if the first left extremity of an inclusion-wise maximal arc (of G, or of the stated dominating set) we encounter when browsing the circle counter clockwise (resp. clockwise) from the left extremity of u is the one of v. In interval graphs, we say that an interval v is at *the left* (resp. at *the right*) of an interval u if the left extremity of v is smaller (resp. larger) than the one of u. Note that since the intervals of the dominating sets are maximal by inclusion, the left and right ordering of these vertices are the same.

Let $u_1 \in D_s$. Let v be the first vertex at the right of u_1 in D_t. We perform the following algorithm, called the Right Sliding Algorithm. By Lemma 3, all the dominating sets of size $|D_s| + 1$ in G_{u_1} contain u_1' and u_1''. Let D_2' be a dominating set of the interval graph G_{u_1} of size $|D_s| + 1$, such that the first vertex at the right of u_1' has the smallest left extremity (we can indeed find such a dominating set in polynomial time). By Theorem 3, there exists a transformation from $(D_s \cup \{u_1', u_1''\}) \setminus \{u_1\}$ to D_2' in G_{u_1}. And by Lemma 3, there exists a transformation from D_s to $D_2 := (D_2' \cup \{u_1\}) \setminus \{u_1', u_1''\}$ in G. We apply this transformation. Informally speaking, it permits to move the token at the left of u_1 closest from u_1, which allows to push the token on u_1 to the right.

We now fix all the vertices of D_2 but u_1 and try to slide the token on u_1 to its right. If we can push it on a vertex at the right of v, we can in particular push it on v (since v is maximal by inclusion) and keep a dominating set. So we set $u_2 = v$ if we can reach v or the rightmost possible vertex maximal by inclusion we can reach otherwise. We now repeat these operations with u_2 instead of u_1, i.e. we apply a reconfiguration sequence towards a dominating set of G in which the first vertex on the left of u_2 is the closest to u_2, then try to slide u_2 to the right, onto u_3. We repeat until $u_i = u_{i+1}$ (i.e. we cannot move to the right anymore) or $u_i = v$. Let $u_1, \ldots, u_\ell$ be the resulting sequence of vertices. Note that this algorithm is polynomial since after at most n steps we reach v or a fixed point.

We can similarly define the Left Sliding Algorithm by replacing the leftmost dominating set of G_{u_i} by the rightmost, and slide u_i to the left for any i. We stop when we cannot slide to the left anymore, or when $u_i = v'$, where v' is the first vertex at the left of u_1 in D_t. Let u_ℓ' be the last vertex of the sequence of vertices given by the Left Sliding Algorithm.

Claim. $(\star)$ We can transform D_s into D_t if and only if $u_\ell = v$ or $u_\ell' = v'$.   □

4   PSPACE-Hardness for Circle Graphs

A *circle graph* $G = (V, E)$ is an intersection graph of chords of a circle C. In other words, we can associate to each vertex of V two points of C, and there is an edge between two vertices if the chords between their pair of points intersect. Equivalently, G can be represented on the real line. We associate to each vertex an interval of the real line and there is an edge between two vertices if their intervals intersect but do not contain each other. In this section, we use the last representation. For any interval I, $\ell(I)$ denotes the left extremity of I, and $r(I)$ its right extremity.

The goal of this section is to show that DSR_{TS} is PSPACE-complete in circle graphs. We provide a polynomial time reduction from SATR to DSR_{TS}. This reduction is inspired from one used in [12] to show that the minimum dominating set problem is NP-complete on circle graphs. The SATR problem is defined as follows:

SATISFIABILITY RECONFIGURATION (SATR)
Input: A Boolean formula F in conjunctive normal form, two variable assignments A_s and A_t that satisfy F.
Output: Does there exist a reconfiguration sequence from A_s to A_t that keeps F satisfied, where the operation consists in a *variable flip*, i.e. the change of the assignment of exactly one variable from $x = 0$ to $x = 1$, or conversely?

Let (F, A_s, A_t) be an instance of SATR. Let $x_1 \ldots, x_n$ be the variables of F. Since F is in conjunctive normal form, it is a conjunction of *clauses* $c_1, \ldots, c_m$ which are disjunctions of literals. A *literal* is a variable or its negation, and we denote by $x_i \in c_j$ (resp. $\overline{x_i} \in c_j$) the fact that x_i (resp. the negation of x_i) is a literal of c_j. Since duplicating clauses does not modify the satisfiability of a formula, we can assume that m is a multiple of 4. We can also assume that for every i, j, x_i or $\overline{x_i}$ is not in c_j (otherwise the clause is satisfied for any assignment and can be removed from F).

4.1   The Reduction

Let us construct an instance $(G_F, D_F(A_s), D_F(A_t))$ of DSR_{TS} from (F, A_s, A_t). We start by constructing the circle graph G_F. In [12], the author gives the coordinates of the endpoints of all the intervals. In this proceeding, we only give the relative position of the intervals. Most of the positions are moreover given graphically. We however outline some of the edges and non edges in G_F that have an impact on the upcoming proofs.

Let us first note that by adding to a circle graph H an interval that starts just before the extremity of another interval u and ends just after, we add one vertex to H, only connected to u. So:

Remark 1. If H is a circle graph and u is a vertex of H, then the graph H plus a new vertex only connected to u is circle graph.

- For each variable x_i, we create m *base intervals* B_j^i where $1 \leq j \leq m$. We also create $\frac{m}{2}$ intervals X_j^i (resp. $\overline{X}_j^i$) called the *positive bridge intervals* (resp. *negative bridge intervals*) of x_i, where $1 \leq j \leq \frac{m}{2}$. The positions of these intervals are illustrated in Fig. 2.
  Base intervals are pairwise non adjacent. On the other hand, every positive (resp. negative) bridge interval is incident to exactly two base intervals; And all the positive (resp. negative) bridge intervals of x_i are incident to pairwise distinct base intervals. In particular, the positive (resp. negative) bridge

intervals dominate the base intervals; And every base interval is adjacent to exactly one positive and one negative bridge interval. All the positive (resp. negative) bridge intervals but X_1^i and $X_{\frac{m}{2}}^i$ have exactly one other positive (resp. negative) bridge interval neighbor. Finally, for every i, every negative bridge interval $\overline{X}_j^i$ has exactly two positive bridge interval neighbors (X_{j-1}^i and X_j^i) except for $\overline{X}_1^i$, which does not have any. Note that a bridge interval of x_i is not adjacent to a bridge interval or a base interval of x_j for $j \neq i$.

- For any clause c_j, we create two identical *clause intervals* C_j and C_j'. As one interval contains the other, C_j and C_j' are not adjacent. The clause intervals are not adjacent to any interval constructed so far.

- For every j such that $x_i \in c_j$ (resp. $\overline{x_i} \in c_j$), we create four intervals T_j^i, U_j^i, V_j^i and W_j^i (resp. $\overline{T}_j^i$, $\overline{U}_j^i$, $\overline{V}_j^i$ and $\overline{W}_j^i$), called the *positive path intervals* (resp. *negative path intervals*) of x_i. See Fig. 2 and Fig. 3 for an illustration. The neighborhood of every clause interval C_j is the set of intervals W_j^i with $x_i \in c_j$ and intervals $\overline{W}_j^i$ with $\overline{x_i} \in c_j$. The interval V_j^i (resp. $\overline{V}_j^i$) is only adjacent to U_j^i and W_j^i (resp. $\overline{U}_j^i$ and $\overline{W}_j^i$). The interval T_j^i (resp. $\overline{T}_j^i$) is only adjacent to B_j^i, U_j^i and one positive bridge interval (resp. B_j^i, $\overline{U}_j^i$ and one negative bridge interval), which is the same one adjacent to B_j^i. Since U_j^i and W_j^i (resp. $\overline{U}_j^i$ and $\overline{W}_j^i$) are not adjacent, B_j^i, T_j^i, U_j^i, V_j^i, W_j^i and C_j (resp. B_j^i, $\overline{T}_j^i$, $\overline{U}_j^i$, $\overline{V}_j^i$, $\overline{W}_j^i$ and C_j) induce a path. Finally, for any two variables x_i and x_i' such that $x_i \neq x_i'$, the only path intervals of respectively x_i and x_i' that can be adjacent are the W and $\overline{W}$ intervals adjacent to different clause intervals.

  The *intervals of* x_i denote the base, bridge and path intervals of x_i.

- For every bridge interval and every U, $\overline{U}$, W and $\overline{W}$ interval, we create a *dead-end interval* only adjacent to it. Then, for any dead-end interval, we create $6mn$ *pending intervals* only adjacent to it. Remark 1 ensures that the resulting graph is a circle graph. Informally speaking, since the dead-end intervals have a lot of pending intervals, they will be forced to be in any dominating set of size at most $6mn$. Thus, in any dominating set, we know that bridge, U, $\overline{U}$, W and $\overline{W}$ intervals (as well as dead-end and pending ones) are already dominated. So the other vertices in the dominating set will only be there to dominate the other vertices of the graph, which are called *important*.

- Finally, we create a *junction interval* J, that starts just before $\ell(C_1)$ and end just after $r(C_m)$. It is adjacent to every W or $\overline{W}$ interval, and to no other interval. This completes the construction of the graph G_F.

4.2   Basic Properties of G_F

Let us first give a couple of properties satisfied by G_F. The following lemma will be used to guarantee that any token can be moved to any vertex of the graph as long as the rest of the tokens form a dominating set.

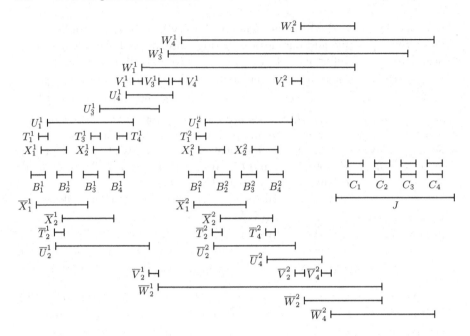

Fig. 2. The intervals obtained for the formula $F = (x_1 \vee x_2) \wedge (\overline{x_1} \vee \overline{x_2}) \wedge (x_1) \wedge (\overline{x_2} \vee x_1)$ with $m = 4$ clauses and $n = 2$ variables. The dead-end intervals and the pending intervals are not represented here.

Lemma 4. $(\star)$ *The graph G_F is connected.*

For any variable assignment A of F, let $D_F(A)$ be the set of intervals of G_F defined as follows. The junction interval J and all the dead-end intervals belong to $D_F(A)$. For any x_i such that $x_i = 1$ in A, the positive bridge, W and $\overline{U}$ intervals of x_i belong to $D_F(A)$. And for any variable x_i such that $x_i = 0$ in A, the negative bridge, $\overline{W}$ and U intervals of x_i belong to $D_F(A)$. The multiplicity of each of these intervals is 1. Thus, $|D_F(A)| = \frac{3mn}{2} + 3\sum_{i=1}^{n} \ell_i + 1$ where for any x_i, ℓ_i is the number of clauses that contain x_i or $\overline{x_i}$.

Lemma 5. $(\star)$ *If A satisfies F, then $D_F(A) \setminus J$ is a dominating set of G_F.*

Let $K := \frac{3mn}{2} + 3\sum_{i=1}^{n} \ell_i + 1$. Since the number $6mn$ of leaves attached on each dead-end interval is strictly more than K (as $\ell_i \leq m$), the following holds.

Remark 2. Any dominating set of size at most K contains all the $(mn+2\sum_{i=1}^{n} \ell_i)$ dead-end intervals.

Thus, in any dominating set of size K, all the pending, dead-end, bridge, U, $\overline{U}$, W and $\overline{W}$ intervals are dominated. So we simply have to focus on the domination of base, T, $\overline{T}$, V, $\overline{V}$ and junction intervals (i.e. the important intervals).

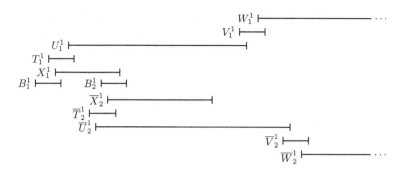

Fig. 3. A zoom on some intervals of the variable x_1.

Lemma 6. $(\star)$ *If D is a dominating set of G then for any x_i, D contains at least ℓ_i intervals dominating the V and $\overline{V}$ intervals of x_i and $\frac{m}{2}$ intervals dominating the base ones. Moreover, these two sets of intervals are disjoint, and they are intervals of x_i.*

Remark 2 and Lemma 6 imply that any dominating set D of size K contains $(mn + 2\sum_{i=1}^{n} \ell_i)$ dead-end intervals and $(\ell_i + \frac{m}{2})$ intervals of x_i for any x_i. Since $K = \frac{3mn}{2} + 3\sum_{i=1}^{n} \ell_i + 1$, there is one remaining token. Thus, for any variable x_i but at most one, there are $(\ell_i + \frac{m}{2})$ intervals of x_i in D. If there exists x_k with more than $(\ell_i + \frac{m}{2})$ intervals of x_k in D, then there are exactly $(\ell_k + \frac{m}{2} + 1)$ of them. The variable x_k is called the *moving variable* of D, denoted by $mv(D)$.

For any x_i, we denote by X_i (resp. $\overline{X_i}$) the set of positive (resp. negative) bridge variables of x_i. Similarly, we denote by W_i (resp. $\overline{W_i}$) the set of W (resp. $\overline{W}$) variables of x_i. Let us give more details about the intervals of x_i in D.

Lemma 7. $(\star)$ *If D is a dominating set of size K, then for any variable $x_i \neq mv(D)$, either $X_i \subseteq D$ and $\overline{X_i} \cap D = \emptyset$, or $\overline{X_i} \subseteq D$ and $X_i \cap D = \emptyset$.*

Lemma 8. $(\star)$ *If D is a dominating set of size K, then for any variable $x_i \neq mv(D)$, if $X_i \subseteq D$ then $\overline{W_i} \cap D = \emptyset$, otherwise $W_i \cap D = \emptyset$.*

4.3 Safeness of the Reduction

Let (F, A_s, A_t) be an instance of SATR, and let $D_s = D_F(A_s)$ and $D_t = D_F(A_t)$. By Lemma 5, (G_F, D_s, D_t) is an instance of DSR$_{\text{TS}}$.

Lemma 9. *If (F, A_s, A_t) is a yes-instance of SATR, then (G_F, D_s, D_t) is a yes-instance of DSR$_{\text{TS}}$.*

Proof. Let (F, A_s, A_t) be a yes-instance of SATR, and $S =< A_1 := A_s, \ldots, A_\ell := A_t >$ be the reconfiguration sequence from A_s to A_t. We construct a sequence S' from D_s to D_t by replacing any flip of variable $x_i \rightsquigarrow \overline{x_i}$ of S from A_r to A_{r+1} by the following sequence of token slides from $D_F(A_r)$ to $D_F(A_{r+1})^2$. In this proceeding, we omit the proof that it maintains a dominating set.

---

[2] And replacing any flip $\overline{x_i} \rightsquigarrow x_i$ by the converse of this sequence.

- We perform a sequence of slides that moves the token on J to $\overline{X}_1^i$.
- For any j such that $x_i \in C_j$, we move the token from W_j^i to V_j^i then U_j^i.
- For j from 1 to $\frac{m}{2} - 1$, we apply the move $X_j^i \rightsquigarrow \overline{X}_{j+1}^i$.
- For any j such that $\overline{x}_i \in C_j$, we move the token from $\overline{U}_j^i$ to $\overline{V}_j^i$ then $\overline{W}_j^i$.
- We perform a sequence of moves that slide the token on $X_{\frac{m}{2}}^i$ to J.   □

Lemma 10. $(\star)$ *If there exists a reconfiguration sequence S from D_s to D_t, then there exists another one S' such that for any two adjacent dominating sets D_r and D_{r+1} of S', if both D_r and D_{r+1} have a moving variable, then it is the same one.*

Lemma 11. *If (G_F, D_s, D_t) is a yes-instance of $\mathrm{DSR_{TS}}$, then (F, A_s, A_t) is a yes-instance of SATR.*

Proof. Let (G_F, D_s, D_t) be a yes-instance of $\mathrm{DSR_{TS}}$. There exists a sequence S' from D_s to D_t. By Lemma 10, we can assume that for any adjacent sets D_r and D_{r+1} of S', if both D_r and D_{r+1} have a moving variable, they are the same.

Let us construct a reconfiguration sequence S from A_s to A_t. To any dominating set D of G_F, we associate a variable assignment $A(D)$ of F as follows. For any $x_i \neq mv(D)$, either $X_i \subset D$ or $\overline{X}_i \subset D$ by Lemma 7. If $X_i \subset D$ we set $x_i = 1$, otherwise $x_i = 0$. Let x_k be such that $mv(D) = x_k$ if it exists. If there exists j such that $W_j^k \in D$ and if for any $x_i \neq x_k$ with $x_i \in c_j$, we have $\overline{X}_i \subset D$, and for any $x_i \neq x_k$ with $\overline{x}_i \in c_j$, we have $X_i \subset D$, then $x_k = 1$. Otherwise $x_k = 0$.

Let S be the sequence of assignments obtained by replacing in S' any dominating set D by $A(D)$. To conclude, we must show that the assignments associated to D_s and D_t are A_s and A_t. Also, for every dominating set D, $A(D)$ has to satisfy F. Finally, for every move in G_F, we must be able to associate a (possibly empty) variable flip. Let us first state a useful claim.

Claim. $(\star)$ For any consecutive dominating sets D_r and D_{r+1} and any x_i that is not $mv(D_r)$ nor $mv(D_{r+1})$, the value of x_i is identical in $A(D_r)$ and $A(D_{r+1})$.

Claim. $(\star)$ We have $A(D_s) = A_s$ and $A(D_t) = A_t$.

Claim. For any dominating set D of S', $A(D)$ satisfies F.

Proof. Since the clause intervals are only adjacent to W and $\overline{W}$ intervals, they are dominated by them, or by themselves in D. But only one clause interval can belong to D. Thus, for any clause interval C_j, if $C_j \in D$, then C_j' must be dominated by a W or a $\overline{W}$ interval, that also dominates C_j. So in any case, C_j is dominated by a W or a $\overline{W}$ interval. We study four possible cases and show that in each case, c_j is satisfied by $A(D)$.

If C_j is dominated in D by an interval W_j^i, where $x_i \neq mv(D)$, then by Lemmas 7 and 8, $X_i \subset D$ and by definition of $A(D)$, $x_i = 1$. Since W_j^i exists, it means that $x_i \in c_j$, thus c_j is satisfied by $A(D)$. Similarly, if C_j is dominated in

D by an interval $\overline{W}_j^i$, where $x_i \neq mv(D)$, then by Lemmas 7 and 8, $\overline{X}_i \subset D$. So $x_i = 0$. Since $\overline{W}_j^i$ exists, $\overline{x}_i \in c_j$, and therefore c_j is satisfied by $A(D)$.

If C_j is only dominated by W_j^k in D, where $x_k = mv(D)$. Then, if there exists $x_i \neq x_k$ with $x_i \in c_j$ and $X_i \subset D$ (resp. $\overline{x}_i \in c_j$ and $\overline{X}_i \subset D$), then $x_i = 1$ (resp. $x_i = 0$) and c_j is satisfied by $A(D)$. So we can assume that, for any $x_i \neq x_k$ with $x_i \in c_j$ we have $X_i \not\subset D$. By Lemma 7, $\overline{X}_i \subset D$. And for any $x_i \neq x_k$ such that $\overline{x}_i \in c_j$ we have $\overline{X}_i \not\subset D$, and thus $X_i \subset D$. So, by definition of $A(D)$, we have $x_k = 1$. Since $x_k \in c_j$ (since W_j^k exists), c_j is satisfied by $A(D)$.

Finally, assume that C_j is only dominated by $\overline{W}_j^k$ in D, where $x_k = mv(D)$. If there exists $x_i \neq x_k$ such that $x_i \in c_j$ and $X_i \subset D$ (resp. $\overline{x}_i \in c_j$ and $\overline{X}_i \subset D$), then $x_i = 1$ (respectively $x_i = 0$) so c_j is satisfied by $A(D)$. Thus, by Lemma 7, we can assume that for any $x_i \neq x_k$ such that $x_i \in c_j$ (resp. $\overline{x}_i \in c_j$), we have $\overline{X}_i \subset D$ (resp. $X_i \subset D$). Let us show that there is no clause interval $C_{j'}$ dominated by a W_i^k interval of x_k in D and that satisfies, for any $x_i \neq x_k$, if $x_i \in c_{j'}$ then $\overline{X}_i \subset D$, and if $\overline{x}_i \in c_{j'}$ then $X_i \subset D$. This will imply $x_k = 0$ by construction and then the fact that c_j is satisfied.

Since D_s has no moving variable, there exists a dominating set before D in S' with no moving variable. Let D_r be the latest in S' amongst them. By assumption, $mv(D_q) = x_k$ for any D_q that comes earlier than D but later than D_r. Thus, by Claim 4.3, for any $x_i \neq x_k$, x_i is the same in $A(D_r)$ and $A(D)$.

Now, by assumption, for any $x_i \neq x_k$ with $x_i \in c_j$ (resp. $\overline{x}_i \in c_j$) we have $\overline{X}_i \subset D$ (resp. $X_i \subset D$). Thus, since x_i has the same value in D and D_r, if $x_i \in c_j$ (resp. $\overline{x}_i \in c_j$) then $\overline{X}_i \subset D_r$ (resp. $X_i \subset D_r$) and then, by Lemma 8, $W_j^i \notin D_r$ (resp. $\overline{W}_j^i \notin D_r$). Therefore, C_j is only dominated by $\overline{W}_j^k$ in D_r. But since D_r has no moving variable, $\overline{X}_k \subset D_r$ by Lemma 7 and Lemma 8. Thus, by Lemma 8, for any $j' \neq j$, $W_{j'}^k \notin D_r$. So for any $j' \neq j$ such that $x_k \in c_{j'}$, $C_{j'}$ is dominated by at least one interval $W_{j'}^i$ or $\overline{W}_{j'}^i$ in D_r, where $x_i \neq x_k$. Lemma 8 ensures that if $C_{j'}$ is dominated by $W_{j'}^i$ (resp. $\overline{W}_{j'}^i$) in D_r then $X_i \subset D_r$ (resp. $\overline{X}_i \subset D_r$), and since x_i has the same value in D and D_r, it gives $X_i \subset D$ (resp. $\overline{X}_i \subset D$). Therefore, by Lemma 7, if a clause interval $C_{j'}$ is dominated by a W interval of x_k in D, then either there exists $x_i \neq x_k$ such that $x_i \in c_{j'}$ and $D(\overline{x}_i) \not\subset D$, or there exists $x_i \neq x_k$ such that $\overline{x}_i \in c'_j$ and $D(x_i) \not\subset D$. By definition of $A(D)$, this implies that $x_k = 0$ in $A(D)$. Since $\overline{W}_j^k$ exists, $\overline{x}_k \in c_j$ thus c_j is satisfied by $A(D)$. ◇

Claim. ($\star$) For any two dominating sets D_r and D_{r+1} of S', either $A(D_{r+1}) = A(D_r)$, or $A(D_{r+1})$ is reachable from $A(D_r)$ with a variable flip move.

Proof (of Theorem 2). Let $D_s = D_F(A_s)$ and $D_t = D_F(A_t)$. Lemma 9 and 11 ensure that (G_F, D_s, D_t) is a yes-instance of DSR_{TS} if and only if (F, A_s, A_t) is a yes-instance of SATR. And SATR is PSPACE-complete [7]. □

5   Open Questions

We left open the following question raised in [3]: is there a class for which MINI-MUM DOMINATING SET is NP-complete but TS-REACHABILITY is polynomial?

More generally, there are many graph classes in which the DSR_{TS} problem remains to be studied. For instance, we are currently working on the complexity of DSR_{TS} in H-free graphs, line graphs and unit disk graphs. It could also be studied in outerplanar graphs. Outerplanar graphs form a natural subclass of circle graphs, of bounded treewidth graph, and of planar graphs on which the complexity of the problem is PSPACE-complete.

A   Appendix

A.1   Proofs of Section 3

Proof of Lemma 1

Lemma. *Let $G = (V, E)$ be a graph, and $u, v \in V$, where $N(u) \subseteq N(v)$. If S is a TS-reconfiguration sequence in G, and S' is obtained by replacing any occurrence of u by v in the dominating sets of S, then S' also is a TS-reconfiguration sequence.*

Proof. Every neighbor of u also is a neighbor of v. Thus, replacing u by v in a dominating set keeps the domination of G. Moreover, any move that involves u can be applied if we replace it by v, which gives the result.   □

Proof of Lemma 2

Lemma. *Let D be a dominating set of G such that $u \in D$, and let $D_u = D \cup \{u', u''\} \setminus \{u\}$. The set D_u is a dominating set of G_u.*

Proof. Every vertex of $N(u)$ in the original graph G is either not in G_u, or is dominated by u' or u''. The neighborhood of all the other vertices have not been modified. Moreover, all the new vertices are dominated since they are all adjacent to u' or u''.   □

Proof of Lemma 3

Lemma. *The following holds:*

(i) *All the dominating sets of G_u of size $|D| + 1$ contain u' and u''.*

(ii) *For every dominating set X of G_u of size $|D| + 1$, $(X \cap V) \cup \{u\}$ is a dominating set of G of size at most $|D|$.*

(iii) *Every reconfiguration sequence in G_u between two dominating sets D_s, D_t of G_u of size at most $|D| + 1$ that does not contain any leaf can be adapted into a reconfiguration sequence in G from $(D_s \setminus \{u', u''\}) \cup \{u\}$ to $(D_t \setminus \{u', u''\}) \cup \{u\}$.*

Proof. **Proof of (i).** The point (i) holds since there are $n + 2$ leaves attached to each of u' and u'' and that $|D| \leq n$.

Proof of (ii). The vertices u' and u'' only dominate vertices of V dominated by u in G and u' and u'' are in any dominating set of size at most $|D| + 1$ of G_u by (i). Moreover no edge between two vertices $x, y \in V(G)$ was created in G_u. Thus $(X \cap V) \cup \{u\}$ is a dominating set of G since the only vertices of $V(G)$ that are not in $V(G_u)$ are vertices whose arcs are strictly included in u and then are dominated by u.

Proof of (iii). By Lemma 1, we can assume that there is no token on u' or u'' at any point. We show that we can adapt the transformation. If the move $x \rightsquigarrow y$ satisfies that $x, y \notin \{u', u''\}$ then the same edge exists in G and by (ii), the resulting set is dominating. So we can assume that x or y are u' or u''. We simply have to slide from or to u since $N(u')$ and $N(u'')$ minus the leaves is equal to $N(u)$. Since there is never a token on the leaves, the conclusion follows.   □

Proof of Claim

Claim. We can transform D_s into D_t if and only if $u_\ell = v$ or $u'_\ell = v'$.

Proof. Firstly, if $u_\ell = v$, then Corollary 1 ensures that there exists a transformation from D_ℓ to D_t and thus from D_s to D_t, and similarly if $u'_\ell = v'$.

Let us now prove the converse direction. If $u_\ell \neq v$ and $u'_\ell \neq v'$, assume for contradiction that there exists a transformation sequence S from D_s to D_t. By Lemma 1 we can assume that all the vertices in any dominating set of S are maximal by inclusion.

Let us consider the first dominating set C of S where the token initially on u_1 is at the right of u_ℓ in G, or at the left of u'_ℓ in G. Such a dominating set exists no token of D_t is between $u'_{\ell'}$ and u_ℓ. Let us denote by C' the dominating set before C in the sequence and $x \rightsquigarrow y$ the move from C to C'. By symmetry, we can assume that y is at the right of u_ℓ. Note that x is at the left of u_ℓ. Note that $C'' = C \setminus \{x\} \cup \{u_\ell\}$ is a dominating set of G since C and $C' = C \setminus \{x\} \cup \{y\}$ are dominating sets and u_ℓ is between x and y.

So $C \setminus \{x\} \cup \{u_\ell, u'_\ell\}$ is a dominating set of G_{u_ℓ} and then for C'' it was possible to move the token on u_ℓ to the right, a contradiction with the fact that u_ℓ was a fixed point.   □

A.2   Proofs of Section 4

A.3   Detailed Construction of G_F

All along this construction, we repeatedly refer to real number as *points*. We say that a point p is *at the left* of a point q (or q is *at the right* of p) if $p < q$. We say that p is *just at the left* of q, (or q is *just at the right* of p) if p is at the left of q, and no interval defined so far has an extremity in $[p, q]$. Note that since we are working with real intervals, given a point p, it is always possible to create a point just at the left (or just at the right) of p. Finally, we say that an interval I

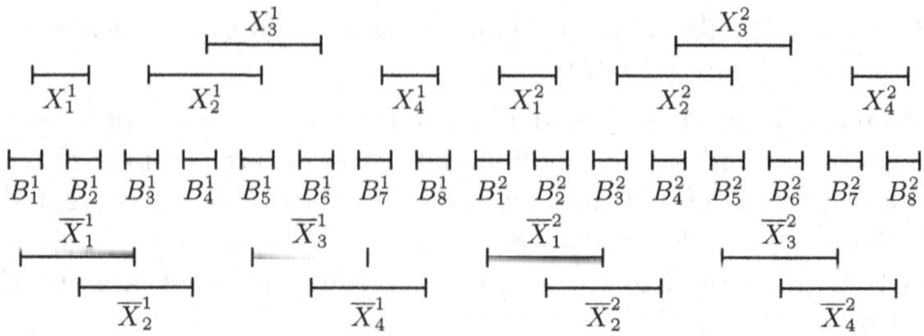

Fig. 4. The base, positive and negative bridge intervals with $n = 2$ and $m = 8$.

frames a set of points P if $\ell(I)$ is just at the left of the minimum of P and $r(I)$ is just at the right of the maximum of P.

The base intervals B_j^i are pairwise disjoint for any i and j, and are ordered by increasing i, then increasing j for a same i.

Let q be such that $m = 4q$. For every i and every $0 \leq r < q$, the interval $\overline{X}_{2r+1}^i$ starts just at the right of $\ell(B_{4r+1}^i)$ and ends just at the right of $\ell(B_{4r+3}^i)$, and $\overline{X}_{2r+2}^i$ starts just at the right of $\ell(B_{4r+2}^i)$ and ends just at the right of $\ell(B_{4r+4}^i)$. The interval X_1^i starts just at the left of $r(B_1^i)$ and ends just at the left of $r(B_2^i)$. For every $1 \leq r < q$, the interval X_{2r}^i starts just at the left of $r(B_{4r-1}^i)$ and ends just at the left of $r(B_{4r+1}^i)$, and X_{2r+1}^i starts just at the left of $r(B_{4r}^i)$ and ends just at the left of $r(B_{4r+2}^i)$. Finally, $X_{\frac{m}{2}}^i$ starts just at the left of $r(B_{m-1}^i)$ and ends just at the left of $r(B_m^i)$.

The positions of these intervals are illustrated in Fig. 4.

The clause intervals C_j are pairwise disjoint and ordered by increasing j, and we have $\ell(C_1) > r(B_m^n)$.

The interval T_j^i frames the right extremity of B_j^i and the extremity of the positive bridge interval that belongs to B_j^i. The interval $\overline{T}_j^i$ frames the left extremity of B_j^i and the extremity of the negative bridge interval that belongs to B_j^i. The interval U_j^i starts just at the left of $r(T_j^i)$, the interval $\overline{U}_j^i$ starts just at the right of $l(\overline{T}_j^i)$, and they both end between the right of the last base interval of the variable x_i and the left of the next base or clause interval. We moreover construct the intervals U_j^i (resp. $\overline{U}_j^i$) in such a way that $r(U_j^i)$ (resp. $r(\overline{U}_j^i)$) is increasing when j is increasing. In other words, the U_j^i (resp. $\overline{U}_j^i$) are pairwise adjacent. The interval V_j^i (resp. $\overline{V}_j^i$) frames the right extremity of U_j^i (resp. $\overline{U}_j^i$). And the interval W_j^i (resp. $\overline{W}_j^i$) starts just at the left of $r(V_j^i)$ (resp. $r(\overline{V}_j^i)$) and ends in an arbitrary point of C_j. Moreover, for any $i \neq i'$, W_j^i (resp. $\overline{W}_j^i$) and $W_j^{i'}$ (resp. $\overline{W}_j^{i'}$) end on the same point of C_j. This ensures that one interval the other and are therefore not adjacent.

The junction interval J frames $\ell(C_1)$ and $r(C_m)$.

Proof of Lemma 4

Lemma. *The graph G_F is connected.*

Proof. Let x_i be a variable. Let us first prove that the intervals of x_i are in the same connected component of G_F. (Recall that they are the base, bridge and path intervals of x_i). Firstly, for any j such that $x_i \in C_j$ (resp. $\overline{x_i} \in C_j$), $B_j^i T_j^i U_j^i V_j^i W_j^i$ (resp. $B_j^i \overline{T}_j^i \overline{U}_j^i \overline{V}_j^i \overline{W}_j^i$) is a path of G_F. Since every base interval of x_i is adjacent to a positive and a negative bridge interval of x_i, it is enough to show that all the bridge intervals of x_i are in the same connected component. Since for every $j \geq 2$, $\overline{X}_j^i$ is adjacent to X_{j-1}^i and X_j^i, we know that $X_1^i \overline{X}_2^i, X_2^i \ldots \overline{X}_{\frac{m}{2}}^i X_{\frac{m}{2}}^i$ is a path of G_F. Moreover, $\overline{X}_1^i$ is adjacent to $\overline{X}_2^i$. So all the intervals of x_i are in the same connected component of G_F.

Now, since the junction interval J is adjacent to every W and $\overline{W}$ interval (and that each variable appears in at least one clause), J is in the connected component of all the path variables, so the intervals of x_i and $x_{i'}$ are in the same connected component for every $i \neq i'$. Since each clause at least one variable, C_j is adjacent to at least one interval W_j^i or $\overline{W}_j^i$. Finally, each dead-end interval is adjacent to a bridge interval or a $U, \overline{U}, W$ or $\overline{W}$ interval, and each pendant interval is adjacent to a dead-end interval. Therefore, G_F is connected. $\qquad\square$

Proof of Lemma 5

Lemma. *If A satisfies F, then $D_F(A) \setminus J$ is a dominating set of G_F.*

Proof. Since every dead-end interval belongs to $D_F(A) \setminus J$, every pending and dead-end interval is dominated, as well as every bridge, $U, \overline{U}, W$ and $\overline{W}$ interval. Since for each variable x_i, the positive (resp. negative) bridge intervals of x_i dominate the base intervals of x_i, the base intervals are dominated. Moreover, the positive (resp. negative) bridge intervals of x_i and the U (resp. $\overline{U}$) intervals of x_i both dominate the T (resp. $\overline{T}$) intervals of x_i. Thus, the T and $\overline{T}$ intervals are all dominated. Moreover, for any variable x_i, the U and W (resp. $\overline{U}$ and $\overline{W}$) intervals of x_i both dominate the V (resp. $\overline{V}$) intervals of x_i. Thus, the V and $\overline{V}$ intervals are all dominated. Finally, since A satisfies F, each clause has at least one of its literal in A. Thus, each C_j and C_j' has at least one adjacent W_j^i or $\overline{W}_j^i$ in $D_F(A) \setminus J$ and are therefore dominated by it, as well as the junction interval. $\qquad\square$

Proof of Lemma 6

Lemma. *If D is a dominating set of G then for any x_i, D at least ℓ_i intervals dominating the V and $\overline{V}$ intervals of x_i and $\frac{m}{2}$ intervals dominating the base ones. Moreover, these two sets of intervals are disjoint, and they are intervals of x_i.*

Proof. For any variable x_i, each interval V_j^i (resp. $\overline{V}_j^i$) can only be dominated by U_j^i, V_j^i or W_j^i (resp. $\overline{U}_j^i$, $\overline{V}_j^i$ or $\overline{W}_j^i$). Indeed V_j^i spans the left extremity of W_j^i and the right extremity of U_j^i and since no interval starts or ends between these two points, the interval V_j^i is only adjacent to U_j^i and W_j^i. And similarly $\overline{V}_j^i$ is only adjacent to $\overline{U}_j^i$ and $\overline{W}_j^i$. Thus, at least ℓ_i intervals dominate the V and $\overline{V}$ intervals of x_i, and they are intervals of x_i. Moreover, only the base, bridge, T and $\overline{T}$ intervals of x_i are adjacent to the base intervals. Since each bridge interval is adjacent to two base intervals, and each T and $\overline{T}$ interval of x_i is adjacent to one base interval of x_i, D must contain at least $\frac{m}{2}$ of such intervals to dominate the m base intervals. □

Proof of Lemma 7

Lemma. *If D is a dominating set of size K, then for any variable $x_i \neq mv(D)$, either $X_i \subseteq D$ and $\overline{X_i} \cap D = \emptyset$, or $\overline{X_i} \subseteq D$ and $X_i \cap D = \emptyset$.*

Proof. Since $x_i \neq mv(D)$, there are exactly $\ell_i + \frac{m}{2}$ variables of x_i in D. Thus, by Lemma 6, exactly $\frac{m}{2}$ intervals of x_i in D dominate the bridge intervals of x_i. Only the bridge, T and $\overline{T}$ intervals of x_i are adjacent to the base intervals. Moreover, bridge intervals are adjacent to two base intervals and T or $\overline{T}$ intervals are adjacent to only one. Since there are m base intervals of x_i, each interval of D must dominate a pair of base intervals (or none of them). So these intervals of D should be some bridge intervals of x_i.

Note that, by cardinality, each pair of bridge intervals of D must dominate pairwise disjoint base intervals. Let us now show by induction that these bridge intervals are either all the positive bridge intervals, or all the negative bridge intervals. We study two cases: either $X_1^i \in D$, or $X_1^i \notin D$.

Assume that $X_1^i \in D$. In D, X_1^i dominates B_1^i and B_2^i. Thus, since $\overline{X}_1^i$ dominates B_1^i and $\overline{X}_2^i$ dominates B_2^i, none of $\overline{X}_1^i, \overline{X}_2^i$ are in D (since their neighborhood in the set of base intervals is not disjoint with X_1^i). But B_3^i (resp B_4^i) is only adjacent to $\overline{X}_1^i$ and X_2^i (resp. $\overline{X}_2^i$ and X_3^i). Thus both X_2^i, X_3^i are in D. Suppose now that for a given j such that j is even and $j \leq \frac{m}{2} - 2$, we have $X_j^i, X_{j+1}^i \in D$. Then, since a base interval dominated by X_j^i (resp. X_{j+1}^i) also is dominated by $\overline{X}_{j+1}^i$ (resp. $\overline{X}_{j+2}^i$), the intervals $\overline{X}_{j+1}^i, \overline{X}_{j+2}^i$ are not in D. But there is a base interval adjacent only to $\overline{X}_{j+1}^i$ and X_{j+2}^i (resp. $\overline{X}_{j+2}^i$ and X_{j+3}^i if $j \neq \frac{m}{2} - 2$, or $\overline{X}_{j+2}^i$ and X_{j+2}^i if $j = \frac{m}{2} - 2$). Therefore, if $j + 2 < \frac{m}{2}$ we have $X_{j+2}^i, X_{j+3}^i \in D$, and $X_{\frac{m}{2}}^i \in D$. By induction, if $X_1^i \in D$ then each of the $\frac{m}{2}$ positive bridge intervals belong to D and thus none of the negative bridge intervals do.

Assume now that $X_1^i \notin D$. Then, to dominate B_1^i and B_2^i, we must have $\overline{X}_1^i, \overline{X}_2^i \in D$. Let us show that if for a given odd j such that $j \leq \frac{m}{2} - 3$ we have $\overline{X}_j^i, \overline{X}_{j+1}^i \in D$, then $\overline{X}_{j+2}^i, \overline{X}_{j+3}^i \in D$. Since $\overline{X}_j^i$ (resp. $\overline{X}_{j+1}^i$) dominates base intervals also dominated by X_{j+1}^i (resp. X_{j+2}^i), we have $X_{j+1}^i, X_{j+2}^i \notin D$.

But there exists a base interval only adjacent to X_{j+1}^i and $\overline{X}_{j+2}^i$ (resp. X_{j+2}^i and $\overline{X}_{j+3}^i$). Thus, $\overline{X}_{j+2}^i, \overline{X}_{j+3}^i \in D$. By induction, if $X_1^i \notin D$ then each of the $\frac{m}{2}$ negative bridge intervals belong to D. Thus, none of the positive bridge intervals belong to D. □

Proof of Lemma 8

Lemma. *If D is a dominating set of size K, then for any variable $x_i \neq mv(D)$, if $X_i \subseteq D$ then $\overline{W}_i \cap D = \emptyset$, otherwise $W_i \cap D = \emptyset$.*

Proof. By Lemma 7, D either X_i or $\overline{X}_i$.

If $X_i \subset D$, Lemma 7 ensures that $\overline{X}_i \cap D = \emptyset$. So the intervals $\overline{T}_j^i$ have to be dominated by other intervals.

By Lemma 6, ℓ_i intervals must dominate the V and $\overline{V}$ intervals of x_i. Since no interval dominates two of them, each $\overline{T}_j^i$ has to be dominated by an interval that is also dominating a V or $\overline{V}$ interval. The only interval that dominates both $\overline{T}_j^i$ and a V or $\overline{V}$ interval is $\overline{U}_j^i$. So all the $\overline{U}$ intervals are in D and $\overline{W} \cap D = \emptyset$ (since the only V or $\overline{V}$ interval dominated by a $\overline{W}$ interval is a $\overline{V}$ interval, which is already dominated).

Similarly if $\overline{X}_i \subset D$, Lemma 7 ensures that $X_i \cap D = \emptyset$. So the intervals T_j^i have to be dominated by other intervals. And one can prove similarly that these intervals should be the U intervals and then the W intervals are not in D. □

Complete Proof of Lemma 9

Lemma. *If (F, A_s, A_t) is a yes-instance of SATR, then (G_F, D_s, D_t) is a yes-instance of DSR$_{\mathrm{TS}}$.*

Proof. Let (F, A_s, A_t) be a yes-instance of SATR, and let $S =< A_1 := A_s, \ldots, A_\ell := A_t >$ be the reconfiguration sequence from A_s to A_t. We construct a reconfiguration sequence S' from D_s to D_t by replacing any flip of variable $x_i \rightsquigarrow \overline{x_i}$ of S from A_r to A_{r+1} by the following sequence of token slides from $D_F(A_r)$ to $D_F(A_{r+1})$.[3]

- We perform a sequence of slides that moves the token on J to $\overline{X}_1^i$. By Lemma 4, G_F is connected, and by Lemma 5, $D_F(A_r) \setminus J$ is a dominating set. So any sequence of moves along a path from J to $\overline{X}_1^i$ keeps a dominating set.
- For any j such that $x_i \in C_j$, we first move the token from W_j^i to V_j^i then from V_j^i to U_j^i. Let us show that this keeps G_F dominated. The important intervals that can be dominated by W_j^i are V_j^i, C_j, and J. The vertex V_j^i is dominated anyway during the sequence since it is also dominated by V_j^i and U_j^i. Moreover, since $x_i \rightsquigarrow \overline{x_i}$ keeps F satisfied, each clause containing x_i has a literal different from x_i that also satisfies the clause. Thus, for each C_j such that $x_i \in C_j$, there exists an interval $W_j^{i'}$ or $\overline{W}_j^{i'}$, with $i' \neq i$, that belongs to $D_F(A_r)$, and then dominates both C_j and J during these two moves.

---

[3] And replacing any flip $\overline{x_i} \rightsquigarrow x_i$ by the converse of this sequence.

- For j from 1 to $\frac{m}{2} - 1$, we apply the move $X_j^i \rightsquigarrow \overline{X}_{j+1}^i$. This move is possible since X_j^i and $\overline{X}_{j+1}^i$ are neighbors in G_F. Let us show that this move keeps a dominating set. For $j = 1$, the important intervals that are dominated by X_1^i are B_1^i, B_2^i, and T_1^i. Since U_1^i is in the current dominating set (by the second point), T_1^i is dominated. Moreover B_1^i is dominated by $\overline{X}_1^i$, and B_2^i is a neighbor of $\overline{X}_2^i$. Thus, $X_1^i \rightsquigarrow \overline{X}_2^i$ maintains a dominating set. For $2 \leq j \leq \frac{m}{2} - 1$, the important intervals that are dominated by X_j^i are B_k^i, B_{k-2}^i and T_j^i where $k = 2j + 1$ if j is even and $k = 2j$ otherwise. Again T_j^i is dominated by the U intervals. Moreover B_{k-2}^i is dominated by $\overline{X}_{j-1}^i$ (on which there is a token since we perform this sequence for increasing j), and B_k^i is also dominated by $\overline{X}_{j+1}^i$.

- For any j such that $\overline{x_i} \in C_j$, we move the token from $\overline{U}_j^i$ to $\overline{V}_j^i$ and then from $\overline{V}_j^i$ to $\overline{W}_j^i$. The important intervals dominated by $\overline{U}_j^i$ are the intervals $\overline{T}_j^i$, $\overline{V}_j^i$. But $\overline{T}_j^i$ is dominated by a negative bridge interval, and $\overline{V}_j^i$ stays dominated by $\overline{V}_j^i$ then $\overline{W}_j^i$.

- The previous moves lead to the dominating set $(D_F(A_{r+1}) \setminus J) \cup X_{\frac{m}{2}}^i$. We finally perform a sequence of moves that slide the token on $X_{\frac{m}{2}}^i$ to J. It can be done since Lemma 4 ensures that G_F is connected. And all along the transformation, we keep a dominating set by Lemma 5. As wanted, it leads to the dominating set $D_F(A_{r+1})$. $\square$

Proof of Lemma 10

Lemma. *If there exists a reconfiguration sequence S from D_s to D_t, then there exists another one S' such that for any two adjacent dominating sets D_r and D_{r+1} of S', if both D_r and D_{r+1} have a moving variable, then it is the same one.*

Proof. Assume that, in S, there exist two adjacent dominating sets D_r and D_{r+1} such that both D_r and D_{r+1} have a moving variable, and $mv(D_r) \neq mv(D_{r+1})$. Let us modify slightly the sequence in order to avoid this move.

Since D_r and D_{r+1} are adjacent in S, we have $D_{r+1} = D_r \cup v \setminus \{u\}$, where uv is an edge of G_F. Since $mv(D_r) \neq mv(D_{r+1})$, u is an interval of $mv(D_r)$, and v an interval of $mv(D_{r+1})$. By construction, the only edges of G_F between intervals of different variables are between their $\{W, \overline{W}\}$ intervals. Thus, both u and v are W or $\overline{W}$ intervals and, in particular they are adjacent to the junction interval J. Moreover, the only important intervals that are adjacent to u (resp. v) are the V or $\overline{V}$ intervals of the same variable as u, W or $\overline{W}$ intervals, clause intervals, or the junction interval J. Since u and v are adjacent, and since they are both W or $\overline{W}$ intervals, they cannot be adjacent to the same clause interval. But the only intervals that are potentially not dominated by $D_r \setminus u = D_{r+1} \setminus v$ should be dominated both by u in D_r and by v in D_{r+1}. So these intervals are included in the set of W or $\overline{W}$ intervals and the junction interval, which are all dominated by J. Thus, $D_r \cup J \setminus u$ is a dominating set of G_F. Therefore, we can add in S the dominating set $D_r \cup J \setminus u$ between D_r and D_{r+1}. This intermediate dominating set has no moving variable. By repeating this procedure while there are adjacent dominating sets in S with different moving variables, we obtain the desired reconfiguration sequence S'.   □

Proof of First Claim

Claim. For any consecutive dominating sets D_r and D_{r+1} and any x_i that is not $mv(D_r)$ nor $mv(D_{r+1})$, the value of x_i is identical in $A(D_r)$ and $A(D_{r+1})$.

Proof. Lemma 7 ensures that for any x_i such that $x_i \neq mv(D_r)$ and $x_i \neq mv(D_{r+1})$, either $X_i \subset D_r$ and $\overline{X_i} \cap D_r = \emptyset$ or $\overline{X_i} \subset D_r$ and $X_i \cap D_r = \emptyset$, and the same holds in D_{r+1}. Since the number of positive and negative bridge intervals is at least 2 (since by assumption m is a multiple of 4), and D_{r+1} is reachable from D_r in a single step, either both D_r and D_{r+1} contain X_i, or both contain $\overline{X_i}$. Thus, by definition of $A(D)$, for any x_i such that $x_i \neq mv(D_r)$ and $x_i \neq mv(D_{r+1})$, x_i has the same value in $A(D_r)$ and $A(D_{r+1})$.   ◇

Proof of Second Claim

Claim. We have $A(D_s) = A_s$ and $A(D_t) = A_t$.

Proof. By definition, $D_s = D_F(A_s)$ and thus D_s the junction interval, which means that it does not have any moving variable. Moreover, $D_s X_i$ for any variable x_i such that $x_i = 1$ in A_s and $\overline{X_i}$ for any variable x_i such that $x_i = 0$ in A_s. Therefore, for any variable x_i, $x_i = 1$ in A_s if and only if $x_i = 1$ in $A(D_s)$. Similarly, $A(D_t) = A_t$.   ◇

Proof of Last Claim

Claim. For any two dominating sets D_r and D_{r+1} of S', either $A(D_{r+1}) = A(D_r)$, or $A(D_{r+1})$ is reachable from $A(D_r)$ with a variable flip move.

Proof. By Claim 4.3, for any variable x_i such that $x_i \neq mv(D_r)$ and $x_i \neq mv(D_{r+1})$, x_i has the same value in $A(D_r)$ and $A(D_{r+1})$. Moreover, by definition of S', if both D_r and D_{r+1} have a moving variable then $mv(D_r) = mv(D_{r+1})$. Therefore, at most one variable changes its value between $A(D_r)$ and $A(D_{r+1})$, which concludes the proof.   ◇

References

1. Bonamy, M., Bousquet, N., Feghali, C., Johnson, M.: On a conjecture of Mohar concerning Kempe equivalence of regular graphs. J. Comb. Theory Ser. B **135**, 179–199 (2019)
2. Bonamy, M., et al.: The perfect matching reconfiguration problem. In: 44th International Symposium on Mathematical Foundations of Computer Science, MFCS 2019, pp. 80:1–80:14 (2019)

3. Bonamy, M., Dorbec, P., Ouvrard, P.: Dominating sets reconfiguration under token sliding. Discret. Appl. Math. **301**, 6–18 (2021)
4. Bousquet, N., Heinrich, M.: A polynomial version of Cereceda's conjecture. arXiv preprint arXiv:1903.05619 (2019)
5. Bousquet, N., Joffard, A., Ouvrard, P.: Linear transformations between dominating sets in the tar-model. In: 31st International Symposium on Algorithms and Computation (ISAAC 2020). Schloss Dagstuhl Leibniz Zentrum für Informatik (2020)
6. Cereceda, L., van den Heuvel, J., Johnson, M.: Finding paths between 3-colorings. J. Graph Theory **67**(1), 69–82 (2011)
7. Gopalan, P., Kolaitis, P.G., Maneva, E., Papadimitriou, C.H.: The connectivity of Boolean satisfiability: computational and structural dichotomies. SIAM J. Comput. **38**(6), 2330–2355 (2009)
8. Haas, R., Seyffarth, K.: The k-dominating graph. Graphs Comb. **30**(3), 609–617 (2014)
9. Haddadan, A., et al.: The complexity of dominating set reconfiguration. Theor. Comput. Sci. **651**, 37–49 (2016)
10. Hearn, R.A., Demaine, E.: PSPACE-completeness of sliding-block puzzles and other problems through the nondeterministic constraint logic model of computation. Theor. Comput. Sci. **343**(1–2) (2005)
11. Ito, T., et al.: On the complexity of reconfiguration problems. Theor. Comput. Sci. **412**(12–14), 1054–1065 (2011)
12. Keil, J.M.: The complexity of domination problems in circle graphs. Discret. Appl. Math. **42**(1), 51–63 (1993)
13. Lokshtanov, D., Mouawad, A.E.: The complexity of independent set reconfiguration on bipartite graphs. ACM Trans. Algorithms (TALG) **15**(1), 1–19 (2018)
14. Nishimura, N.: Introduction to reconfiguration. Algorithms **11**(4), 52 (2018)
15. Suzuki, A., Mouawad, A.E., Nishimura, N.: Reconfiguration of dominating sets. J. Comb. Optim. **32**(4), 1182–1195 (2016)
16. van den Heuvel, J.: The complexity of change. Surveys in Comb. **409**(2013), 127–160 (2013)

Bipartite 3-Regular Counting Problems with Mixed Signs

Jin-Yi Cai, Austen Z. Fan[iD], and Yin Liu[✉]

University of Wisconsin-Madison, Madison, WI 53706, USA
{jyc,afan,yinl}@cs.wisc.edu

Abstract. We prove a complexity dichotomy for a class of counting problems expressible as bipartite 3-regular Holant problems. For every problem of the form Holant $(f \mid =_3)$, where f is any integer-valued ternary symmetric constraint function on Boolean variables, we prove that it is either P-time computable or #P-hard, depending on an explicit criterion of f. The constraint function can take both positive and negative values, allowing for cancellations. In addition, we discover a new phenomenon: there is a set $\mathcal{F}$ with the property that for every $f \in \mathcal{F}$ the problem Holant $(f \mid =_3)$ is planar P-time computable but #P-hard in general, yet its planar tractability is by a *combination* of a holographic transformation by $\left[\begin{smallmatrix} 1 & 1 \\ 1 & -1 \end{smallmatrix}\right]$ to FKT *together* with an independent global argument.

Keywords: Dichotomy theorem · Holant problem · Bipartite graph

1 Introduction

Holant problems encompass a broad class of counting problems [1–3,9–12,17,18, 20,24,25,27]. For symmetric constraint functions this is also equivalent to edge-coloring models [21,22]. These problems extend counting constraint satisfaction problems. Freedman, Lovász and Schrijver proved that some prototypical Holant problems, such as counting perfect matchings, cannot be expressed as vertex-coloring models known as graph homomorphisms [16,19]. To be more precise, counting CSP are precisely Holant problems where EQUALITIES of all arities are assumed to be part of the constraint functions. Meanwhile, Holant problems can be viewed as counting CSP with read-twice variables. The classification program of counting problems is to classify as broad a class of these problems as possible into either #P-hard or P-time computable. This paper is an investigation of a class of restricted Holant problems.

While much progress has been made for the classification of counting CSP [4,6,7,14], and some progress for Holant problems [5], classifying Holant problems on regular bipartite graphs is particularly challenging. In a very recent paper [15] we initiated the study of Holant problems in the simplest setting of 3-regular bipartite graphs with *nonnegative* constraint functions. Admittedly, this

Supported by NSF CCF-1714275.

E. Bampis and A. Pagourtzis (Eds.): FCT 2021, LNCS 12867, pp. 135–148, 2021.
https://doi.org/10.1007/978-3-030-86593-1_9

is a severe restriction, because nonnegativity of the constraint functions rules out cancellation, which is a source of non-trivial P-time algorithms. Cancellation is in a sense the *raison d'être* for the Holant framework following Valiant's holographic algorithms [24–26]. The (potential) existence of P-time algorithms by cancellation is exciting, but at the same time creates obstacles if we want to classify every problem in the family into either P-time computable or #P-hard. At the same time, restricting to nonnegative constraints makes the classification theorem easier to prove. In this paper, we remove this nonnegativity restriction.

More formally, a Holant problem is defined on a graph where edges are variables and vertices are constraint functions. The aim of a Holant problem is to compute its partition function, which is a sum over all $\{0,1\}$-edge assignments of the product over all vertices of the constraint function evaluations. E.g., if every vertex has the EXACT-ONE function (which evaluates to 1 if exactly one incident edge is 1, and evaluates to 0 otherwise), then the partition function gives the number of perfect matchings. In this paper we consider Holant problems on 3-regular bipartite graphs $G = (U, V, E)$, where the Holant problem Holant $(f \mid =_3)$ computes the following partition function[1]

$$\text{Holant}(G) = \sum_{\sigma: E \to \{0,1\}} \prod_{u \in U} f\left(\sigma|_{E(u)}\right) \prod_{v \in V} (=_3)\left(\sigma|_{E(v)}\right),$$

where $f = [f_0, f_1, f_2, f_3]$ at each $u \in U$ is an integer-valued constraint function that evaluates to f_i if σ assigns exactly i among 3 incident edges $E(u)$ to 1, and $(=_3) = [1, 0, 0, 1]$ is the EQUALITY function on 3 variables (which is 1 iff all three inputs are equal). E.g., if we take the EXACT-ONE function $f = [0, 1, 0, 0]$ in the 3-regular bipartite problem Holant $(f \mid =_3)$, the right hand side (RHS) $(=_3)$ represents 3-element subsets and each element appears in 3 such subsets, and Holant(G) counts the number of exact-3-covers; if f is the OR function $[0, 1, 1, 1]$ then Holant(G) counts the number of all set covers.

The main theorem in this paper is a complexity dichotomy (Theorem 6): for any rational-valued function f of arity 3, the problem Holant $(f \mid =_3)$ is either #P-hard or P-time computable, depending on an explicit criterion on f. The main advance is to allow f to take both positive and negative values, thus cancellations in the sum $\sum_{\sigma: E \to \{0,1\}}$ can occur.

A major component of the classification program is to account for some algorithms, called holographic algorithms, that were initially discovered by Valiant [24]. These algorithms introduce quantum-like cancellations as the main tool. In the past 10 to 15 years we have gained a great deal of understanding of these mysteriously looking algorithms. In particular, it was proved in [12] that for all counting CSP with arbitrary constraint functions on Boolean variables, there is a precise 3-way division of problem types: (1) P-time computable in general, (2) P-time computable on planar structures but #P-hard in general, and (3) #P-hard even on planar structures. Moreover, every problem in type

---

[1] If we replace f by a set $\mathcal{F}$ of constraint functions, each $u \in U$ is assigned some $f_u \in \mathcal{F}$, and replace $(=_3)$ by $\mathcal{EQ}$, the set of Equality of all arities, then Holant $(\mathcal{F} \mid \mathcal{EQ})$ can be taken as the definition of counting CSP.

(2) is so by Valiant's holographic reduction to the Fisher-Kasteleyn-Temperley algorithm (FKT) for planar perfect matchings. In [8] for (non-bipartite) Holant problems with symmetric constraint functions, the 3-way division above persists, but problems in (2) include one more subtype unrelated to Valiant's holographic reduction. In this paper, we have a surprising discovery. We found a new set of functions $\mathcal{F}$ which fits into type (2) problems above, but the planar P-time tractability is *neither* by Valiant's holographic reduction alone, *nor* entirely independent of it. Rather it is by a combination of a holographic reduction together with a global argument. Here is an example of such a problems: We say $(X, \mathcal{S})$ is a 3-regular k-uniform set system, if $\mathcal{S}$ consists of a family of subsets $S \subset X$ each of size $|S| = k$, and every $x \in X$ is in exactly 3 sets. If $k = 2$ this is just a 3-regular graph. We consider 3-regular 3-uniform set systems. We say $\mathcal{S}'$ is a *leafless partial cover* if every $x \in \bigcup_{S \in \mathcal{S}'} S$ belongs to more than one set $S \in \mathcal{S}'$. We say x is *lightly covered* if $|\{S \in \mathcal{S}' : x \in S\}|$ is 2, and *heavily covered* if it is 3.

Problem: Weighted-Leafless-Partial-Cover.
Input: A 3-regular 3-uniform set system $(X, \mathcal{S})$.
Output: $\sum_{\mathcal{S}'} (-1)^l 2^h$, where the sum is over all leafless partial covers $\mathcal{S}'$, and l (resp. h) is the number of $x \in X$ that are lightly covered (resp. heavily covered).

One can show that this problem is just Holant $(f \mid=_3)$, where $f = [1, 0, -1, 2]$. This problem is a special case of a set of problems of the form $f = [3a + b, -a - b, -a + b, 3a - b]$. We show that all these problems belong to type (2) above, although they are not directly solvable by a holographic algorithm since they are provably not matchgates-transformable.

In this paper, we use Mathematica™ to perform symbolic computation. In particular, the procedure CylindricalDecomposition in Mathematica™ is an implementation (of a version) of Tarski's theorem on the decidability of the theory of real-closed fields. Some of our proof steps involve heavy symbolic computation. This stems from the bipartite structure. In order to preserve this structure, one has to connect each vertex from the left hand side (LHS) to RHS when constructing subgraph fragments called gadgets. In 3-regular bipartite graphs, it is easy to show that any gadget construction produces a constraint function that has the following restriction: the difference of the arities between the two sides is 0 mod 3. This severely limits the possible constructions within a moderate size, and a reasonable sized construction tends to produce gigantic polynomials. To "solve" some of these polynomials seems beyond direct manipulation by hand.

We believe our dichotomy (Theorem 6) is valid even for (algebraic) real or complex-valued constraint functions. However, in this paper we can only prove it for rational-valued constraint functions. There are two difficulties of extending our proof beyond $\mathbb{Q}$. The first is that we use the idea of interpolating degenerate straddled functions, for which we need to ensure that the ratio of the eigenvalues of the interpolating gadget matrix is not a root of unity. With rational-valued constraint functions, the only roots of unity that can occur are in a degree 2 extension field. For general constraint functions, they can be arbitrary roots of unity. Another difficulty is that some Mathematica™ steps showing the nonexistence of some cases are only valid for $\mathbb{Q}$.

2 Preliminaries

We use graph fragments called *gadgets* in our constructions. As illustrated in Fig. 1, 2, 3, 4 and 5, a gadget is a bipartite graph $G = (U, V, E_{\text{int}}, E_{\text{ext}})$ with internal edges E_{int} and dangling edges E_{ext}. There can be m (respectively n) dangling edges internally incident to vertices from U (respectively V). These $m + n$ dangling edges represent input Boolean variables $x_1, \ldots, x_m, y_1, \ldots, y_n$ and the gadget defines a constraint function, a.k.a. a *signature*

$$f(x_1, \ldots, x_m, y_1, \ldots, y_n) = \sum_{\sigma: E_{\text{int}} \to \{0,1\}} \prod_{u \in U} f\left(\hat{\sigma}|_{E(u)}\right) \prod_{v \in V} (=_3)\left(\hat{\sigma}|_{E(v)}\right),$$

where $\hat{\sigma}$ denotes the extension of σ by the assignment on the dangling edges. As indicated before, in any gadget construction we must be careful to keep the bipartite structure. In particular, we must keep track whether an input variable is on the LHS (labeled by f), or it is on the RHS (labeled by $(=_3)$). In each figure of gadgets presented later, we use a blue square to represent a signature from LHS, which under most of the cases will be $[f_0, f_1, f_2, f_3]$, a green circle to represent the ternary equality $(=_3)$, and a black triangle to represent a unary signature whose values depend on the context.

A *symmetric* signature is a function that is invariant under any permutation of its variables. The value of such a signature depends only on the Hamming weight of its input. We denote a ternary symmetric signature f by the notation $f = [f_0, f_1, f_2, f_3]$, where f_i is the value on inputs of Hamming weight i. The EQUALITY of arities 3 is $(=_3) = [1, 0, 0, 1]$. A symmetric signature f is called (1) *degenerate* if it is the tensor power of a unary signature; (2) *Generalized Equality*, or Gen-Eq, if it is zero unless all inputs are equal. Affine signatures were discovered in the dichotomy for counting constraint satisfaction problems (#CSP) [5]. A (real valued) ternary symmetric signature is *affine* if it has the form $[1, 0, 0, \pm 1], [1, 0, 1, 0], [1, 0, -1, 0], [1, 1, -1, -1]$ or $[1, -1, -1, 1]$, or by reversing the order of the entries, up to a constant factor. If f is degenerate, Gen-Eq, or affine, then the problem $\# \text{CSP}(f)$ and thus Holant $(f \mid =_3)$ is in P (for a more detailed exposition of this theory, see [5]). Our dichotomy asserts that, for all signatures f with $f_i \in \mathbb{Q}$, these three classes are the only tractable cases of the problem Holant $(f \mid =_3)$; all other signatures lead to #P-hardness.

By a slight abuse of terminology, we say f is in P, or respectively #P-hard, if the problem Holant $(f \mid (=_3))$ is computable in P, or respectively #P-hard. We will use the following theorem [20] on spin systems when proving our results:

Theorem 1. *Let* $a, b \in \mathbb{C}$, *and* $X = ab$, $Z = \left(\frac{a^3 + b^3}{2}\right)^2$. *Holant* $([a, 1, b] \mid (=_3))$ *is #P-hard except in the following cases when the problem is in P.*

1. $X = 1$;
2. $X = Z = 0$;
3. $X = -1$ *and* $Z = 0$;
4. $X = -1$ *and* $Z = -1$.

Fig. 1. G_1: the square node is labeled $f = [1, a, b, c]$, the circle node is $(=_3)$.

An important observation is that in the context of Holant $(f \mid (=_3))$, every gadget construction produces a signature with $m \equiv n \bmod 3$, where m and n are the numbers of input variables (arities) from the LHS and RHS respectively. Thus, any construction that produces a signature purely on either the LHS or the RHS will have arity a multiple of 3. In order that our constructions are more manageable in size, we will make heavy use of *straddled gadgets* with $m = n = 1$ that do not belong to either side and yet can be easily iterated. The signatures of the iterated gadgets are represented by matrix powers.

Consider the binary straddled gadget G_1 in Fig. 1. Its signature is $G_1 = \begin{bmatrix} 1 & b \\ a & c \end{bmatrix}$, where $G_1(i, j)$ (at row i column j) is the value of this gadget when the left dangling edge (from the "square") and the right dangling edge (from the "circle" $(=_3)$) are assigned i and j respectively, for $i, j \in \{0, 1\}$. Iterating G_1 sequentially k times is represented by the matrix power G_1^k. It turns out that it is very useful either to produce directly or to obtain by interpolation a rank *deficient* straddled signature, which would in most cases allow us to obtain unary signatures on either side. With unary signatures we can connect to a ternary signature to produce binary signatures on one side and then apply Theorem 1. The proof idea of Lemma 1 is the same as in [15] for nonnegative signatures.

Lemma 1. *Given the binary straddled signature* $G_1 = \begin{bmatrix} 1 & b \\ a & c \end{bmatrix}$, *we can interpolate the degenerate binary straddled signature* $\begin{bmatrix} y & xy \\ 1 & x \end{bmatrix}$, *provided that* $c \neq ab$, $a \neq 0$, $\Delta = \sqrt{(1-c)^2 + 4ab} \neq 0$ *and* $\frac{\lambda}{\mu}$ *is not a root of unity, where* $\lambda = \frac{-\Delta + (1+c)}{2}$, $\mu = \frac{\Delta + (1+c)}{2}$ *are the two eigenvalues, and* $x = \frac{\Delta - (1-c)}{2a}$ *and* $y = \frac{\Delta + (1-c)}{2a}$.

Proof. We have $x + y = \Delta/a \neq 0$ and so $\begin{bmatrix} -x & y \\ 1 & 1 \end{bmatrix}^{-1}$ exists, and the matrix G_1 has the Jordan Normal Form

$$G_1 = \begin{pmatrix} 1 & b \\ a & c \end{pmatrix} = \begin{pmatrix} -x & y \\ 1 & 1 \end{pmatrix} \begin{pmatrix} \lambda & 0 \\ 0 & \mu \end{pmatrix} \begin{pmatrix} -x & y \\ 1 & 1 \end{pmatrix}^{-1}.$$

Here the matrix G_1 is non-degenerate since $c \neq ab$, and so λ and μ are nonzero. Consider

$$D = \frac{1}{x+y} \begin{pmatrix} y & xy \\ 1 & x \end{pmatrix} = \begin{pmatrix} -x & y \\ 1 & 1 \end{pmatrix} \begin{pmatrix} 0 & 0 \\ 0 & 1 \end{pmatrix} \begin{pmatrix} -x & y \\ 1 & 1 \end{pmatrix}^{-1}.$$

Given any bipartite graph Ω where the binary degenerate straddled signature D appears n times, we form gadgets G_1^s where $0 \leq s \leq n$ by iterating the G_1 gadget s times and replacing each occurrence of D with G_1^s. (For $s = 0$ we replace each copy of D by an edge.) Denote the resulting bipartite graph as Ω_s. We stratify the assignments in the Holant sum for Ω according to assignments to $\begin{bmatrix} 0 & 0 \\ 0 & 1 \end{bmatrix}$ as:

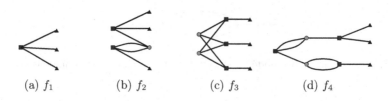

(a) f_1 (b) f_2 (c) f_3 (d) f_4

Fig. 2. Four gadgets where each triangle represents the unary gadget $[1, x]$

- $(0,0)$ i times;
- $(1,1)$ j times;

with $i + j = n$; all other assignments will contribute 0 in the Holant sum for Ω. The same statement is true for each Ω_s with the matrix $\begin{bmatrix} \lambda^s & 0 \\ 0 & \mu^s \end{bmatrix}$. Let $c_{i,j}$ be the sum, in Ω, over all such assignments of the products of evaluations of all other signatures other than that represented by the matrix $\begin{bmatrix} 0 & 0 \\ 0 & 1 \end{bmatrix}$, including the contributions from $\begin{bmatrix} -x & y \\ 1 & 1 \end{bmatrix}$ and its inverse. The same quantities c_{ij} appear for each Ω_s, independent of s, with the substitution of the matrix $\begin{bmatrix} \lambda^s & 0 \\ 0 & \mu^s \end{bmatrix}$. Then, for $0 \le s \le n$, we have

$$\text{Holant}_{\Omega_s} = \sum_{i+j=n} \left(\lambda^i \mu^j \right)^s \cdot c_{i,j} \tag{2.1}$$

and $\text{Holant}_\Omega = c_{0,n}$. Since λ/μ is not a root of unity, the quantities $\lambda^i \mu^{n-i}$ are pairwise distinct, thus (2.1) is a full ranked Vandermonde system. Thus we can compute Holant_Ω from Holant_{Ω_s} by solving the linear system in polynomial time. Thus we can interpolate D in polynomial time.

The next lemma allows us to get unary signatures.

Lemma 2. *For* $\text{Holant}([1, a, b, c] \mid =_3)$, $a, b, c \in \mathbb{Q}$, $a \neq 0$, *with the availability of binary degenerate straddled signature* $\begin{bmatrix} y & xy \\ 1 & x \end{bmatrix}$ *(here* $x, y \in \mathbb{C}$ *can be arbitrary), in polynomial time*

1. *we can interpolate* $[y, 1]$ *on the LHS, i.e.,* $\text{Holant}(\{[1, a, b, c], [y, 1]\} \mid =_3) \le_T$ $\text{Holant}([1, a, b, c] \mid =_3)$;
2. *we can interpolate* $[1, x]$ *on the RHS, i.e.,* $\text{Holant}([1, a, b, c] \mid \{(=_3), [1, x]\}) \le_T$ $\text{Holant}([1, a, b, c] \mid =_3)$, *except for two cases:* $[1, a, a, 1]$, $[1, a, -1 - 2a, 2 + 3a]$.

Proof. For the problem $\text{Holant}(\{[1, a, b, c], [y, 1]\} \mid =_3)$, the number of occurrences of $[y, 1]$ on LHS is 0 mod 3, say $3n$, since the other signatures are both of arity 3. Now, for each occurrence of $[y, 1]$, we replace it with the binary straddled signature $\begin{bmatrix} y & xy \\ 1 & x \end{bmatrix}$, leaving $3n$ dangling edges on RHS yet to be connected to LHS, each of which represents a unary signature $[1, x]$. We build a gadget to connect every triple of such dangling edges. We claim that at least one of the connection gadgets in Figs. 2a, 2b, 2c and 2d creates a nonzero global factor. The factors of these four gadgets are $f_1 = cx^3 + 3bx^2 + 3ax + 1$, $f_2 = (ab + c)x^3 + (3bc + 2a^2 + b)x^2 + (2b^2 + ac + 3a)x + ab + 1$, $f_3 = (a^3 + b^3 + c^3)x^3 + 3(a^2 + 2ab^2 + bc^2)x^2 +$

(a) g_1 (b) g_2 (c) g_3

Fig. 3. Three gadgets where each triangle represents the unary gadget $[y, 1]$

$3(a + 2a^2b + b^2c)x + 1 + 2a^3 + b^3$ and $f_4 = (ab + 2abc + c^3)x^3 + (2a^2 + b + 2a^2c + 3ab^2 + bc + 3b^2c)x^2 + (3a + 3a^2b + ac + 2b^2 + 2b^2c + ac^2)x + 1 + 2ab + abc$ respectively. By setting the four formulae to be 0 simultaneously, with $a \neq 0$, $a, b, c \in \mathbb{Q}$ and $x \in \mathbb{C}$, we found that there is no solution. This verification uses Mathematica™. Thus, we can always "absorb" the left-over $[1, x]$'s at the cost of some easily computable nonzero global factor.

For the other claim on $[1, x]$ on RHS, i.e.,

$$\text{Holant}([1, a, b, c] \mid \{(=_3), [1, x]\}) \leq_T \text{Holant}([1, a, b, c] \mid =_3)$$

we use a similar strategy to "absorb" the left-over copies of $[y, 1]$ on the LHS by connecting them to $(=_3)$ in the gadgets in the Figs. 3a, 3b or 3c. These gadgets produce factors $g_1 = y^3 + 1$, $g_2 = y^3 + by^2 + ay + c$ and $g_3 = y^3 + 3a^2y^2 + 3b^2y + c^2$ respectively. It can be directly checked that, for complex y, all these factors are 0 iff $y = -1$, and the signature has the form $[1, a, a, 1]$ or $[1, a, -2a - 1, 3a + 2]$. □

A main thrust in our proof is we want to be assured that such degenerate binary straddled signature can be obtained, and the corresponding unary signatures in Lemma 2 can be produced. We now first consider the two exceptional cases $[1, a, a, 1]$ and $[1, a, -2a - 1, 3a + 2]$ where this is not possible.

Lemma 3. *The problem* $[1, a, a, 1]$ *is #P-hard unless* $a \in \{0, \pm 1\}$ *in which case it is in* P.

Proof. If $a = 0$ or $a = \pm 1$, then it is either Gen-Eq or degenerate or affine, and thus Holant($[1, a, a, 1] \mid =_3$) is in P. Now assume $a \neq 0$ and $a \neq \pm 1$.

Using the gadget G_1, we have $\Delta = |2a|$ and $x = y = \Delta/2a = \pm 1$ depending on the sign of a. So we get the signature $[y, 1] = [\pm 1, 1]$ on LHS by Lemmas 1 and 2. Connecting two copies of $[y, 1]$ to $[1, 0, 0, 1]$ on RHS, we get $[1, 1]$ on RHS regardless of the sign. Connecting $[1, 1]$ to $[1, a, a, 1]$ on LHS, we get $[1 + a, 2a, 1 + a]$ on LHS. The problem Holant($[1 + a, 2a, 1 + a] \mid =_3$) is #P-hard by Theorem 1 unless $a = 0, \pm 1$ or $-\frac{1}{3}$, thus we only need to consider the signature $[3, -1, -1, 3]$.

If $a = -\frac{1}{3}$, we apply holographic transformation with the Hadamard matrix $H = \begin{bmatrix} 1 & 1 \\ 1 & -1 \end{bmatrix}$. Note that $[3, -1, -1, 3] = 4((1, 0)^{\otimes 3} + (0, 1)^{\otimes 3}) - (1, 1)^{\otimes 3}$. Here each tensor power represents a truth-table of 8 entries, or a vector of dimension 8; the linear combination is the truth-table for the symmetric signature $[3, -1, -1, 3]$, which is in fact a short hand for the vector $(3, -1, -1, -1, -1, -1, -1, 3)$. Also note that, $(1, 0)H = (1, 1)$, $(0, 1)H = (1, -1)$ and $(1, 1)H = (2, 0)$, thus we have

$[3, -1, -1, 3]H^{\otimes 3} = 4((1,1)^{\otimes 3} + (1,-1)^{\otimes 3}) - (2,0)^{\otimes 3} = 4[2,0,2,0] - [8,0,0,0] = [0,0,8,0]$, which is equivalent to $[0,0,1,0]$ by a global factor. So, we get

$$\text{Holant} ([3,-1,-1,3] \mid (=_3)) \equiv_T \text{Holant} \left([3,-1,-1,3]H^{\otimes 3} \mid (H^{\otimes 3})^{-1}[1,0,0,1]\right)$$
$$\equiv_T \text{Holant} ([0,0,1,0] \mid [1,0,1,0])$$
$$\equiv_T \text{Holant} ([0,0,1,0] \mid [0,0,1,0])$$
$$\equiv_T \text{Holant} ([0,1,0,0] \mid [0,1,0,0])$$

where the first reduction is by Valiant's Holant theorem [25], the third reduction comes from the following observation: given a bipartite 3-regular graph $G = (V, U, E)$ where the vertices in V are assigned the signature $[0,0,1,0]$ and the vertices in U are assigned the signature $[1,0,1,0]$, every nonzero term in the Holant sum must correspond to a mapping $\sigma : E \to \{0,1\}$ where exactly two edges of any vertex are assigned 1. The fourth reduction is by simply flipping 0's and 1's. The problem Holant $([0,1,0,0] \mid [0,1,0,0])$ is the problem of counting perfect matchings in 3-regular bipartite graphs, which Dagum and Luby proved to be #P-complete (Theorem 6.2 in [13]). □

Lemma 4. *The problem* $[1, a, -2a - 1, 3a + 2]$ *is #P-hard unless* $a = -1$ *in which case it is in* P.

Proof. Observe that the truth-table of the symmetric signature $[1, a, -2a-1, 3a+2]$ written as an 8-dimensional column vector is just

$$2(a+1) \left(\begin{bmatrix} 1 \\ 0 \end{bmatrix}^{\otimes 3} + \begin{bmatrix} 0 \\ 1 \end{bmatrix}^{\otimes 3} \right) - \frac{a+1}{2} \left(\begin{bmatrix} 1 \\ 1 \end{bmatrix}^{\otimes 3} + \begin{bmatrix} 1 \\ -1 \end{bmatrix}^{\otimes 3} \right) - a \begin{bmatrix} 1 \\ -1 \end{bmatrix}^{\otimes 3}.$$

Here again, the tensor powers as 8-dimensional vectors represent truth-tables, and the linear combination of these vectors "holographically" reconstitute a truth-table of the symmetric signature $[1, a, -2a - 1, 3a + 2]$. We apply the holographic transformation with the Hadamard matrix $H = \begin{bmatrix} 1 & 1 \\ 1 & -1 \end{bmatrix}$, and we get

$$\text{Holant} ([1, a, -2a - 1, 3a + 2] \mid (=_3))$$
$$\equiv_T \text{Holant} \left([1, a, -2a - 1, 3a + 2]H^{\otimes 3} \mid (H^{\otimes 3})^{-1}[1,0,0,1]\right)$$
$$\equiv_T \text{Holant} ([0, 0, a + 1, -3a - 1] \mid [1,0,1,0])$$
$$\equiv_T \text{Holant} ([0, 0, a + 1, 0] \mid [0,0,1,0]),$$

where the last equivalence follows from the observation that for each nonzero term in the Holant sum, every vertex on the LHS has at least two of three edges assigned 1 (from $[0, 0, a+1, -3a-1]$), meanwhile every vertex on the RHS has at most two of three edges assigned 1 (from $[1, 0, 1, 0]$). The graph being bipartite and 3-regular, the number of vertices on both sides must equal, thus every vertex has exactly two incident edges assigned 1.

Then by flipping 0's and 1's,

$$\text{Holant}\,([0,0,a+1,0] \mid [0,0,1,0]) \equiv_T \text{Holant}\,([0,a+1,0,0] \mid [0,1,0,0])\,.$$

For $a \neq -1$, this problem is equivalent to counting perfect matchings in bipartite 3-regular graphs, which is #P-complete. If $a = -1$, the signature $[1,-1,1,-1] = [1,-1]^{\otimes 3}$ is degenerate, and thus in P. The holographic reduction also reveals that, not only the problem is in P, but the Holant sum is 0.   □

We can generalize Lemma 4 to get the following corollary.

Corollary 1. *The problem* $\text{Holant}\,(f \mid =_3)$, *where* $f = [3a+b, -a-b, -a+b, 3a-b]$, *is computable in polynomial time on planar graphs for all* a, b, *but is #P-hard on general graphs for all* $a \neq 0$.

Proof. The following equivalence is by a holographic transformation using H:

$$\begin{aligned}
\text{Holant}\,(f \mid (=_3)) &\equiv_T \text{Holant}\,\left(fH^{\otimes 3} \mid (H^{-1})^{\otimes 3}(=_3)\right) \\
&\equiv_T \text{Holant}\,([0,0,a,b] \mid [1,0,1,0]) \\
&\equiv_T \text{Holant}\,([0,0,a,0] \mid [0,0,1,0]) \\
&\equiv_T \text{Holant}\,([0,a,0,0] \mid [0,1,0,0])
\end{aligned}$$

where the third reduction follows the same reasoning as in the proof of Lemma 4. When $a \neq 0$, $\text{Holant}\,([0,a,0,0] \mid [0,1,0,0])$ is (up to a global nonzero factor) the perfect matching problem on 3-regular bipartite graphs. This problem is computable in polynomial time on planar graphs and the reductions are valid for planar graphs as well. It is #P-hard on general graphs (for $a \neq 0$).   □

Remark: The planar tractability of the problem $\text{Holant}\,(f \mid =_3)$, for $f = [3a + b, -a - b, -a + b, 3a - b]$, is a remarkable fact. It is neither accomplished by a holographic transformation to matchgates alone, nor entirely independent from it. One can prove that the signature f is not matchgates-transformable (for nonzero a, b; see [5] for the theory of matchgates and the realizability of signatures by matchgates under holographic transformation). In previous complexity dichotomies, we have found that for the entire class of counting CSP problems over Boolean variables, all problems that are #P-hard in general but P-time tractable on planar graphs are tractable by the following universal algorithmic strategy—a holographic transformation to matchgates followed by the FKT algorithm [12]. On the other hand, for (non-bipartite) Holant problems with arbitrary symmetric signature sets, this category of problems (planar tractable but #P-hard in general) is completely characterized by two types [8]: (1) holographic transformations to matchgates, and (2) a separate kind that depends on the existence of "a wheel structure" (unrelated to holographic transformations and matchgates). Here in Corollary 1 we have found the first instance where a new type has emerged.

Proposition 1. *For* $G_1 = \left[\begin{smallmatrix} 1 & b \\ a & c \end{smallmatrix}\right]$, *with* $a, b, c \in \mathbb{Q}$, *if it is non-singular (i.e.,* $c \neq ab$), *then it has two nonzero eigenvalues* λ *and* μ. *The ratio* λ/μ *is not a root of unity unless at least one of the following conditions holds:*

$$\begin{cases} c + 1 = 0 \\ ab + c^2 + c + 1 = 0 \\ 2ab + c^2 + 1 = 0 \\ 3ab + c^2 - c + 1 = 0 \\ 4ab + c^2 - 2c + 1 = 0 \end{cases} \tag{2.2}$$

Now we introduce two more binary straddled signatures—G_2 and G_3 in Fig. 4. The signature matrix of G_2 is $\left[\begin{smallmatrix} w & b' \\ a' & c' \end{smallmatrix}\right]$, where $w = 1 + 2a^3 + b^3$, $a' = a + 2a^2b + b^2c$, $b' = a^2 + 2ab^2 + bc^2$ and $c' = a^3 + 2b^3 + c^3$. The signature matrix of G_3 is $\left[\begin{smallmatrix} 1+ab & a^2+bc \\ a+b^2 & ab+c^2 \end{smallmatrix}\right]$. In this case we define $w = 1 + ab$, $a' = a + b^2$, $b' = a^2 + bc$ and $c' = ab + c^2$. Similar to Proposition 1, we have the following on G_2 and G_3.

(a) G_2 (b) G_3

Fig. 4. Two binary straddled gadgets

Proposition 2. *For each gadget* G_2 *and* G_3 *respectively, if the signature matrix is non-degenerate, then the ratio* λ'/μ' *of its eigenvalues is not a root of unity unless at least one of the following conditions holds, where* $A = w + c'$, $B = (c' - w)^2 + 4a'b'$.

$$\begin{cases} A = 0 \\ B = 0 \\ A^2 + B = 0 \\ A^2 + 3B = 0 \\ 3A^2 + B = 0 \end{cases} \tag{2.3}$$

Lemma 5. *Suppose* $a, b, c \in \mathbb{Q}$, $a \neq 0$ *and* $c \neq ab$ *and* a, b, c *do not satisfy any condition in (2.2). Let* $x = \frac{\Delta - (1-c)}{2a}$, $y = \frac{\Delta + (1-c)}{2a}$ *and* $\Delta = \sqrt{(1-c)^2 + 4ab}$. *Then for* Holant($[1, a, b, c] \mid =_3$),

1. *we can interpolate* $[y, 1]$ *on LHS;*
2. *we can interpolate* $[1, x]$ *on RHS except for 2 cases:* $[1, a, a, 1]$, $[1, a, -1 - 2a, 2 + 3a]$.

We have similar statements corresponding to G_2 (resp. G_3). When the signature matrix is non-singular and does not satisfy any condition in (2.3), we can interpolate the corresponding $[y', 1]$ on LHS, and we can also interpolate the corresponding $[1, x']$ on RHS except when $y' = -1$.

Definition 1. *For* Holant($[1, a, b, c] \mid =_3$)*, with* $a, b, c \in \mathbb{Q}$*,* $a \neq 0$*, we say a binary straddled gadget* G *works if the signature matrix of* G *is non-degenerate and the ratio of its two eigenvalues* λ/μ *is not a root of unity.*

Remark: Explicitly, the condition that G_1 *works* is that $c \neq ab$ and a, b, c do not satisfy any condition in (2.2), which is just the assumptions in Lemma 5. G_1 *works* implies that it can be used to interpolate $[y, 1]$ on LHS, and to interpolate $[1, x]$ on RHS with two exceptions for which we already proved the dichotomy. The x, y are as stated in Lemma 5. Similar remarks are valid for the binary straddled gadgets G_2 and G_3.

The unary signatures $\Delta_0 = [1, 0]$ and $\Delta_1 = [0, 1]$ are called the pinning signatures because they "pin" a variable to 0 or 1. One good use of having unary signatures is that we can use Lemma 7 to get the two pinning signatures. Pinning signatures are helpful as the following lemma shows.

Lemma 6. *If* Δ_0 *and* Δ_1 *are available on the RHS in* Holant($[1, a, b, c] \mid =_3$)*, where* $a, b, c \in \mathbb{Q}$*,* $ab \neq 0$*, then the problem is* #*P-hard unless* $[1, a, b, c]$ *is affine or degenerate, in which cases it is in P.*

The following lemma lets us interpolate arbitrary unary signatures on RHS, in particular Δ_0 and Δ_1, from a binary gadget with a straddled signature and a suitable unary signature s on RHS. Mathematically, the proof is essentially the same as in [23], but technically Lemma 7 applies to binary straddled signatures.

Lemma 7. *Let* $M \in \mathbb{R}^{2 \times 2}$ *be a non-singular signature matrix for a binary straddled gadget which is diagonalizable with distinct eigenvalues, and* $s = [a, b]$ *be a unary signature on RHS that is not a row eigenvector of* M*. Then* $\{s \cdot M^j\}_{j \geq 0}$ *can be used to interpolate any unary signature on RHS.*

The ternary gadget G_4 in Fig. 5 will be used in the paper.

Fig. 5. G_4

3   Main Theorem

Our main theorem is Theorem 6. The main part of its proof is to deal with the generic case $f = [f_0, f_1, f_2, f_3]$ where f_0, f_1, f_2, f_3 are all nonzero. Roughly speaking, we argue that *either* the simplest gadget G_1 *works*, in which case we get suitable unary signatures with some small number of exceptional cases which we can handle separately, *or* the gadget G_1 does not work, in which case we gain a polynomial condition in (2.2). When G_1 works we can construct signatures from f to show #P-hardness which implies that f is also #P-hard, or we get additional stringent conditions on f_0, f_1, f_2, f_3 which we can analyse separately. When G_1 does not work, armed by the polynomial condition from (2.2), we can afford to use more complicated gadgets G_2 and G_3. We also use G_4 to produce new ternary signatures. It turns out that we need to iterate this theme more than once, but every time we either succeed because the current gadget *works*, or it does not, in which case we gain an additional polynomial condition similar to the one in (2.2), and from that we can afford to use a more complicated gadget, every time staying just one step ahead of a symbolic computational explosion due to the high complexity of `CylindricalDecomposition` for Tarski's theorem. The exact proof differs somewhat from the outline above because of some technical complications.

We need the following preliminary dichotomies:

Theorem 2. *The problem* $[1, a, b, 0]$ *for* $a, b \in \mathbb{Q}$ *is #P-hard unless it is degenerate or affine, and thus in P.*

Theorem 3. *The problem* $[1, a, 0, c]$ *with* $a, c \in \mathbb{Q}$ *is #P-hard unless* $a = 0$, *in which case it is Gen-Eq and thus in P.*

Theorem 4. *The problem* $[0, a, b, 0]$ *with* $a, b \in \mathbb{Q}$ *is #P-hard unless* $a = b = 0$, *in which case the Holant value is 0.*

Theorem 5. *The problem* $[1, a, b, c]$ *with* $a, b, c \in \mathbb{Q}$, $abc \neq 0$, *is #P-hard unless it is degenerate, Gen-Eq or affine.*

The logical structure of our proof of the main dichotomy theorem is illustrated in the following flowchart.

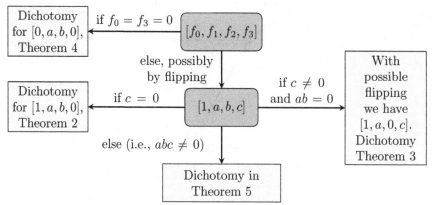

We are now ready to prove our main theorem.

Theorem 6. *The problem* Holant$\{[f_0, f_1, f_2, f_3] \,|\, (=_3)\}$ *with* $f_i \in \mathbb{Q}$ ($i = 0, 1, 2, 3$) *is #P-hard unless the signature* $[f_0, f_1, f_2, f_3]$ *is degenerate, Gen-Eq or belongs to the affine class.*

Proof. First, if $f_0 = f_3 = 0$, by Theorem 4, we know that it is #P-hard unless it is $[0, 0, 0, 0]$ which is degenerate. Note that in all other cases, $[0, f_1, f_2, 0]$ is not Gen-Eq, degenerate or affine.

Assume now at least one of f_0 and f_3 is not 0. By flipping the role of 0 and 1, we can assume $f_0 \neq 0$, then the signature becomes $[1, a, b, c]$ after normalization.

If $c = 0$, the dichotomy for $[1, a, b, 0]$ is proved in Theorem 2.

If in $[1, a, b, c]$, $c \neq 0$, then a and b are symmetric by flipping. Now if $ab = 0$, we can assume $b = 0$ by the afore-mentioned symmetry, i.e., the signature becomes $[1, a, 0, c]$. By Theorem 3, it is #P-hard unless $a = 0$, in which case it is Gen-Eq. In all other cases, it is not Gen-Eq or degenerate or affine.

Finally, for the problem $[1, a, b, c]$ where $abc \neq 0$, Theorem 5 proves the dichotomy that it is #P-hard unless the signature is degenerate or Gen-Eq or affine. $\qquad\square$

References

1. Backens, M.: A new Holant dichotomy inspired by quantum computation. In: 44th International Colloquium on Automata, Languages, and Programming, ICALP, pp. 16:1–16:14 (2017)
2. Backens, M.: A complete dichotomy for complex-valued Holant^c. In: 45th International Colloquium on Automata, Languages, and Programming, ICALP, pp. 12:1–12:14 (2018)
3. Backens, M., Goldberg, L.A.: Holant clones and the approximability of conservative Holant problems. ACM Trans. Algorithms **16**(2), 23:1–23:55 (2020)
4. Bulatov, A.A.: A dichotomy theorem for constraint satisfaction problems on a 3-element set. J. ACM (JACM) **53**(1), 66–120 (2006)
5. Cai, J., Chen, X.: Complexity Dichotomies for Counting Problems: Boolean Domain, vol. 1. Cambridge University Press, Cambridge (2017)
6. Cai, J., Chen, X.: Complexity of counting CSP with complex weights. J. ACM **64**(3), 19:1–19:39 (2017)
7. Cai, J., Chen, X., Lu, P.: Nonnegative weighted# CSP: an effective complexity dichotomy. SIAM J. Comput. **45**(6), 2177–2198 (2016)
8. Cai, J., Fu, Z., Guo, H., Williams, T.: A Holant dichotomy: is the FKT algorithm universal? In: IEEE 56th Annual Symposium on Foundations of Computer Science, FOCS, pp. 1259–1276 (2015)
9. Cai, J., Guo, H., Williams, T.: A complete dichotomy rises from the capture of vanishing signatures. SIAM J. Comput. **45**(5), 1671–1728 (2016)
10. Cai, J., Lu, P.: Holographic algorithms: from art to science. J. Comput. Syst. Sci. **77**(1), 41–61 (2011)
11. Cai, J., Lu, P., Xia, M.: Holant problems and counting CSP. In: Proceedings of the Forty-First Annual ACM Symposium on Theory of Computing, pp. 715–724 (2009)
12. Cai, J., Lu, P., Xia, M.: Holographic algorithms with matchgates capture precisely tractable planar #CSP. In: 51th Annual IEEE Symposium on Foundations of Computer Science, FOCS, pp. 427–436 (2010)

13. Dagum, P., Luby, M.: Approximating the permanent of graphs with large factors. Theor. Comput. Sci. **102**(2), 283–305 (1992)
14. Dyer, M., Richerby, D.: An effective dichotomy for the counting constraint satisfaction problem. SIAM J. Comput. **42**(3), 1245–1274 (2013)
15. Fan, A.Z., Cai, J.: Dichotomy result on 3-regular bipartite non-negative functions. arXiv preprint arXiv:2011.09110 (2020)
16. Freedman, M., Lovász, L., Schrijver, A.: Reflection positivity, rank connectivity, and homomorphism of graphs. J. Am. Math. Soc. **20**(1), 37–51 (2007)
17. Guo, H., Huang, S., Lu, P., Xia, M.: The complexity of weighted Boolean #CSP modulo K. In: 28th International Symposium on Theoretical Aspects of Computer Science, STACS, pp. 249–260 (2011)
18. Guo, H., Lu, P., Valiant, L.G.: The complexity of symmetric Boolean parity Holant problems. SIAM J. Comput. **42**(1), 324–356 (2013)
19. Hell, P., Nesetril, J.: Graphs and Homomorphisms, Oxford Lecture Series in Mathematics and its Applications, vol. 28. Oxford University Press (2004)
20. Kowalczyk, M., Cai, J.: Holant problems for regular graphs with complex edge functions. In: 27th International Symposium on Theoretical Aspects of Computer Science, STACS, pp. 525–536 (2010)
21. Szegedy, B.: Edge coloring models and reflection positivity. J. Am. Math. Soc. **20**(4), 969–988 (2007)
22. Szegedy, B.: Edge coloring models as singular vertex coloring models. In: Katona, G.O.H., Schrijver, A., Szőnyi, T., Sági, G. (eds.) Fete of Combinatorics and Computer Science. BSMS, pp. 327–336. Springer, Heidelberg (2010). https://doi.org/10.1007/978-3-642-13580-4_12
23. Vadhan, S.P.: The complexity of counting in sparse, regular, and planar graphs. SIAM J. Comput. **31**(2), 398–427 (2001)
24. Valiant, L.G.: Accidental algorthims. In: 47th Annual IEEE Symposium on Foundations of Computer Science, FOCS, pp. 509–517. IEEE (2006)
25. Valiant, L.G.: Holographic algorithms. SIAM J. Comput. **37**(5), 1565–1594 (2008)
26. Valiant, L.G.: Some observations on holographic algorithms. Comput. Complex. **27**(3), 351–374 (2017). https://doi.org/10.1007/s00037-017-0160-4
27. Xia, M.: Holographic reduction: a domain changed application and its partial converse theorems. In: Automata, Languages and Programming, pp. 666–677 (2010)

The Satisfiability Problem
for a Quantitative Fragment of PCTL

Miroslav Chodil and Antonín Kučera[✉]

Masaryk University, Brno, Czechia
tony@fi.muni.cz

Abstract. We give a sufficient condition under which every finite-satisfiable formula of a given PCTL fragment has a model with at most doubly exponential number of states (consequently, the finite satisfiability problem for the fragment is in **2-EXPSPACE**). The condition is semantic and it is based on enforcing a form of "progress" in non-bottom SCCs contributing to the satisfaction of a given PCTL formula. We show that the condition is satisfied by PCTL fragments beyond the reach of existing methods.

Keywords: Probabilistic temporal logics · Satisfiability · PCTL

1 Introduction

Probabilistic CTL (PCTL) [19] is a temporal logic applicable to discrete-time probabilistic systems with Markov chain semantics. PCTL is obtained from the standard CTL (see, e.g., [14]) by replacing the existential/universal path quantifiers with the probabilistic operator $P(\Phi) \bowtie r$. Here, Φ is a path formula, $\bowtie$ is a comparison such as $\geq$ or $<$, and r is a numerical constant. A formula $P(\Phi) \bowtie r$ holds in a state s if the probability of all runs initiated in s satisfying Φ is $\bowtie$-bounded by r. The *satisfiability problem for PCTL*, asking whether a given PCTL formula has a model, is a long-standing open question in probabilistic verification resisting numerous research attempts.

Unlike CTL and other non-probabilistic temporal logics, PCTL does not have a small model property guaranteeing the existence of a bounded-size model for every satisfiable formula. In fact, one can easily construct satisfiable PCTL formulae without *any* finite model (see, e.g., [8]). Hence, the PCTL satisfiability problem is studied in two basic variants: (1) *finite satisfiability*, where we ask about the existence of a finite model, and (2) *general satisfiability*, where we ask about the existence of an unrestricted model.

For the *qualitative fragment* of PCTL, where the range of admissible probability constraints is restricted to $\{=0, >0, =1, <1\}$, both variants of the satisfiability problem are **EXPTIME**-complete, and a finite description of a model for a satisfiable formula is effectively constructible [8]. Unfortunately, the underlying proof techniques are not applicable to general PCTL with unrestricted (quantitative) probability constraints such as ≥ 0.25 or <0.7.

© Springer Nature Switzerland AG 2021
E. Bampis and A. Pagourtzis (Eds.): FCT 2021, LNCS 12867, pp. 149–161, 2021.
https://doi.org/10.1007/978-3-030-86593-1_10

To solve the *finite* satisfiability problem for some PCTL fragment, it suffices to establish a computable upper bound on the size (the number of states) of a model for a finite-satisfiable formula of the fragment[1]. At first glance, one is tempted to conjecture the existence of such a bound for the whole PCTL because there is no apparent way how a finite-satisfiable PCTL formula φ can "enforce" the existence of $F(\varphi)$ distinct states in a model of φ, where F grows faster than any computable function. Interestingly, this conjecture is *provably wrong* in a slightly modified setting where we ask about finite PCTL satisfiability in a *subclass* of Markov chains $\mathcal{M}^k$ where every state has at most $k \geq 2$ immediate successors (the k is an arbitrarily large fixed constant). This problem is *undecidable* and hence no computable upper bound on the size of a finite model in $\mathcal{M}^k$ exists [8] (see [9] for a full proof). So far, all attempts at extending the undecidability proof of [8] to the class of unrestricted Markov chains have failed; it is not yet clear whether the obstacles are invincible.

Regardless of the ultimate decidability status of the (finite) PCTL satisfiability, the study of PCTL fragments brings important insights into the structure and expressiveness of PCTL formulae. The existing works [12,22] identify several fragments where every (finite) satisfiable formula has a model of bounded size and specific shape. In [12], it is shown that every formula φ of the *bounded fragment* of PCTL, where the validity of φ in a state s depends only on a bounded prefix of a run initiated in s, has a bounded-size tree model. In [22], seven PCTL fragments based on F and G operators are studied. For each of these fragments, it is shown that every satisfiable (or finite-satisfiable) formula has a bounded-size model such that every non-bottom SCC is a singleton. It is also shown that there are finite-satisfiable PCTL formulae without a model of this shape. An example of such formula is

$$\psi \quad \equiv \quad G_{=1}\big(F_{\geq 0.5}(a \wedge F_{\geq 0.2}\neg a) \vee a\big) \ \wedge \ F_{=1}\,G_{=1}\,a \ \wedge \ \neg a$$

In [22], it is shown that ψ is finite satisfiable[2], but every finite model of ψ has a non-bottom SCC with at least two states, such as the Markov chain M of Fig. 1.

Our Contribution. A crucial step towards solving the finite satisfiability problem for PCTL is understanding the role of non-bottom SCCs. Intuitively, if a given PCTL formula φ enforces a model with a non-bottom SCC, then the top SCC must achieve some sort of "progress" in satisfying φ, and successor SCCs are required to satisfy only some "simpler" formulae. In this paper, we develop this intuition into an algorithm deciding finite satisfiability for various PCTL fragments beyond the reach of existing methods.

More concretely, we design a sufficient condition under which every formula in a given PCTL fragment has a bounded model with at most doubly

---

[1] Although there are uncountably many Markov chains with n states, the edge probabilities can be represented symbolically by variables, and the satisfiability of a given PCTL formula in a Markov chain with n states can then be encoded in the existential fragment of first-order theory of the reals. This construction is presented in [13].

[2] In [22], the formula ψ has the same structure but uses qualitative probability constraints.

exponential number of states. Consequently, the finite satisfiability problem for the fragment is in **2-EXPSPACE**. The condition says that a progress in satisfying φ is achievable by a SCC C where C has a bounded number states and takes the form of a loop with one exit state of bounded outdegree (see Fig. 2). Furthermore, the successor states are required to satisfy PCTL formulae strictly simpler than φ in a precisely defined sense. Hence, bounded models for these formulae exist by induction hypothesis, and thus we complete the construction of a bounded model for φ.

The above sufficient condition is "semantic" and it is satisfied by various mutually incomparable syntactic fragments of PCTL that are not covered by the methods of [22] (two of these fragments contain the formula ψ presented above). Hence, our semantic condition can be seen as a "unifying principle" behind these new decidability results.

In our construction, we had to address fundamental issues specific to quantitative PCTL. The basic observation behind the small model property proofs for non-probabilistic temporal logics (and also *qualitative* PCTL) is that the satisfaction of a given formula in a given state s is determined by the satisfaction of φ and its subformulae in the successor states of s. For quantitative PCTL, this is not true. For example, knowing whether immediate successors of a state s satisfy the formula $F_{\geq 0.2}\,\varphi$ does not necessarily allow to determine the satisfaction of $F_{\geq 0.2}\,\varphi$ in s. What we need is a *precise probability* of satisfying the path formula $F\,\varphi$ in the successors of s. Clearly, it makes no sense to filter a model according to the satisfaction of infinitely many formulae of the form $F_{\geq r}\,\varphi$. In our proof, we invent a method for extending the set of "relevant formulae" so that it remains bounded and still captures the crucial properties of states.

The methodology presented in this paper can be extended by considering SCCs with increasingly complex structure and analyzing the achievable "progress in satisfaction" of PCTL formulae (see Sect. 4 for more comments). We believe that this effort may eventually result in a decidability proof for the whole PCTL.

Related Work. The satisfiability problem for non-probabilistic CTL is known to be **EXPTIME**-complete [15]. The same upper bound is valid also for a richer logic of the modal μ-calculus [3,18]. The probabilistic extension of CTL (and also CTL*) was initially studied in its qualitative form [20,23]. The satisfiability problem is shown decidable in these works. A precise complexity classification of general and finite satisfiability, together with a construction of (a finite description of) a model is given in [8]. In the same paper, it is also shown that the satisfiability and the finite satisfiability problems are undecidable when the class of admissible models is restricted to Markov chains with a k-bounded branching degree, where $k \geq 2$ is an arbitrary constant. A variant of the bounded satisfiability problem, where transition probabilities are restricted to $\{\frac{1}{2}, 1\}$, is proven **NP**-complete in [4]. The decidability of finite satisfiability for fragments of quantitative PCTL is established in the works [12,22] discussed above.

The *model-checking* problem for PCTL has been studied both for finite Markov chains (see, e.g., [1,2,5,21]) and for infinite Markov chains generated by probabilistic pushdown automata and their subclasses [11,16,17]. PCTL

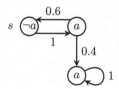

Fig. 1. A Markov chain M such that $s \models \psi$.

formulae have also been used as *objectives* in Markov decision processes (MDPs) and stochastic games, where the players controlling non-deterministic states strive to satisfy/falsify a given PCTL formula. Positive decidability results exist for finite MDPs and qualitative PCTL formulae [10]. For quantitative PCTL and finite MDPs, the problem becomes undecidable [7]. Let us note that the aforementioned undecidability results for the (finite) PCTL satisfiability problem in subclasses of Markov chains with bounded branching degree follow by utilizing proof techniques of [7].

2   Preliminaries

We use $\mathbb{N}$, $\mathbb{Q}$, $\mathbb{R}$ to denote the sets of non-negative integers, rational numbers, and real numbers, respectively. We use the standard notation for writing intervals of real numbers, e.g., $[0, 1)$ denotes the set of all $r \in \mathbb{R}$ such that $0 \le r < 1$.

The logic PCTL [19] is a probabilistic version of Computational Tree Logic [14] obtained by replacing the existential and universal path quantifiers with the probabilistic operator $P(\Phi) \bowtie r$, where Φ is a path formula, $\bowtie$ is a comparison, and $r \in [0, 1]$ is a constant.

In full PCTL, the syntax of path formulae is based on the X and U ('next' and 'until') operators. In this paper, we consider a variant of PCTL based on F and G operators.

Definition 1 (PCTL). *Let AP be a set of atomic propositions. The syntax of PCTL state and path formulae is defined by the following abstract syntax equations:*

$$\varphi \quad ::= \quad a \mid \neg a \mid \varphi \wedge \varphi \mid \varphi \vee \varphi \mid P(\Phi) \triangleright r$$
$$\Phi \quad ::= \quad F\,\varphi \mid G\,\varphi$$

Here, $a \in AP$, $\triangleright \in \{\ge, >\}$, and $r \in [0, 1]$.

For the sake of simplicity, the trivial probability constraints '≥ 0' and '>1' are syntactically forbidden. Since the formula Φ in the probabilistic operator $P(\Phi) \triangleright r$ is always of the form $F\,\varphi$ or $G\,\varphi$, we often write just $F_{\triangleright r}\,\varphi$ and $G_{\triangleright r}\,\varphi$ instead of $P(F\,\varphi) \triangleright r$ and $P(G\,\varphi) \triangleright r$, respectively. The probability constraint '≥ 1' is usually written as '$=1$'. The set of all state subformulae of a given state formula φ is denoted by $sub(\varphi)$. For a set X of PCTL formulae, we use $sub(X)$ to denote $\bigcup_{\varphi \in X} sub(\varphi)$.

Observe that the negation is applicable only to atomic propositions and the comparison ranges only over $\{\geq, >\}$. This causes no loss of generality because negations can be pushed inside, and formulae such as $F_{\leq r}\,\varphi$ and $G_{< r}\,\varphi$ are equivalent to $G_{\geq 1-r}\,\neg\varphi$ and $F_{> 1-r}\,\neg\varphi$, respectively.

PCTL formulae are interpreted over Markov chains where every state s is assigned a subset $v(s) \subseteq AP$ of atomic propositions valid in s.

Definition 2 (Markov chain). *A Markov chain is a triple $M = (S, P, v)$, where S is a finite or countably infinite set of* states, $P \colon S \times S \to [0,1]$ *is a function such that $\sum_{t \in S} P(s,t) = 1$ for every $s \in S$, and $v \colon S \to 2^{AP}$.*

A *path* in M is a finite sequence $w = s_0 \ldots s_n$ of states such that $P(s_i, s_{i+1}) > 0$ for all $i < n$. A *run* in M is an infinite sequence $\pi = s_0 s_1 \ldots$ of states such that every finite prefix of π is a path in M. We also use $\pi(i)$ to denote the state s_i of π.

A *strongly connected component (SCC)* of M is a maximal $U \subseteq S$ such that, for all $s, t \in U$, there is a path from s to t. A *bottom SCC (BSCC)* of M is a SCC U such that for every $s \in U$ and every path $s_0 \ldots s_n$ where $s = s_0$ we have that $s_n \in U$.

For every path $w = s_0 \ldots s_n$, let $Run(w)$ be the set of all runs starting with w, and let $\mathbb{P}(Run(w)) = \prod_{i=0}^{n-1} P(s_i, s_{i+1})$. To every state s, we associate the probability space $(Run(s), \mathcal{F}_s, \mathbb{P}_s)$, where $\mathcal{F}_s$ is the σ-field generated by all $Run(w)$ where w starts in s, and $\mathbb{P}_s$ is the unique probability measure obtained by extending $\mathbb{P}$ in the standard way (see, e.g., [6]).

The *validity* of a PCTL state/path formula for a given state/run of M is defined inductively as follows:

$$
\begin{aligned}
s &\models a && \text{iff} && a \in v(s), \\
s &\models \neg a && \text{iff} && a \notin v(s), \\
s &\models \varphi_1 \wedge \varphi_2 && \text{iff} && s \models \varphi_1 \text{ and } s \models \varphi_2, \\
s &\models \varphi_1 \vee \varphi_2 && \text{iff} && s \models \varphi_1 \text{ or } s \models \varphi_2, \\
s &\models P(\Phi) \triangleright r && \text{iff} && \mathbb{P}_s(\{\pi \in Run(s) \mid \pi \models \Phi\}) \triangleright r, \\[4pt]
\pi &\models F\,\varphi && \text{iff} && \pi(i) \models \varphi \text{ for some } i \in \mathbb{N}, \\
\pi &\models G\,\varphi && \text{iff} && \pi(i) \models \varphi \text{ for all } i \in \mathbb{N}.
\end{aligned}
$$

For a set X of PCTL state formulae, we write $s \models X$ iff $x \models \varphi$ for every $\varphi \in X$.

We say that M is a *model* of φ if $s \models \varphi$ for some state s of M. The *(finite) PCTL satisfiability problem* is the question whether a given PCTL formula has a (finite) model.

A *PCTL fragment* is a set of PCTL state formulae $\mathcal{L}$ closed under state subformulae and changes in probability constraints, i.e., if $F_{\triangleright r}\,\varphi \in \mathcal{L}$ (or $G_{\triangleright r}\,\varphi \in \mathcal{L}$), then $F_{\geq r'}\,\varphi \in \mathcal{L}$ (or $G_{\geq r'}\,\varphi \in \mathcal{L}$) for every $r' \in (0,1]$.

3   Results

In this section, we formulate our main results. As a running example, we use the formula ψ of Sect. 1 and its model of Fig. 1.

Definition 3. *Let ψ be a PCTL formula and s a state in a Markov chain such that $s \models \psi$. The closure of ψ in s, denoted by $C_s(\psi)$, is the least set K satisfying the following conditions:*

- *$\psi \in K$;*
- *if $\varphi_1 \vee \varphi_2 \in K$ and $s \models \varphi_1$, then $\varphi_1 \in K$;*
- *if $\varphi_1 \vee \varphi_2 \in K$ and $s \models \varphi_2$, then $\varphi_2 \in K$;*
  if $\varphi_1 \wedge \varphi_2 \subset K$, then $\varphi_1, \varphi_2 \in K$;
- *if $F_{\rhd r}\, \varphi \in K$ and $s \models \varphi$, then $\varphi \in K$;*

Furthermore, for a finite set X of PCTL formulae such that $s \models X$, we put $C_s(X) = \bigcup_{\psi \in X} C_s(\psi)$.

Observe that $C_s(\psi)$ contains some but not necessarily *all* subformulae of ψ that are valid in s. In particular, there is no rule saying that if $G_{\rhd r}\, \varphi \in K$, then $\varphi \in K$. As we shall see, the subformulae within the scope of $G_{\rhd r}$ operator need special treatment.

Example 1. For the formula ψ and the state s of our running example, we obtain

$$C_s(\psi) \;=\; \{\psi, \quad G_{=1}\left(F_{\geq 0.5}(a \wedge F_{\geq 0.2}\, \neg a) \vee a\right), \quad F_{=1}\, G_{=1}\, a, \quad \neg a\}$$

Observe that although $F_{=1}\, G_{=1}\, a \in C_s(\psi)$, the formula $G_{=1}\, a$ is not included into $C_s(\psi)$ because $s \not\models G_{=1}\, a$.

The set $C_s(\psi)$ does not give a precise information about the satisfaction of relevant path formulae. Therefore, we allow for "updating" the closure with precise quantities.

Definition 4. *Let X be a set of PCTL formulae and s a state in a Markov chain such that $s \models \varphi$ for every $\varphi \in X$. The update of X in s, denoted by $U_s(X)$, is the set of formulae obtained by replacing every formula of the form $P(\Phi) \rhd r$ in X with the formula $P(\Phi) \geq r'$, where $r' = \mathbb{P}_s(\{\pi \in Run(s) \mid \pi \models \Phi\})$.*

Observe that $r' \geq r$, and the formulae of X which are *not* of the form $P(\Phi) \rhd r$ are left unchanged by U_s. The UC_s operator is defined by $UC_s(X) = U_s(C_s(X))$. Observe that UC_s is idempotent, i.e., $UC_s(UC_s(X)) = UC_s(X)$.

Example 2. In our running example, we have that

$$UC_s(\psi) \;=\; \{\psi, \quad G_{=1}\left(F_{\geq 0.5}(a \wedge F_{\geq 0.2}\, \neg a) \vee a\right), \quad F_{=1}\, G_{=1}\, a, \quad \neg a\}$$

because the probability constraint r in the two formulae of the form $P(\Phi) \rhd r$ is equal to 1 and cannot be enlarged.

In our next definition, we introduce a sufficient condition under which the finite satisfiability problem is decidable in a PCTL fragment $\mathcal{L}$.

Definition 5. *We say that a PCTL fragment $\mathcal{L}$ is progressive if for every finite set X of PCTL formulae and every state s of a finite Markov chain such that*

- $s \models X$,
- X is closed and updated (i.e., $X = UC_s(X)$),
- $X \subseteq \mathcal{L}$

there exists a progress loop, i.e., a finite sequence $\mathscr{L} = L_0, \ldots, L_n$ of subsets of $sub(X)$ satisfying the following conditions:

(1) $X \subseteq L_i$ for some $i \in \{0, \ldots, n\}$;
(2) $L_0, \ldots, L_n$ are pairwise different (this induces an upper bound on n);
(3) for every $i \in \{0, \ldots, n\}$, we have that
- if $a \in L_i$, then $\neg a \notin L_i$;
- if $\varphi_1 \wedge \varphi_2 \in L_i$, then $\varphi_1, \varphi_2 \in L_i$;
- if $\varphi_1 \vee \varphi_2 \in L_i$, then $\varphi_1 \in L_i$ or $\varphi_2 \in L_i$;
- if $G_{\triangleright r} \varphi \in L_i$, then $\varphi \in L_j$ for every $j \in \{0, \ldots, n\}$.

Furthermore, let $\Delta(\mathscr{L})$ be the set of all $\varphi \in L_0 \cup \cdots \cup L_n$ such that one of the following conditions holds:

- $\varphi \equiv G_{\triangleright r} \psi$;
- $\varphi \equiv F_{\triangleright r} \psi$ and $\psi \notin L_0 \cup \cdots \cup L_n$;
- $\varphi \equiv F_{=1} \psi$ and $F_{=1} \psi \in L_i$ for some i such that $\psi \notin L_i \cup \cdots \cup L_n$.

We require that

(4) $s \models \Delta(\mathscr{L})$;
(5) $s \not\models \psi$ for every formula of form $F_{\triangleright r} \psi$ such that $F_{\triangleright r} \psi \in \Delta(\mathscr{L})$;
(6) $deg_s(\Delta(\mathscr{L})) \subset deg_s(X)$ or $cf_s(\Delta(\mathscr{L})) \subseteq cf_s(X)$.
  Here, the set $deg_s(Y)$ consists of formulae $G \varphi$ such that $sub(Y)$ contains a formula of the form $G_{\triangleright r} \varphi$ and $s \not\models G_{=1} \varphi$. The set $cf_s(Y)$ consists of formulae $F \varphi$ such that Y contains a formula of the form $F_{\triangleright r} \varphi$, $s \not\models \varphi$, and there is a finite path from s to a state t where $t \models \varphi$ and $deg_t(Y) = deg_s(Y)$.

Example 3. In our running example, consider $X = UC_s(\psi)$. Then L_0, L_1, where

$$L_0 = \{\psi, \; G_{=1}\big(F_{\geq 0.5}(a \wedge F_{\geq 0.2} \neg a) \vee a\big), \; F_{\geq 0.5}(a \wedge F_{\geq 0.2} \neg a) \vee a,$$
$$F_{\geq 0.5}(a \wedge F_{\geq 0.2} \neg a), \; F_{=1} G_{=1} a, \; \neg a\}$$
$$L_1 = \{F_{\geq 0.5}(a \wedge F_{\geq 0.2} \neg a) \vee a, \; F_{\geq 0.5}(a \wedge F_{\geq 0.2} \neg a), a \wedge F_{\geq 0.2} \neg a, \; a, \; F_{\geq 0.2} \neg a\}$$

is a progress loop for X and s. Observe that

$$\Delta(\mathscr{L}) = \{G_{=1}\big(F_{\geq 0.5}(a \wedge F_{\geq 0.2} \neg a) \vee a\big), \; F_{=1} G_{=1} a\}.$$

Furthermore, $X \subseteq L_0$ and $\Delta(\mathscr{L}) \subseteq X$.

Intuitively, a progress loop allows to prove the existence of a bounded-size model for a finite-satisfiable formula ψ where $\psi \in \mathcal{L}$. Let us fix some (unspecified) finite Markov chain M and a state s of M such that $s \models \psi$. Initially, we put $X = UC_s(\psi)$. Then, we consider a progress loop $\mathscr{L} = L_0, \ldots, L_n$ for X and s, and construct the graph of Fig. 2. The states $\ell_0, \ldots, \ell_n$ correspond to $L_0, \ldots, L_n$, and, as we shall see, $\ell_i \models L_i$ for every $i \in \{0, \ldots, n\}$ after completing our

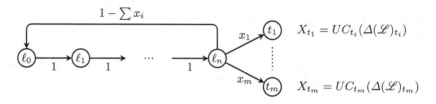

Fig. 2. A graph for a progress loop $L_0, \ldots, L_n$.

construction. Intuitively, the set $\Delta(\mathscr{L})$ contains formulae whose satisfaction is not ensured by the loop itself, and must be "propagated" to the successors of ℓ_n. The probabilities $x_1, \ldots, x_m$ are chosen so that $1 - \sum x_i$ is larger than the maximal $r \neq 1$ appearing in formulae of the form $F_{\triangleright r}\, \varphi \in L_0 \cup \cdots \cup L_n$. This ensures that every ℓ_i visits every ℓ_j with high-enough probability.

The loop is required not to spoil relevant formulae of the form $G_{\triangleright r}\, \psi$ (see the last condition in (3)) and to satisfy almost all of the "new" formulae of the form $F_{\triangleright r}\, \psi$ that do not appear in X and have been added to the loop because of the last condition in (3). In Example 3, such a "new" formula is, e.g., $F_{\geq 0.5}(a \wedge F_{\geq 0.2}\, \neg a)$. Formulae in $\Delta(\mathscr{L})$ must satisfy the technical condition (6) which ensures progress with respect to the measure defined in Sect. 3.1 even in the presence of the "new" formulae added to $\Delta(\mathscr{L})$.

Now we explain how the successors $t_1, \ldots, t_m$ of ℓ_n are constructed, and what is the bound on m. Recall that

$$\Delta(\mathscr{L}) = \big\{ P(\Phi_1) \triangleright r_1, \ \ldots, \ P(\Phi_v) \triangleright r_v, \ \ldots, \ P(\Phi_u) \triangleright r_u \big\}$$

where $\Phi_i \equiv F\, \varphi_i$ for $1 \leq i \leq v$, and $\Phi_i \equiv G\, \varphi_i$ for $v < i \leq u$, respectively. Clearly, u is bounded by the number of subformulae of the considered formula ψ. For every state t of M, let α_t be the u-dimensional vector such that

$$\alpha_t(i) = \mathbb{P}_t\big(\{\pi \in Run(t) \mid \pi \models \Phi_i\}\big).$$

Furthermore, let B be the set of all states t of M such that t either belongs to a BSCC of M or $t \models \varphi_i$ for some $1 \leq i \leq v$. Since M is finite, B is also finite, but there is no upper bound on the size of B. For every $t \in B$, let y_t be the probability of all runs initiated in s visiting the state t so that all states preceding the first visit to t are not contained in B. It is easy to see that

$$\alpha_s \ \leq \ \sum_{t \in B} y_t \cdot \alpha_t \tag{1}$$

Observe that the two vectors of (1) are equal on the first v components. For the remaining components, the inequality can be strict because some of the φ_i formulae, where $i > v$, can become invalid along a path from s before visiting a state of B (the runs initiated by such a path do not satisfy $G\, \varphi_i$).

Since $\sum_{t \in B} y_t = 1$, we can apply Carathéodory's convex hull theorem and thus obtain a subset $B' \subseteq B$ with at most $u + 1$ elements such that $\sum_{t \in B} y_t \cdot \alpha_t$ lies in the convex hull of α_t, $t \in B'$. That is,

$$\alpha_s \;\; \leq \;\; p_1 \cdot \alpha_{t_1} + \cdots + p_m \cdot \alpha_{t_m}$$

where $m \leq u + 1$, $0 < p_i \leq 1$ for all $i \in \{1, \ldots, m\}$, $\sum_{i=1}^{m} p_i = 1$, and $B' = \{t_1, \ldots, t_m\}$. Let $\varrho > 0$ be a constant such that $1 - \varrho > r$ for every $r \neq 1$ appearing in formulae of the form $F_{\rhd r}\, \varphi \in L_0 \cup \cdots \cup L_n$. For every $i \in \{1, \ldots, m\}$, the probability x_i (see Fig. 2) is defined by $x_i = p_i \cdot \varrho$. Furthermore, for every $t_i \in B'$, we construct the set[3]

$$\Delta(\mathscr{L})_{t_i} = \big\{ P(\Phi_1) \geq \alpha_{t_i}(1), \;\; \ldots, \;\; P(\Phi_v) \geq \alpha_{t_i}(v), \;\; \ldots, \;\; P(\Phi_u) \geq \alpha_{t_i}(u) \big\}$$

and then the set $X_{t_i} = UC_{t_i}(\Delta(\mathscr{L})_{t_i})$. We have that $X_{t_i} \subseteq \mathcal{L}$, and X_{t_i} is *smaller* than X with respect to the measure defined in Sect. 3.1. Hence, we proceed by induction, and construct a finite model of bounded size for X_{t_i} by considering a progress loop for X_{t_i} and t_i. Thus, we obtain the following theorem:

Theorem 1. *Let $\mathcal{L}$ be a progressive PCTL fragment. Then every finite-satisfiable formula $\psi \in \mathcal{L}$ has a model with at most $a^{a^{a+5}}$ states where $a = |sub(\psi)|$, such that every non-bottom SCC is a simple loop with one exit state (see Fig. 2). Consequently, the finite satisfiability problem for $\mathcal{L}$ is in* **2-EXPSPACE**.

A full technical proof of Theorem 1 formalizing the above sketch is given in [13].

The **2-EXPSPACE** upper bound is obtained by encoding the bounded satisfiability into existential theory of the reals. This encoding is recalled in [13].

Theorem 1 can be applied to various PCTL fragments by demonstrating their progressivity, and can be interpreted as a "unifying principle" behind these concrete decidability results. To illustrate this, we give examples of progressive fragments in Sect. 3.2.

3.1 Progress Measure

A crucial ingredient of our result is a function measuring the complexity of PCTL formulae. The value of this function, denoted by $\| \cdot \|_s$, is strictly decreased by every progress loop, i.e., $\|X_{t_i}\|_{t_i} < \|X\|_s$ for every X_{t_i}. Now we explain the definition of $\|\dot{X}\|_s$. We start by introducing some auxiliary notions.

Let X be a set of PCTL state formulae. Recall that $sub(X)$ denotes the set all state subformulae of all $\varphi \in X$. The set $psub(X)$ consists of all path formulae Φ such that $sub(X)$ contains a state formula of the form $P(\Phi) \rhd r$.

Let Φ be a path formula of the form $F\, \varphi$ or $G\, \varphi$. The *size* of Φ, denoted by $\|\Phi\|$, is defined as follows:

$$\|\Phi\| \;\; = \;\; 1 + \sum_{\Psi \in psub(\varphi)} \|\Psi\|$$

---

[3] We do not include formulae with the trivial "≥ 0" probability constraint.

Here, the empty sum denotes 0. Note that this definition is correct because the nesting depth of F and G is finite in every path formula, and the above equality makes sense also for $psub(\varphi) = \emptyset$.

Definition 6 (Progress measure $\|\cdot\|_s$). *Let X be a finite set of PCTL formulae and s a state in a Markov chain. We put*

$$\|X\|_s = 1 + |deg_s(X)| \cdot \left(1 + \sum_{\Phi \in psub(X)} \|\Phi\|\right) + \sum_{\Phi \in cf_s(X)} \|\Phi\|$$

The progress measure of Definition 6 appears technical, but it faithfully captures the simplification achieved by a progress loop.

Example 4. Let $X = UC_s(\psi)$ for the ψ and s of our running example, i.e.,

$$X = \{\psi, \quad G_{=1}\left(F_{\geq 0.5}(a \wedge F_{\geq 0.2}\neg a) \vee a\right), \quad F_{=1} G_{=1} a, \quad \neg a\}$$

We have that

– $deg_s(X) = \{G\, a\}$,
– $psub(X) = \{G\left(F_{\geq 0.5}(a \wedge F_{\geq 0.2}\neg a) \vee a\right), F\, G_{=1}\, a\}$,
– $cf_s(X) = \emptyset$.

Since $\|G\left(F_{\geq 0.5}(a \wedge F_{\geq 0.2}\neg a) \vee a\right)\| = 3$ and $\|F\, G_{=1}\, a\| = 2$, we obtain $\|X\|_s = 7$.

3.2 Progressive PCTL Fragments

In this section, we give examples of several progressive PCTL fragments. The constraint $\triangleright r$ has the same meaning as in Definition 1, and $\triangleright w$ stands for an arbitrary constraint except for '=1'.

Fragment $\mathcal{L}_1$

$$\varphi ::= a \mid \neg a \mid \varphi_1 \wedge \varphi_2 \mid \varphi_1 \vee \varphi_2 \mid F_{\triangleright r}\, \varphi \mid G_{\triangleright r}\, \psi$$
$$\psi ::= a \mid \neg a \mid \psi_1 \wedge \psi_2 \mid \psi_1 \vee \psi_2 \mid G_{\triangleright r}\, \psi$$

Fragment $\mathcal{L}_2$

$$\varphi ::= a \mid \neg a \mid \varphi_1 \wedge \varphi_2 \mid \varphi_1 \vee \varphi_2 \mid F_{\triangleright r}\, \varphi \mid G_{=1}\, \psi$$
$$\psi ::= a \mid \neg a \mid \psi_1 \wedge \psi_2 \mid \psi_1 \vee \psi_2 \mid F_{\triangleright w}\, \psi$$

Fragment $\mathcal{L}_3$

$$\varphi ::= a \mid \neg a \mid \varphi_1 \wedge \varphi_2 \mid \varphi_1 \vee \varphi_2 \mid F_{\triangleright r}\, \varphi \mid G_{=1}\, \psi \mid G_{=1}\, \varrho$$
$$\psi ::= a \mid \neg a \mid \psi_1 \wedge \psi_2 \mid \psi_1 \vee \psi_2 \mid F_{\triangleright w}\, \psi$$
$$\varrho ::= \varrho_1 \wedge \varrho_2 \mid \varrho_1 \vee \varrho_2 \mid F_{\triangleright w}\, \psi \mid G_{=1}\, \psi \mid G_{=1}\, \varrho$$

Fragment $\mathcal{L}_4$

$$\varphi \quad ::= \quad a \mid \neg a \mid \varphi_1 \wedge \varphi_2 \mid \varphi_1 \vee \varphi_2 \mid F_{\triangleright r}\,\varphi \mid G_{=1}\,\psi$$
$$\psi \quad ::= \quad a \mid \neg a \mid \psi_1 \wedge \psi_2 \mid \psi_1 \vee \psi_2 \mid F_{>0}\,\psi \mid G_{=1}\,\psi$$

Observe that $\mathcal{L}_2$ and $\mathcal{L}_3$ contain the formula ψ of our running example. The above fragments are chosen so that they are not covered by the results of [22] and illustrate various properties of Definition 5. Fragments $\mathcal{L}_2$, $\mathcal{L}_3$, and $\mathcal{L}_4$ contain formulae requiring non-bottom SCCs with more than one state.

To demonstrate the applicability of Theorem 1, we explicitly show that $\mathcal{L}_2$ is progressive.

Proposition 1. *Fragment $\mathcal{L}_2$ is progressive.*

Proof. Let $X \subseteq \mathcal{L}_2$ be a finite set of formulae and s a state of a finite Markov chain such that $s \models X$ and $X = UC_s(X)$. We show that there exists a progress loop for X and s. To achieve that, we inductively construct a finite sequence $L_0, \ldots, L_n$, where every L_i is associated to some state t_i reachable from s such that $t_i \models L_i$. The set L_0 is the least set M satisfying the following conditions:

- $X \subseteq M$;
- if $\varphi_1 \wedge \varphi_2 \in M$, then $\varphi_1, \varphi_2 \in M$;
- if $\varphi_1 \vee \varphi_2 \in M$ and $s \models \varphi_1$, then $\varphi_1 \in M$;
- if $\varphi_1 \vee \varphi_2 \in M$ and $s \models \varphi_2$, then $\varphi_2 \in M$;
- if $G_{\triangleright r}\,\varphi \in M$, then $\varphi \in M$.
- if $F_{\triangleright r}\,\varphi \in M$ and $s \models \varphi$, then $\varphi \in M$.

We put $t_0 = s$ (observe $s \models L_0$). Furthermore, let N be the set of all formulae ξ such that $G_{=1}\,\xi \in L_0$.

Suppose that $L_0, \ldots, L_n$ are the sets constructed so far where $t_i \models L_i$ for every $i \in \{0, \ldots, n\}$. Now we distinguish two possibilities.

- If for every formula of the form $F_{\triangleright r}\,\xi \in L_0 \cup \ldots \cup L_n$ where $F_{\triangleright r}\,\varphi \notin X$ there exists $i \in \{0, \ldots, n\}$ such that $\xi \in L_i$, then the construction terminates.
- Otherwise, let $F_{\triangleright r}\,\xi \in L_i$ be a formula such that $F_{\triangleright r}\,\xi \notin X$ and $\xi \notin L_0 \cup \ldots \cup L_n$. It follows from the definition of the fragment $\mathcal{L}_2$ that $r \neq 1$. Furthermore, $t_i \not\models \xi$ (this is guaranteed by the closure rules defining L_0 and L_{n+1}, see below). Since $t_i \models F_{\triangleright r}\,\xi$, there exists a state t reachable from t_i (and hence also from s) such that $t \models \xi$. Furthermore, $t \models N$. Now, we construct L_{n+1}, which is the least set M satisfying the following conditions:
  - $\xi \in M$;
  - $N \subseteq M$;
  - if $\varphi_1 \wedge \varphi_2 \in M$, then $\varphi_1, \varphi_2 \in M$;
  - if $\varphi_1 \vee \varphi_2 \in M$ and $t \models \varphi_1$, then $\varphi_1 \in M$;
  - if $\varphi_1 \vee \varphi_2 \in M$ and $t \models \varphi_2$, then $\varphi_2 \in M$;
  - if $F_{\triangleright r}\,\varphi \in M$ and $t \models \varphi$, then $\varphi \in M$.

Observe that if $G_{=1}\,\varphi \in M$, then $G_{\triangleright r}\,\varphi \in N$ because ξ does not contain any subformula of the form $G_{=1}\,\varphi$ (see the definition of $\mathcal{L}_2$). Furthermore, $t \models L_{n+1}$.

Note that if $F_{\triangleright r} \varphi \in \Delta(\mathcal{L})$, then this formula belongs also to X. Now it is not hard to see that the constructed $L_0, \ldots, L_n$ is a progress loop.   □

Let us note that arguments justifying the progressiveness of $\mathcal{L}_1$ are simple, arguments for $\mathcal{L}_3$ are obtained by extending the ones for $\mathcal{L}_2$, and arguments for $\mathcal{L}_4$ already involve the technical condition (6) in Definition 5 in a non-trivial manner.

4   Conclusions

We have shown that the finite satisfiability problem is decidable in doubly exponential space for all PCTL fragments where a progress loop is guaranteed to exist. A natural continuation of our work is to generalize the shape of a progress SCC and the associated progress measure. Natural candidates are loops with several exit states, and SCCs with arbitrary topology but one exit state. Here, increasing the probability of satisfying $F \varphi$ subformulae can be "traded" for decreasing the probability of satisfying $G \varphi$ formulae, and understanding this phenomenon is another important step towards solving the finite satisfiability problem for the whole PCTL.

Let us note that the technique introduced in this paper can also be used to tackle the decidability of *general* satisfiability for PCTL fragments containing formulae that are not finitely satisfiable. By unfolding progress loops into infinite-state Markov chains and arranging the probabilities appropriately, formulae of the form $G \varphi$ can be satisfied with arbitrarily large probability by the progress loop itself, although the loop is still exited with positive probability. Elaborating this idea is another interesting challenge for future work.

Acknowledgement. The work is supported by the Czech Science Foundation, Grant No. 21-24711S.

References

1. Baier, C., Katoen, J.P.: Principles of Model Checking. The MIT Press, Cambridge (2008)
2. Baier, C., Kwiatkowska, M.: Model checking for a probabilistic branching time logic with fairness. Distrib. Comput. **11**(3), 125–155 (1998)
3. Banieqbal, B., Barringer, H.: Temporal logic with fixed points. In: Banieqbal, B., Barringer, H., Pnueli, A. (eds.) Temporal Logic in Specification. LNCS, vol. 398, pp. 62–74. Springer, Heidelberg (1989). https://doi.org/10.1007/3-540-51803-7_22
4. Bertrand, N., Fearnley, J., Schewe, S.: Bounded satisfiability for PCTL. In: Proceedings of CSL 2012. Leibniz International Proceedings in Informatics, vol. 16, pp. 92–106. Schloss Dagstuhl-Leibniz-Zentrum für Informatik (2012)
5. Bianco, A., de Alfaro, L.: Model checking of probabilistic and nondeterministic systems. In: Thiagarajan, P.S. (ed.) FSTTCS 1995. LNCS, vol. 1026, pp. 499–513. Springer, Heidelberg (1995). https://doi.org/10.1007/3-540-60692-0_70
6. Billingsley, P.: Probability and Measure. Wiley, Hoboken (1995)

7. Brázdil, T., Brožek, V., Forejt, V., Kučera, A.: Stochastic games with branching-time winning objectives. In: Proceedings of LICS 2006, pp. 349–358. IEEE Computer Society Press (2006)
8. Brázdil, T., Forejt, V., Křetínský, J., Kučera, A.: The satisfiability problem for probabilistic CTL. In: Proceedings of LICS 2008, pp. 391–402. IEEE Computer Society Press (2008)
9. Brázdil, T., Forejt, V., Křetínský, J., Kučera, A.: The satisfiability problem for probabilistic CTL. Technical report FIMU-RS-2008-03, Faculty of Informatics, Masaryk University (2008)
10. Brázdil, T., Forejt, V., Kučera, A.: Controller synthesis and verification for Markov decision processes with qualitative branching time objectives. In: Aceto, L., Damgård, I., Goldberg, L.A., Halldórsson, M.M., Ingólfsdóttir, A., Walukiewicz, I. (eds.) ICALP 2008. LNCS, vol. 5126, pp. 148–159. Springer, Heidelberg (2008). https://doi.org/10.1007/978-3-540-70583-3_13
11. Brázdil, T., Kučera, A., Stražovský, O.: On the decidability of temporal properties of probabilistic pushdown automata. In: Diekert, V., Durand, B. (eds.) STACS 2005. LNCS, vol. 3404, pp. 145–157. Springer, Heidelberg (2005). https://doi.org/10.1007/978-3-540-31856-9_12
12. Chakraborty, S., Katoen, J.: On the satisfiability of some simple probabilistic logics. In: Proceedings of LICS 2016, pp. 56–65 (2016)
13. Chodil, M., Kučera, A.: The satisfiability problem for a quantitative fragment of PCTL. arXiv:2107.03794 [cs.LO] (2021)
14. Emerson, E.: Temporal and modal logic. In: Handbook of Theoretical Computer Science B, pp. 995–1072 (1991)
15. Emerson, E., Halpern, J.: Decision procedures and expressiveness in the temporal logic of branching time. In: Proceedings of STOC 1982, pp. 169–180. ACM Press (1982)
16. Esparza, J., Kučera, A., Mayr, R.: Model-checking probabilistic pushdown automata. Logical Methods Comput. Sci. **2**(1:2), 1–31 (2006)
17. Etessami, K., Yannakakis, M.: Model checking of recursive probabilistic systems. ACM Trans. Comput. Logic **13** (2012)
18. Fischer, M., Ladner, R.: Propositional dynamic logic of regular programs. J. Comput. Syst. Sci. **18**, 194–211 (1979)
19. Hansson, H., Jonsson, B.: A logic for reasoning about time and reliability. Formal Aspects Comput. **6**, 512–535 (1994)
20. Hart, S., Sharir, M.: Probabilistic temporal logic for finite and bounded models. In: Proceedings of POPL 1984, pp. 1–13. ACM Press (1984)
21. Huth, M., Kwiatkowska, M.: Quantitative analysis and model checking. In: Proceedings of LICS 1997, pp. 111–122. IEEE Computer Society Press (1997)
22. Křetínský, J., Rotar, A.: The satisfiability problem for unbounded fragments of probabilistic CTL. In: Proceedings of CONCUR 2018. Leibniz International Proceedings in Informatics, vol. 118, pp. 32:1–32:16. Schloss Dagstuhl-Leibniz-Zentrum für Informatik (2018)
23. Lehman, D., Shelah, S.: Reasoning with time and chance. Inf. Control **53**, 165–198 (1982)

Beyond the BEST Theorem: Fast Assessment of Eulerian Trails

Alessio Conte[1] , Roberto Grossi[1,2] , Grigorios Loukides[3] ,
Nadia Pisanti[1,2] , Solon P. Pissis[2,4,5] , and Giulia Punzi[1(✉)]

[1] Università di Pisa, Pisa, Italy
{conte,grossi,pisanti}@di.unipi.it, giulia.punzi@phd.unipi.it
[2] ERABLE Team, Lyon, France
[3] King's College London, London, UK
grigorios.loukides@kcl.ac.uk
[4] CWI, Amsterdam, The Netherlands
solon.pissis@cwi.nl
[5] Vrije Universiteit, Amsterdam, The Netherlands

Abstract. Given a directed multigraph $G = (V, E)$, with $|V| = n$ nodes and $|E| = m$ edges, and an integer z, we are asked to assess whether the number $\#ET(G)$ of node-distinct Eulerian trails of G is at least z; two trails are called *node-distinct* if their *node* sequences are different. This problem has been formalized by Bernardini et al. [ALENEX 2020] as it is the core computational problem in several string processing applications. It can be solved in $\mathcal{O}(n^\omega)$ arithmetic operations by applying the well-known BEST theorem, where $\omega < 2.373$ denotes the matrix multiplication exponent. The algorithmic challenge is: *Can we solve this problem faster for certain values of m and z?* Namely, we want to design a combinatorial algorithm for assessing whether $\#ET(G) \geq z$, which does not resort to the BEST theorem and has a predictably bounded cost as a function of m and z. We address this challenge here by providing a combinatorial algorithm requiring $\mathcal{O}(m \cdot \min\{z, \#ET(G)\})$ time.

1 Introduction

Eulerian trails (or Eulerian paths) were introduced by Euler in 1736: Given a multigraph $G = (V, E)$, an Eulerian trail traverses every edge in E exactly once, allowing for revisiting nodes in V. An Eulerian cycle is an Eulerian trail that starts and ends on the same node in V. The perhaps most fundamental algorithmic question related to Eulerian trails is whether we can efficiently identify one of them. Hierholzer's paper ([9], [5, 1B]) can be employed for this purpose to get a linear-time algorithm. A related question is *counting* Eulerian trails: this is $\#P$-complete for undirected graphs [6], while for directed graphs the number of Eulerian trails can be computed in polynomial time using the BEST theorem [1], named after de Bruijn, van Aardenne-Ehrenfest, Smith and Tutte.

Two trails are called *node-distinct* if their *node* sequences are different. Bernardini et al. formalized the following basic problem in [3], which surprisingly

© Springer Nature Switzerland AG 2021
E. Bampis and A. Pagourtzis (Eds.): FCT 2021, LNCS 12867, pp. 162–175, 2021.
https://doi.org/10.1007/978-3-030-86593-1_11

had not been previously posed to the best of our knowledge: Given a directed multigraph $G = (V, E)$, with $|V| = n$ nodes and $|E| = m$ edges, two nodes $s, t \in V$, and a positive integer z, assess whether the number $\#ET(G)$ of node-distinct Eulerian trails of G with source s and target t is at least z. This is the core computational problem in several string processing applications [3,11,12].

This problem can be solved in $\mathcal{O}(n^\omega)$ arithmetic operations as follows [3,11], where $\omega < 2.373$ denotes the matrix multiplication exponent [2,8,16]. (The underlying assumption is that G is Eulerian[1], that is, the indegree equals the outdegree in each node, possibly except for the source and the target of the trail.) Let $A = (a_{uv})$ be the adjacency matrix of G allowing both $a_{uv} > 1$ (multi-edges) and $a_{uu} > 0$ (self-loops). Let $r_u = d^+(u)$ for $u \neq t$, $r_t = d^+(t) + 1$, where $d^+(u)$ denotes the outdegree of u, and the edges are counted with multiplicity. We can apply the BEST theorem using its formulation for directed multigraphs [10]:

$$\#ET(G) = (\det L) \cdot \left(\prod_{u \in V} (r_u - 1)! \right) \cdot \left(\prod_{(u,v) \in E} (a_{uv})! \right)^{-1} \qquad (1)$$

where $L = (l_{uv})$ is the $n \times n$ matrix with $l_{uu} = r_u - a_{uu}$ and $l_{uv} = -a_{uv}$. The original BEST theorem states that the number of Eulerian trails of a (directed) graph can be obtained by multiplying the number of arborescences rooted at any node of the graph (given by $\det L$) by the number of permutations of the edges outgoing from each node ($\prod_{u \in V}(r_u - 1)!$). In the multigraph version given in Eq. (1), the formula is further divided by the number of permutations of multi-edges ($\prod_{(u,v) \in E} a_{uv}!$), in order to only count node-distinct trails.

Beyond the BEST Theorem: We address the following algorithmic challenge [3, Final Remarks]: design a combinatorial algorithm for assessing $\#ET(G) \geq z$, which does not resort to the BEST theorem and has a predictably bounded cost as a function of m and z (as we do not need $\#ET(G)$ to provide an answer).

We first illustrate how this challenge is well-founded *without* matrix multiplication. In Eq. (1), we can single out two factors: the determinant $\det L$ and the ratio of factorials $F = \prod_{u \in V}(r_u - 1)!(\prod_{(u,v) \in E} a_{uv}!)^{-1}$. First, the assessment based on Eq. (1) cannot rely on the assumption that $F \geq z$ (which would imply that $\#ET(G) \geq z$), since we can have $F \ll 1$. As an example, consider a directed multi-cycle with n nodes, $u_1, \ldots, u_n$, each connected to the next (and u_n back to u_1) with k multi-edges: we have only one node-distinct Eulerian trail, so $\#ET(G) = 1$, but there are k^{n-1} arborescences, so $\det L = k^{n-1}$. In particular, we have that $F = \frac{k!(k-1)!^{n-1}}{k!^n} = \frac{1}{k^{n-1}} \ll 1$, for any choice of $s = t$. Second, enumerating arborescences [7,15], progressively bounding $\det L$ to check whether $\det L \geq z/F$, might be costly. This is because the number of arborescences could be exponential in $\#ET(G)$, as in our example. Therefore, in this paper, we follow a fundamentally different approach, which takes $\mathcal{O}(m) = \mathcal{O}(nk)$ time in our example, and can be generalized to more involved graphs.

---

[1] If G is not Eulerian, the answer is trivially negative for any non-zero z.

We remark that exploiting Eq. (1) *with* matrix multiplication could be costly too. Avoiding matrix multiplication makes a difference of several orders of magnitude as these arithmetic operations can be costly. In typical instances in experiments, computing $\det L$ with state-of-the-art (sparse) matrix multiplication libraries and other tricks can still take several hours (see [3] for more details).

Our Results and Techniques: Our main contribution is to introduce an approach that does not merely employ the structure of the BEST theorem, but it actually goes beyond that: we design an efficient algorithm which directly provides an assessment by looking directly at Eulerian trails, without considering the different factors of Eq. (1). We first present a natural (but non-trivial) algorithm to facilitate the reader's comprehension. The main idea consists in providing a lower bound on $\#ET(G)$, based on the product of the lower bounds for the node-distinct trails of its strongly connected components. This lower bound is then progressively refined by considering any arbitrarily chosen component, and its contribution is improved by employing some novel structural properties of strongly connected components of Eulerian graphs. Our method conceptually provides a recursive enumeration approach whose calls enumerate the first z node-distinct Eulerian trails in $\Theta(m^2 \cdot \min\{z, \#ET(G)\})$ time. However, we improve upon that, as our lower-bound driven algorithm does not necessarily perform all the recursive calls to assess whether or not $\#ET(G) \geq z$.

The above algorithm may require quadratic time per Eulerian trail because each call might require $\mathcal{O}(m)$ time. We refine it to bring its complexity down by a double numbering on the edges, which guarantees that every call generates *at least two distinct calls*. This double numbering gives us insight on the interior connectivity structure of the graph; namely, how strongly connected components change when we start removing edges, which is the source of the quadratic time. With this double numbering, we manage to instantly retrieve edges that generate new trails, no longer needing to iterate for $\mathcal{O}(m)$ unsuccessful steps. We thus reduce the time by a factor of m, which gives a time-optimal algorithm for $z = \mathcal{O}(1)$ or for $\#ET(G) = \mathcal{O}(1)$. Our main result is formalized as follows.[2]

Theorem 1. *Given a directed multigraph $G = (V, E)$, with $|E| = m$, and an integer z, assessing $\#ET(G) \geq z$ can be done in $\mathcal{O}(m \cdot \min\{z, \#ET(G)\})$ time.*

Let us remark that our algorithms can potentially run asymptotically faster than the worst-case bounds given above. Under suitable assumptions and values of z, it is possible that they run in less than Constant Amortized Time (CAT) per solution (cf. [13]). In principle, this implies that, under suitable assumptions, assessment might be intrinsically more efficient than counting.

Paper Organization: Sect. 2 introduces the basic definitions and notation used throughout. In Sect. 3, we prove the combinatorial properties of Eulerian graphs which form the basis of our technique. In Sect. 4, we present the simple $\mathcal{O}(m^2 \cdot$

---

[2] We assume throughout that basic arithmetic operations take constant time, which is the case when $z = \mathcal{O}(\text{poly}(m))$.

$\min\{z, \#ET(G)\}$)-time algorithm. This algorithm is then refined to our main result, which is described in Sect. 5.

2   Definitions and Notation

Consider a directed graph $G = (V, E)$ with multi-edges and self-loops, and let $|V| = n$ and $|E| = m$; the edges are counted with multiplicity. A *trail* over G is a sequence of adjacent distinct edges. Two trails are *node-distinct* if their *node* sequences are different. An *Eulerian trail* of G is a trail that traverses every edge exactly once. We consider *node-distinct* Eulerian trails. The set of node-distinct Eulerian trails of G is denoted by $ET(G)$ and its size is denoted by $\#ET(G)$. We may omit the term "node-distinct" when it is clear from its context.

Given a node $u \in V$, we define its *outdegree* (resp. *indegree*) as the number of edges of the form (u, v) (resp. (v, u)), counting multiplicity and self-loops. We then denote by $\Delta(u)$ the difference outdegree(u) − indegree(u). Furthermore, we define the set of *out-neighbors* of u as $N^+(u) = \{v \in V \mid (u, v) \in E\}$. Finally, we use the notation $N_C^+(u) = N^+(u) \cap C$, when referring only to the out-neighbors inside some subgraph C of G.

G is called *strongly connected* if there is a trail in each direction between each pair of the graph nodes. A *strongly connected component* (SCC) of G is a strongly connected subgraph of G. G is called *weakly connected* if replacing all of its edges by undirected edges produces a *connected* graph: it has at least one node and there is a trail between every pair of nodes.

Definition 1. *A directed graph $G = (V, E)$ is Eulerian with source s and target t, where $s, t \in V$, if it is weakly connected and (i) $\Delta(s) = 1$, $\Delta(t) = -1$, and $\Delta(u) = 0$ for all $u \in V \setminus \{s, t\}$; or (ii) $\Delta(u) = 0$ for all $u \in V$. In Case (i), G has an Eulerian trail from s to t. In Case (ii), G has an Eulerian cycle: an Eulerian trail that starts and ends on $s = t$.*

3   Structure and Properties of Directed Eulerian Graphs

The SCCs of a directed Eulerian graph G induce a directed acyclic graph G_{SCC}. Considering this graph, we derive some non-trivial and useful properties, upon which we will heavily rely to design our algorithms for assessing the number of node-distinct Eulerian trails. Let us start with the following crucial lemma whose proof is deferred to the full version of the paper.

Lemma 1. *Let G be an Eulerian graph, with SCCs $C_0, \ldots, C_k$, source $s \in C_0$, and target $t \in C_k$. The corresponding G_{SCC} is a chain graph of the form $C_0 \to C_1 \to \ldots \to C_k$, where the arrow between C_i and C_{i+1} represents a single edge $(t_i, s_{i+1}) \in E$, called bridging edge. Furthermore, each C_i is Eulerian with source s_i and target t_i, where $s_0 = s, t_k = t$.*

It follows from Lemma 1 that every trail from s to t must traverse all edges of $C_0, \ldots, C_i$ before crossing the bridging edge (t_i, s_{i+1}). As a consequence, we obtain the following.

Corollary 1. *Let G be an Eulerian graph with SCCs $C_0, \ldots, C_k$. Then we have that $ET(G) = \prod_{i=0}^{k} ET(C_i)$, where $\prod$ denotes the cartesian product. It follows that the number of trails of G is the product of the number of trails of its SCCs.*

We can thus focus on an individual SCC or, equivalently, assume, wlog, that the Eulerian graph is strongly connected. The following lemma, whose proof is deferred to the full version of the paper, forms the basis of our technique.

Lemma 2. *Let C be a strongly connected Eulerian graph with source s and target t. For every edge (s, u), there is an Eulerian trail of C whose first two traversed nodes are s and u. Moreover, the residual graph $C \setminus (s, u)$ remains Eulerian with new source u.*

Corollary 2. *Let C_i be any SCC of an Eulerian graph with source s_i. Then:*

$$ET(C_i) = \bigcup_{u \in N_{C_i}^+(s_i)} (s_i, u) \cdot ET(C_i \setminus (s_i, u)),$$

that is, the Eulerian trails of C_i are given by concatenating each possible start of the trail (s_i, u) with all its possible continuations, i.e., the trails in $ET(C_i \setminus (s_i, u))$. Thus the number of trails of C_i is the sum of the number of trails of the subgraphs with edges (s_i, u) removed, for every $u \in N_{C_i}^+(s_i)$ distinct out-neighbor of s_i in C_i, with u as the new source.

Proof. This follows from Lemma 2 applied to the SCCs: we know that each distinct out-neighbor of s_i leads to at least one trail; furthermore, no two of these trails can be equal since they begin with distinct edges. Lastly, all trails are accounted for, since we consider every trail starting from every distinct out-neighbor of s_i, and s_i is the source of C_i. $\qquad\square$

Note the subtle point in the statement of Corollary 2, where we use $N_{C_i}^+(s_i)$ instead of $N^+(s_i)$: if the latter two differ, it is because s_i has an outgoing bridging edge, and this should be traversed *after* all other edges in C_i.

4   Assessment Algorithm for #ET(G)

We present ASSESSET, a simple but non-trivial algorithm for assessing the number of node-distinct Eulerian trails on a given directed graph, which will be refined in Sect. 5. ASSESSET takes the following input parameters: (i) a weakly connected Eulerian graph $G = (V, E)$ with source s and target t; (ii) a positive integer threshold z, and (iii) a function $lb(\cdot)$, which outputs a lower bound on the number of the node-distinct Eulerian trails in G. To achieve the desired complexity, $lb(\cdot)$ must be computable in $\mathcal{O}(m)$ time and $lb(\cdot) \geq 1$ must hold.

Proposition 1. *Given graph G, nodes s and t, integer z, and $lb(\cdot)$, ASSESSET assesses $\#ET(G) \geq z$ in $\mathcal{O}(m^2 \cdot \min\{z, \#ET(G)\})$ time using $\mathcal{O}(mz)$ space.*

Main Idea: Let $C_0, \ldots, C_k$ be the set of SCCs of an Eulerian graph G as illustrated in Lemma 1. ASSESSET exploits Corollary 1, Lemma 2, and Corollary 2, to provide a lower bound on the number of node-distinct Eulerian trails of graph G, denoted by $lb_{ET}(G)$, where $lb_{ET}(G) \leq \#ET(G)$. Initially, we set $lb_{ET}(G) = \prod_{i=0}^{k} lb(C_i)$, based on the product of the lower bounds for the number of node-distinct Eulerian trails of the SCCs of G by Corollary 1. Then $lb_{ET}(G)$ is progressively refined by considering any arbitrarily chosen component, say C_i, and in turn replacing its lower bound $lb(C_i)$ with a new lower bound $lb_{ET}(C_i)$ that exploits Lemma 2 and its Corollary 2. That is, we remove each different outgoing edge from the source s_i of C_i, and after computing the $lb(\cdot)$ function on all of the resulting graphs, we sum these lower bounds to obtain $lb_{ET}(C_i)$, and update $lb_{ET}(G)$. We proceed in this way until either $lb_{ET}(G) \geq z$, or we compute the actual number of trails: $lb_{ET}(G) = \#ET(G)$.

The requirements for the lower bound function are trivially satisfied by the constant function $lb(\cdot) \equiv 1$. However, we use a better lower bound given by Lemma 3 below, whose proof is deferred to the full version of the paper.

Lemma 3. *For any Eulerian graph G, the function*

$$lb(G) = 1 + \sum_{v \in V(G):|N_G^+(v)| \geq 3} (|N_G^+(v)| - 2). \tag{2}$$

is a lower bound for the number $\#ET(G)$ of node-distinct Eulerian trails of G.

Function COMPUTESCC: Our algorithm relies on a function COMPUTESCC(G), which computes the SCCs of a given input graph G. This function only outputs the non-trivial components (i.e., comprised of multiple nodes), and it requires $\mathcal{O}(m)$ time to achieve this (specifically, we make use of [14]).

Frontier Data Structure: In order to efficiently explore the different SCCs as discussed above, we introduce the *Frontier Data Structure*, denoted by $\mathcal{F} = \{f_1, \ldots, f_{|\mathcal{F}|}\}$, representing the frontier of the recursive tree we are *implicitly* constructing when traversing a component. At any moment of the computation, an element $f_j \in \mathcal{F}$ is a tuple $\langle C_0^j, \ldots, C_{h_j}^j \rangle$, where $C_0, \ldots, C_{h_j}$ are the non-trivial SCCs of some Eulerian subgraph $G_j \subset G$. A component is considered *trivial* if it is comprised of a single node. A trivial component is omitted because it contributes to the product in Eq. (3) below by a factor of one. Different G_j's are obtained from G by removing different edges that are outgoing from the source of a component, as per Lemma 2; thus G_j differs from any other G_l by at least one removed edge. In this way, each element of the frontier represents at least one node-distinct Eulerian trail of G. Furthermore, our data structure $\mathcal{F}$ retains an important invariant: at any moment, the elements of $\mathcal{F}$ are the SCC decompositions of the subgraphs which realize the current bound. That is,

$$lb_{ET}(G) = \sum_{j=1}^{|\mathcal{F}|} lb_{ET}(f_j) = \sum_{j=1}^{|\mathcal{F}|} \prod_{i=0}^{h_j} lb(C_i^j). \tag{3}$$

Each component $f_j[i] = C_i^j$, with source s_i^j and target t_i^j, is represented in $f \in \mathcal{F}$ as a tuple of the form $(V[C_i^j], E[C_i^j], s_i^j, t_i^j, lb(C_i^j))$. In what follows, we consider $\mathcal{F}$ implemented as a *stack*: both removing and inserting elements requires $\mathcal{O}(1)$ time with POP and PUSH operations. Performing these operations also modifies the size of $\mathcal{F}$, which is accounted for. We can thus answer whether the stack is empty in $\mathcal{O}(1)$ time.

Algorithm ASSESSET. The algorithm maintains a running bound lb_{ET}, induced by the components currently forming the elements of the stack, according to Eq. (3), where lb_{ET} is the current value of $lb_{ET}(G)$. We proceed as follows:

1. Compute the (non-trivial) SCCs of graph G. If there is none, we only have one trail, and $lb_{ET} = 1$. Otherwise, we initialize the stack with the tuple $\langle C_0, \ldots, C_k \rangle$ of these SCCs, and also initialize the bound accordingly setting $lb_{ET} \leftarrow \prod_{j=0}^{k} lb(C_j)$.
2. While $lb_{ET} < z$, we perform the following:
   (a) If the stack is empty, we output NO. Since non-trivial components are never added into the stack, the stack is empty if and only if $lb_{ET} = \#ET(G)$ and $lb_{ET} < z$.
   (b) Otherwise, we pop an element f from the stack, and remove its contribution from the current bound: $lb_{ET} \leftarrow lb_{ET} - lb_{ET}(f)$, where $lb_{ET}(f) = \prod_{i=1}^{|f|} lb(f[i])$.
   (c) We pick an arbitrary component $C_i = f[i]$ of tuple f, and let s_i be its source. We remove the component from f.
   (d) For all distinct out-neighbors $u \in N_{C_i}^+(s_i)$:
      i. We compute the SCCs $\mathcal{C}$ of C_i with edge (s_i, u) removed.
      ii. If f with the added new components $\mathcal{C}$ (i.e. $f \cdot \mathcal{C}$) is non-empty, we add it into the stack and increase the running bound accordingly as $lb_{ET} \leftarrow lb_{ET} + lb_{ET}(f \cdot \mathcal{C})$. If $f \cdot \mathcal{C}$ is empty, it corresponds to a single Eulerian trail, so we increase the bound lb_{ET} by one.
3. If we exit from the while loop in Step 2, then $lb_{ET} \geq z$ and we output YES.

When $lb(\cdot)$ always returns 1, ASSESSET makes $\mathcal{O}(mz)$ calls to compute the SCCs, of $\mathcal{O}(m)$ time each, as it essentially enumerates z Eulerian trails one by one. However, when $lb(\cdot) > 1$, a lot of these calls are avoided as lower bounds are multiplied. The pseudocode of ASSESSET is provided in Algorithm 1.

The correctness of ASSESSET follows from Corollary 1, Lemma 2 and Corollary 2. The analysis of time and space complexity of ASSESSET, which completes the proof of Proposition 1, is deferred to the full version of the paper.

5   Improved Assessment Algorithm

We may think of ASSESSET as a recursive computation (handled explicitly with pop/push on a stack) having the drawback that it makes $\mathcal{O}(mz)$ recursive calls. To try and speed up the process, one could resort to existing decremental SCC

Algorithm 1. (AssessET)

1: **procedure** ASSESSET($G = (V, E)$, $z = \mathcal{O}(\mathrm{poly}(|E|))$, $lb(\cdot)$)
2:     $C_0, \ldots, C_k \leftarrow$ COMPUTESCC(G)   ▷ Only considers non-trivial SCCs
3:     $f \leftarrow \langle C_0, \ldots, C_k \rangle$
4:     **if** f is empty **then** $lb_{ET} \leftarrow 1$
5:     **else** STACK.PUSH(f)   ▷ Initialization
6:         $lb_{ET} \leftarrow \prod_{j=0}^{k} lb(C_j)$
7:     **while** $lb_{ET} < z$ **do**
8:         **if** STACK.ISEMPTY() **then Output NO**
9:         $f \leftarrow$ STACK.POP()
10:        $lb_{ET} \leftarrow lb_{ET} - lb_{ET}(f)$ ▷ $lb_{ET}(f) = \prod_{i=1}^{|f|} lb(f[i])$
11:        Choose any i; let $C_i = f[i]$ and s_i be its source   ▷ $f[i]$ is the i-th SCC of f
12:        Remove C_i from f
13:        **for all** $u \in N_{C_i}^{+}(s_i)$ **do**
14:            $\mathcal{C} \leftarrow$ COMPUTESCC($C_i \setminus (s_i, u)$)
15:            **if** $f \cdot \mathcal{C}$ is not empty **then**   ▷ $f \cdot \mathcal{C}$: f with each SCC of $\mathcal{C}$ appended
16:                STACK.PUSH($f \cdot \mathcal{C}$)
17:                $lb_{ET} \leftarrow lb_{ET} + lb_{ET}(f \cdot \mathcal{C})$
18:            **else** $lb_{ET} \leftarrow lb_{ET} + 1$
19:    **Output YES**

algorithms [4]. However, these tend to add (poly)logarithmic factors, and do not immediately yield improvements unless further amortization is suitably designed.

We use a different approach, reducing the number of calls to $\mathcal{O}(z)$ by guaranteeing that each call generates at least two further calls or immediately halts when one Eulerian trail is found. In this section, we show how to attain this goal with an efficient combinatorial procedure.

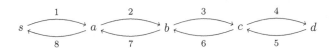

Fig. 1. This graph has a single node-distinct Eulerian trail from s to itself, even though all nodes except for s and d are branching.

5.1   Introducing Function BranchingSource

Consider the SCC C_i chosen in ASSESSET, and its source s_i. We call a node $u \in C_i$ *branching* if it has at least two distinct out-neighbors in C_i, that is, $|N_{C_i}^{+}(u)| \geq 2$. Thus, if s_i is branching, we have at least two calls by Lemma 2. The issue comes when s_i has just one out-neighbor, as illustrated in Fig. 1: some of the remaining nodes could be branching but, unfortunately, only one node-distinct Eulerian trail exists. Thus, the existence of branching nodes when the source s_i is not branching does not guarantee that we attain our goal.

One first solution comes to mind, as it is exploited in our lower bound $lb(\cdot)$ of Eq. 2. Consider a trail $T \in ET(C_i)$, which is nonempty as C_i is Eulerian: a node u gives rise to at least $|N_{C_i}^+(u)| - 2$ further Eulerian trails by Lemma 2 as, when u becomes a source for the first time, one out-neighbor of u is part of T and at most one out-neighbor of u leads to a bridging edge; thus the remaining $|N_{C_i}^+(u)| - 2$ out-neighbors can be traversed in any order by so many other Eulerian trails. While this helps for $|N_{C_i}^+(u)| \geq 3$, it is not so useful in the situation illustrated in Fig. 1, where all branching nodes have $|N_{C_i}^+(u)| = 2$.

Main Idea: A better solution is obtained by introducing a function BRANCHING-SOURCE to be applied to any tuple f of SCCs from the frontier data structure $\mathcal{F}$. If any of these SCCs has a branching source, then BRANCHINGSOURCE returns f itself. Otherwise, it examines each SCC C in f: if $\#ET(C) = 1$, it removes C from f as it is trivial; otherwise, it finds the longest common prefix P of all trails in $ET(C)$, and computes the SCCs of $C \setminus P$, which take the place of C in f. Among these SCCs, one is guaranteed to have a branching source, so BRANCHINGSOURCE returns f updated in this way. Note that only trivial SCCs are removed by BRANCHINGSOURCE, and hence the number of Eulerian trails cannot change. If f is empty, then we have a single Eulerian trail as there is no choice. BRANCHINGSOURCE can be implemented in $\mathcal{O}(m^2)$ time as it simulates what ASSESSET does until a branching source is found. The challenge is to implement it in $\mathcal{O}(m)$ time. Armed with that, we can modify ASSESSET and get IMPROVEDASSESSET, where we guarantee in $\mathcal{O}(m)$ time that *the source s_i is always branching.* The modification is just a few lines, once BRANCHINGSOURCE is available, so we do not provide a detailed description of the pseudocode.

5.2 Linear-Time Computation of BranchingSource

Suppose that tuple f in the frontier data structure $\mathcal{F}$ contains only SCCs with non-branching sources (otherwise, BRANCHINGSOURCE returns f unchanged). Consider any SCC C in the tuple f. The main idea is to fix any trail $T \in ET(C)$, which can be found in $\mathcal{O}(|E(C)|)$ time, and traverse T asking at each node u whether there is an alternative trail T' branching at u.

Swap Edges: Let us start with the following definition.

Definition 2. *Given an Eulerian trail T of an SCC C, let T_u be the prefix of T from its source s to the first time u is met, and let (u, v) be the next edge traversed by T. An edge (u, v') in C is a* swap edge *if $T_u \cdot (u, v')$ is prefix of another Eulerian trail $T' \neq T$ and $v' \neq v$. We say that u admits a swap edge and $T_u = T'_u$ is the longest common prefix of T and T'.*

The discovery of swap edges in C is key to BRANCHINGSOURCE: although different Eulerian trails of C may give rise to different swap edges in C, these trails all share P, so the node u at the end of P can be identified by T_u (Definition 2), for *any* trail $T \in ET(C)$. The proof of the following lemma is deferred to the full version of the paper.

Lemma 4. *Suppose that all swap edges are known in an SCC C of f for any given trail $T \in ET(C)$. Then (i) $\#ET(C) = 1$ if and only if there are no swap edges in C; moreover, (ii) if $\#ET(C) > 1$, let u be the first node that is met traversing T and that admits a swap edge. Then $P = T_u$ is the longest common prefix of all the trails in $ET(C)$.*

Using Swap Edges in BranchingSource: Based on Lemma 4, BRANCHINGSOURCE examines each $C \in f$: it tests whether C is trivial ($\#ET(C) = 1$), or it finds the longest common prefix $P = T_u$ of all the trails. If all SCCs are trivial, it returns an empty f. Otherwise, it deletes from f the trivial SCCs found so far, and for the current non-trivial SCC C, it computes the set $\mathcal{C} = \text{COMPUTESCC}(C \setminus T_u)$ of SCCs. Note that u becomes the source of an SCC in $\mathcal{C}$ and u is branching as it admits a swap edge (u, v'), along with (u, v) from its trail T. Thus, u keeps at least two out-neighbors v and v' in $\mathcal{C}$. At this point, BRANCHINGSOURCE stops its computation, updates f by replacing C with the SCCs from $\mathcal{C}$, and returns f. Since only trivial SCCs are removed from f, and the number of Eulerian trails in C is the product of those in the SCCs of $\mathcal{C}$, the overall number of Eulerian trails in f does not change before and after its update. This proves the following.

Lemma 5. *Given any tuple f in $\mathcal{F}$ and the set of swap edges in the SCCs of f, the function BRANCHINGSOURCE takes $\mathcal{O}(m)$ time to update f, so that either f is empty (a single Eulerian trail exists in f), or f contains at least one SCC with branching source.*

Remark 1. Since every swap edge generates *at least one new Eulerian trail*, if we can find all swap edges in $\mathcal{O}(m)$ time, we can employ $lb(G) = 1+$ "number of swap edges" in our algorithm. Any node with three different out-neighbors generates at least a swap edge. Thus, this new choice for the lower bound function necessarily performs better than the one shown in Eq. (2). For example, in Fig. 2, there are 5 swap edges whereas $lb(\cdot) = 1$.

Finding Swap Edges in Linear Time: We are thus interested in finding all the swap edges in linear time. We need the following property, whose proof is deferred to the full version of the paper, to characterize them for an SCC C of f.

Lemma 6. *Let C be an SCC, and let T with prefix $T_u \cdot (u, v)$ be one of its Eulerian trails. Edge (u, v'), for $v' \neq v$, is a swap edge if and only if there is a trail from v' to u (i.e., u is reachable from v') in $C \setminus T_u$.*

In order to find the swap edges, we need to traverse T in reverse order and assign each edge $e \in E(C)$ two integers, as illustrated in the example of Fig. 2: (i) the *Eulerian trail numbering* $etn(e)$, which represents the position of e inside T and is immediate to compute, and (ii) the *disconnecting index* $di(e)$, which is discussed in the next paragraph as its computation is a bit more involved. As we will see (Lemma 7), comparing these integers allows us to check if a given edge is a swap edge in constant time.

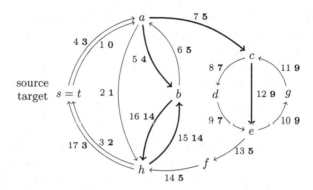

Fig. 2. Example of an Eulerian graph with $s = t$. The black (left) numbers on the edges are the Eulerian trail numbers *etn* for a given trail T; the orange (right) ones are the disconnecting indices *dis*. Swap edges are in bold.

Disconnecting Indices: We introduce the notion of disconnecting index relatively to a given trail $T \in ET(C)$, according to the following rationale. We observe that Lemma 5 characterizes a swap edge (u, v') by stating that u and v' must belong to the same SCC after T_u is removed from C. Suppose that we want to traverse T to discover the swap edges. Equivalently, we take the edges according to their *etn* order in T. Fix any edge (u, v'). At the beginning, u and v' are in the same SCC C. Next, we start to conceptually remove, from C, the edges traversed by an increasingly long prefix of T: how long will u and v' stay in the same SCC? In this scenario, the disconnecting index of (u, v') corresponds to the maximum *etn* (hence prefix of T) for which u and v' will stay in the same SCC, i.e., removing any prefix of T longer than this one from C disconnects v' from u.

For any $\ell \in [0, m]$, we denote by $T_{\leq \ell}$ the prefix T_u of T such that $|T_u| = \ell$. When $\ell = 0$, it is the empty prefix; when $\ell = m$, it is T itself.

Definition 3. *Given an edge* $(u, v') \in E(C)$*, its* disconnecting index *is*

$$di(u, v') = \max \left\{ 0 \leq \ell < etn(u, v') \mid u, v' \text{ are inside an SCC of } C \setminus T_{\leq \ell} \right\}.$$

Figure 2 illustrates an example where the following property, whose proof is deferred to the full version of the paper, can be checked by inspection.

Lemma 7. *For any edge* $(u, v') \in E(C)$*, we have that* (u, v') *is a swap edge for a given trail if and only if* $di(u, v') \geq etn(u, v) - 1$ *for some* $v \neq v'$*.*

Linear-Time Computation of Disconnecting Indices: Consider an SCC C from $f \in \mathcal{F}$, and any arbitrary trail $T \in ET(C)$ (computable in $\mathcal{O}(|E(C)|)$ time). Assign the Eulerian trail numbering $etn(e)$ to each edge $e \in E(C)$. We discuss how to assign the disconnecting index $di(e)$ to each edge e in $\mathcal{O}(|E(C)|)$ time.

We proceed by reconstructing T backwards. That is, we conceptually start from an empty graph, and we add edges from T, one at a time from last to first (i.e., in decreasing order of their *etn* values), until all edges from T are added back obtaining again the SCC C. During this task, along with disconnecting indices, we also assign a *flag* $tr(u) = true$ to the nodes u touched by the edges that have been added. We keep a stack, BRIDGES, for the edges that have been added but do not yet have a disconnecting index, i.e., they are not in an SCC of the current partial graph. More formally, we will guarantee the following invariants:

I1 The edges in BRIDGES have increasing *etn* values, starting from the top.
I2 The edges in BRIDGES are all and only the bridging edges of the current graph.
I3 Given any two consecutive edges e, e' in BRIDGES, the edges with *etn* values in $[etn(e) + 1, etn(e') - 1]$ (which, observe, are not in BRIDGES) make up an SCC of the current graph.
I4 The flag $tr(u)$ is *true* if and only if u is incident to an edge of the current graph.

We describe the algorithm, prove its correctness and all invariants.

For $\ell = m, m - 1, \ldots, 1$, step ℓ adds back to the current graph the edge (u, v) such that $etn(u, v) = \ell$. Let u be the tail and v be the head of the edge.

- If the tail u has not been explored yet (i.e., $tr(u) = false$), we add (u, v) to BRIDGES and set $tr(u) = true$. If $\ell = m$, then v is the last node of the trail and we also set $tr(v) = true$.
- Otherwise, u has been traversed before, and there must be at least an edge incoming in u in our current graph; let (z, u) be the one such edge with highest *etn* value, say, $etn(z, u) = x$. We assign $di(u, v) = etn(u, v) - 1$, and pop all edges e from BRIDGES such that $etn(e) \leq x$, assigning $di(e) = etn(u, v) - 1$ to all of these too.

Lemma 8. *Given an SCC C from f in $\mathcal{F}$, and any arbitrary trail $T \in ET(C)$, the disconnecting indices of T can be computed in $\mathcal{O}(|E(C)|)$ time and space.*

The proof of Lemma 8 is deferred to the full version of the paper. We thus arrive at our main result.

Theorem 1. *Given a directed multigraph $G = (V, E)$, with $|E| = m$, and an integer z, assessing $\#ET(G) \geq z$ can be done in $\mathcal{O}(m \cdot \min\{z, \#ET(G)\})$ time.*

Other than being easily extensible to the *edge-distinct* case, the algorithm underlying Theorem 1 has an attractive property: its number of $\mathcal{O}(m)$-time steps is z in the worst case, but can be significantly smaller in practice, thanks to suitable lower bounding techniques. This property means that our assessment algorithm can potentially run in less than CAT per solution on favorable instances.

Acknowledgments. We wish to thank Luca Versari for useful discussions on a previous version of the main algorithm. This paper is part of the PANGAIA project that has received funding from the European Union's Horizon 2020 research and innovation programme under the Marie Skłodowska-Curie grant agreement No 872539. This paper is also part of the ALPACA project that has received funding from the European Union's Horizon 2020 research and innovation programme under the Marie Skłodowska-Curie grant agreement No 956229.

References

1. van Aardenne-Ehrenfest, T., de Bruijn, N.G.: Circuits and Trees in Oriented Linear Graphs, pp. 149–163. Birkhäuser Boston, Boston, MA (1987). https://doi.org/10. 1007/978-0-8176-4842-8_12
2. Alman, J., Williams, V.V.: A refined laser method and faster matrix multiplication. In: Marx, D. (ed.) Proceedings of the 2021 ACM-SIAM Symposium on Discrete Algorithms, SODA 2021, Virtual Conference, 10–13 January 2021, pp. 522–539. SIAM (2021). https://doi.org/10.1137/1.9781611976465.32
3. Bernardini, G., Chen, H., Fici, G., Loukides, G., Pissis, S.P.: Reverse-safe data structures for text indexing. In: Blelloch, G.E., Finocchi, I. (eds.) Proceedings of the Symposium on Algorithm Engineering and Experiments, ALENEX 2020, Salt Lake City, UT, USA, 5–6 January 2020. pp. 199–213. SIAM (2020). https://doi. org/10.1137/1.9781611976007.16
4. Bernstein, A., Probst, M., Wulff-Nilsen, C.: Decremental strongly-connected components and single-source reachability in near-linear time. In: Proceedings of the 51st Annual ACM SIGACT Symposium on Theory of Computing. p. 365–376. STOC 2019, Association for Computing Machinery, New York, NY, USA (2019). https://doi.org/10.1145/3313276.3316335
5. Biggs, N.L., Lloyd, E.K., Wilson, R.J.: Graph Theory 1736–1936. Clarendon Press (1976)
6. Brightwell, G.R., Winkler, P.: Counting Eulerian circuits is #P-complete. In: Demetrescu, C., Sedgewick, R., Tamassia, R. (eds.) Proceedings of the Seventh Workshop on Algorithm Engineering and Experiments and the Second Workshop on Analytic Algorithmics and Combinatorics, ALENEX/ANALCO 2005, Vancouver, BC, Canada, 22 January 2005, pp. 259–262. SIAM (2005), http://www.siam. org/meetings/analco05/papers/09grbrightwell.pdf
7. Gabow, H.N., Myers, E.W.: Finding all spanning trees of directed and undirected graphs. SIAM J. Comput. **7**(3), 280–287 (1978)
8. Gall, F.L.: Powers of tensors and fast matrix multiplication. In: Nabeshima, K., Nagasaka, K., Winkler, F., Szántó, Á. (eds.) International Symposium on Symbolic and Algebraic Computation, ISSAC 2014, Kobe, Japan, July 23–25, 2014, pp. 296–303. ACM (2014). https://doi.org/10.1145/2608628.2608664
9. Hierholzer, C., Wiener, C.: Über die möglichkeit, einen linienzug ohne wiederholung und ohne unterbrechung zu umfahren. Math. Ann. **6**(1), 30–32 (1873)
10. Hutchinson, J.P., Wilf, H.S.: On Eulerian circuits and words with prescribed adjacency patterns. J. Combin. Theor. Ser. A **18**(1), 80–87 (1975)
11. Kingsford, C., Schatz, M.C., Pop, M.: Assembly complexity of prokaryotic genomes using short reads. BMC Bioinform. **11**, 21 (2010). https://doi.org/10.1186/1471-2105-11-21
12. Patro, R., Mount, S.M., Kingsford, C.: Sailfish: alignment-free isoform quantification from RNA-seq reads using lightweight algorithms. Nature Biotechnol. **32**, 462–464 (2014). https://doi.org/10.1038/nbt.2862, https://www.nature.com/articles/nbt.2862
13. Ruskey, F.: Combinatorial generation. Preliminary working draft. University of Victoria, Victoria, BC, Canada 11, 20 (2003)

14. Tarjan, R.E.: Depth-first search and linear graph algorithms. SIAM J. Comput. **1**(2), 146–160 (1972). https://doi.org/10.1137/0201010, https://doi.org/10.1137/0201010

15. Uno, Takeaki: A new approach for speeding up enumeration algorithms and its application for matroid bases. In: Asano, Takano, Imai, Hideki, Lee, D.. T.., Nakano, Shin-ichi, Tokuyama, Takeshi (eds.) COCOON 1999. LNCS, vol. 1627, pp. 349–359. Springer, Heidelberg (1999). https://doi.org/10.1007/3-540-48686-0_35

16. Williams, V.V.: Multiplying matrices faster than coppersmith-winograd. In: Karloff, H.J., Pitassi, T. (eds.) Proceedings of the 44th Symposium on Theory of Computing Conference, STOC 2012, New York, NY, USA, 19–22 May 2012, pp. 887–898. ACM (2012). https://doi.org/10.1145/2213977.2214056

Linear-Time Minimal Cograph Editing

Christophe Crespelle$^{(\boxtimes)}$

University of Lyon, UCB Lyon 1, ENS de Lyon, CNRS, Inria, LIP UMR 5668,
15 Parvis René Descartes, 69342 Lyon, France
`christophe.crespelle@univ-lyon1.fr`

Abstract. We present an algorithm for computing a minimal editing of an arbitrary graph G into a cograph, i.e. a set of edits (additions and deletions of edges) that turns G into a cograph and that is minimal for inclusion. Our algorithm runs in linear time in the size of the input graph, that is $O(n + m)$ time where n and m are the number of vertices and the number of edges of G, respectively.

1 Introduction

We consider the problem of *editing* an arbitrary graph into a *cograph*, i.e. a graph with no induced path on 4 vertices. This lies within the general framework of *graph modification problems*, in which one wants to perform elementary modifications to an input graph, typically adding and removing edges and vertices, in order to obtain a graph belonging to a given target class of graphs, which satisfies some additional property compared to the input. Ideally, one would like to do so by performing a minimum number of elementary modifications. This is a fundamental problem in graph algorithms, which answers the question to know how far is a given graph from satisfying a target property.

Here, we consider the *edge modification problem* called *editing*, where two operations are allowed: adding an edge and deleting an edge. In other words, given a graph $G = (V, E)$, we want to find a set $M \subseteq \{\{x, y\} \mid x, y \in V\}$ of pairs of vertices, called *edits*, such that the edited graph $H = (V, E \Delta M)$ belongs to the target class. In this case, the quantity to be minimised, called the *cost* of the editing, is the number $|M|$ of adjacencies that are modified, i.e. the number of edges that are added plus the number of edges that are deleted. There exist two other edge modification problems, called *completion* and *deletion*, which are particular cases of editing where only addition of edges or only deletion of edges is allowed, respectively. Edge modification problems are essential in algorithmic graph theory, where they are closely related to some important graph parameters, such as treewidth [1]. They also naturally appear in many problems arising in computer science [3,23], molecular biology [6] and genomics, where they played a key role in the mapping of the human genome [14,22]. Recently,

This project has received funding from the European Union's Horizon 2020 research and innovation programme under the Marie Sklodowska-Curie grant agreement No 749022.

© Springer Nature Switzerland AG 2021
E. Bampis and A. Pagourtzis (Eds.): FCT 2021, LNCS 12867, pp. 176–189, 2021.
https://doi.org/10.1007/978-3-030-86593-1_12

edge modification problems into the class of cographs and some of its subclasses has become a powerful approach to solve problems in complex networks analysis, such as inference of phylogenomics [19,20], identification of groups in social networks [21,28] and measures of centrality of nodes in networks [5,31]. For these applications, the need to treat real-world datasets, whose size is often huge and constantly growing, asks for extremely efficient algorithms with regard to the running time that provide solutions of good quality (number of edits).

Unfortunately, finding the minimum number of edits to be performed in an editing problem is NP-hard for most of the target classes of interest (see, e.g., the thesis of Mancini [26] for further discussion and references). Moreover, the classic approaches developped to deal with this difficulty of computation, such as exact exponential algorithms (see e.g. [4]) and parameterised algorithms (see e.g. [2,7]), are of no help in practice to deal with real-world networks of hundreds of thousands of edges or more (up to billions in some cases). The reason is that their complexity exponentially (or super-polynomially) depends on some parameter, either the size of the graph or the number of edits, that is almost always very large in practice, at least some thousands, which forbids to run these algorithms on such data.

As an alternative, the approach based on the inclusion-minimal relaxation of the problem, called *minimal editing*, allows to design very fast polynomial time algorithms that still provide good solutions. Instead of asking for the minimum number of edits, the minimal editing problem only asks for a set of edits which is minimal for inclusion, i.e. which does not contain any proper subset of edits that also results in a graph in the target class. This approach has been extensively used for completion problems, for the class of cographs itself [11,25], as well as for many other graph classes, including chordal graphs [17], interval graphs [13, 29], proper interval graphs [30], split graphs [18], comparability graphs [16] and permutation graphs [12]. The main reason for the success of inclusion-minimal completion is that, for all these classes, it provides a heuristic for minimum-cardinality completion that runs in a very low polynomial complexity, often $O(nm)$ or $O(n^2)$ time.

Rather surprisingly, the inclusion-minimal approach has never been used for editing problems, where both addition and deletion are allowed, which constitutes a big lack for the domain. Indeed, from a practical point of view, the general version of editing has much more interest as the number of edits obtained is usually much lower when both operations are used. There even exist examples (see Fig. 1) where n edits are enough to turn the graph into a cograph while $\Omega(n^2)$ modifications are required both for pure deletion and for pure completion. Moreover, we show in this paper that dealing with both operations may be beneficial not only for the number of edits output, but also for the time complexity of the algorithm.

Related Work. Edge modification problems into the class of cographs have already received a great amount of attention, including in the parameterized complexity framework [15,24,27]. Concerning the inclusion-minimal approach, unlike the minimal editing problem, minimal cograph completion has already

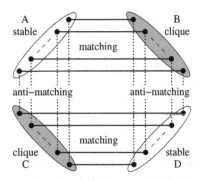

Fig. 1. Graph G_n on $n = 4k$ vertices, with $k \geq 1$. The edges between A and B and between C and D induce matchings while the edges between A and C and between B and D induce anti-matchings (i.e. the complement of a matching). There are no edges between A and D and between B and C. Any cograph deletion (resp. completion) of G_n requires $\Omega(n^2)$ deletions (resp. additions) of edges, but n edits are enough to turn G_n into a cograph.

been studied. [25] designed an incremental algorithm that gives an inclusion-minimal cograph completion in time $O(n+m')$, where n is the number of vertices and m' the number of edges in the output cograph. Later, [11] improved the running time to $O(n + m \log^2 n)$ for the inputs where the number of edges m is small and m' is large. They also show that within the $O(n+m')$ time complexity, it is possible to determine the minimum number of edges to be added at each step of the incremental algorithm.

[21] and [19] designed heuristics, for cograph deletion and cograph editing respectively, that are not intended to output a minimal set of modifications. In the worst case, the algorithm of [21] runs in time at least $O(m^2)$ and the algorithm of [19] in time greater than $O(n^2)$. Finally, let us mention that for the subclass of quasi-threshold graphs, there exist two heuristics dedicated to the editing problem. The one in [28] runs in cubic time in the worst case while [5] obtains a complexity which is often close to linear in practice, but which remains quadratic in the worst case.

Our Results. We design an algorithm that computes a minimal cograph editing of an arbitrary graph in linear time in the size of the input graph, i.e. $O(n+m)$ time. This is the first algorithm for a cograph edge modification problem that runs in linear time. Even compared to the $O(n + m')$-time algorithm of [25] for the pure completion problem, the $O(n + m)$ complexity we obtain here for the editing problem is a significant improvement since, as shown in [11], many instances having only $m = O(n)$ edges actually require $m' = \Omega(n^2)$ edges in any of their completions. Note that, as a particular case, our minimal cograph editing algorithm solves the cograph recognition problem in linear time, and the technique we use can be seen as an extension of the seminal work of [9]. Like the algorithms in [11,25], our algorithm is incremental on the vertices and as the one in [11], it is able to provide a minimum number of edits to be performed at each incremental step and is even able to list all the editings having this minimum cardinality, within the same complexity. This implies that the editing output at the end of the algorithm never contains more than m edits.

2 Preliminaries

All graphs considered here are finite and simple. In the following, G is a graph, V (or $V(G)$) is its vertex set and E (or $E(G)$) is its edge set. We use the notation $G = (V, E)$ and n stands for the cardinality $|V|$ of V and m for $|E|$. An edge between vertices x and y will be arbitrarily denoted by xy or yx. The (open) neighbourhood of x is denoted by $N(x)$ (or $N_G(x)$). The subgraph of G induced by the subset of vertices $X \subseteq V$ is denoted by $G[X] = (X, \{xy \in E \mid x, y \in X\})$.

For a rooted tree T, we employ the usual terminology for *children, parent, ancestors* and *descendants* of a node u in T (the two later notions including u itself). We denote by $\mathcal{C}(u)$ the set of children of u and by $parent(u)$ its parent. The subtree of T rooted at u, denoted T_u, is the tree induced by the descendants of node u in T (which include u itself). We denote $lca(u, v)$ the lowest common ancestor of nodes u and v in T.

Cographs. One of the characterisations of the class of *cographs* (see [8] for a more detailed introduction to the class) is that they are the graphs obtained from a single vertex under the closure of the parallel composition and the series composition. The parallel composition of two graphs $G_1 = (V_1, E_1)$ and $G_2 = (V_2, E_2)$ is the disjoint union of G_1 and G_2, i.e., the graph $G_{par} = (V_1 \cup V_2, E_1 \cup E_2)$. The series composition of G_1 and G_2 is the disjoint union of G_1 and G_2 plus all possible edges from a vertex of G_1 to one of G_2, i.e., the graph $G_{ser}(V_1 \cup V_2, E_1 \cup E_2 \cup \{xy \mid x \in V_1, y \in V_2\})$. These operations are naturally extended to a finite number of graphs.

This gives a nice representation of a cograph G by a tree whose leaves are the vertices of G and whose internal nodes (non-leaf nodes) are labelled P, for parallel, or S, for series, corresponding to the operations used in the construction of G. It is always possible to find such a labelled tree T representing G such that every internal node has at least two children, no two parallel nodes are adjacent in T and no two series nodes are adjacent. This tree T is unique [8] and is called the *cotree* of G. Note that the subtree T_u rooted at some node u of cotree T also defines a cograph whose vertex set, denoted $V(u)$, is the set of leaves of T_u. The adjacencies between the vertices of a cograph can easily be read off its cotree, in the following way.

Remark 1. Two vertices x and y of a cograph G having cotree T are adjacent iff the lowest common ancestor u of leaves x and y in T is a series node. Otherwise, if u is a parallel node, x and y are not adjacent.

Let us emphasize that the class of cographs is *hereditary*, i.e., an induced subgraph of a cograph is also a cograph (since the class also admits a definition by forbidden induced subgraphs).

Incremental Minimal Cograph Editing. Our approach for computing a minimal cograph editing of an arbitrary graph G is incremental, in the sense that we take the vertices of G one by one, in an arbitrary order $(x_1, \ldots, x_n)$, and at step i we compute a minimal cograph editing H_i of $G_i = G[\{x_1, \ldots, x_i\}]$ from a minimal cograph editing H_{i-1} of G_{i-1}, by modifying only adjacencies involving x_i. This is possible thanks to the following observation that is general to all hereditary graph classes that are stable under the addition of universal vertices and isolated vertices, like cographs.

Lemma 1. (see e.g. [29]). *Let G be an arbitrary graph and let H be a minimal cograph editing (resp. completion or deletion) of G. Consider a new graph $G' = G + x$, obtained by adding to G a new vertex x adjacent to an arbitrary set $N(x)$ of vertices of G. There is a minimal cograph editing (resp. completion or deletion) H' of G' such that $H' - x = H$.*

The new problem. Thanks to Lemma 1 above, in all the rest of this article, we consider the following problem and use the following (slightly modified) notations: G is a cograph, $G + x$ is the graph obtained by adding to G a new vertex x adjacent to some set $N(x)$ of vertices of G and our goal is to compute a minimal cograph editing H of $G + x$ such that $H - x = G$. We sometimes write $G + (x, N(x))$ instead of $G + x$, when we want to make the neighbourhood of x in $G + x$ explicit, and we denote $d = |N(x)|$.

Definition 1 (Full, hollow, mixed). *A subset $S \subseteq V(G)$ is full if $S \subseteq N(x)$, hollow if $S \cap N(x) = \varnothing$ and mixed if S is neither full nor hollow. We use the same vocabulary for nodes u of T, referring to their associated set of vertices $V(u)$.*

From [9,10], we have the following characterisation of the case where the insertion of x in cograph G yields a cograph $G + x$, in the case where the root of T is mixed. Note that if the root of T is not mixed, then x is a universal vertex or an isolated vertex in $G + x$ and $G + x$ is necessarily a cograph (because cographs are closed under both the addition of a universal vertex and the addition of an isolated vertex).

Theorem 1. (Reformulated from [9,10]). *If the root of T is mixed, then $G + x$ is a cograph iff there exists a mixed node u of T such that:*

1. all the children of u are full or hollow and
2. for all vertices $y \in V(G) \setminus V(u)$, $y \in N(x)$ iff $lca(y, u)$ is a series node.

Moreover, when such a node u exists, it is unique and it is called the insertion node.

In addition, when $G + x$ is a cograph, its cotree can be obtained from the cotree T of G by inserting x as a grand child of the insertion node u in T, see [9,10] for a full description of the modifications of the cotree under the insertion of x.

3 Characterisation of Minimal Cograph Editings of G + x

In this section, we build upon Theorem 1 to get a characterisation of all the minimal cograph editings of $G+x$ (Lemmas 3 and 4 below), extending the work of [11] for pure completion. Note that from Theorem 1, any minimal editing defines a unique insertion node.

Definition 2 (Minimal insertion node). *A node u of T is a* minimal insertion node *iff there exists a minimal editing H of $G+x$ such that u is the insertion node associated to H.*

Definition 3 (Consistent and settled). *An editing H of $G+x$ is* consistent *with a node u of T iff the three following conditions are satisfied:*

1. *H makes x adjacent to all the vertices $y \notin V(u)$ such that $lca(y,u)$ is a series node and non-adjacent to all the vertices $z \notin V(u)$ such that $lca(z,u)$ is a parallel node, and*
2. *all the children of u are full or hollow in H, and*
3. *all the hollow (resp. full) children of u in $G+x$ are hollow (resp. full) in H as well.*

If, in addition, u is mixed in H, we say that H is settled *at u. A* minimum consistent editing *(resp.* minimum settled editing*) is an editing consistent with some node u (resp. settled at u) and having minimum cost among editings consistent with u (resp. settled at u).*

Lemma 2 below states that any minimal cograph editing of $G+x$ is an editing settled at some mixed node of T.

Lemma 2. *If u is a minimal insertion node and H a minimal editing whose insertion node is u, then u is mixed and H is an editing settled at u.*

We will need the following definitions in order to characterise the minimal editings of $G+x$.

Definition 4. (Completion-forced [11] and deletion-forced). *A comple* tion-forced *node u is inductively defined as a node satisfying at least one of the three following conditions:*

1. *u is full, or*
2. *u is a parallel node with all its children non-hollow, or*
3. *u is a series node with all its children completion-forced.*

A node u is deletion-forced*, in $G + (x, N(x))$, iff u is completion-forced in $\overline{G} + (x, V(G) \setminus N(x))$, where $\overline{G}$ is the complement graph of G.*

Remark 2. The complement graph $\overline{G}$ of a cograph G is a cograph and the cotree of $\overline{G}$ is obtained from the cotree T of G by flipping the labels of the internal nodes: series nodes become parallel nodes and vice-versa.

The rationale for these notions is that for u a completion-forced (resp. deletion-forced) node, $G[V(u)] + x$ admits a unique cograph completion (resp. deletion), which is the trivial one that makes x universal (resp. isolated) in $G[V(u)] + x$.

Definition 5 (Clean). *A mixed node u of T is* clean *iff either:*

- *the parent v of u is a parallel node and u is the unique non-hollow child of v, or*
- *the parent v of u is a series node and u is the unique non-full child of v.*

We can now state our characterisation. Lemma 3 identifies the minimal insertion nodes u and Lemma 4 gives all the editings settled at u that are minimal for inclusion.

Lemma 3. *A mixed node u of T is a minimal insertion node iff one of these conditions holds:*

1. *u is parallel and either u has at least 3 children and no clean non-completion-forced child or u has exactly 2 children, at least one of which is completion-forced; or*
2. *u is series and either u has at least 3 children and no clean non-deletion-forced child or u has exactly 2 children, at least one of which is deletion-forced.*

Lemma 4. *Let u be a mixed node of T and let H be an editing settled at u. H is a minimal editing iff exactly one of the two following conditions is satisfied:*

- *u is a parallel node and either at least two children of u are full in H or exactly one child of u is full in H and this child is completion-forced.*
- *u is a series node and either at least two children of u are hollow in H or exactly one child of u is hollow in H and this child is deletion-forced.*

Using the characterisations provided by Lemmas 3 and 4, one can show the following theorem, whose proof is omitted due to space restriction.

Theorem 2. *There exists an $O(n)$-time algorithm that, given a cograph G, its cotree T and a new vertex x together with its neighbourhood, determines all the minimal insertion nodes u of T and for each of them determines an editing of minimum cost, denoted $mincost(u)$, among those settled at u, called a minimum settled editing.*

4   An $O(n + m)$-time Algorithm for Minimal Cograph Editing

We now design an incremental algorithm in which one step runs in $O(d)$ time, where $d = |N(x)|$ is the degree of the new vertex x in $G + (x, N(x))$, resulting in an overall $O(n + m)$ time complexity for the whole algorithm. The difficulty to do so is that the minimal insertion nodes can be far from each other in the tree, at distance $\Omega(n)$, so we cannot explore entirely the part of T connecting them, as done by the algorithm of Theorem 2. Therefore, the main idea to achieve a linear complexity is to avoid to search all the minimal insertion nodes of T by discarding those for which the best settled editing has cost greater than d, i.e. is more costly than the trivial delete-all editing (in which all the edges incident to x are deleted). Thanks to this, we can limit ourselves to search a sub-forest of T that has size $O(d)$. In particular, note that we call the $O(n)$-time algorithm of Theorem 2 only on some disjoint subtrees T_u rooted at nodes u that satisfy $|V(u) \cap N(x)| \geq |V(u) \setminus N(x)|$, which we call the *preponderant nodes*, ensuring that all of these calls need no more than $O(\sum_u |T_u|) = O(\sum_u 2|V(u) \cap N(x)|) = O(d)$ total computation time.

A preponderant node is maximal if none of its strict ancestors in T is preponderant. We denote by Γ the set of parents of the maximal preponderant nodes of T. The algorithm colours the neighbours of x in black and leaves the other vertices white. We denote $B(u) = |V(u) \cap N(x)|$ the number of black leaves of T_u and $W(u) = |V(u) \setminus N(x)|$ its number of white leaves. For any node u of T, we denote cost-above(u) the number of edits that is to be performed on the adjacencies between x and the vertices of $V(G) \setminus V(u)$ in any editing settled at u. And we denote diff-above$(u) = \text{cost-above}(u) - (d - B(u))$, which is the difference of cost, restricted to the adjacencies between x and $V(G) \setminus V(u)$, between any editing settled at u and the delete-all editing. For a node $u \in T$, we denote $\mathcal{C}_{prep}(u)$ the set of its preponderant children and we denote $B_{prep}(u) = \bigcup_{v \in \mathcal{C}_{prep}(u)} B(v)$ and $W_{prep}(u) = \bigcup_{v \in \mathcal{C}_{prep}(u)} W(v)$. The general scheme of the algorithm is in three steps:

1. determine the maximal preponderant nodes of T,
2. for every parent u of some maximal preponderant node, i.e. $u \in \Gamma$, decide whether diff-above$(u) \leq B_{prep}(u)$ and determine the exact value of diff-above(u) when u satisfies this condition,
3. for such nodes u, for each of their preponderant children u' and for each minimal insertion node $v \in T_{u'}$, determine one editing of minimum cost among those settled at v.

The rationale behind this approach is as follows. We try to discover all the minimum editings of $G + x$ by listing the minimum settled editings of the nodes $u \in T$, whose cost has been denoted $mincost(u)$. Doing so, we can safely disregard the nodes u such that $mincost(u) > d$ (the cost of the delete-all editing) or $mincost(u) > mincost(v)$ for another node v. For this reason, we can focus only on the nodes that are in the subtree or are the parent of some maximal preponderant node, since all other insertion nodes are more costly. Let $u \in \Gamma$, let u' be a preponderant child of u and let $v \in T_{u'}$. Another necessary condition so that $mincost(v) \leq d$ is that diff-above$(v) \leq B(v)$, because any editing can save at most $B(v)$ edits in T_v compared to the delete-all editing. This implies the same condition for all the ancestors of v, including u', and an even stronger condition for u, namely diff-above$(u) \leq B_{prep}(u)$. This is why our algorithm first determines the nodes $u \in \Gamma$ that satisfy this necessary condition, then determines their preponderant children u' that satisfy diff-above$(u') \leq B(u')$ and searches completely their subtrees $T_{u'}$, thanks to the algorithm of Theorem 2, to obtain a minimum settled editing for each minimal insertion node v of $T_{u'}$ (and for u as well).

Theorem 3 below, whose proof is omitted due to page limit, states that the set of all maximal preponderant nodes can be found in $O(d)$ time.

Theorem 3. *There exists an algorithm that, given a cograph G, its cotree T and a new vertex x together with its neighbourhood, determines all the maximal preponderant nodes of T in time $O(d)$, where d is the degree of x.*

4.1   Determining Diff-Above

For the rest of the algorithm, we slightly modify the tree T as follows: for every parent u of some maximal preponderant node, i.e. $u \in \Gamma$, we modify the list of children of u by cutting off all preponderant children of u. The set of nodes of the tree T' obtained in this way is exactly the set of non-preponderant nodes of T that have no preponderant ancestor. In particular, T' contains only white leaves, since black leaves are preponderant nodes. We keep track of the parts of T that have been removed by storing, for each node $u \in \Gamma$, the numbers $B_{prep}(u)$ and $W_{prep}(u)$ of the black leaves and white leaves, respectively, in the subtrees of the preponderant children of u.

Routine Search-tree. In order to determine diff-above we use an auxiliary routine, called **Search-tree**, which is our main tool to ensure that we search a part of T' that has size $O(d)$. It performs a limited search of the subtree of a node u which has the following fundamental property.

Lemma 5. *After an $O(d)$-time preprocessing of T', for any non-preponderant node $u \in T'$ and any positive integer s, a call to Routine **Search-tree**(u, s) determines whether $W(u) - B(u) \leq s$ in time $O(\min\{s, W(u) - B(u)\})$.*

The preprocessing step is also performed by Routine **Search-tree**, of which we now give a coarse-grain description. Routine **Search-tree**(u, s) performs a depth-first search of T'_u with budget s. The budget s of the search is converted into a *ttl* (standing for *time to live*) which is initially set to $2 + 5s$ and which is decreased by 1 everytime an edge of T' is traversed (either upward or downward). During the search of T'_u, thanks to the numbers $B_{prep}(v)$ and $W_{prep}(v)$ stored in the nodes $v \in T'_u \cap \Gamma$, we maintain a counter *cpt* of the difference between the number of white leaves and the number of black leaves in the part of T_u that corresponds to the part of T'_u that has been searched so far. The search of T'_u stops when either the subtree of u has been entirely searched (*cpt* then contains the exact value of $W(u) - B(u)$) or when the *ttl* becomes negative for the first time, if this happens before. Therefore, the search never takes more than $O(s)$ time.

The preprocessing step consists in assigning a weighted shorcut to each node $u \in \Gamma$, which points to the node of T'_u on which stops the call to $\texttt{Search-tree}(u, exc(u))$, where $exc(u) = \sum_{u_i \in \mathcal{C}_{prep}(u)}(B(u_i) - W(u_i))$, and whose weight is the value of cpt when the call terminates. The interest of these shorcuts is that the calls to $\texttt{Search-tree}(v, s)$, with v an ancestor of u in T, that are made later do not need to perform this part of the search again, they simply follow the shorcut and update their own counter with the weight cpt of the shorcut. This allows to achieve the complexity mentionned in Lemma 5 and to keep the overall complexity of the algorithn linear. During the search of T'_u if a node $v \in \Gamma$ that has not been assigned a shorcut yet is encountered, then we first assign its shorcut to v before we continue the search of T'_u. In this way, each part of T' is searched at most once and the complexity of the preprocessing step is $O(\sum_{u \in \Gamma} exc(u)) = O(d)$.

Finally, one can prove that once all the shorcuts have been assigned, the number of edge traversals and shorcut traversals needed to entirely search the subtree T'_u of a non-preponderant node u is at most $2 + 5(W(u) - B(u))$. Therfeore, if $W(u) - B(u) \leq s$, $\texttt{Search-tree}(u, s)$ entirely searches the subtree T'_u, in $O(W(u) - B(u))$ time, which proves Lemma 5.

Searching the Branch of Node u. Now, we show how to decide, for u the parent of some maximal preponderant node, whether diff-above$(u) \leq B_{prep}(u)$ and we determine the exact value of diff-above(u) when this condition holds. To this purpose, we search the branch from u up to the root of T, using a budget bud initially set at $bud(u) = B_{prep}(u)$. Along the search, we maintain the current value $bud(v)$ of bud on the ancestor v of u so that $bud(v) = B_{prep}(u) -$ diff-above$_v(u)$, where diff-above$_v(u)$ is defined as diff-above(u) but in the subtree T_v (instead of the whole cotree T). Let v be the current node of the search and p its parent. The search stops when either:

1. v is the root of T, or
2. $bud(p) = B_{prep}(u) -$ diff-above$_p(u) < 0$, or
3. p has some preponderant child and $bud(p) = B_{prep}(u) -$ diff-above$_p(u) \geq 0$ but $bud(p) < B_{prep}(p)$.

In the first case, u satisfies the condition diff-above$(u) \leq B_{prep}(u)$, as the search reached the root with a non-negative budget. In the second case, $mincost(u) > mincost(v)$: we stop the search and discard it. In the third case, for complexity reasons, we stop the search initiated at u and make it a child of the search initiated at p in an auxiliary forest F, as this latter search will stop further up in T. Afterwards, when all searches have been performed, we use F to decide whether u satisfies the condition diff-above$(u) \leq B_{prep}(u)$.

The key for searching the branch of u is to be able to update $bud(v) = B_{prep}(u) -$ diff-above$_v(u)$ when moving from the current node v to its parent p. If p is a parallel node, this is easy as diff-above$_p(u) =$ diff-above$_v(u)$ and so $bud(p) = bud(v)$. When p is a series node, we obtain $bud(p)$ by adding to $bud(v)$ the quantity diff-above$_p(u) -$ diff-above$_v(u) = (B_{prep}(p) - W_{prep}(p)) - \sum_{q \in \mathcal{C}_{non}(p) \setminus \{v\}}(W(q) - B(q))$, where $\mathcal{C}_{non}(p)$ denotes the set of non-preponderant children of p. If p has no preponderant children, then we continue the search iff $bud(p) \geq 0$ (Condition 2 above), which writes in this case $\sum_{q \in \mathcal{C}_{non}(p) \setminus \{v\}}(W(q) -$

$B(q)) \leq bud(v)$. This can be tested by a call to Routine Search-tree$(q, .)$ on each node $q \in \mathcal{C}_{non}(p) \setminus \{v\}$ with a global budget of $bud(v)$ for all the calls (each call starts with the budget remaining from the previous calls). If p has some preponderant child, then we continue the search iff $bud(p) \geq B_{prep}(p)$ (Condition 3 above). Instead of checking directly this condition, we check a weaker condition, namely that $bud(p) \geq B_{prep}(p) - W_{prep}(p)$, which also writes $\sum_{q \in \mathcal{C}_{non}(p) \setminus \{v\}}(W(q) - B(q)) < bud(v)$. As previously, we check this condition thanks to Routine Search-tree: if it does not hold, then we also have $bud(p) < B_{prep}(p)$ so we stop the search initiated at u and make it a child in F of the search initiated at p; otherwise, if $\sum_{q \in \mathcal{C}_{non}(p) \setminus \{v\}}(W(q) - B(q)) \leq bud(v)$, then Routine Search-tree has searched all the subtrees rooted at the nodes $q \in \mathcal{C}_{non}(p) \setminus \{v\}$. Therefore, we can determine directly the exact value of $bud(p)$, compare it to $B_{prep}(p)$ and either continue the search initiated at u or stop it and make it a child in F of the search initiated at p.

The main idea for getting an $O(d)$ time complexity is to ensure that each black leaf will participate to the budget of the search of only one branch. This is true for the initial budgets $B_{prep}(u)$ of the searches. But unfortunately, when the next node p of the search is a series node with some preponderant child, an additional quantity $B_{prep}(p) - W_{prep}(p)$ is added to the current budget of the search. This means that this quantity $B_{prep}(p) - W_{prep}(p)$ will contribute to the budget of all the searches that were initiated at some node in the subtree of p and that continue after p, which threatens the linear complexity we aim at. Therefore, for preserving the complexity, instead of continuing all the searches that should continue beyond p, we only continue one of them, say f, that reached p with maximum remaining budget and we make the other searches children of this search f in F. In this way, the quantity $B_{prep}(p) - W_{prep}(p)$ contributes only to search f and the total time complexity of all the searches is bounded by $O(\sum_{u \in \Gamma} B_{prep}(u)) = O(d)$. There is an additional difficulty to be solved here: we have to be sure that at the time we determine the search reaching p with maximum remaining budget, all the searches that will eventually reach p have done so already. To ensure this, we launch the searches in an order σ that guarantees that all the searches initiated at some strict descendant of any $p \in \Gamma$ are launched before the search initiated at p itself is launched and we determine the search reaching p with maximum remaining budget at the time when the search initiated at p is launched. It is possible to determine such an order σ in $O(d)$ time by a careful partial bottom-up search of T starting from all the nodes in Γ. The difficulty is again that we cannot afford to search all T but only an $O(d)$-size subpart of it.

Once all the searches have been performed, we obtain a forest F on these searches. We parse each tree F_i in F, starting from its root. For each search f in F_i, we determine whether the node u at which search f was initiated satisfies the condition diff-above$(u) \leq B_{prep}(u)$ (this is easy for the search $\hat{f}$ being the root of F_i, as we just need to check whether $\hat{f}$ reached the root of T). If u does not satisfy the condition, we discard all the descendant searches f' of f in F_i as the nodes they were initiated at also fail to satisfy the condition. If u satisfies the condition, we can determine the exact value of diff-above(u). We then determine which of the child searches f' of f in F_i have their initiator node v that satisfies the condition. We can do this thanks to diff-above(u) and the value $bud(p)$ of the remaining budget at node p of the search f', where p is the node of T where the searches f and f' met and where f' was made a child of f. When the parse of F is over, we obtain the set of all nodes $u \in \Gamma$ that satisfy the condition diff-above(u) $\leq B_{prep}(u)$ and we have the exact value of diff-above(u) for each of them. Overall, this takes time $O(d)$.

4.2   Final Stage of the Algorithm and Overall Complexity

We first find all the maximal preponderant nodes and determine for each of their parents u whether the condition diff-above(u) $\leq B_{prep}(u)$ holds (and we get the value of diff-above(u) in the positive), as shown above. This takes $O(d)$ time. As noted earlier, we know that all the nodes $v' \in T$ such that $mincost(v') \leq d$ either belong to the subset of nodes of Γ that satisfy the condition above, or are in the subtree of some preponderant child of such a node. Then, for all the nodes $u \in \Gamma$ that satisfy diff-above(u) $\leq B_{prep}(u)$, we determine the set of their preponderant children v that satisfy diff-above(v) $\leq B(v)$, which is also a necessary condition so that T_v contains some node v' with $mincost(v') \leq d$. This can be done thanks to Routine Search-tree and takes $O(d)$ time. Then, for each child v of u satisfying this condition, we call the algorithm of Theorem 2 on $G[V(v)] + x$ to get an editing of minimum cost among those settled in T_v. This takes time $O(|V(v)|) = O(B(v))$ since v is a preponderant node. Finally, we select one editing of minimum cost among the editings obtained in each tree T_v and the editings settled at nodes $u \in \Gamma$ themselves: this is a minimum editing of $G + x$, say settled at node w, and it takes $O(d)$ time to find it. Once this insertion node w has been determined, we update the cotree T and its factorising permutation π (which we use to find maximal preponderant nodes) as shown in [9,10], as well as the reciprocal pointers between the nodes of T and the cells of π. This takes $O(d)$ time and we can update all the information our algorithm needs on the tree within the same time. It is worth noting that our algorithm represents and manipulates the current cograph G only via its cotree. Nevertheless, it can straightforwardly be augmented to maintain the adjacency lists of G as well, within the same time complexity, by observing that at each incremental step, the modified degree of the new vertex x is always less than $2d$.

5   Conclusion and Perspectives

We designed an $O(n + m)$-time algorithm for minimal cograph editing using the fact that there always exists an editing of cost at most m, which is not true for pure completion. Our result, which, to the best of our knowledge, is the first algorithm for an inclusion-minimal editing problem, suggests that considering minimal editing instead of minimal completion may allow to design faster algorithms. Since the editing problem is likely to provide smaller sets of edits than the completion problem, this possibility should be explored for other target classes of graphs.

References

1. Arnborg, S., Corneil, D.G., Proskurowski, A.: Complexity of finding embeddings in a k-tree. SIAM J. Algebraic Discrete Methods **8**(2), 277–284 (1987)
2. Bliznets, I., Fomin, F.V., Pilipczuk, M., Pilipczuk, M.: Subexponential parameterized algorithm for interval completion. In: SODA 2016, pp. 1116–1131. SIAM (2016)
3. Böcker, S., Baumbach, J.: Cluster editing. In: Bonizzoni, P., Brattka, V., Löwe, B. (eds.) CiE 2013. LNCS, vol. 7921, pp. 33–44. Springer, Heidelberg (2013). https://doi.org/10.1007/978-3-642-39053-1_5
4. Böcker, S., Briesemeister, S., Klau, G.W.: Exact algorithms for cluster editing: evaluation and experiments. Algorithmica **60**(2), 316–334 (2011)
5. Brandes, Ulrik, Hamann, Michael, Strasser, Ben, Wagner, Dorothea: Fast quasi-threshold editing. In: Bansal, Nikhil, Finocchi, Irene (eds.) ESA 2015. LNCS, vol. 9294, pp. 251–262. Springer, Heidelberg (2015). https://doi.org/10.1007/978-3-662-48350-3_22
6. Bruckner, S., Hüffner, F., Komusiewicz, C.: A graph modification approach for finding core-periphery structures in protein interaction networks. Algorithms Mol. Biol. **10**(1), 1–13 (2015)
7. Cao, Y.: Unit interval editing is fixed-parameter tractable. Inf. Comput. **253**, 109–126 (2017)
8. Corneil, D., Lerchs, H., Burlingham, L.: Complement reducible graphs. Discret. Appl. Math. **3**(3), 163–174 (1981)
9. Corneil, D., Perl, Y., Stewart, L.: A linear time recognition algorithm for cographs. SIAM J. Comput. **14**(4), 926–934 (1985)
10. Crespelle, C., Paul, C.: Fully dynamic recognition algorithm and certificate for directed cographs. Discret. Appl. Math. **154**(12), 1722–1741 (2006)
11. Crespelle, C., Lokshtanov, D., Phan, T.H.D., Thierry, E.: Faster and enhanced inclusion-minimal cograph completion. In: Gao, X., Du, H., Han, M. (eds.) COCOA 2017. LNCS, vol. 10627, pp. 210–224. Springer, Cham (2017). https://doi.org/10.1007/978-3-319-71150-8_19
12. Crespelle, C., Perez, A., Todinca, I.: An $o(n^2)$-time algorithm for the minimal permutation completion problem. Discret. Appl. Math. **254**, 80–95 (2019)
13. Crespelle, C., Todinca, I.: An $O(n^2)$-time algorithm for the minimal interval completion problem. Theor. Comput. Sci. **494**, 75–85 (2013)
14. Goldberg, P., Golumbic, M., Kaplan, H., Shamir, R.: Four strikes against physical mapping of DNA. J. Comput. Biol. **2**, 139–152 (1995)
15. Guillemot, S., Havet, F., Paul, C., Perez, A.: On the (non-)existence of polynomial kernels for P_l-free edge modification problems. Algorithmica **65**(4), 900–926 (2012)
16. Heggernes, P., Mancini, F., Papadopoulos, C.: Minimal comparability completions of arbitrary graphs. Discret. Appl. Math. **156**(5), 705–718 (2008)
17. Heggernes, P., Telle, J.A., Villanger, Y.: Computing minimal triangulations in time $O(n^{\alpha \log n}) = o(n^{2.376})$. SIAM J. Discrete Math. 19(4), 900–913 (2005)
18. Heggernes, P., Mancini, F.: Minimal split completions. Discret. Appl. Math. **157**(12), 2659–2669 (2009)

19. Hellmuth, M., Fritz, A., Wieseke, N., Stadler, P.F.: Techniques for the cograph editing problem: Module merge is equivalent to editing P4s. CoRR abs/1509.06983 (2015)
20. Hellmuth, M., Wieseke, N., Lechner, M., Lenhof, H.P., Middendorf, M., Stadler, P.F.: Phylogenomics with paralogs. PNAS **112**(7), 2058–2063 (2015)
21. Jia, S., et al.: Defining and identifying cograph communities in complex networks. New J. Phys. **17**(1), 013044 (2015)
22. Karp, R.: Mapping the genome: some combinatorial problems arising in molecular biology. In: 25th ACM Symposium on Theory of Computing (STOC 1993), pp. 278–285. ACM (1993)
23. Liu, K., Terzi, E.: Towards identity anonymization on graphs. In: ACM SIGMOD International Conference on Management of Data (SIGMOD 2008), pp. 93–106. ACM (2008)
24. Liu, Y., Wang, J., Guo, J., Chen, J.: Complexity and parameterized algorithms for cograph editing. Theoret. Comput. Sci. **461**, 45–54 (2012)
25. Lokshtanov, D., Mancini, F., Papadopoulos, C.: Characterizing and computing minimal cograph completions. Discrete Appl. Math. **158**(7), 755–764 (2010)
26. Mancini, F.: Graph Modification Problems Related to Graph Classes. Ph.D. Thesis, University of Bergen, Norway (2008)
27. Nastos, J., Gao, Y.: Bounded search tree algorithms for parametrized cograph deletion: Efficient branching rules by exploiting structures of special graph classes. Discrete Math., Alg. and Appl. **4** (2012)
28. Nastos, J., Gao, Y.: Familial groups in social networks. Social Networks **35**(3), 439–450 (2013)
29. Ohtsuki, T., Mori, H., Kashiwabara, T., Fujisawa, T.: On minimal augmentation of a graph to obtain an interval graph. J. Comput. Syst. Sci. **22**(1), 60–97 (1981)
30. Rapaport, I., Suchan, K., Todinca, I.: Minimal proper interval completions. Inf. Process. Lett. **106**(5), 195–202 (2008)
31. Schoch, D., Brandes, U.: Stars, neighborhood inclusion and network centrality. In: SIAM Workshop on Network Science (2015)

Regular Model Checking with Regular Relations

Vrunda Dave[1], Taylor Dohmen[2(✉)], Shankara Narayanan Krishna[1],
and Ashutosh Trivedi[2]

[1] IIT Bombay, Mumbai, India
{vrunda,krishnas}@cse.iitb.ac.in
[2] Univeristy of Colorado, Boulder, USA
{taylor.dohmen,ashutosh.trivedi}@colorado.edu

Abstract. Regular model checking is an exploration technique for infinite state systems where state spaces are represented as regular languages and transition relations are expressed using rational relations over infinite (or finite) strings. We extend the regular model checking paradigm to permit the use of more powerful transition relations: the class of regular relations, of which the rational relations are a strict subset. We use the language of monadic second-order logic (MSO) on infinite strings to specify such relations and adopt streaming string transducers (SSTs) as a suitable computational model. We introduce nondeterministic SSTs over infinite strings (ω-NSSTs) and show that they precisely capture the relations definable in MSO. We further explore theoretical properties of ω-NSSTs required to effectively carry out regular model checking. In particular, we establish that the regular type checking problem for ω-NSSTs is decidable in PSPACE. Since the post-image of a regular language under a regular relation may not be regular (or even context-free), approaches that iteratively compute the image can not be effectively carried out in this setting. Instead, we utilize the fact that regular relations are closed under composition, which, together with our decidability result, provides a foundation for regular model checking with regular relations.

1 Introduction

Regular model checking [2,3,13,24,31] is a symbolic exploration and verification technique where sets of configurations are expressed as regular languages and transition relations are encoded as rational relations [27–29] in the form of generalized sequential machines. A generalized sequential machine (GSM) is essentially a finite state machine with output capability; on every transition an input symbol is read, the state changes, and a finite string is appended to an output string (see Fig. 1, for instance, where the label α/s indicates that the machine reads the symbol α and writes the string s on any such transition). While regular model checking is undecidable in general, a number of approximation schemes and heuristics [1,8,12,13,18,22,23,30] have made it a practical verification approach. It has, for example, been applied to verify programs with

© Springer Nature Switzerland AG 2021
E. Bampis and A. Pagourtzis (Eds.): FCT 2021, LNCS 12867, pp. 190–203, 2021.
https://doi.org/10.1007/978-3-030-86593-1_13

unbounded data structures such as lists and stacks [3,13]. Moreover, since infinite strings over a finite alphabet can be naturally interpreted as real numbers in the unit interval, regular model checking over infinite strings provides a framework [7,9,10,14,25,26] to analyze properties of dynamical systems.

This paper generalizes the regular model checking approach so that transition relations can be expressed using *regular relations* over infinite strings. We propose the computational model of nondeterministic streaming string transducers on infinite strings (ω-NSST), and explore theoretical properties of ω-NSSTs required to effectively carry out regular model checking.

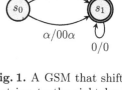

Fig. 1. A GSM that shifts a string to the right by 1 or 2, or equivalently realizing division of the binary encoding of real numbers in $[0,1]$ by 2 or 4.

Regular Relations. While rational relations are capable of modelling a rich set of transition systems, their limitations can be observed by noting their inability to express common transformations such as copy $\overset{\text{def}}{=} w \mapsto ww$ and reverse $\overset{\text{def}}{=} w \mapsto \overleftarrow{w}$, where the string $\overleftarrow{w}$ is the reverse of the string w. Courcelle [16,17] initiated the use of monadic second-order logic (MSO) in defining deterministic and nondeterministic graph-to-graph transformations which are known to include some non-rational transformations like copy and reverse. Engelfriet and Hoogeboom [20] showed that deterministic MSO-definable transformations (DMSOT) over finite strings coincide exactly with the transformations that can be realized by generalizations of GSMs that can read inputs in two directions (2GSM). Furthermore, they showed that this correspondence does not extend to the set of nondeterministic MSO-definable transformations (NMSOT) and nondeterministic 2GSMs (N2GSM).

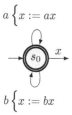

Fig. 2. SST implementing reverse. Here, x is a string variable and input strings ending in the final state s_0 output variable x (as shown by the label on the outgoing arrow from s_0.)

Alur and Černý [4] proposed a one-way machine capable of realizing the same transformations as DMSOTs. These machines, known as *streaming string transducers* (SST), work by storing and combining partial outputs in a finite set of variables, and enjoy a number of appealing properties including decidability of functional equivalence and type-checking (see Fig. 2 for an SST realization of reverse). Alur and Deshmukh followed up this work by introducing nondeterministic streaming string transducers (NSST) as a natural generalization [5] and proved this model captures precisely the same set of relations as NMSOTs. Since the connection between automata and logic is often used as a yardstick for regularity, MSO-definable functions and relations over finite strings are often called regular functions and regular relations.

Regular Relations over Infinite Strings. The expressiveness of SSTs and MSO-definable transformations also coincide when representing functions over infinite strings [6]. Deterministic SSTs operating on infinite strings are known as ω-DSSTs, however, for regular relations of infinite strings, no existing computational model exists. We combine and generalize results in the literature on NSSTs and ω-DSSTs to propose the computational model of nondeterministic streaming ω-string transducers (ω-NSST) capturing regular relations of ω strings.

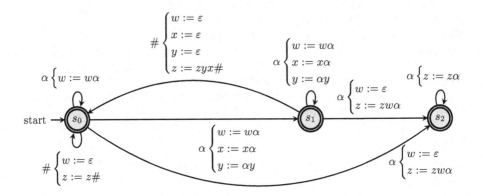

Fig. 3. An ω-NSST implementing the relation $R_{\overleftarrow{u}u}$ from Example 1. Let α denote all symbols in A, excluding $\#$. Variable w remembers the string since the last $\#$, while x and y store the chosen suffix and its reverse. The output variable is z.

Example 1. Let A be a finite alphabet and $\#$ be a special separator not in A. For $u, v \in A^*$, we say that $v \preceq u$ if v is a suffix of u. Consider a relation $R_{\overleftarrow{u}u}$ that transforms strings in $(A \cup \{\#\})^\omega$ such that each maximal $\#$-free finite substring u occurring in the input string is transformed into $\overleftarrow{v}v$ for some suffix v of u. Formally, $R_{\overleftarrow{u}u}$ is defined as

$$\{(u_1\# \cdots \#u_n\#w, \overleftarrow{v_1}v_1\# \cdots \#\overleftarrow{v_n}v_n\#w) : u_i, v_i \in A^*, w \in A^\omega, \text{ and } v_i \preceq u_i\}$$
$$\cup \{(u_1\#u_2\# \ldots, \overleftarrow{v_1}v_1\#\overleftarrow{v_2}v_2\# \ldots) : u_i, v_i \in A^* \text{ and } v_i \preceq u_i\},$$

and can be implemented as an ω-NSST with Büchi acceptance condition (accepting states are visited infinitely often for accepting strings) as shown in Fig. 3.

Contributions and Outline. In Sect. 2 we introduce ω-NSSTs and their semantics as a computational model for regular relations. In Sect. 3 we prove that the ω-NSST-definable relations coincide exactly with MSO-definable relations of infinite strings. In Sect. 4 we consider regular model checking with regular relations. To enable regular model checking with regular relations, we study the following key verification problem. The *type checking problem* for ω-NSSTs asks to decide, given two ω-regular languages L_1, L_2 and an ω-NSST, whether $[\![T]\!](L_1) \subseteq L_2$, where $[\![T]\!]$ is the regular relation implemented by T. We show that type checking for ω-NSSTs is decidable in PSPACE.

2   Regular Relations for Infinite Strings

An alphabet A is a finite set of letters. A *string* w over an alphabet A is a finite sequence of symbols in A. We denote the empty string by ε. We write A^* for the set of all finite strings over A, and for $w \in A^*$ we write $|w|$ for its length. A language L over A is a subset of A^*. An *ω-string* x over A is a function $x : \mathbb{N} \to A$, and written as $x = x(0)x(1) \cdots$. We write A^ω for the set of all ω-strings over A, and A^∞ for $A^* \cup A^\omega$. An ω-language L over A is a subset of A^ω.

2.1   MSO Definable Relations

Strings may be viewed as ordered structures encoded over the signature $\mathcal{S}_A = \{(a)_{a \in A}, <\}$ and interpreted with respect to A^* or A^ω. The domain of a string in this context refers to the set of valid positions in the string, and the relation $<$ in $\mathcal{S}_A$ ranges over this domain. The expression $a(x)$ holds true if the symbol at position x is a, and $x < y$ holds if x is a lesser index than y.

Formulae in MSO over $\mathcal{S}_A$ are defined relative to a countable set of first-order variables $x, y, z, \ldots$ that range over individual elements of the domain and a countable set of second-order variables $X, Y, Z, \ldots$ that range over subsets of the domain. The syntax for well-formed formulae is given as:

$$\phi ::= \exists X.\ \phi \mid \exists x.\ \phi \mid \phi \wedge \phi \mid \phi \vee \phi \mid \neg\phi \mid a(x) \mid x < y \mid x \in X$$

MSO transducers are particular specifications in this logic that define transformations between strings. Intuitively, each such transducer copies each input string some fixed number of times and treats the positions in each copy as nodes in a graph, which are then relabeled and and rearranged in accordance with the formulae of the transducer to produce an output.

Definition 1. *A deterministic* MSO *ω-string transducer (ω-DMSOT) is a tuple*

$$\left(A, B, \mathsf{dom}, N, (\phi_b^n(x))_{b \in B}^{n \in N}, (\psi^{n,m}(x, y))^{n,m \in N}\right),$$

where A and B are input and output alphabets, $N = \{1, \ldots, n\}$ is a set of copy indices, dom is an MSO sentence that defines an input language, the node formulae $(\phi_b^n(x))_{b \in B}^{n \in N}$ specify the labels of positions in the output, and the edge formulae $(\psi^{n,m}(x, y))^{n,m \in N}$ specify which positions in the output will be adjacent.

A ω-DMSOT operates over N disjoint copies of the string graph of an input. Each formula ϕ_b^n has a single free variable and should be interpreted such that if a position satisfies ϕ_b^n, then that position will be labeled by the symbol b in the nth disjoint string graph comprising the output. Each formula $\psi^{(n,m)}$ has two free variables and a satisfying pair of indices indicates that there is a link between the former index in copy n and the latter index in copy m.

Nondeterminism is introduced through additional set variables $X_1, \ldots, X_k$ called *parameters*. Fixing a valuation—sets of positions of the input graph satisfying the domain formula—of these parameters determines an output graph,

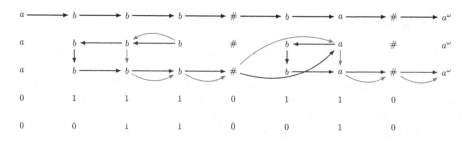

Fig. 4. Two possible outputs of the relation given in Example 1 constructed according ot the ω-NMSOT from Example 2.

just as in the deterministic case. Each possible valuation may result in a different output graph for the same input graph, and thus nondeterminism arises from the choice of valuation.

Definition 2. *A nondeterministic MSO ω-string transducer (ω-NMSOT) with k free set variables $\mathbf{X}_k = (X_1, \ldots, X_k)$ is given as a tuple*

$$\left(A, B, \mathrm{dom}(\mathbf{X}_k), N, (\phi_b^n(x, \mathbf{X}_k))_{b \in B}^{n \in N}, (\psi^{n,m}(x, y, \mathbf{X}_k))^{n,m \in N}\right),$$

where all formulae are parameterized by the free second-order variables in addition to the required first-order parameters.

A relation between strings is a *regular relation* if it is definable by a ω-NMSOT. Since ω-DMSOTs can map each input to at most one output, the relations definable by ω-DMSOTs are called the *regular functions*.

Example 2. We now describe a ω-NMSOT capturing the relation given in Example 1. Set $A = \{a, b, \#\} = B$, $N = \{1, 2\}$, and consider a single parameter $\mathbf{X}_1 = \{X_1\}$. The domain of the relation is simply A^ω, so we omit the formula. For all symbols $\beta \in B$ and copy indices $n \in N$, the node formulae labels each position with the same symbol as the corresponding position in the input string: $\phi_\beta^n(x, X_1) \overset{\mathrm{def}}{=} \beta(x)$. We omit formal specifications of the edge formulae (which can be found in the extended version of this work [19]) and describe them informally. The formula for edges from copy 1 to copy 1 connects adjacent non-$\#$ positions that belong to X_1 in the reverse order. The formula for edges from copy 1 to copy 2 connects non-$\#$ positions to themselves when the predecessor position is not in X_1. The formula for edges from copy 2 to copy 2 links the right-most sequence of positions in X_1 that preceed a $\#$ symbol and also connect all those positions coming after the final $\#$ if required. Finally, the formula for edges from copy 2 to copy 1 links $\#$ symbols to the last position in X_1 left of the next $\#$.

Two possible outputs from the relation of Example 1 are displayed in Fig. 4 which shows how the above ω-NMSOT constructs an output string for two different valuations of X_1. A 1 in the blue (resp. green) row signifies that the position at that column is in X_1, while a 0 indicates that it is not in X_1.

2.2   Nondeterministic Streaming String Transducers

Definition 3. *A nondeterministic streaming string transducer T over ω-strings (ω-NSST) is a tuple $(A, B, S, I, \mathsf{Acc}, \Delta, f, X, U)$, where*

- *A and B are finite input and output alphabets,*
- *S is a finite set of states,*
- *$I \subseteq Q$ is a set of initial states,*
- *Acc is an acceptance condition,*
- *X is a finite set of string variables,*
- *U is a finite set of variable update functions of type $X \to (X \cup B)^*$,*
- *Δ is a transition function of type $(S \times A) \to 2^{U \times S}$, and*
- *$f \in X$ is an append-only output variable.*

Such a machine is deterministic (a ω-DSST) if $|\Delta(s, a)| = 1$, for all states $s \in S$ and symbols $a \in A$, and $|I| = 1$; it is nondeterministic otherwise.

On each transition $s_k \xrightarrow[u_k]{a_k} s_{k+1}$, the transducer changes state and applies the update u_k to each variable of X in parallel. An ω-NSST is *copyless* if every variable in X occurs at most once in the image $\mathrm{im}(u)$ of every update $u \in U$. Alternately stated, an update $u \in U$ is copyless if the string $u(x_0)u(x_1)\ldots u(x_{n-1})$ has at most one occurrence of each $x \in X$, and an ω-NSST is copyless if all of its updates are copyless.

A run of an ω-NSST on an infinite string $a_1 a_2 \cdots \in A^\omega$ is an infinite sequence of states and transitions $s_0 \xrightarrow[u_0]{a_0} s_1 \xrightarrow[u_1]{a_1} \ldots$ where $s_0 \in I$ and $(s_{k+1}, u_k) \in \Delta(s_k, a_k)$ for all $k \in \mathbb{N}$. Let $\mathsf{Runs}_T(w)$ be the set of all runs in T, given input w. An update function $u : X \to (X \cup B)^*$ can easily be extended to $\widehat{u} : (X \cup B)^* \to (X \cup B)^*$ such that $\widehat{u}(w) \overset{\mathrm{def}}{=} \varepsilon$ if $w = \varepsilon$, $\widehat{u}(w) \overset{\mathrm{def}}{=} b\widehat{u}(w')$ if $w = bw'$, and $u(x)\widehat{u}(w')$ if $w = xw'$. The effect of two updates $u_1, u_2 \in U$ in sequence can be *summarized* by the function composition $\widehat{u}_1 \circ \widehat{u}_2$; likewise a sequence of updates of arbitrary length would be summarized by $\widehat{u}_0 \circ \widehat{u}_1 \circ \ldots \circ \widehat{u}_{n-1}$. For notational convenience, we often omit the hats when the extension is clear from context. Notice that if all updates in a sequence of compositions are copyless, then so is the entire summary.

A valuation is a function $X \to B^*$ mapping each variable to a string value. The initial valuation val_ε of all variables is the empty string ε. A valuation is well-defined after any finite prefix r_n of a run r and is computed as a composition of updates occurring on this prefix: $\mathsf{val}_{r_n} = \mathsf{val}_\varepsilon \circ u_0 \circ u_1 \circ \cdots \circ u_{n-1}$. The output $T(r) \overset{\mathrm{def}}{=} \lim_{n \to \infty} \mathsf{val}_{r_n}(f)$ of T on r is well-defined only if r is accepted by T. Since the output variable f is only ever appended to and never prepended, this limit exists and is an ω-string whenever r is accepted, otherwise we set $T(r) = \bot$. The relation $[\![T]\!]$ realized by an ω-NSST T is given by $[\![T]\!] \overset{\mathrm{def}}{=} \{(w, T(r)) : r \in \mathsf{Runs}_T(w)\}$. An ω-NSST T is *functional* if for every w the set $\{w' : (w, w') \in [\![T]\!]\}$ has cardinality at most 1.

We consider both Büchi and Muller acceptance conditions for ω-NSSTs and reference these classes of machines by the initialisms NBT and NMT (DBT and

DMT for their deterministic versions), respectively. For a run $r \in \mathsf{Runs}_T(w)$, let $\mathsf{Inf}(r) \subseteq S$ denote the set of states visited infinitely often.

1. A *Büchi acceptance condition* is given by a set of states $F \subseteq S$ and is interpreted such that a NBT is defined on an input $w \in A^\omega$ if there exists a run $r \in \mathsf{Runs}_T(w)$ for which $\mathsf{Inf}(r) \cap F \neq \emptyset$.
2. A *Muller acceptance condition* is given as a set of sets $\mathbb{F} = \{F_0, \ldots, F_n\} \subset 2^S$, interpreted such that a NMT is defined on input $w \in A^\omega$ if there exists a run $r \in \mathsf{Runs}_T(w)$ for which $\mathsf{Inf}(r) \in \mathbb{F}$.

Proposition 1. *A relation is* NBT *definable if, and only if, it is* NMT *definable.*

The equivalence of NBT and NMT -definable relations follows from a straightforward application of the equivalence of nondeterministic Büchi automata and nondeterministic Muller automata. Equivalence of these acceptance conditions in transducers allows us to switch between them whenever convenient.

Remark 1. Observe that DMTs and functional NMTs, both of which were introduced in [6], have a slightly different output mechanism, which is defined as a function $\Omega : 2^S \rightharpoonup X^*$ such that the output string $\Omega(S')$ is copyless and of the form $x_1 \ldots x_n$, for all $S' \subseteq S$ for which $\Omega(S') \neq \perp$. Furthermore, there is the condition that if $s, s' \in S'$ and $a \in A$ s.t. $(u, s') \in \Delta(s, a)$, then (1) $u(x_k) = x_k$ for all $k < n$ and (2) $u(x_n) = x_n w$ for some $w \in (X \cup B)^*$.

In contrast, our definition has a unique append-only output variable $f \in X$. However, our model with the Muller acceptance is as expressive as that studied in [6]. One can use nondeterminism to guess a position in the input after which states in a Muller accepting set S' will be visited infinitely often. The output function can be defined by guessing a Muller set, and keeping an extra variable for the output. Upon making the guess, it will move the contents of $x_1 \ldots x_n$ to the variable f and make a transition to a copy $T_{S'}$ of the transducer where $\mathsf{Acc} = \{S'\}$. If any state outside the set S' is visited, or the variables $x_1 \ldots, x_{n-1}$ are updated, or the variable f is assigned in non-appending fashion, then $T_{S'}$ makes a transition to a rejecting sink state. Alur, Filiot, and Trivedi [6] showed the equivalence of functional NMT with DMT. This implies that the transductions definable using *functional* NMTs or *functional* NBTs (in our definition) are precisely those definable by ω-DMSOT.

3   Equivalence of ω-NMSOT and ω-NSST

Alur and Deshmukh [5] showed that relations over finite strings definable by nondeterministic MSO transducers coincide with those definable by nondeterministic streaming string transducers. We generalize this result by proving that a relation is definable by an ω-NMSOT if, and only if, it is definable by an ω-NSST. We provide symmetric arguments to connect ω-NSST, ω-DSST and ω-NMSOT, ω-DMSOT, resulting in a simple proof.

Our arguments use the concept of a relabeling relation, following Engelfriet and Hoogeboom [20]. A relation $\rho \subseteq A^\omega \times B^\omega$ is a *relabeling*, if there exists another relation $\rho' \subseteq A \times B$ such that $(aw, bv) \in \rho$ iff $(a, b) \in \rho'$ and $(w, v) \in \rho$. In other words, ρ is obtained by lifting the letter-to-letter relation ρ', in a straightforward manner, to ω-strings. Let $\mathsf{Let}(\rho)$ denote the letter to letter relation $\rho' \subseteq A \times B$ corresponding to ρ and let RL be the set of all such relabelings.

Theorem 1. $\omega\text{-NMSOT} = \omega\text{-NSST}$.

The proof of Theorem 1 proceeds in two stages. In the first part (Lemma 1), we show that every ω-NSST is equivalent to the composition of a nondeterministic relabeling and a ω-DSST. In the second part (Lemma 2), we show that every ω-NMSOT is equivalent to the composition of a nondeterministic relabeling and a ω-DMSOT. These two lemmas, in conjunction with the equivalence of DMTs and functional NMTs [6], allow us to equate these two models of transformation via a simple assignment.

Lemma 1. $\omega\text{-NSST} = \omega\text{-DSST} \circ \text{RL}$

Proof. We first show $\omega\text{-DSST} \circ \text{RL} \subseteq \omega\text{-NSST}$ by proving that for every DMT $T \stackrel{\text{def}}{=} (B, C, S, I, \mathbb{F}, \Delta, f, X, U)$ and nondeterministic relabeling $\rho \subseteq A^\omega \times B^\omega$, there is a NMT $T' \stackrel{\text{def}}{=} (A, C, S, I, \mathbb{F}, \Delta_\rho, f, X, U)$ such that $[\![T']\!] = [\![T]\!] \circ \rho$. As indicated by the tuple given to specify T', the only distinct components between the two machines are their input alphabets and their transition functions Δ and Δ_ρ. The latter is given as $\Delta_\rho \stackrel{\text{def}}{=} (s, a) \mapsto \bigcup\limits_{(a,b) \in \mathsf{Let}(\rho)} \Delta(s, b)$. The nondeterminism of ρ is therefore captured in Δ_ρ. This results in a unique run through T', for every possible relabeling of inputs for T. Since the remaining pieces of T are untouched in the process of constructing T', it is clear that $[\![T']\!] = [\![T]\!] \circ \rho$.

What remains to be shown is the inclusion $\omega\text{-NSST} \subseteq \omega\text{-DSST} \circ \text{RL}$: for any NMT $T \stackrel{\text{def}}{=} (A, B, S, I, \mathbb{F}, \Delta, f, X, U)$, there exists a DMT T' and a nondeterministic relabeling ρ such that $[\![T]\!] = [\![T']\!] \circ \rho$. From T, we can construct a nondeterministic, letter-to-letter relation $\rho' \subseteq A \times (U \times S)$ as follows: $\rho' \stackrel{\text{def}}{=} \{(a, (u, s')) : (u, s') \in \Delta(s, a)\}$. Now let $\rho \subseteq A^\omega \times (U \times S)^\omega$ be the extension of ρ' as described previously. The relation ρ contains the set of all possible runs through T for any possible input in A^ω.

Next, we construct a DMT $T' \stackrel{\text{def}}{=} (U \times S, B, S, I, \mathbb{F}, \Delta_\rho, f, X, U)$ with transition function $\Delta_\rho \stackrel{\text{def}}{=} (s, (u, s')) \mapsto \{(u, s') : (u, s') \in \Delta(s, a)$ for some $a \in A\}$. Consequently, T' retains only the pairs in ρ which correspond to valid runs T and encodes them as ω-strings over the alphabet $S \times U$. The DMT T' then simply follows the instructions encoded in its input and thereby simulates only legitimate runs through T. Thus, we may conclude that $[\![T]\!] = [\![T']\!] \circ \rho$. $\square$

Lemma 2. $\omega\text{-NMSOT} = \omega\text{-DMSOT} \circ \text{RL}$.

Proof. We begin by showing the inclusion $\omega\text{-NMSOT} \subseteq \omega\text{-DMSOT} \circ \text{RL}$: for any ω-NMSOT T, there exists an ω-DMSOT T' and a relabeling ρ such that $[\![T]\!] = [\![T']\!] \circ \rho$. Nondeterministic choice in T is determined by the choice of assignment to free variables in $\mathbf{X}_k$. Alternatively, the job of facilitating nondeterminism can be placed upon a relabeling relation, thereby allowing us to remove the parameter variables. Define a letter-to-letter relation $\rho' \subseteq A \times (A \times \{0,1\}^k)$ as follows: $\rho' \stackrel{\text{def}}{=} \{(a,(a,b)) : b \in \{0,1\}^k\}$, and let the relabeling $\rho \subseteq A^\omega \times (A \times \{0,1\}^k)^\omega$ be its extension. This relabeling essentially gives us a new alphabet such that each symbol from A is tagged with encodings of its membership status for each set parameter from $\mathbf{X}_k$. Now, we can construct an ω-DMSOT T' that is identical to T, apart from two distinctions. Firstly, T' is deterministic (i.e. it has no free set variables), and every occurrence of a subformula $x \in X_i$ in T is replaced by a subformula $\bigvee\limits_{b \in \{0,1\}^k \wedge b[i]=1} (a,b)(x)$ in T'. As a result of this encoding, the equality $[\![T]\!] = [\![T']\!] \circ \rho$ holds.

The converse inclusion, ω-DMSOT$\circ$RL $\subseteq \omega$-NMSOT, is much simpler. Every relabeling ρ in RL is ω-NMSOT definable: consider $\rho' = \text{Let}(\rho) \subseteq A \times B$. The ω-NMSOT specifying ρ is similar to identity/copy, except that here we have that the output label is b iff the input label is a and $(a,b) \in \rho'$. This can be implemented using second-order variables X_b for all $b \in B$. Let $\mathbf{X}_B$ represent this set. Only a single copy is required to produce the output. Node formulae are given by $\phi_b^1(x, \mathbf{X}_B) \stackrel{\text{def}}{=} \bigvee\limits_{a \in A} \bigvee\limits_{(a,b) \in \rho'} (a(x) \wedge x \in X_b)$, and the edge formulae by $\psi^{1,1}(x, y, \mathbf{X}_B) \stackrel{\text{def}}{=} x < y$. It is known that ω-NMSOT are closed under composition [17]. Thus, we conclude that any composition of a nondeterministic relabeling and a ω-DMSOT is definable by a ω-NMSOT and that ω-MSOT$\circ$RL $\subseteq \omega$-NMSOT. $\qquad\square$

In conjunction Lemmas 1 and 2 along with the results of [6] allow us to write the following equation, thereby proving Theorem 1.

$$\omega\text{-NMSOT} = \omega\text{-DMSOT} \circ \text{RL} = \text{DMT} \circ \text{RL} = \text{NMT} = \omega\text{-NSST}$$

4   MSO-Definable Regular Model Checking

In this section, we explain how algorithms for deciding properties of regular relations can be used to perform regular model checking. Given two relations T_1 and T_2, their *sequential composition* is $[\![T_2 \circ T_1]\!] \stackrel{\text{def}}{=} \{(x,z) : (x,y) \in [\![T_1]\!], (y,z) \in [\![T_2]\!]\}$. Let T^k denote the k-fold composition of a relation T with itself. Let T^* denote the transitive closure of T.

Suppose that INIT and BAD are regular languages representing sets of states in some system that are initial, and unsafe, respectively. Given a generic transition relation T which captures the dynamics of the system, the *regular model checking problem* asks to decide whether any element of BAD is reachable from any element of INIT via repeated applications of T. In precise terms, the regular model checking problem asks to decide whether the equation $[\![T^*]\!](\text{INIT}) \cap \text{BAD} = \emptyset$ holds. Bounded model checking, in this setting, asks to decide, given $n \in \mathbb{N}$, whether $[\![T^k]\!](\text{INIT}) \cap \text{BAD} = \emptyset$ holds, for all $k \leq n$. Unbounded model checking is undecidable (cf. [19] for a proof), so we focus on bounded model checking.

When T is a rational relation, its image is always a regular language, and this permits the approach of iteratively applying T from INIT and checking whether this set intersects with BAD by standard automata-theoretic methods. If T is a regular relation, its image may not be a regular language, and we must iteratively compute compositions of T with itself and test whether these compositions enter the BAD language. To allow this, we establish decidability of the *type checking problem* for ω-NSSTs: given two ω-regular languages L_1, L_2 and an ω-NSST T, decide if the inclusions $L_1 \subseteq \mathrm{dom}(T)$ and $[\![T]\!](L_1) \subseteq L_2$ hold.

Theorem 2. *The type checking problem for ω-NSSTs is decidable in* PSPACE.

Proof. Suppose that $T \stackrel{\mathrm{def}}{=} (A, B, S, I, F, \Delta, f, X, U)$ is an NBT and $L_1 \subseteq A^\omega$ and $L_2 \subseteq B^\omega$ are ω-regular languages, encoded, respectively, as deterministic Muller automata (DMA) M_1 and M_2. We first check whether T is defined for all ω-strings $w \in L_1$, i.e. whether $L_1 \subseteq \mathrm{dom}(T)$. A nondeterministic Büchi automaton (NBA) $\mathcal{C}$ that recognizes the domain of T can be constructed in linear time by ignoring variables and output mechanism. The inclusion $L_1 \subseteq \mathrm{dom}(T)$ can be decided in PSPACE by checking emptiness of $M_1' \cap \mathcal{C}$ where M_1' is the NBA equivalent to M_1 and $\mathcal{C}$ is the NBA representing the complement language of $\mathrm{dom}(T)$. It is known that an NBA can be constructed from a DMA with exponential blowup in the number of states [11]. A complement automaton can be constructed for an NBA with exponential increase in the number of states as well [11]. Hence $\mathcal{C}$ has exponentially many states relative to T and M_1. Intersection of M_1' and $\mathcal{C}$ is a standard product construction with a flag so that both M_1' and $\mathcal{C}$ visit good states infinitely often. Thus the intersection NBA $M_1' \cap \mathcal{C}$ has exponentially many states relative to T and M_1. Thanks to the fact that emptiness of NBA can be checked in NLOGSPACE [11], the emptiness of this product automaton, can be decided in NPSPACE = PSPACE.

We now assume that T is well-defined on L_1 and construct a nondeterministic Muller automaton (NMA) $\mathcal{A}$ such that the language of $\mathcal{A}$ is defined as $\{w \in L_1 : \exists w' \in [\![T]\!](w) \text{ s.t. } w' \notin L_2\}$. Next, we construct a DMA $\overline{M_2}$ for $\overline{L_2}$ by complementing the Acc set. The automaton $\mathcal{A}$ simulates M_1, T and $\overline{M_2}$ in parallel. Next, we construct an NMT T' corresponding to the NBT T in order to homogenize the acceptance condition accross these machines. Let us fix the definition for all three machines: (i) $M_1 \stackrel{\mathrm{def}}{=} (A, S_1, p_0, \mathbb{F}_1, \Delta_1)$, (ii) $T' \stackrel{\mathrm{def}}{=} (A, B, S, I, \mathbb{F}', \Delta, f, X, U)$, (iii) $\overline{M_2} \stackrel{\mathrm{def}}{=} (B, S_2, r_0, \mathbb{F}_2, \Delta_2)$.

The NMA $\mathcal{A}$ is defined as the product of M_1 and T' (without the output mechanism), and it stores a state summary map—i.e. the effect of running current valuation of each variable starting from all states of $\overline{M_2}$—in each of its own states. Formally, the states of $\mathcal{A}$ comprise a finite subset of $S_1 \times S \times (S_2 \times X \to S_2 \cup \{\bot\})$. A state (q, p, g) with $g(r, x) = r'$ represents that, starting from state r, if we read the current value of variable x, then we reach state r'. If $g(r, x) = \bot$, it indicates that there is no run on valuation of x starting from r. This information can be updated along the run of $\mathcal{A}$. For instance, if a transition of T updates x as $aybx$, then the summary map g is updated to g' such that $g'(r, x) = g(\Delta_2(g(\Delta_2(r, a), y), b), x)$, and summarizes the effect of reading $x = aybx$ in $\overline{M_2}$ starting from state r.

The set of states of $\mathcal{A}$ is $S_{\mathcal{A}} = S_1 \times S \times (S_2 \times X \to S_2 \cup \{\bot\})$, in which S_1, S, and S_2 represent the state sets of M_1, T', and $\overline{M_2}$, respectively. The transition relation $\Delta_{\mathcal{A}}$ is defined such that $(q', p', g') \in \Delta_{\mathcal{A}}((q, p, g), a)$ iff (i) $\Delta_1(q, a) = q'$, (ii) $(u, p') \in \Delta_1(p, a)$, and (iii) $g'(r, x) = r'$ and $\Delta_2(r, \mathsf{val}_{u(x)}) = r'$, for all $x \in X$ and $r \in S_2$,. Initial states are the product of initial states i.e. a set $I_{\mathcal{A}} = \{(q_0, p_0, r_0) : q_0 \in I\}$. The Muller accepting set of $\mathcal{A}$ is defined as the collection of all $P \subseteq S_{\mathcal{A}}$ such that (i) $\pi_1(P) \in \mathbb{F}_1$, (ii) $\pi_2(P) \in \mathbb{F}$, and (iii) $(\pi_3(P))(r_0, f) \in \mathbb{F}_2$, where π_i is the ith projection. The size of NMA $\mathcal{A}$ is exponential in the number variables of T, polynomial in the number of states of M_1 and T. Thanks to the fact that emptiness of an NMA can be determined in NLOGSPACE [11], emptiness of $\mathcal{A}$ having exponential states in the inputs T, M_1 and M_2, can be decided in NPSPACE and thus, by Savitch's theorem, also in PSPACE.   □

Since regular relations are definable in MSO, they are closed under sequential composition. In combination with Theorems 1 and 2, this establishes the necessary conditions for bounded regular model checking with regular relations to be possible. Thus, we have the following corollary.

Corollary 1. *Bounded model checking with regular relations is decidable.*

Despite the fact that unbounded regular model checking is undecidable, bounded regular model checking provides a refutation procedure. That is, it allows us to search for a witness for proving the system unsafe. Unfortunately, we cannot use bounded model checking of this kind to decide if the system does satisfy the desired property. On the other hand, we identify several special cases of the problem which permit the safety of the system to be verified in finite time. In general, we assume that INIT $\subseteq \overline{\text{BAD}}$, where $\overline{\text{BAD}}$ is the complement of BAD.

Functional Fixed Points. The first instance applies when T is functional, i.e. $[\![T]\!]$ is a function, and relies on the following result of Alur, Filiot, and Trivedi [6].

Theorem 3. *Given an ω-NSST T, it is decidable if $[\![T]\!]$ is a function. Given a pair of functional ω-NSSTs T_1 and T_2, it is decidable if $[\![T_1]\!] = [\![T_2]\!]$.*

At every step of the bounded regular model checking procedure, one can check if T^k is functional, if T^{k+1} is functional, and if $[\![T^k]\!] = [\![T^{k+1}]\!]$. If these three conditions hold, then, for all $m \geq 0$, we have that $[\![T^k]\!] = [\![T^{k+m}]\!]$. When this occurs and $[\![T^k]\!](\text{INIT}) \subseteq \overline{\text{BAD}}$ holds, it follows that $[\![T^k]\!] = [\![T^*]\!]$ and therefore that $[\![T^*]\!](\text{INIT}) \subseteq \overline{\text{BAD}}$ which implies $[\![T^*]\!](\text{INIT}) \cap \text{BAD} = \emptyset$. Note that T^k can be functional even when T is not. To see this, consider a non-functional ω-NSST T such that $[\![T]\!](a^\omega) = \{b^\omega, c^\omega\}$, and $[\![T]\!](b^\omega) = d^\omega = [\![T]\!](c^\omega)$. If $a^\omega \in \text{INIT}$ and $|[\![T]\!](w)| = 1$ for every other input w and $a^\omega \notin \text{im}(T)$, then T^2 is functional.

Inductive Invariants. An alternative approach involves showing that $[\![T]\!]$ satisfies some inductive invariant. Select, as a candidate invariant, a regular or ω-regular language L which is contained in the set of safe states $L \subseteq \overline{\text{BAD}}$. Now, L provides a witness to the unbounded safety of the system if the following pair of conditions

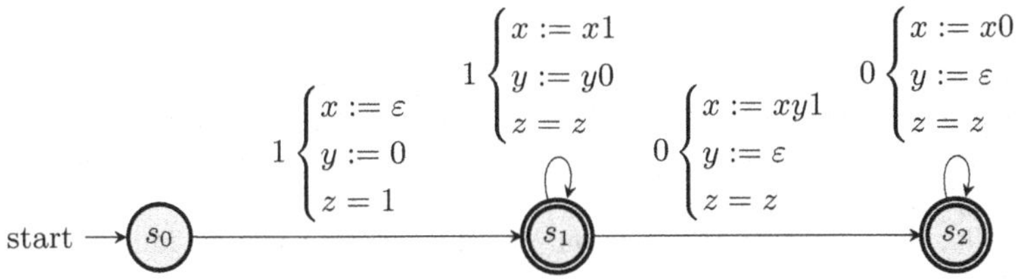

$$\text{start} \rightarrow \boxed{s_0} \xrightarrow{1\begin{cases} x := \varepsilon \\ y := 0 \\ z = 1 \end{cases}} \boxed{s_1} \xrightarrow{0\begin{cases} x := xy1 \\ y := \varepsilon \\ z = z \end{cases}} \boxed{s_2}$$

with self-loops:

s_1: $1\begin{cases} x := x1 \\ y := y0 \\ z = z \end{cases}$

s_2: $0\begin{cases} x := x0 \\ y := \varepsilon \\ z = z \end{cases}$

Fig. 5. An ω-SST squaring a number with binary expansion of the form $1^n 0^\omega$. The output at s_1 and s_2 is x. Notice that this function can not be expressed as a GSM.

are met: (i) INIT $\subseteq L$ and (ii) $[\![T]\!](L) \subseteq L$. Together, (i) and (ii) imply that $[\![T^*]\!](\text{INIT}) \subseteq L$, and in combination with the assumption that $L \subseteq \overline{\text{BAD}}$ this yields that $[\![T^*]\!](\text{INIT}) \cap \text{BAD} = \emptyset$. The necessary inclusions can be formulated as instances of the type checking problem, and so, given an appropriately chosen inductive invariant in the form of an ω-regular language, the global safety of such a system may be verified in polynomial space. This method is easily generalized by searching for k-inductive invariants: ω-regular languages for which there is a $k \in \mathbb{N}$ such that $[\![T^k]\!](L) \subseteq L$. The k-inductive approach complements bounded regular model checking, since, for a given k, bounded regular model checking lets us decide if the system is safe for up to k transitions while k-induction lets us decide if it is safe after at least k transitions.

5   Conclusion

We introduced ω-NSSTs as a computational model for regular relations over infinite strings, and showed that the relations definable by ω-NSST coincide exactly with those definable in MSO. Motivated by potential applications in formal verification, we studied algorithmic properties of these objects and established the minimal theoretical results required for bounded regular model checking to be possible with regular transition relations.

Regular functions and relations provide an intriguing class of models for real valued functions, see Fig. 5 for example. In [15, 21] analytic properties such as continuity and differentiability of real functions encoded by ω-automata have been studied. Extending this line of research by going beyond standard ω-automata is both theoretically interesting and could be leveraged towards applications involving verification and control of dynamical systems. The present work indicates the viability of generalizing the automata-theoretic approach to modeling real functions. With this application in mind, it would be worthwhile to study the approximation techniques developed for traditional regular model checking to see if they generalize to handle regular relations.

References

1. Abdulla, P.A., Jonsson, B., Nilsson, M., d'Orso, J.: Regular model checking made simple and effcient. In: Brim, L., Křetínský, M., Kučera, A., Jančar, P. (eds.) CONCUR 2002. LNCS, vol. 2421, pp. 116–131. Springer, Heidelberg (2002). https://doi.org/10.1007/3-540-45694-5_9

2. Abdulla, P.A., Jonsson, B., Nilsson, M., d'Orso, J., Saksena, M.: Regular model checking for LTL(MSO). Int. J. Softw. Tools Technol. Transf. **14**(2), 223–241 (2012). https://doi.org/10.1007/s10009-011-0212-z

3. Abdulla, P.A., Jonsson, B., Nilsson, M., Saksena, M.: A survey of regular model checking. In: Gardner, P., Yoshida, N. (eds.) CONCUR 2004. LNCS, vol. 3170, pp. 35–48. Springer, Heidelberg (2004). https://doi.org/10.1007/978-3-540-28644-8_3

4. Alur, R., Cerný, P.: Expressiveness of streaming string transducers. In: IARCS Annual Conference on Foundations of Software Technology and Theoretical Computer Science, FSTTCS. LIPIcs, vol. 8, pp. 1–12. Schloss Dagstuhl - Leibniz-Zentrum für Informatik (2010). https://doi.org/10.4230/LIPIcs.FSTTCS.2010.1

5. Alur, R., Deshmukh, J.V.: Nondeterministic streaming string transducers. In: Aceto, L., Henzinger, M., Sgall, J. (eds.) ICALP 2011. LNCS, vol. 6756, pp. 1–20. Springer, Heidelberg (2011). https://doi.org/10.1007/978-3-642-22012-8_1

6. Alur, R., Filiot, E., Trivedi, A.: Regular transformations of infinite strings. In: Proceedings of the 27th Annual IEEE Symposium on Logic in Computer Science, LICS, pp. 65–74. IEEE Computer Society (2012). https://doi.org/10.1109/LICS.2012.18

7. Boigelot, B., Jodogne, S., Wolper, P.: An effective decision procedure for linear arithmetic over the integers and reals. ACM Trans. Comput. Log. **6**(3), 614–633 (2005). https://doi.org/10.1145/1071596.1071601

8. Boigelot, B., Legay, A., Wolper, P.: Iterating Transducers in the Large. In: Hunt, W.A., Somenzi, F. (eds.) CAV 2003. LNCS, vol. 2725, pp. 223–235. Springer, Heidelberg (2003). https://doi.org/10.1007/978-3-540-45069-6_24

9. Boigelot, B., Legay, A., Wolper, P.: Omega-regular model checking. In: Jensen, K., Podelski, A. (eds.) TACAS 2004. LNCS, vol. 2988, pp. 561–575. Springer, Heidelberg (2004). https://doi.org/10.1007/978-3-540-24730-2_41

10. Boigelot, B., Wolper, P.: Representing arithmetic constraints with finite automata: an overview. In: Stuckey, P.J. (ed.) ICLP 2002. LNCS, vol. 2401, pp. 1–20. Springer, Heidelberg (2002). https://doi.org/10.1007/3-540-45619-8_1

11. Boker, U.: Why these automata types? In: LPAR-22. 22nd International Conference on Logic for Programming, Artificial Intelligence and Reasoning. EPiC Series in Computing, vol. 57, pp. 143–163. EasyChair (2018). https://easychair.org/publications/paper/G5dD

12. Bouajjani, A., Habermehl, P., Vojnar, T.: Abstract regular model checking. In: Alur, R., Peled, D.A. (eds.) CAV 2004. LNCS, vol. 3114, pp. 372–386. Springer, Heidelberg (2004). https://doi.org/10.1007/978-3-540-27813-9_29

13. Bouajjani, A., Jonsson, B., Nilsson, M., Touili, T.: Regular model checking. In: Emerson, E.A., Sistla, A.P. (eds.) CAV 2000. LNCS, vol. 1855, pp. 403–418. Springer, Heidelberg (2000). https://doi.org/10.1007/10722167_31

14. Bouajjani, A., Legay, A., Wolper, P.: Handling liveness properties in (omega-)regular model checking. In: Proceedings of the 6th International Workshop on Verification of Infinite-State Systems, INFINITY. Electronic Notes in Theoretical Computer Science, vol. 138, pp. 101–115. Elsevier (2004). https://doi.org/10.1016/j.entcs.2005.02.061

15. Chaudhuri, S., Sankaranarayanan, S., Vardi, M.Y.: Regular real analysis. In: 28th Annual ACM/IEEE Symposium on Logic in Computer Science, LICS, pp. 509–518. IEEE Computer Society (2013). https://doi.org/10.1109/LICS.2013.57

16. Courcelle, B.: Monadic second-order definable graph transductions: a survey. Theor. Comput. Sci. **126**(1), 53–75 (1994). https://doi.org/10.1016/0304-3975(94)90268-2

17. Courcelle, B., Engelfriet, J.: Graph Structure and Monadic Second-Order Logic - A Language-Theoretic Approach, Encyclopedia of mathematics and its applications, vol. 138. Cambridge University Press (2012). http://www.cambridge.org/fr/knowledge/isbn/item5758776/?site_locale=fr_FR

18. Dams, D., Lakhnech, Y., Steffen, M.: Iterating transducers. In: Berry, G., Comon, H., Finkel, A. (eds.) CAV 2001. LNCS, vol. 2102, pp. 286–297. Springer, Heidelberg (2001). https://doi.org/10.1007/3-540-44585-4_27

19. Dave, V., Dohmen, T., Krishna, S.N., Trivedi, A.: Regular model checking with regular relations. CoRR abs/1910.09072 (2019). http://arxiv.org/abs/1910.09072

20. Engelfriet, J., Hoogeboom, H.J.: MSO definable string transductions and two-way finite-state transducers. ACM Trans. Comput. Log. **2**(2), 216–254 (2001). https://doi.org/10.1145/371316.371512

21. Gorman, A.B., et al.: Continuous regular functions. Log. Methods Comput. Sci. **16**(1) (2020). https://doi.org/10.23638/LMCS-16(1:17)2020

22. Habermehl, P., Vojnar, T.: Regular model checking using inference of regular languages. In: Proceedings of the 6th International Workshop on Verification of Infinite-State Systems, INFINITY. Electronic Notes in Theoretical Computer Science, vol. 138, pp. 21–36. Elsevier (2004). https://doi.org/10.1016/j.entcs.2005.01.044

23. Jonsson, B., Nilsson, M.: Transitive closures of regular relations for verifying infinite-state systems. In: Graf, S., Schwartzbach, M. (eds.) TACAS 2000. LNCS, vol. 1785, pp. 220–235. Springer, Heidelberg (2000). https://doi.org/10.1007/3-540-46419-0_16

24. Kesten, Y., Maler, O., Marcus, M., Pnueli, A., Shahar, E.: Symbolic model checking with rich assertional languages. Theor. Comput. Sci. **256**(12), 93–112 (2001). https://doi.org/10.1016/S0304-3975(00)00103-1

25. Legay, A.: Extrapolating (omega-)regular model checking. Int. J. Softw. Tools Technol. Transf. **14**(2), 119–143 (2012). https://doi.org/10.1007/s10009-011-0209-7

26. Legay, A., Wolper, P.: On (omega-)regular model checking. ACM Trans. Comput. Log. **12**(1), 2:1-2:46 (2010). https://doi.org/10.1145/1838552.1838554

27. Löding, C., Spinrath, C.: Decision problems for subclasses of rational relations over finite and infinite words. Discret. Math. Theor. Comput. Sci. 21(3) (2019). http://dmtcs.episciences.org/5141

28. Sakarovitch, J.: Elements of Automata Theory. Cambridge University Press, Cambridge (2009). https://doi.org/10.1017/CBO9781139195218

29. Schützenberger, M.: Sur les relations rationelles entre monoïdes libres. Theor. Comput. Sci. 243–259 (1976)

30. Touili, T.: Regular model checking using widening techniques. Electron. Notes Theor. Comput. Sci. **50**(4), 342–356 (2001). https://doi.org/10.1016/S1571-0661(04)00187-2

31. Wolper, P., Boigelot, B.: Verifying systems with infinite but regular state spaces. In: Hu, A.J., Vardi, M.Y. (eds.) CAV 1998. LNCS, vol. 1427, pp. 88–97. Springer, Heidelberg (1998). https://doi.org/10.1007/BFb0028736

Minimum Consistent Subset Problem
for Trees

Sanjana Dey[1]($^{(\boxtimes)}$), Anil Maheshwari[2], and Subhas C. Nandy[1]

[1] ACM Unit, Indian Statistical Institute, Kolkata, India
info4.sanjana@gmail.com, nandysc@isical.ac.in
[2] School of Computer Science, Carleton University, Ottawa, Canada
anil@scs.carleton.ca

Abstract. In the minimum consistent subset (MCS) problem, a connected simple undirected graph $G = (V, E)$ is given in which each vertex is colored by one of the possible colors $\{c_1, c_2, \ldots, c_k\}$; the objective is to compute a minimum size subset $\mathcal{C} \subseteq V$ such that for each vertex $v \in V$, one of its nearest neighbors in $\mathcal{C}$, with respect to the hop-distance, is of the same color as the color of v. The decision version of the MCS problem is NP-complete even for planar graphs. We propose a polynomial-time algorithm for computing a minimum consistent subset of a bi-chromatic tree.

Keywords: Consistent subset · Graphs · Trees · Optimal algorithm

1 Introduction

The consistent subset problem was first introduced by Hart [6] in the context of reducing the size of the learning set. In pattern recognition, classification is done using the nearest neighbor rule: given the learning set O of objects already classified, each new object is classified into the same class as its closest object from O. A subset O' of the learning set O is a consistent subset if, for each object o from O, the object o and its closest neighbor in O' are in the same class. Ritter [9] introduced the problem of finding consistent subsets of minimum size. Some other important applications are in the field of speech recognition, handwriting recognition, object recognition in vision research etc., see [8].

A geometric variant of the consistent subset problem is as follows. Let P be a set of colored points in the plane. A consistent subset of P is a subset $S \subseteq P$ such that for every point $p \in P \setminus S$, its closest point among the points in S has the same color as that of p. In the minimum consistent subset (MCS) problem, the objective is to find a consistent subset of P of minimum cardinality. In [8], it is shown that the decision version of this problem is NP-complete for 3-colored point sets in $\mathbf{R}^2$. In [7], NP-hardness is shown for 2-colored point sets in $\mathbf{R}^2$. Recently, a sub-exponential time algorithm for the MCS problem in $\mathbf{R}^2$

Anil Maheshwari—Research supported by NSERC.

© Springer Nature Switzerland AG 2021
E. Bampis and A. Pagourtzis (Eds.): FCT 2021, LNCS 12867, pp. 204–216, 2021.
https://doi.org/10.1007/978-3-030-86593-1_14

is proposed in [2]. It is also shown in [2] that in $O(n \log n)$ time, one can test whether the size of the minimum consistent subset of a bi-colored point set in $\mathbf{R}^2$ is of size 2 or not. In the same paper, an $O(n)$ time algorithm is presented for the special case of collinear points.

We study the following graph-theoretic version of the MCS problem in this paper:

Minimum consistent subset problem in a graph

Let $G = (V, E)$ be a graph whose vertices are partitioned into k classes (i.e., colors), namely $V_1, V_2, \ldots, V_k$. The objective is to choose subsets $V_i' \subseteq V_i$, $i = 1, 2, \ldots, k$ such that for each member $v \in V$, if $v \in V_i$ then among its nearest neighbors in $\cup_{i=1}^k V_i'$ there is a vertex of V_i', and $\sum_{i=1}^k |V_i'|$ is minimum.

The distance between a pair of vertices u and v is the number of edges in the shortest path from u to v, and is referred to as $dist(u, v)$. The nearest neighbor is defined with respect to this distance measure in G.

The decision version of the MCS problem is NP-Hard for general graphs $G = (V, E)$. The hardness reduction is from the minimum dominating set of an undirected graph [1]. The same result also holds for planar graphs as the dominating set problem for planar graphs is NP-Hard. To the best of our knowledge, surprisingly, only little is known about the algorithmic complexity of minimum consistent subset problem in graphs. In [5], polynomial-time algorithms are proposed for some simple graph classes, namely (i) paths, (ii) caterpillars, (iii) spiders and (iv) combs. A related problem is recently studied in [3,4], where the *inverse Voronoi diagram* (IVD) in graphs is defined as follows. A graph $G = (V, E)$ with positive edge weights, and a sequence of subsets $\{V_1, V_2, \ldots, V_k\}$, $V_i \subseteq V$ for $i = 1, 2, \ldots, k$, are given where each subset V_i is connected in G, and $\cup_{i=1}^k V_i = V$. The objective is to identify the existence of a subset $X = \{x_1, x_2, \ldots, x_k\}$, where each $x_i \in V_i$ is such that for every element $v \in V_i$ its nearest neighbor in X is x_i. Here, by distance of a pair of vertices $u, v \in V$, we mean the shortest path distance with respect to the edge weights in G. In [3,4], it is shown that the IVD problem for planar graphs is NP-complete. For trees, the IVD problem can be solved in $O(N + n \log^2 n)$ time [4], where $N = n + \sum_{i=1}^k |V_i|$.

In this paper, we propose a polynomial-time algorithm for computing a MCS of a bi-chromatic tree $\mathcal{T} = (V, E)$ in which each vertex is colored *red* or *blue*. Note that this problem is different from the IVD problem in [3,4]. For example, even if we assume that the connected components of the vertices of the same color belong to the same subset, we may need to choose none, one or more than one vertex from each subset to make all the vertices in $\mathcal{T}$ consistent.

2   Preliminaries

We use $\mathcal{C} \subseteq V$ to denote a consistent subset of the tree $\mathcal{T} = (V, E)$ of minimum cardinality.

Observation 1. *If all the vertices in $\mathcal{T}$ are of the same color, then $\mathcal{C}$ consists of a single vertex (any vertex forms an MCS) of $\mathcal{T}$. If $\mathcal{T}$ can be arranged as a*

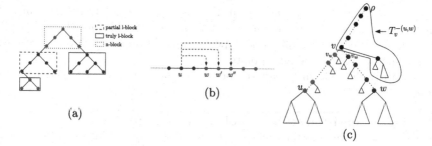

Fig. 1. (a) Illustration of n_block and ℓ_block, (b) Illustration of gates $\Gamma(u, w)$, $\Gamma(u, w')$, $\Gamma(u, w'')$, and (c) the covered tree $T_v^{-(u,w)} = T \setminus (T_{v_u} \cup T_{v_w})$

rooted tree with an appropriate vertex, say $\rho \in V$, as the root, and the vertices in each alternate level of T are of a different color then $C = V$.

In linear time we can determine whether T satisfies Observation 1 and report an appropriate MCS. Thus, for the rest of the paper, we assume that T doesn't satisfy Observation 1.

We will use the following notations to describe our algorithm. Let $u \longrightarrow v$ denotes the path between u and v in T, and let $dist(u, v)$ denotes the length of the path $u \longrightarrow v$ in T. A path $u \longrightarrow v$ in which all the vertices are of the same color is referred to as a *run*. Let v and x be two vertices in T, and let v_x be the neighbor of v such that if we delete the edge (v, v_x) from T, then T is split into two sub-trees, one (rooted at v_x) containing the vertex x, and the remaining part contains the vertex v. The sub-tree containing v will be referred to as T_v^{-x}. Similarly, for a triple of vertices v, x and y, $T_v^{-(x,y)}$ is obtained by deleting the sub-trees rooted at v_x and v_y by removing the edges (v, v_x) and (v, v_y). From now onwards, the term *covered* for a set of vertices implies that those vertices are *consistently covered*, i.e., the vertices satisfy the property of being consistent with some vertex in C.

Definition 1 (Blocks). A *block* in a tree is defined as a connected set of vertices of the same color. We have two types of blocks, namely *leaf blocks* and *non-leaf blocks* (see Fig. 1(a)). A *leaf block* is a block including at least one leaf of T, and is denoted as ℓ_block. An ℓ_block having exactly one vertex connected with a vertex of another color, through which this ℓ-block is connected with the rest of the tree, is called a *true ℓ_block*; if an ℓ_block has more than one vertex connected with vertices of another color, then it is called a *partial ℓ_block*. A *non-leaf block* does not contain any leaf vertex of T, and is called an n_block.

Definition 2 (Gates). A *gate* $\Gamma(u, w)$ is defined by a tuple (u, w), $u, w \in V$, such that (a) u and w are of different colors, (b) there exist *exactly* two runs of different colors on the path $u \longrightarrow w$ in T, and (c) the difference in the number of vertices in these two runs is at most 1 (see Fig. 1(b)). If the number of vertices on the path $u \longrightarrow w$ is odd, then the middle-most vertex v on this path is

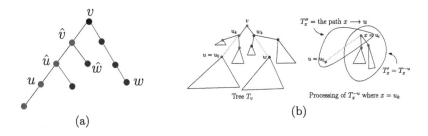

Fig. 2. (a) Nested sibling gate and (b) Processing of a useful sibling gate

referred to as the *anchor vertex* of $\Gamma(u, w)$. If the number of vertices on the path $u \longrightarrow w$ is even, then there is no anchor vertex in that gate. However, then the middle-most edge is referred to as the *anchor edge* of $\Gamma(u, w)$.

Observation 2. *Every bi-colored tree $\mathcal{T}$ of size greater than three that does not satisfy Observation 1, has at least one gate.*

Observation 3. *Let the number of vertices in the path $u \longrightarrow w$ of a gate $\Gamma(u, w)$ be odd. Let v be the anchor vertex of $\Gamma(u, w)$, $N(v)$ be the set of neighbors of v, and for the pair of neighbors $v_u, v_w \in N(v)$, the subtrees T_{v_u} and T_{v_w}, obtained by deleting the edge (v, v_u) and (v, v_w) from $\mathcal{T}$, contain u (in T_{v_u}) and w (in T_{v_w}), respectively. If $\{u, w\}$ are in $\mathcal{C}$, and no other vertex at distance less than $dist(u, v)$ from v is in $\mathcal{C}$, then all the vertices in $\mathcal{T} \setminus (T_{v_u} \cup T_{v_w})$ are consistently covered (see Fig. 1(c)).*

Proof. Given the premise, for each vertex $z \in T_v^{-(u,w)}$, $dist(u, z) = dist(w, z)$. Since the colors of u and w are distinct, z is consistently covered by the inclusion of $\{u, w\}$ in $\mathcal{C}$. $\qquad\qquad\square$

Definition 3 (Sibling Gates). In a rooted tree $\mathcal{T}$, a gate $\Gamma(u, w)$ will be referred to as a *sibling gate* if the number of vertices in the path $u \longrightarrow w$ is odd, and the anchor v of $\Gamma(u, w)$ is a predecessor of both u and w in $\mathcal{T}$. We denote this type of gate as $\Gamma_{sib}(u, v, w)$.

As mentioned earlier, if $\{u, w\}$ of $\Gamma_{sib}(u, v, w)$ is included in $\mathcal{C}$ then all the vertices of the tree $T_v^{-(u,w)}$ are consistently covered.

Observation 4. *Let $\Gamma_{sib}(u, v, w)$ be a sibling gate in $\mathcal{T}$; v_u and v_w are two children of v such that the sub-trees rooted at v_u and v_w contain u and w, respectively. Now, if there exists a sibling gate $\Gamma_{sib}(\hat{u}, \hat{v}, \hat{w})$ in any of the sub-trees rooted at v_u and v_w (see Fig. 2(a)), then the set of vertices consistently covered by $\{u, w\}$ is a subset of vertices consistently covered by $\{\hat{u}, \hat{w}\}$.*

Definition 4 (Useful Sibling Gates). A sibling gate $\Gamma_{sib}(u, v, w)$ is said to be a *useful sibling gate* if the two sub-trees of v containing the vertices u and w respectively do not contain any other sibling gates.

Assume that $\mathcal{T}$ does not satisfy Observation 1. Let us consider an MCS $\mathcal{C}$ of a tree $\mathcal{T}$. Let us consider a pair of bi-chromatic vertices $\{u, w\}$ in $\mathcal{C}$ such that u is red and w is blue and among all such bichromatic pairs in $\mathcal{C}$ they have the minimum distance in $\mathcal{T}$. Observe that this pair of vertices forms a gate $\Gamma(u, w)$. There are two cases depending on whether the number of vertices in the path $u \to w$ is even or odd.

First, consider the case it is odd. By Observation 3, the middlemost vertex v in the path $u \to w$ is an anchor vertex. Let T_v denote the tree $\mathcal{T}$ rooted at v. Now T_v has a minimum consistent subset $\mathcal{C}$ such that $\mathcal{C}$ contains a pair of vertices constituting a sibling gate with the root of T_v as the anchor. Now consider the case that the number of vertices in the path $u \to w$ is even. Now we have an anchor edge (see Definition 2). We introduce a *fictitious* vertex v on the anchor edge and root the tree $\mathcal{T}$ at this vertex (as if the tree is 'rooted at the anchor edge'). In this case, T_v has a minimum consistent subset $\mathcal{C}$ such that $\mathcal{C}$ contains a pair of vertices constituting a sibling gate with the root of T_v as the anchor. Thus we have,

Lemma 1. *Assume that $\mathcal{T}$ does not satisfy Observation 1. There exists a vertex v (real or fictitious), such that if $\mathcal{T}$ is rooted at v, $\mathcal{T}$ has a minimum consistent subset $\mathcal{C}$ such that $\mathcal{C}$ contains a pair of vertices constituting a sibling gate with v as their anchor.*

We now explain the main structure of our algorithm for the computation of MCS for a bi-colored tree $\mathcal{T}$. See Algorithm 1 for a pseudocode. If $\mathcal{T}$ satisfies Observation 1, then the computation of MCS is straightforward. Otherwise, we compute MCS of all the rooted trees at the anchor vertices of the gates as stated in Lemma 1. Among all the rooted trees, the MCS of $\mathcal{T}$ will be the one that is of the smallest size.

Next, we briefly discuss each of the steps of Algorithm 1. Step 1 can be accomplished by rooting $\mathcal{T}$ at an arbitrary vertex and then checking whether the conditions for Observation 1 are met. Step 2 requires rooting tree $\mathcal{T}$ at n vertices of $\mathcal{T}$ and $n - 1$ fictitious vertices corresponding to each edge of $\mathcal{T}$. In Step 3, we need to check whether the root v of T_v is an anchor of a useful sibling gate. Step 4 is discussed in detail in Sect. 3. In Sect. 4, we show that Algorithm 1 is correct and runs in polynomial time.

3   Computing MCS of a Tree Rooted at an Anchor

In this section, we show how to compute minimum consistent set $\mathcal{C}(T_v)$ for a tree T_v rooted at an anchor vertex v. Note that v is an anchor vertex (either a vertex of $\mathcal{T}$ or a fictitious vertex) corresponding to a sibling gate $\Gamma(u, v, w)$ (see Definition 3). Moreover, we can assume that $\Gamma(u, v, w)$ is a useful sibling gate (see Definition 4). This follows from Observation 4 as the two children v_u and v_w of v, where the sub-trees rooted at v_u and v_w contain u and w, respectively, do not contain any sibling gates.

Algorithm 1: An MCS $\mathcal{C}$ of $\mathcal{T}$

Input: A tree $\mathcal{T} = (V, E)$
Output: An MCS $\mathcal{C}$ for the tree $\mathcal{T}$
//STEP 1;
if $\mathcal{T}$ *is monochromatic or* $\mathcal{T}$ *is alternating* **then**
  | report MCS and return;
end
//STEP 2;
$\mathcal{R} = \emptyset$;
for *each vertex v of $\mathcal{T}$* **do**
  | root $\mathcal{T}$ at v, and let T_v be the resulting rooted tree;
  | $\mathcal{R} = \mathcal{R} \cup \{T_v\}$;
end
for *each edge e of $\mathcal{T}$* **do**
  | Root $\mathcal{T}$ at a fictitious vertex v on e, let T_v be the rooted tree;
  | $\mathcal{R} = \mathcal{R} \cup \{T_v\}$;
end
//STEP 3;
for *each rooted tree $T_v \in \mathcal{R}$* **do**
  | **if** *the root v is not an anchor of a useful sibling gate* **then**
  | | $\mathcal{R} = \mathcal{R} \setminus \{T_v\}$;
  | **end**
end
//STEP 4;
for *each rooted tree $T_v \in \mathcal{R}$* **do**
  | Compute $\mathcal{C}(T_v)$ of T_v;
end
//STEP 5;
$\mathcal{C} = \mathcal{C}(T_z)$ such that $T_z \in \mathcal{R}$ corresponds to $\min\limits_{T_v \in \mathcal{R}} |\mathcal{C}(T_v)|$;

return $\mathcal{C}$;

Lemma 2. *If there exists a sibling gate $\Gamma_{sib}(\hat{u}, \hat{v}, \hat{w})$ in any of the sub-trees rooted at v_u and v_w, then the set of vertices consistently covered by $\{u, w\}$ is a subset of vertices consistently covered by $\{\hat{u}, \hat{w}\}$. In that case, the size of the consistent set of T_v cannot be smaller than that of $T_{\hat{v}}$, and hence there is no point in computing $\mathcal{C}(T_v)$.*

We first note that there may be several useful sibling gates that are anchored at the root v of T_v. Traverse each pair of bi-colored runs incident at v in T_v. Let Π_{red} and Π_{blue} be a pair of such runs, and $k = \min\{|\Pi_{red}|, |\Pi_{blue}|\}$. Let $u \in \Pi_{red}$ and $w \in \Pi_{blue}$ be two vertices having hop-distance k from v along the paths Π_{red} and Π_{blue}. Now, every pair of vertices (u_i, w_i) forms a useful sibling gate $\Gamma_{sib}(u_i, v, w_i)$, where $i = 1, 2, \ldots, k$, $u_i \in \Pi_{red}$, $w_i \in \Pi_{blue}$. Next we outline the computation of MCS with respect to one of these useful sibling gates. Overall MCS of T_v is the minimum among MCS for all these sibling gates.

The MCS $\mathcal{C}(T_v)$ for a useful sibling gate $\Gamma_{sib}(u, v, w)$ is computed using the following equation:

$$\mathcal{C}(T_v) = \{u, w\} \bigcup \cup_{u_i \in U} \mathcal{C}_u(T_{u_i}^{-u}) \bigcup \cup_{w_i \in W} \mathcal{C}_w(T_{w_i}^{-w}), \qquad (1)$$

where $U = \{u_0 = u, u_1, \ldots, u_k = v_u\}$, and $W = \{w_0 = w, w_1, \ldots, w_k = v_w\}$ (see Fig. 2(b)), and $\mathcal{C}_a(T_x^{-a})$ for a pair of vertices $a, x \in T_v$, where a and x belong to the same block, is defined as follows.

Definition 5. Let a and x be two vertices in a block. We use $\mathcal{C}_a(T_x^{-a})$ to denote the MCS of a subtree T_x^{-a}, assuming that the vertex x is consistently covered by the vertex $a \in \mathcal{C}_a(T_x^{-a})$ (see Fig. 3(a)). In other words, there does not exist $y \in \mathcal{C}_a(T_x^{-a})$ with $color(y) = color(x) = color(a)$ such that $dist(x, y) < dist(x, a)$. Surely, in order to maintain the consistent covering of x, there does not exist any vertex $\zeta \in \mathcal{C}_a(T_x^{-a})$ of $color(\zeta) \neq color(a)$ (i.e., ζ in an adjacent block $z \longrightarrow z'$ of the block containing x) with $dist(x, \zeta) < dist(a, x)$.

Remark 1. There may not exist any such consistent set of the tree T_x^{-a}. This occurs when the run of different color $z \longrightarrow z'$ in T_x^{-a} closest to x satisfies $dist(x, z') < dist(x, a)$. Here, in order to consistently cover the vertices on the path $z \longrightarrow z'$ one needs to choose a vertex in the run $\{z \longrightarrow z'\}$ in $\mathcal{C}_a(T_x^{-a})$, which will make the vertex x inconsistent (see Fig. 3(b)).

Remark 2. There may exist a situation where the length of the run $x \longrightarrow x'$ (of $color(a)$) is greater than the length of its adjacent run (of other color) $z \longrightarrow z'$ (see Fig. 3(c)). In such a case, one chooses another vertex $a' \in \{x \longrightarrow x'\}$ (satisfying $dist(x, a') \geq dist(x, a)$, i.e., maintaining x to cover by a) to include in $\mathcal{C}_a(T_x^{-a})$, such that a vertex $\zeta \in \{z \longrightarrow z'\}$ may be chosen to have a gate (a', ζ), and the vertices in $z \longrightarrow z'$ be covered by including $\zeta \in \mathcal{C}_a(T_x^{-a})$.

Lemma 3. *For a useful sibling gate $\Gamma_{sib}(u, v, w)$, following holds:*

(a) *Let $z, y \in U$, and $z \neq y$. The computation of the consistent subset $\mathcal{C}_u(T_z^{-u})$ does not affect the computation of $\mathcal{C}_u(T_y^{-u})$, and vice-versa.*

(b) *Similarly, if $z \in U$ and $y \in W$, then the computation of the consistent subset $\mathcal{C}_u(T_z^{-u})$ does not affect the computation of $\mathcal{C}_w(T_y^{-w})$, and vice-versa.*

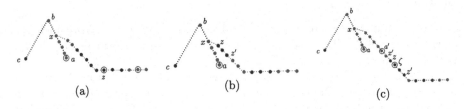

(a)   (b)   (c)

Fig. 3. Demonstration of $\mathcal{C}_a(T_x^{-a})$: (a) $dist(x, z) > dist(a, x)$, (b) $dist(x, z') < dist(x, a)$, and (c) $dist(x, x') \gg dist(z, z')$; we choose another vertex $a' \in x \longrightarrow x'$ (satisfying $dist(x, a') \geq dist(x, a)$) to have a next feasible gate $\Gamma(a, \zeta)$, $\zeta \in z \longrightarrow z'$.

3.1   Computation of $\mathcal{C}(T_z)$

Recall Eq. 1, and consider the processing of the uncovered sub-trees T_z rooted at every vertex z on the path $v \longrightarrow u$ of the sibling gate $\Gamma_{sib}(u, v, w)$. The processing of the uncovered sub-trees rooted at the vertices on the path $v \longrightarrow w$ is analogous. While processing the vertex z on the path $v \longrightarrow u$, we will consider $T_z = T'_z \cup T''_z$, where $T'_z = T_z^{-u}$, and $T''_z =$ the path $z \longrightarrow u$. Note that the set $\mathcal{C}(T_v)$ contains u, and z is covered by the vertex u (see Fig. 4(a)). We know that T_z does not contain any sibling gate, as $\Gamma_{sib}(u, v, w)$ is a useful sibling gate. However, it may have ordinary *gates* (see Definition 2). If Observation 1 holds for T'_z, then $\mathcal{C}(T_z)$ is easy to compute. Otherwise, the following characterizations are required to formulate the algorithm for computing $\mathcal{C}(T_z)$.

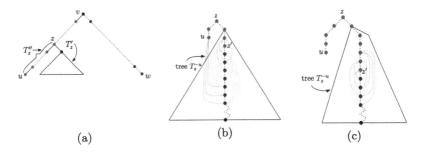

(a)   (b)   (c)

Fig. 4. (a) Computation of $\mathcal{C}(T_z)$: $T_z = T'_z \cup T''_z$, and demonstration of the graph G_Π for the path $\Pi = u \longrightarrow z \longrightarrow \tau$ for processing the tree T_z^{-u} – (b) where G_Π is connected, and (c) where G_Π is disconnected as the red run $z \longrightarrow z'$ is much longer than the next blue run so that there is no edge from u (Color figure online)

Lemma 4. $\mathcal{C}(T_z)$ *contains at least one vertex that belongs to a leaf block of* T_z.

Lemma 5. *The set* $\mathcal{C}(T_z)$ *contains a subset of vertices* $\mathcal{C}'$ *whose elements can be arranged in increasing order of their names, say* $\{\chi_1, \chi_2, \ldots, \chi_m\}$, *such that (i)* χ_m *is in a leaf block of* T_z, *(ii) at least one vertex from each block on the path from* z *to* χ_m *is present in* $\mathcal{C}'$, *and (iii)* χ_i *is the predecessor (not necessarily immediate predecessor) of* χ_{i+1} *in* T_z *for each* $i = 1, 2, \ldots, m - 1$.

Proof. Part (i) follows from Lemma 4. Part (ii) follows trivially since if there exists a block on the path from z to χ_m which has no representative in $\mathcal{C}'$, then the nearest neighbor of each vertex (of color, say red) in that block is of color blue, and hence it becomes inconsistent.

We prove part (iii) by contradiction. Assume that, $\chi \in \mathcal{C}(T_z)$ is the representative of a leaf block. Consider a sequence of vertices $\Psi = \{\psi_1, \psi_2, \ldots, \psi_{m'} = \chi\} \in \mathcal{C}'$ such that for each pair (ψ_j, ψ_{j+1}) either they are in the same block, or they are in the adjacent block, $j = 1, 2, \ldots, m' - 1$. Let (ψ_i, ψ_{i+1}) be the first pair observed in the sequence Ψ such that ψ_i is not the predecessor of ψ_{i+1}. Here, if ψ_i lies in a partial leaf block, the path from z to ψ_i

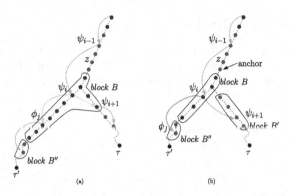

Fig. 5. Illustration of part (iii) of Lemma 5

satisfies the lemma. If ψ_i lies in a non-leaf block, say B, two cases may happen: (a) $color(\psi_{i+1}) = color(\psi_i)$, i.e., $\psi_i, \psi_{i+1} \in B$ (see Fig. 5(a)), and (b) $color(\psi_{i+1}) \neq color(\psi_i)$ ($\psi_i \in B$ and $\psi_{i+1} \in B'$, where B and B' are adjacent blocks) (see Fig. 5(b)). In either case, as B is a non-leaf block, there exists another block B'' adjacent to B which can be reached from B using a gate $\Gamma(\psi', \phi')$ (where $dist(z, \psi_i) < dist(z, \psi')$) and ψ' is reachable from ψ_i using the successor links in $\mathcal{T}$. So instead of considering the path Ψ, we will consider the path $\Psi' = \{\psi_1, \ldots, \psi_i, \ldots \psi' \ldots \phi' \ldots\}$ (see the thick edges in Fig. 5(a,b)). Proceeding in this way, we will reach a leaf block. Thus, the result follows.   □

As demonstrated in Fig. 3(c), there may exist multiple vertices of a block in $\mathcal{C}(T_z)$. Moreover, from the definition of the gate (of odd length), it is also clear that the representative of all the blocks may not have representatives in $\mathcal{C}(T_z)$. Thus, we consider each leaf τ of T_z and compute the size of the MCS assuming that the ℓ_block λ containing τ has a representative in $\mathcal{C}(T_z)$.

Consider the path $\Pi = u \longrightarrow z \longrightarrow \tau$, where τ is a leaf of T_z; τ belongs to the leaf block λ. We formulate the problem as a shortest path problem in a directed weighted graph $G_\Pi = (X, E)$, called the *consistency graph*. Here X contains the vertices on the path Π and two vertices $\{u, t\}$, where u defines the sibling gate $\Gamma_{sib}(u, v, w)$ under process, and t is a dummy sink vertex, and z is a vertex on the path $u \longrightarrow v$. We assume that $u \in \mathcal{C}(T_z)$ and define the edge set $E = E_z^0 \cup E_z^1 \cup E_z^2 \cup E_z^3$, as follows (see Fig. 4(b,c)).

E_z^0: It consists of consistency edges (colored orange) from u to the next run (of color different from that of u) in Π. We may have at most three such edges in E_z^0 depending on the size of that run. See Figs. 4(b) and 4(c) for two different situations depending on the length of the run $u \longrightarrow z \longrightarrow z'$.

E_z^1: It consists of consistency edges (colored orange) between every pair of adjacent runs in Π.

E_z^2: It consists of the edges (colored pink) of a complete graph among the vertices of each run on the path Π.

E_z^3: The vertex t is connected with every vertex of the leaf block λ (these edges are not shown in Fig. 4).

For an edge $\overrightarrow{ab} \in E_z^1$, if the number of vertices on the path $a \longrightarrow b$ is odd, then the subtree $T_c^{-(a,b)}$ rooted at the anchor c of $gate(a,b)$ are consistently covered. Thus, the weight of the edge (a,b) is

$$w(a,b) = \sum_{x=a}^{c_a} |\mathcal{C}_a(T_x^{-a})| + \sum_{x=c_b}^{b'} |\mathcal{C}_c(T_z^{-b})|, \tag{2}$$

where b' is the neighbor of b on the path $a \longrightarrow b$, c_a and c_b are the neighbors of c along the path $a \longrightarrow c$ and $c \longrightarrow b$ respectively[1].

For an edge $\overrightarrow{ab} \in E_z^1$, if the number of vertices on the path $a \longrightarrow b$ is even, then

$$w(a,b) = \sum_{x=a}^{c_a} |\mathcal{C}_a(T_x^{-a})| + \sum_{x=c_b}^{b'} |\mathcal{C}_b(T_x^{-b})|, \tag{3}$$

where c_a and c_b are the pair of middle-most vertices along the path segment $a \longrightarrow b$. To avoid the confusion, we mention that $T_a^{-a} = T_a$.

The weight computation of the edges $(u,b) \in E_z^0$ are done with a minor change in the first sum in Eqs. 2 and 3; here the range of the first sum is from vertex z to c_a instead of u to c_a.

For an edge $(a,b) \in E_z^2$,

$$w(a,b) = \sum_{x=a}^{c_a} |\mathcal{C}_a(T_x^{-a})| + \sum_{x=c_b}^{b'} |\mathcal{C}_b(T_x^{-b})|. \tag{4}$$

Here, if the number of vertices on the path $a \longrightarrow b$ is odd then we assume $c_a = c$ and c_b is the immediate successor of c along the path $c \longrightarrow b$; and if it is even then c_a and c_b are as defined in Eq. 3. Each edge $(a,t) \in E_z^3$ will have weight $w(a,t) = 0$.

The graph G_{Π} may not be connected as the vertex u may not be connected to a vertex $b \in X$ in the next run on the path Π. This situation happens when the path $u \longrightarrow c$ is much longer than the path $c \longrightarrow b$. In such a case the path Π (or the corresponding leaf τ) contributes ∞ in $\mathcal{C}(T_z)$. Thus, it remains to explain the computation of $\mathcal{C}_a(T_x^{-a})$, where a and x are in the same block on Π.

We use the bottom-up dynamic programming to compute $\mathcal{C}(T_x)$ for all vertices $x \in \Pi$. Let $M(x)$ denote the members in the run containing vertex x, and $m(x) = |M(x)|$. These vertices are named as $M(x) = \{b_1 = x, b_2, \ldots, b_{m(x)}\}$ in order. The vertex x is attached with an array A_x of size $M(x)$. For each element $b_i \in M(x)$, $A_x(b_i)$ contains $\mathcal{C}_{b_i}(T_x^{-b_i})$. While processing a vertex $x \in T_z$, we assume that the $A_x(b_i)$ parameters of all the vertices $b_i \in M(x)$ in the run containing x are available; otherwise we recurse. We initialize $A_x(b_i) = \infty$ for all

---

[1] The subtree rooted at b will be considered when another edge from b to a successor vertex will be considered.

$b_i \in M(x)$. Next, we consider every leaf vertex θ of the tree $T_x^{-b_i}$, and construct the graph G_Φ for the path $\Phi = \beta \longrightarrow b_i \longrightarrow \theta$, where β is the $m(x)$-th vertex in the run containing b_i. The edges in the graph G_Φ are similar to those in G_Π constructed while processing the vertex z on the path $\Pi = u \longrightarrow z$ as described earlier[2]. If for every vertex y on the path Φ the $A_y(.)$ values are available, then the edge costs of those edges can be computed using Eqs. 2 and 3 as described for z. Otherwise, this will lead to a further recursive call. Finally, $A_x(b_i)$ is updated by comparing the existing value of $A_x(b_i)$ and the shortest path cost of G_Φ.

4   Analysis of Algorithm

Theorem 1. *Algorithm 1 correctly computes a minimum consistent subset of a bi-colored tree T on n vertices in $O(n^4)$ time.*

Proof. From the property of sibling gates, it follows that the presence of $\{u, w\}$ in $\mathcal{C}(T_v)$ of any sibling gate, say $\Gamma_{sib}(u, v, w)$ anchored at v, will consistently cover all the vertices of $T \setminus (T_{v_u} \cup T_{v_w})$. We have chosen the one of minimum size among all possible useful sibling gates anchored at v. Now, it remains to prove the correctness and minimality of computing $\mathcal{C}(T_v)$.

Again $\{u, w\}$ consistently covers the vertices on the paths $v \longrightarrow u$ and $v \longrightarrow w$. We added the MCS' $\mathcal{C}_u(T_x^{-u})$ for each $x \in \{v \longrightarrow u\}$ and $\mathcal{C}_w(T_y^{-w})$ for each $y \in \{v \longrightarrow w\}$. The computation of consistent subsets for the sub-trees rooted at the vertices on the path $v \longrightarrow u$ and $v \longrightarrow w$, under the condition that $u, w \in \mathcal{C}(T_v)$, can be done independently (see Lemma 3). Now, we prove $\mathcal{C}_u(T_x^{-(u)})$ is correctly computed. By Lemmas 4 and 5, there is a path Π from vertex x to a leaf of the tree T_x^{-u} such that the vertices on a path of the consistency graph G_Π are in the consistent subset $\mathcal{C}_u(T_x^{-u})$, and $\{u\} \cup \mathcal{C}_u(T_x^{-u})$ covers all the vertices on Π. The recursive argument of computing the MCS for the uncovered sub-trees of T_x justifies the correctness of computing the $\mathcal{C}(T_v)$. The minimality is ensured from the fact that for each leaf τ of T_x, we considered the path $\Pi = u \longrightarrow x \longrightarrow \tau$, and considered the shortest path of the graph G_Π, and have chosen the result for a leaf that produces the minimum cost.

Now, we will analyze the time complexity. Step 1 of Algorithm 1 can be implemented in $O(n)$ time. Step 2 requires $O(n^2)$ time as we are constructing $O(n)$ rooted trees. Step 3, for each tree T_v, can be implemented in $O(n)$ time. Now we analyze Step 4.

While processing a vertex z on the path $u \longrightarrow v$ of a sibling gate $\Gamma_{sib}(u, v, w)$, assuming that the array $A_x(.)$ of every vertex on $x \in T_v$ are available, the time of processing a path $\Pi = u \longrightarrow z \longrightarrow \tau$, where τ is a leaf of T_z^{-u}, needs computation of edge costs of G_Π and the computation of shortest path in G_Π. In the graph G_Π, the vertex u and each vertex of the path $z \longrightarrow \tau$ has at most three orange edges to its successor run in Π. Thus the total number of orange edges in G_Π is $O(m_\Pi)$, where m_Π is the length of Π. Moreover, the span of an edge (a_i, b_j) covers the span of another edge (a_{i+1}, b_{j-1}) (if it exists). Thus, the

---

[2] for vertex u defining the sibling gate $\Gamma_{sib}(u, v, w)$.

total time needed for computation of edge costs of all these pink edges in G_Π is $O(m_\Pi)$. However, we may have $O(m_\Pi^2)$ pink edges in G_Π. Processing each leaf vertex τ in T_z^{-u} incurs $O(m_\Pi^2)$ time. The shortest path computation of a directed graph needs time proportional to its number of edges. As the number of leaves of the tree T_z is $O(n_z)$ in the worst case, where n_z is the number of vertices in T_z, the total time of processing the vertex z is $O(n_z^3)$. Since the trees for the vertices along the path $u \longrightarrow z$ are disjoint, the time complexity of processing these vertices are additive. Again, as the vertices in the subtree rooted at the anchor of useful sibling gates in $\mathcal{T}$ are also disjoint, the total time for processing all the sibling gates is $O(n^3)$ in the worst case, provided the $A_x(.)$ values of every vertex on $x \in \mathcal{T}$ are available.

Now, we consider the computation of $A_x(.)$ values of every vertex $x \in T_v$. We consider the vertices in each level of the rooted tree T_v separately, and compute their $A_x(.)$ values. While processing the vertices in its predecessor level, we will use those without recomputing. Similar to the processing of the vertex $z \in \{u \longrightarrow v\}$ discussed earlier, the processing of every vertex $x \in T_z$ for the computation of their $A_x(.)$ values requires $O(n_x^3)$ time, where n_x is the number of vertices in T_x. The sub-trees rooted at the vertices in a particular level are disjoint, and the total computation time for the vertices in a level is additive. Thus, the overall time complexity of the algorithm is $O(n^3 h)$, where h is the maximum number of levels among the sub-trees rooted at the anchor of all possible useful sibling gates in $\mathcal{T}$. Though we will consider the anchor of every gate as a sibling gate and do the above computation, this will not increase the time complexity since the result of a sibling gate once computed can be used later when it is needed. $\square$

5   Conclusion

In this paper, we present a polynomial-time algorithm for computing a minimum consistent subset of bi-chromatic trees. The problem remains unsolved for trees with more than two colors and outerplanar graphs.

References

1. Banerjee, S., Bhore, S., Chitnis, R.: Algorithms and hardness results for nearest neighbor problems in bicolored point sets. In: Proceedings LATIN 2018: Theoretical Informatics, pp. 80–93 (2018)
2. Biniaz, A., et al.: On the minimum consistent subset problem. To appear in Algorithmica, Preliminary version in Proceedings of the WADS 2019, pp. 155–167 (2019)
3. Bonnet, É., Cabello, S., Mohar, B., Pérez-Rosés, H.: The inverse Voronoi problem in graphs I: hardness. Algorithmica **82**(10), 3018–3040 (2020)
4. Bonnet, É., Cabello, S., Mohar, B., Pérez-Rosés, H.: The inverse Voronoi problem in graphs II: trees. Algorithmica **83**(5), 1165–1200 (2021)
5. Dey, S., Maheshwari, A., Nandy, S.C.: Minimum consistent subset of simple graph classes. In: Proceedings of the CALDAM 2021, pp. 471–484 (2021)

6. Hart, P.: The condensed nearest neighbor rule (corresp.). IEEE Trans. Inf. Theory, 14(3), 515–516 (1968)
7. Khodamoradi, K., Krishnamurti, R., Roy, B.: Consistent subset problem with two labels. In: Proceedings of the CALDAM 2018, pp. 131–142 (2018)
8. Wilfong, G.: Nearest neighbor problems. In: Proceedings of the SCG 1991, pp. 224–233 (1991)
9. Ritter, G.L., Woodruff, H.B., Lowry, S.R., Isenhour, T.L.: An algorithm for a selective nearest neighbor decision rule (corresp.). IEEE Trans. Inform. Theor. **21**(6), 665–669 (1975)

Parameterized Complexity of Finding Subgraphs with Hereditary Properties on Hereditary Graph Classes

David Eppstein[1], Siddharth Gupta[2], and Elham Havvaei[1(✉)]

[1] Department of Computer Science, University of California Irvine, Irvine, USA
{eppstein,ehavvaei}@uci.edu
[2] Department of Computer Science, Ben-Gurion University of the Negev,
Be'er Sheva, Israel
siddhart@post.bgu.ac.il

Abstract. We investigate the parameterized complexity of finding subgraphs with hereditary properties on graphs belonging to a hereditary graph class. Given a graph G, a non-trivial hereditary property Π and an integer parameter k, the general problem $P(G, \Pi, k)$ asks whether there exists k vertices of G that induce a subgraph satisfying property Π. This problem, $P(G, \Pi, k)$ has been proved to be NP-complete by Lewis and Yannakakis. The parameterized complexity of this problem is shown to be W[1]-complete by Khot and Raman, if Π includes all trivial graphs (graphs with no edges) but not all complete graphs and vice versa; and is fixed-parameter tractable, otherwise. As the problem is W[1]-complete on general graphs when Π includes all trivial graphs but not all complete graphs and vice versa, it is natural to further investigate the problem on restricted graph classes.

Motivated by this line of research, we study the problem on graphs which also belong to a hereditary graph class and establish a framework which settles the parameterized complexity of the problem for various hereditary graph classes. In particular, we show that:

- $P(G, \Pi, k)$ is solvable in polynomial time when the graph G is co-bipartite and Π is the property of being planar, bipartite or triangle-free (or vice-versa).
- $P(G, \Pi, k)$ is fixed-parameter tractable when the graph G is planar, bipartite or triangle-free and Π is the property of being planar, bipartite or triangle-free, or graph G is co-bipartite and Π is the property of being co-bipartite.
- $P(G, \Pi, k)$ is W[1]-complete when the graph G is C_4-free, $K_{1,4}$-free or a unit disk graph and Π is the property of being either planar or bipartite.

Keywords: Hereditary properties · Ramsey's theorem · Fixed-parameter tractable · W-hardness

1 Introduction

In this paper, we study the parameterized complexity of finding k-vertex induced subgraphs in a given hereditary class of graphs, within larger graphs belonging to

© Springer Nature Switzerland AG 2021
E. Bampis and A. Pagourtzis (Eds.): FCT 2021, LNCS 12867, pp. 217–229, 2021.
https://doi.org/10.1007/978-3-030-86593-1_15

a different hereditary class of graphs. A prototypical instance of the induced subgraph problem is the k-clique problem, which asks whether a given graph G has
a clique of size k. Although k-clique is W[1]-complete for general graphs [15], and
NP-complete even when the input graph is constrained to be a multiple-interval
graph, [6], it is fixed-parameter tractable[1] in this special case [20]. This example,
of a W[1]-complete problem for general graphs which becomes fixed-parameter
tractable on constrained inputs, motivates us to seek additional examples of this
phenomenon, and more broadly to attempt a classification of induced subgraph
problems which can determine in many cases whether a constrained induced
subgraph problem is tractable or remains hard. We denote by FPT the class of
all fixed-parameter tractable problems.

We formalize a graph property as a set Π of the graphs that have the property. A property is *nontrivial* if it neither is empty nor contains all the graphs,
and more strongly it is *interesting* if infinitely many graphs have the property
and infinitely many graphs do not have the property. A nontrivial graph property Π is *hereditary* if it is closed under taking induced subgraphs. That is, if
Π is hereditary and a graph G belongs to Π, then every induced subgraph of G
also belongs to Π. Given a hereditary property Π, let $\overline{\Pi}$ be the complementary
property, the set of graphs which do not belong to Π. The *forbidden* set $\mathcal{F}_\Pi$
of Π is the set of graphs that are minimal for $\overline{\Pi}$: they belong to $\overline{\Pi}$, but all of
their proper induced subgraphs belong to Π. For a hereditary property Π, a
graph G belongs to Π if and only if G has no induced subgraph in $\mathcal{F}_\Pi$. Khot
and Raman [26] studied the parameterized complexity of the following unified
formulation of the induced-subgraph problem, without constraints on the input
graph: Given a graph G, an interesting hereditary property Π and a positive
integer k, the problem $P(G, \Pi, k)$ asks whether there exists an induced subgraph of G of size k that belongs to Π. They proved a dichotomy theorem for
this problem: If Π includes all trivial graphs (graphs with no edges) but not all
complete graphs, or vice-versa, then the problem is W[1]-complete. However, in
all remaining cases, the problem is fixed-parameter tractable.

Our work studies the parameterized complexity of the problem $P(G, \Pi, k)$,
in cases for which it is W[1]-complete for general graphs, under the constraint
that the input graph G belongs to a hereditary graph class Π_G. (Note that Π_G
should be a different class than Π, for otherwise the problem is trivial: just return
any k-vertex induced subgraph of the input.) Given a graph G, the interesting
hereditary properties Π_G and Π, and an integer k, we denote our problem by
$P(G, \Pi_G, \Pi, k)$. The main tool that we use for finding efficient algorithms for
$P(G, \Pi_G, \Pi, k)$ is Ramsey's theorem, which allows us to prove the existence of
either large cliques or large independent sets in arbitrary graphs, allowing some
combinations of input graph size and parameter to be answered immediately
without performing a search. For the cases where we find hardness results, we
do so by reductions from $P(G, \Pi_G, \mathsf{IS}, k)$ to $P(G, \Pi_G, \Pi, k)$, where IS is the
property of being an independent set. We believe our framework has interest
in its own right, as a way to settle a wide class of induced-subgraph properties

---

[1] For basic notions in parameterized complexity, see Sect. 2.

while avoiding the need to develop many tedious hardness proofs for individual problems.

1.1 Our Contributions

We partition interesting hereditary properties into four classes named AA, AS, SA, and SS as follows. A hereditary property Π belongs to:

- AA, if it includes all complete graphs and all independent sets.
- AS, if it includes all complete graphs but excludes some independent sets.
- SA, if it excludes some complete graphs but includes all independent sets.
- SS, if it excludes some complete graphs as well as some independent sets.

By Ramsey's theorem, an interesting hereditary property cannot belong to SS. The interesting cases for the problem $P(G, \Pi_G, \Pi, k)$ with respect to Π are either $\Pi \in$ SA or $\Pi \in$ AS. In the other two cases, when $\Pi \in$ AA or $\Pi \in$ SS the problem $P(G, \Pi_G, \Pi, k)$ is known to be fixed-parameter tractable regardless of Π_G [26]. We prove the following results related to the problem $P(G, \Pi_G, \Pi, k)$, for these interesting cases:

- If $\Pi_G \in$ AS and $\Pi \in$ SA or vice versa, then the problem $P(G, \Pi_G, \Pi, k)$ is solvable in polynomial time (Theorem 1). Although the exponent of the polynomial depends in general on Π, some classes Π_G for which subgraph isomorphism is in FPT also have polynomial-time algorithms for $P(G, \Pi_G, \Pi, k)$ whose exponent is fixed independently of Π (Theorem 2). The key insight for these problems is that these assumptions cause $\Pi_G \cap \Pi$ to be a finite set, limiting the value of k and making it possible to perform a brute-force search for an induced subgraph while remaining within polynomial time.

  A class of problems of this form that have been extensively studied involve finding cliques in sparse graphs or sparse classes such as planar graphs; beyond being polynomial for any fixed hereditary sparse AS or class of graphs, it is fixed-parameter tractable for general graphs when parameterized by degeneracy, a parameter describing the sparsity of the given graph [19]. Another example problem of this type that is covered by this result is finding planar induced subgraphs of co-bipartite graphs; here, Π is the property of being planar, in SA, and Π_G is the property of being co-bipartite, in AS. Similarly, this result covers finding a k-vertex bipartite or triangle-free induced subgraph of a co-bipartite graph, or finding a k-vertex co-bipartite induced subgraph of a planar, bipartite, or triangle-free graph.

- If both Π_G and Π belong either to AS or both belong to SA, then the problem $P(G, \Pi_G, \Pi, k)$ is in FPT (Theorem 3). The insight that leads to this result is that large-enough graphs in Π_G necessarily contain k-vertex cliques (for properties in AS) or independent sets (for properties in SA), which also belong to Π. Therefore, the only instances for which a more complicated search is needed are those for which k is large enough relative to G that the existence of a k-vertex clique or independent set cannot be guaranteed. For that range of the parameter k, the search complexity is in FPT.

Problems of this type that have been studied previously include finding independent sets in sparse graph families, as well as finding planar induced subgraphs of sparse classes of graphs [4]. Finding a k-vertex graph that belongs to one of the four classes of forests, planar graphs, bipartite graphs, or triangle-free graphs, as an induced subgraph of a graph G that belongs to another of these three classes, belongs to the problems of this type.

- If $\Pi_G \in$ SS, then the problem $P(G, \Pi_G, \Pi, k)$ is solvable in polynomial time (Theorem 4). This case is trivial: there can be only finitely many graphs in Π_G and we can precompute the answers to each one.
- In the remaining cases, $\Pi_G \in$ AA, while Π belongs to AS or to SA. These cases include both problems known to be polynomial, such as finding independent sets in various classes of perfect graphs, problems known to be fixed-parameter tractable, including several other cases of independent sets [13], and problems known to be hard for parameterized computation, such as finding independent sets in unit disk graphs [28]. Therefore, we cannot expect definitive results that apply to all cases of this form, as we obtained in the previous cases. Instead, we provide partial results suggesting that in many natural cases the complexity of $P(G, \Pi_G, \Pi, k)$ is controlled by the complexity of the simpler problem of finding independent sets:
  - If Π_G is closed under duplication of vertices (strong products with complete graphs), and Π contains the graphs $n \cdot K_{\chi(\Pi)}$ (disjoint unions of complete graphs with the maximum chromatic number for Π), then $P(G, \Pi_G, \Pi, k)$ is as hard as $P(G, \Pi_G, \mathsf{IS}, k)$ (Theorem 5).
    Families Π_G that meet these conditions, for which finding independent sets is W[1]-complete, include the property of being a unit disk graph, the property of being C_4-free, and the property of being $K_{1,4}$-free. Families Π that meet these conditions include the property of being either planar or bipartite. Therefore, $P(G, \Pi_G, \Pi, k)$ is also W[1]-complete in these families.
  - If $\Pi_G \in$ AA and is closed under joins with disjoint unions of cliques, and if Π contains all joins of an independent set with a disjoint union of cliques that have size at most $\omega(\Pi) - 1$, then $P(G, \Pi_G, \Pi, k)$ is as hard as $P(G, \Pi_G, \mathsf{IS}, k)$ (Theorem 6).

1.2   Other Related Work

Before the investigation of the parameterized complexity of $P(G, \Pi, k)$, Lewis and Yannakakis had studied the dual of this problem, the NODE DELETION problem, for interesting hereditary properties, which is defined as follows: Given a graph G and an interesting hereditary property Π, find the minimum number of nodes to delete from G such that the resulting graph belongs to Π. They proved that the NODE DELETION problem is NP-complete [27]. Cai [7] studied the parameterized version of NODE DELETION and proved that the problem is fixed-parameter tractable, parameterized by the number of deleted vertices, for an interesting hereditary property with a finite forbidden set.

Related to our line of work on the parameterized complexity of hereditary properties, finding an independent set with the maximum cardinality (MIS) on a general graph, has been proved to be NP-hard even for planar graphs of degree at most three [21], unit disk graphs [11], and C_4-free graphs [1]. Fellows et al. proved that finding a k-Independent Set is W[1]-hard for 2-interval graphs while its complementary problem, k-clique, as mentioned before is fixed-parameter tractable for multiple-interval graphs [20].

2 Preliminaries

Throughout the paper, we consider finite undirected graphs. Given a graph G, we denote its vertex set and edge set by $V(G)$ and $E(G)$, respectively. For a vertex $v \in V(G)$, we denote the set of all adjacent vertices of v in G by $N_G(v)$, i.e. $N_G(v) = \{u \in V(G) \mid \{u,v\} \in E(G)\}$. The degree of a vertex $v \in V(G)$ in G is denoted by $\deg_G(v)$. Given a vertex set $S \subseteq V(G)$, $G[S]$ represents the subgraph of G induced by S. The maximum clique size of G is denoted by $\omega(G)$. The maximum clique size of a graph property Π, denoted by $\omega(\Pi)$, is the maximum clique size of any graph $G \in \Pi$. The *chromatic number*, $\chi(G)$, of G is the minimum number of colors needed to color the vertices such that no two adjacent vertices get the same color. The chromatic number, $\chi(\Pi)$, of a graph property Π is the maximum chromatic number of any graph $G \in \Pi$.

Let Π be a hereditary graph property. If $\Pi \in$ AS or $\Pi \in$ SS, then we denote the size of the smallest independent set that does not belong to Π by i_Π. Similarly, if $\Pi \in$ SA or $\Pi \in$ SS, then we denote the number of vertices in the smallest clique that does not belong to Π by c_Π. Observe that, $c_\Pi = \omega(\Pi) + 1$. We denote the property of being an independent set (the family of all all independent sets) as IS.

The use of parameterized complexity has been growing remarkably, in recent decades. What has emerged is a very extensive collection of techniques in diverse areas on numerous parameters. A problem L is a *parameterized* problem if each problem instance of L is associated with a *parameter k*. For simplicity, we represent an instance of a parameterized problem L as a pair (I, k) where k is the parameter associated with input I. Formally, we say that L is *fixed-parameter tractable* if any instance (I, k) of L is solvable in time $f(k) \cdot |I|^{\mathcal{O}(1)}$, where $|I|$ is the number of bits required to specify input I and f is a computable function of k. We remark that this framework also provides methods to show that a parameterized problem is unlikely to be fixed-parameter tractable. The main technique is the one of parameterized reductions analogous to those employed in classical complexity, with the concept of W[1]-hardness replacing NP-hardness. For problems whose solution is a set (for instance of vertices or edges), the size of this set is a natural parameter for the study of the parameterized complexity of the problem. Various problems such as k-vertex cover [5,8,9], k-directed feedback vertex set [10] have been studied under this definition of natural parameter. There are numerous examples of other studies not solely parameterized by the size of the solution [2,18,22,32]. In this paper, we study our problems under their natural parameter, the number of vertices of the subgraph we are seeking. For more information on parameterized complexity, we refer the reader to [12,16].

3   Tractability Results

In this section, we identify pairs of hereditary properties Π_G and Π for which the problem $P(G, \Pi_G, \Pi, k)$ is either in P or FPT. Our proofs use Ramsey numbers which we begin by defining. For any positive integers r and s, there exists a minimum positive integer $R(r, s)$ such that any graph on at least $R(r, s)$ vertices contains either a clique of size r or an independent set of size s. It is well known that $R(r, s) \leq \binom{r+s-2}{r-1}$ [23]. It will also be convenient in our analysis to have a notation for the time to test whether a given k-vertex graph (typically, a subgraph of our given graph G) has property Π; we let $t_\Pi(k)$ denote this time complexity.

Theorem 1. *If $\Pi_G \in$ AS and $\Pi \in$ SA or vice versa, then the problem $P(G, \Pi_G, \Pi, k)$ is solvable in polynomial time.*

Proof. We give a proof for the case when $\Pi_G \in$ AS and $\Pi \in$ SA. The proof for the other case is symmetric under reversal of the roles of cliques and independent sets. Recall that every graph on $R(c_\Pi, i_{\Pi_G})$ vertices contains either a clique of size c_Π, too large to have property Π, or it contains an independent set of size i_{Π_G}, too large to have property Π_G. Therefore, when $k \geq R(c_\Pi, i_{\Pi_G})$, it is impossible for a k-vertex induced subgraph of a graph G in Π_G to also have property Π, because such a subgraph would either have a large clique (contradicting the membership of the subgraph in Π) or a large independent set (contradicting the membership of G in Π_G). Therefore, for such large values of k, an algorithm for $P(G, \Pi_G, \Pi, k)$ can simply answer No without doing any searching.

When $k < R(c_\Pi, i_{\Pi_G})$, then we can use a brute force search to test if there exists a k-vertex induced subgraph having property Π. Specifically, we enumerate all k-vertex subsets of the vertices of G, construct the induced subgraph for each subset, and test whether any of these induced subgraphs belongs to Π. Given a representation of G for which we can test adjacency in constant time, the time to construct each subgraph is $O(k^2)$, so the total time taken by this search is

$$\binom{n}{k} \left(O(k^2) + t_\Pi(k) \right) \leq n^r \left(O(r^2) + t_\Pi(r) \right),$$

where $r = R(c_\Pi, i_{\Pi_G}) - 1$. As the right hand side of this time bound is a polynomial of n without any dependence on k, this is a polynomial time algorithm. Thus, the problem $P(G, \Pi_G, \Pi, k)$ is solvable in polynomial time.   □

Although polynomial, the time bound of Theorem 1 has an exponent r that depends on Π and Π_G, and may be large. An alternative approach, which we outline next, may lead to better algorithms for properties Π_G for which the induced subgraph isomorphism problem is in FPT, as it is for instance for planar graphs [17] or more generally for nowhere-dense families of graphs [30].

Theorem 2. *If $\Pi_G \in$ AS and $\Pi \in$ SA or vice versa, and induced subgraph isomorphism is in* FPT *in Π_G with time $t_{sgi}(n, k)$ to find k-vertex induced subgraphs of n-vertex graphs, then the problem $P(G, \Pi_G, \Pi, k)$ is solvable in polynomial time $O(t_{sgi}(n, r))$, for the same constant r (depending on Π and Π_G but not on k or G) as in Theorem 1.*

Proof. If $k > r$, we answer No immediately as in Theorem 1. Otherwise, we generate all k-vertex graphs, test each of them for having property Π, and if so apply the subgraph isomorphism algorithm for graphs with property Π_G to G and the generated graph. There are $2^{O(r^2)}$ graphs to generate, testing for property Π takes time $t_\Pi(r)$ for each one, and testing for being an induced subgraph of G takes time $t_{sgi}(n, r)$ for each one, so the time is as stated.   □

In particular, these problems can be solved in linear time for planar graphs.

Theorem 3. *If both Π_G and Π belong to* AS, *or if both belong to* SA, *then the problem $P(G, \Pi_G, \Pi, k)$ is in* FPT.

Proof. We give a proof for the case when both Π_G and Π belong to AS. The proof for the other case is again symmetric under reversal of the roles of cliques and independent sets. For a graph $G \in \Pi_G$ that is large enough that $|V(G)| \geq R(k, i_{\Pi_G})$, it must be the case that G contains a clique C of size k, for it cannot contain an independent set of size i_{Π_G} without violating the assumption that it belongs to Π_G. Because Π is assumed to be in AS, it contains all cliques, so this k-vertex clique belongs to Π. Therefore, for graphs with this many vertices, it is safe to answer Yes. There is a small subtlety here, in that we do not know an efficient method to calculate $R(k, i_{\Pi_G})$, and an inefficient method would unnecessarily increase the dependence of our time bounds on the parameter k. However, we can use the inequality

$$R(k, i_{\Pi_G}) \leq \binom{k + i_{\Pi_G} - 2}{k - 1}$$

to get a bound on this number that is easier to calculate. Our algorithm can simply test whether $|V| \geq \binom{k+i_{\Pi_G}-2}{k-1}$, and if so we return Yes without doing any searching.

If $|V(G)| < \binom{k+i_{\Pi_G}-2}{k-1}$, then constructing and checking all induced subgraphs of G of size k to detect whether there exists such a subgraph belonging to Π takes time

$$\binom{k + i_{\Pi_G} - 2}{k - 1}^k \left(O(k^2) + t_\Pi(k) \right),$$

a time complexity that is bounded by a function of k but independent of n. As the times for both cases are of the appropriate form, the problem $P(G, \Pi_G, \Pi, k)$ is in FPT.   □

The following corollaries can be directly obtained from Theorem 1 and Theorem 3.

Table 1. Summary of Theorems 1, 3 and 4.

	$\Pi \in$ SA	$\Pi \in$ AS
$\Pi_G \in$ AS	If $k < R(c_\Pi, i_{\Pi_G})$ check all induced subgraphs of size k, otherwise return No	If $|V(G)| < \binom{k+i_{\Pi_G}-2}{k-1}$ check all induced subgraphs of size k, otherwise return Yes
$\Pi_G \in$ SA	If $|V(G)| < \binom{k+c_{\Pi_G}-2}{k-1}$ check all induced subgraphs of size k, otherwise return Yes	If $k < R(c_{\Pi_G}, i_\Pi)$ check all induced subgraphs of size k, otherwise return No
$\Pi_G \in$ SS	$|V(G)| < R(c_{\Pi_G}, i_{\Pi_G})$, precompute all possible inputs	

Corollary 1. *If Π_G is the property of being co-bipartite and Π is the property of being a forest, planar, bipartite or triangle-free (or vice versa), then the problem $P(G, \Pi_G, \Pi, k)$ is solvable in polynomial time.*

Corollary 2. *If Π_G and Π are the properties of being planar, bipartite or triangle-free, then the problem $P(G, \Pi_G, \Pi, k)$ is fixed-parameter tractable.*

For completeness, we state the following (trivial) theorem:

Theorem 4. *If $\Pi_G \in$ SS, then the problem $P(G, \Pi_G, \Pi, k)$ is solvable in polynomial time.*

Proof. We have $|V(G)| < R(c_{\Pi_G}, i_{\Pi_G})$, because otherwise G has either a clique of size c_{Π_G} or a trivial graph of size i_{Π_G}, a contradiction. Because $V(G)$ is bounded, there are only finitely many valid inputs to the problem $P(G, \Pi_G, \Pi, k)$ and we can precompute the solutions to each one.   □

Table 1 briefly summarizes the results of Theorems 1, 3 and 4.

4   Hardness from Strong Products

In this section, we prove some hardness results for the problem $P(G, \Pi_G, \Pi, k)$, when $\Pi_G \in$ AA and $\Pi \in$ SA.

4.1   Hardness from Strong Products with Cliques

To formulate the first of these results in full generality, we need some definitions. The *strong product* $G \boxtimes H$ is defined as a graph whose vertex set $V(G) \times V(H)$ consists of the ordered pairs of a vertex in G and a vertex in H, with two of these ordered pairs (u, v) and (u', v') adjacent if u and u' are adjacent or equal, and v and v' are adjacent or equal. In particular, the strong product with a complete graph, $G \boxtimes K_i$, can be thought of as making i copies of each vertex in G, with two copies of the same vertex always adjacent, and with adjacency between copies of different vertices remaining the same as in G. We use the notation $n \cdot K_i$ to denote the disjoint union of n copies of an i-vertex complete graph; this is the strong product of an n-vertex independent set with an i-vertex clique.

Observation 1. *Given a graph G on n vertices, there exists an independent set of G of size at least $n/\chi(G)$.*

Namely, the large independent set of the observation can be chosen as the largest color class of any optimal coloring of G.

Theorem 5. *Let $\Pi_G \in$ AA be a hereditary property which is closed under strong products with complete graphs, and let $\Pi \in$ SA be a hereditary property such that, for all n, the graph $n \cdot K_{\chi(\Pi)}$ belongs to Π. Then, the problem $P(G, \Pi_G, \Pi, k)$ is as hard as $P(G, \Pi_G, \mathsf{IS}, k)$.*

Proof. We describe a polynomial-time parameterized reduction from instances of $P(G, \Pi_G, \mathsf{IS}, k)$ to equivalent instance of $P(G', \Pi_G, \Pi, k')$, where k' depends only on k (and not on G). The reduction transforms the graph G of the instance into a new graph $G' = G \boxtimes K_{\chi(\Pi)}$, and transforms the parameter k into a new parameter value $k' = k \cdot \chi(\Pi)$. As $\Pi \in$ SA and for all n, the graph $n \cdot K_{\chi(\Pi)}$ belongs to Π, $\chi(\Pi)$ is exactly equal to $\omega(\Pi)$ and hence constantly bounded.

As we have assumed that Π_G is closed under strong products with complete graphs, it follows that $G' \in \Pi_G$, so the reduction produces a valid instance of $P(G, \Pi_G, \Pi, k')$. To show that this instance is equivalent to the starting instance, we show that G has an independent set of size k if and only if G' has an induced subgraph of size k' belonging to Π.

($\Rightarrow$) Let I be an independent set of G of size k, and let $X = I \boxtimes K_{\chi(\Pi)}$ be the subgraph of G' induced by the set of all copies of vertices in I. Then $|V(X)| = k'$ and, as a graph of the form $k \cdot K_{\chi(\Pi)}$, X belongs to Π by assumption.

($\Leftarrow$) Let $H \in \Pi$ be an induced subgraph of G' of size k'. By Observation 1, it has an independent set I' of size $k'/\chi(\Pi) \geq k$. This independent set can include at most one copy of each vertex in G, so the set I of vertices in G whose copies are used in I' must also have size $\geq k$. Further, I is independent, for any edge between its vertices would be copied as an edge in G', contradicting the assumption that we have an independent set in G'. Therefore, I is an independent set of size $\geq k$ in G, as desired. $\square$

The families of unit-disk graphs, C_4-free graphs, and $K_{1,4}$-free graphs all belong to AA, and are closed under strong products with complete graphs. Finding independent sets is also known to be W[1]-complete for unit-disk graphs [28,29], C_4-free graphs [3], and $K_{1,4}$-free graphs [24]. Moreover, the families of planar graphs and of bipartite graphs both have the property that $n \cdot K_{\chi(\Pi)} \in \Pi$. For instance, in planar graphs, the graph $n \cdot K_{\chi(\Pi)}$ consists of n disjoint copies of K_4, a planar graph, and forming disjoint unions preserves planarity. Therefore, we have the following corollary:

Corollary 3. *If Π_G is the property of being (a) unit-disk, (b) C_4-free, or (c) $K_{1,4}$-free , and Π is the property of being either planar or bipartite, then the problem $P(G, \Pi_G, \Pi, k)$ is W[1]-complete.*

4.2 Hardness from Joins with Cliques

The *join* of two graphs $G + H$ is a graph formed from the disjoint union of G and H by adding edges from each vertex of G to each vertex of H. The reduction that we consider in this section involves the join with a disjoint union of cliques, $G + t \cdot K_c$. That is, starting from G we add t cliques of size c, with each vertex in G connected to all vertices in these cliques.

Observation 2. *Given a graph G and two positive integers t and c, the maximum clique size of $G + t \cdot K_c$ is $\omega(G) + c$.*

Theorem 6. *Let $\Pi_G \in$ AA be a hereditary property which is closed under joins with disjoint unions of cliques, and $\Pi \in$ SA be a hereditary property which includes all subgraphs $I + n \cdot K_{\omega(\Pi)-1}$ for an independent set I and positive integer n. Then the problem $P(G, \Pi_G, \Pi, k)$ is as hard as $P(G, \Pi_G, \mathsf{IS}, k)$.*

Proof. We first construct a new graph $G' = G + r \cdot K_c$, where $r = R(\omega(\Pi)+1, k)$ and $c = \omega(\Pi) - 1$, and a new parameter value $k' = k + rc$. By the assumption that Π_G is closed under joins with disjoint unions of cliques, $G' \in \Pi_G$. Now, we show that G has an independent set of size k if and only if G' has an induced subgraph of size k' belonging to Π.

($\Rightarrow$) Let I be an independent set of G of size k. Consider the induced subgraph $I + r \cdot K_c$ of G', formed by including all vertices that were added to G. This subgraph has size $k' = k + rc$, and by assumption it belongs to Π.

($\Leftarrow$)] Let $H \in \Pi$ be an induced subgraph of G' of size k'. The vertices of H can be partitioned into two sets $S_1 \subset V(G)$ and $S_2 \subset r \cdot K_c$. The following two cases can occur:

- If S_1 is not an independent set, let uv be an edge in S_1. Then S_2 must have at most $c - 1$ vertices in each clique of $r \cdot K_c$, for if it contained all c vertices of one of these cliques, then these c vertices together with u and v would form a clique of size $\omega(\Pi) + 1$, which is disallowed in Π. Therefore, S_2 has at most $r(c-1)$ vertices, and to obtain total size k', S_1 must have at least $k + r$ vertices. By the definition of r and by Ramsey's theorem, S_1 has either a clique of size $\omega(\Pi) + 1$ (again, an impossibility) or an independent set of size k, as desired.
- If S_1 is an independent set, we observe that, even if S_2 includes all of the vertices added to G to form G', it has only rc vertices. Therefore, to obtain total size k', S_1 must have at least k vertices, and contains an independent set of size k, as desired. $\square$

There are many families Π_G that meet the requirements on Π_G in this theorem, but do not meet the requirements of Theorem 5: this will be true, for instance, when the forbidden subgraphs of Π_G do not include disjoint unions of cliques, and are co-connected (so they cannot be formed by joins, which produce co-disconnected graphs) but at least one of these graphs contains two adjacent twin vertices (with the same neighbors other than each other). The requirement on Π in this theorem is met, for instance, by the family Π of bipartite graphs. In this case, $\omega(\Pi) = 2$, so the graphs $I + n \cdot K_{\omega(\Pi)-1}$ are just complete bipartite graphs, which are of course bipartite.

As an example, finding k-independent sets in $\overline{K_{1,3}}$-free graphs (the complements of claw-free graphs) is known to be NP-complete, from the completeness of the same problem in triangle-free graphs [31]. Theorem 6 then shows that finding k-vertex bipartite induced subgraphs of $\overline{K_{1,3}}$-free graphs is also NP-complete. However, we cannot use this method to prove parameterized hardness for this example, because the k-independent set problem in $\overline{K_{1,3}}$-free graphs can be solved in FPT by applying a fixed-parameter tractable algorithm for $(k-1)$-independent sets in triangle-free graphs [13] to the sets of non-neighbors of all vertices.

5   Conclusion

We have further narrowed down the parameterized complexity of the problem $P(G, \Pi, k)$ for the case when it is W[1]-complete. In particular, restricting the input graph G to belong to a hereditary graph class Π_G helps us to settle parameterized complexity of numerous graph classes circumventing long and tedious reduction proofs. It remains an open problem to determine the parameterized complexity of the problem $P(G, \Pi_G, \Pi, k)$ when $\Pi_G \in \mathsf{AA}$ with fewer restrictions. It would be also interesting to investigate this problem under other graph parameters beyond the size of the solution. Further, related to our line of work, the problem of counting all induced subgraphs of size k in a graph G that satisfy the property Π has been introduced and shown to be W[1]-hard by Jerrum and Meeks [25]. Recently, it is shown that, given any graph property Π that is closed under the removal of vertices and edges, and that is non-trivial for bipartite graphs, the problem is W[1]-hard [14]. It would be interesting to revisit this counting problem for the case when both the graph and the property belong to hereditary graph classes.

Acknowledgments. The second author is supported in part by the Zuckerman STEM Leadership Program.

References

1. Alekseev, V.E.: The effect of local constraints on the complexity of determination of the graph independence number. Combin.-Algebraic Methods Appl. Math. 3–13 (1982)
2. Bannister, M.J., Eppstein, D.: Crossing minimization for 1-page and 2-page drawings of graphs with bounded treewidth. J. Graph Algorithms Appl. **22**(4), 577–606 (2018)
3. Bonnet, É., Bousquet, N., Charbit, P., Thomassé, S., Watrigant, R.: Parameterized complexity of independent set in h-free graphs. In: 13th International Symposium on Parameterized and Exact Computation, IPEC 2018, 20–24 August 2018, Helsinki, Finland, pp. 17:1–17:13 (2018)

4. Borradaile, G., Eppstein, D., Zhu, P.: Planar induced subgraphs of sparse graphs. J. Graph Algorithms Appl. **19**(1), 281–297 (2015). https://doi.org/10.7155/jgaa.00358
5. Buss, J.F., Goldsmith, J.: Nondeterminism within P. SIAM J. Comput. **22**(3), 560–572 (1993)
6. Butman, A., Hermelin, D., Lewenstein, M., Rawitz, D.: Optimization problems in multiple-interval graphs. In: Bansal, N., Pruhs, K., Stein, C. (eds.) Proceedings of the Eighteenth Annual ACM SIAM Symposium on Discrete Algorithms, SODA 2007, New Orleans, Louisiana, USA, 7–9 January 2007, pp. 268–277. SIAM (2007)
7. Cai, L.: Fixed-parameter tractability of graph modification problems for hereditary properties. Inf. Process. Lett. **58**(4), 171–176 (1996)
8. Chen, J., Kanj, I.A., Jia, W.: Vertex cover: further observations and further improvements. J. Algorithms **41**(2), 280–301 (2001)
9. Chen, J., Kanj, I.A., Xia, G.: Improved parameterized upper bounds for vertex cover. In: Královič, R., Urzyczyn, P. (eds.) MFCS 2006. LNCS, vol. 4162, pp. 238–249. Springer, Heidelberg (2006). https://doi.org/10.1007/11821069_21
10. Chen, J., Liu, Y., Lu, S., O'Sullivan, B., Razgon, I.: A fixed-parameter algorithm for the directed feedback vertex set problem. In: Dwork, C. (ed.) Proceedings of the 40th Annual ACM Symposium on Theory of Computing, Victoria, British Columbia, Canada, 17–20 May 2008, pp. 177–186. ACM (2008)
11. Clark, B.N., Colbourn, C.J., Johnson, D.S.: Unit disk graphs. Discret. Math. **86**(1–3), 165–177 (1990)
12. Cygan, M., et al.: Parameterized Algorithms. Springer, Heidelberg (2015). https://doi.org/10.1007/978-3-319-21275-3
13. Dabrowski, K., Lozin, V., Müller, H., Rautenbach, D.: Parameterized algorithms for the independent set problem in some hereditary graph classes. In: Iliopoulos, C.S., Smyth, W.F. (eds.) IWOCA 2010. LNCS, vol. 6460, pp. 1–9. Springer, Heidelberg (2011). https://doi.org/10.1007/978-3-642-19222-7_1
14. Dörfler, J., Roth, M., Schmitt, J., Wellnitz, P.: Counting induced subgraphs: an algebraic approach to #w[1]-hardness. In: Rossmanith, P., Heggernes, P., Katoen, J. (eds.) 44th International Symposium on Mathematical Foundations of Computer Science, MFCS 2019, 26–30 August 2019, Aachen, Germany. LIPIcs, vol. 138, pp. 26:1–26:14. Schloss Dagstuhl - Leibniz-Zentrum für Informatik (2019). https://doi.org/10.4230/LIPIcs.MFCS.2019.26
15. Downey, R.G., Fellows, M.R.: Fixed-parameter tractability and completeness II: on completeness for W[1]. Theor. Comput. Sci. **141**(1 & 2), 109–131 (1995)
16. Downey, R.G., Fellows, M.R.: Fundamentals of Parameterized Complexity. Texts in Computer Science, Springer, London (2013). https://doi.org/10.1007/978-1-4471-5559-1
17. Eppstein, D.: Subgraph isomorphism in planar graphs and related problems. J. Graph Algorithms Appl. **3**(3), 1–27 (1999)
18. Eppstein, D., Havvaei, E.: Parameterized leaf power recognition via embedding into graph products. Algorithmica **82**(8), 2337–2359 (2020)
19. Eppstein, D., Strash, D., Löffler, M.: Listing all maximal cliques in large sparse real-world graphs in near-optimal time. J. Exp. Algorithmics **18**(3), 3.1 (2013). https://doi.org/10.1145/2543629
20. Fellows, M.R., Hermelin, D., Rosamond, F.A., Vialette, S.: On the parameterized complexity of multiple-interval graph problems. Theor. Comput. Sci. **410**(1), 53–61 (2009)
21. Garey, M.R., Johnson, D.S.: Computers and Intractability: A Guide to the Theory of NP-Completeness. W. H. Freeman (1979)

22. Gomes, G.C.M., dos Santos, V.F., da Silva, M.V.G., Szwarcfiter, J.L.: FPT and kernelization algorithms for the induced tree problem. In: Calamoneri, T., Corò, F. (eds.) CIAC 2021. LNCS, vol. 12701, pp. 158–172. Springer, Cham (2021). https:// doi.org/10.1007/978-3-030-75242-2_11
23. Harary, F.: Graph Theory. Addison-Wesley, Boston (1991)
24. Hermelin, D., Mnich, M., van Leeuwen, E.J.: Parameterized complexity of induced graph matching on claw-free graphs. Algorithmica **70**(3), 513–560 (2014)
25. Jerrum, M., Meeks, K.: The parameterised complexity of counting connected subgraphs and graph motifs. J. Comput. Syst. Sci. **81**(4), 702–716 (2015). https:// doi.org/10.1016/j.jcss.2014.11.015
26. Khot, S., Raman, V.: Parameterized complexity of finding subgraphs with hereditary properties. Theor. Comput. Sci. **289**(2), 997–1008 (2002)
27. Lewis, J.M., Yannakakis, M.: The node-deletion problem for hereditary properties is NP-complete. J. Comput. Syst. Sci. **20**(2), 219–230 (1980)
28. Marx, D.: Efficient approximation schemes for geometric problems? In: Brodal, G.S., Leonardi, S. (eds.) ESA 2005. LNCS, vol. 3669, pp. 448–459. Springer, Heidelberg (2005). https://doi.org/10.1007/11561071_41
29. Marx, D.: Parameterized complexity of independence and domination on geometric graphs. In: Bodlaender, H.L., Langston, M.A. (eds.) IWPEC 2006. LNCS, vol. 4169, pp. 154–165. Springer, Heidelberg (2006). https://doi.org/10.1007/11847250_14
30. Nešetřil, J., de Mendez, P.O.: 18.3 the subgraph isomorphism problem and boolean queries. In: Nešetřil, J., de Mendez, P.O. (eds.) Sparsity: Graphs, Structures, and Algorithms, Algorithms and Combinatorics, vol. 28, pp. 400–401. Springer, Heidelberg (2012). https://doi.org/10.1007/978-3-642-27875-4
31. Poljak, S.: A note on stable sets and colorings of graphs. Comment. Math. Univ. Carolinae **15**, 307–309 (1974)
32. Szeider, S.: On fixed-parameter tractable parameterizations of SAT. In: Giunchiglia, E., Tacchella, A. (eds.) SAT 2003. LNCS, vol. 2919, pp. 188–202. Springer, Heidelberg (2004). https://doi.org/10.1007/978-3-540-24605-3_15

The Space Complexity of Sum Labelling

Henning Fernau[1] and Kshitij Gajjar[2(✉)]

[1] Universität Trier, FB 4 – Informatikwissenschaften, Trier, Germany
fernau@uni-trier.de
[2] National University of Singapore, Singapore, Singapore

Abstract. A graph is called a *sum graph* if its vertices can be labelled by distinct positive integers such that there is an edge between two vertices if and only if the sum of their labels is the label of another vertex of the graph. Most papers on sum graphs consider combinatorial questions like the minimum number of isolated vertices that need to be added to a given graph to make it a sum graph. In this paper, we initiate the study of sum graphs from the viewpoint of computational complexity. Note that every n-vertex sum graph can be represented by a sorted list of n positive integers where edge queries can be answered in $\mathscr{O}(\log n)$ time. Thus, limiting the size of the vertex labels upper-bounds the space complexity of storing the graph.

We show that every n-vertex, m-edge, d-degenerate graph can be made a sum graph by adding at most m isolated vertices to it such that the size of each vertex label is at most $\mathscr{O}(n^2 d)$. This enables us to store the graph using $\mathscr{O}(m \log n)$ bits of memory. For sparse graphs (graphs with $\mathscr{O}(n)$ edges), this matches the trivial lower bound of $\Omega(n \log n)$. Since planar graphs and forests have constant degeneracy, our result implies an upper bound of $\mathscr{O}(n^2)$ on their label size. The previously best known upper bound on the label size of general graphs with the minimum number of isolated vertices was $\mathscr{O}(4^n)$, due to Kratochvíl, Miller & Nguyen [23]. Furthermore, their proof was existential whereas our labelling can be constructed in polynomial time.

Keywords: Sum labelling · Exclusive labelling · Graph databases · Sparse graphs · Graph representations · Space complexity

1 Introduction

There is a vast body of literature on graph labelling, testified by a dynamic survey on the topic maintained by Gallian [11]. The 553-page survey (last updated in 2020) mentions over 3000 papers on different ways of labelling graphs. We focus on a type of labelling introduced by Harary [16] in 1990, called sum labelling.

Part of this work was done when the author was a researcher at Technion, Israel. This project has received funding from the European Union's Horizon 2020 research and innovation programme under grant agreement No. 682203-ERC-[Inf-Speed-Tradeoff].

E. Bampis and A. Pagourtzis (Eds.): FCT 2021, LNCS 12867, pp. 230–244, 2021.
https://doi.org/10.1007/978-3-030-86593-1_16

Definition 1. *A simple, undirected, unweighted graph G is a sum graph if there exists an injective function $\lambda : V(G) \rightarrow \mathbb{N}$ such that for all vertices $v_1, v_2 \in V(G)$,*

$$(v_1, v_2) \in E(G) \quad \Longleftrightarrow \quad \exists v_3 \in V(G) \;\; s.t. \;\; \lambda(v_1) + \lambda(v_2) = \lambda(v_3).$$

Then we say that λ is a sum labelling *of (the vertices of) G.*

Note that Definition 1 implies that the edge set of a sum graph G can be recovered completely using the labels on its vertex set given by λ.

Gould & Rödl [14] showed that every n-vertex graph can be made a sum graph by adding at most n^2 isolated vertices to it. In fact, certain graphs can be encoded much more succinctly with sum labelling than with the more traditional methods of storing a graph (e.g., adjacency matrix, incidence matrix, adjacency list). This makes sum labelling an intriguing concept not just to mathematicians but to computer scientists, as well. Sum labelling could also be of interest in graph databases [1,2,24] and in collections of benchmark graphs [6,18,25]. However, no systematic study of this question has been undertaken so far. With this paper, we intend to start such a line of research, bringing sum labellings closer to the research in *labelling schemes* [19]. To the best of our knowledge, the only known application of sum labelling before our work is in secret sharing schemes [38].

The idea of using sum labelling to efficiently store graphs was already considered by Sutton [40]. However, Sutton focused on the number of additional isolated vertices needed to store a given graph, whereas our focus is on the number of *bits* that are needed to store the graph.

In other words, while Sutton's work attempts to minimize the number of additional vertices, it does not take into account the *size* of the vertex labels required to do so. This is crucial because it is known that there are several graph families for which the size of the vertex labels grows exponentially with the number of vertices. One popular example is the sum labelling scheme for trees presented by Ellingham [9]. Another example is the more esoteric graph family known as the generalised friendship graph [10].

Another parameter associated with sum graphs is the difference between the largest and smallest label, called *spum* (also called *range* in [23]). Interestingly, while the concept of spum was around for quite some time (Gallian's survey [11] refers to an unpublished manuscript by a group of six students), the first publication that studies spum for different basic classes of graphs is a very recent one [37]. Unfortunately, this measure also does not reflect the whole truth about storing graphs, as it neglects the number of additional vertices that need to be stored. Moreover, spum is somewhat dependent on the definition of the sum number (see below for a formal definition), which might be slightly unnatural for the purpose of storing a graph.

In this paper, we introduce a new graph parameter $\sigma_{\mathbf{store}}$ that takes into account both the number of additional vertices and their label size. We explain this formally in the next section.

2   Definitions and Main Result

Let us now fix some notation in order to formally introduce the concepts in this paper.

As isolated vertices (i.e., vertices of degree zero) are usually irrelevant in applications, $\lambda(V)$ can be also viewed as the description of $G - I$, where I collects all isolates of G. Then, $\lambda(V)$ is called the *sum number encoding* of $G - I$. Conversely, given a graph G without isolates, the minimum number of isolates needed to be added in order to turn G into a sum graph is called the *sum number* of G, written $\sigma(G)$, i.e., $G + \overline{K_{\sigma(G)}}$ is a sum graph. Here, $+$ is used to denote the disjoint union of graphs, $\overline{H}$ denotes the complement of graph H, and K_n is the complete graph on n vertices, so that $\overline{K_n}$ is the null (i.e., edgeless) graph on n vertices. The *spum* of G, written spum(G), is defined as the minimum over all sum labellings of $G + \overline{K_{\sigma(G)}}$ of the difference between the maximum and minimum labels.

A labelling function λ can be also seen as operating on edges by the summability condition: $\lambda(e)$ for an edge $e = xy \in E$ is defined as $\lambda(x) + \lambda(y)$. A labelling of a sum graph $G = (V, E)$ is called an *exclusive* sum labelling [27,29,34,44] if for every $e \in E$, $\lambda(e) = \lambda(i)$ for some $i \in I$, where I is the set of isolated vertices of G. Accordingly, $\epsilon(G)$ denotes the *exclusive sum number* of G.

Are substantial savings possible when considering sum number encodings of graphs? As most research in the area of sum labellings went into studying quite specific families of graphs, some partial answers are possible. For instance, analyzing the expositions in [33,45], one sees that for the complete bipartite graph $K_{n,n+1}$, with n vertices in one partition and $n + 1$ vertices in the other, $\sigma(K_{n,n+1}) = 2n - 1$. In other words, in order to represent $K_{n,n+1}$, we need $4n$ numbers. Ignoring the size of these numbers, this is a clear advantage over any traditional way to store the complete bipartite graph $K_{n,n+1}$, which would need $\mathcal{O}(n^2)$ bits. However, after a closer look at the labelling presented in [33], it becomes clear that the numbers needed to label a $K_{m,n}$ are of size $\mathcal{O}(nm)$. Therefore, it is clear that storing the complete bipartite graph $K_{n,n+1}$ needs $\mathcal{O}(n \log n)$ bits only, if we take its sum graph encoding. As we will see later, this is in fact storage-optimal in a certain sense.

Similarly, $\sigma(K_n) = 2n - 3$ is known for $n \geq 4$, i.e., $3n - 3$ numbers are necessary to store the information about the complete graph K_n, while again traditional methods would need $\mathcal{O}(n^2)$ bits. As mentioned in [39], this can be obtained by labelling vertex x_i with $4i - 3$, with $1 \leq i \leq n$, leading to isolate labels $4j + 2$ for $1 \leq j \leq 2n - 3$. Hence, the sizes of the labels are in fact linear in n, which is, in a sense, even better than what is known for complete bipartite graphs. We will continue our discussions on storage issues in the next section. It is known that the sum number of general graphs will grow with the order of its edges, see [31]. In fact, this can happen even with sparse graphs, see [17,42].

As we have seen so far, neither the sum number of a graph nor the spum of a graph models the storage requirements of storing graphs with the help of sum numberings in a faithful manner. Therefore, we suggest another graph parameter, based on

$$\textbf{storage}(\lambda, G) = \sum_{v \in V} \lceil \log_2(\lambda(v)) \rceil \le |V| \max_{v \in V} \lceil \log_2(\lambda(v)) \rceil \qquad (1)$$

for a labelling $\lambda : V \to \mathbb{N}$ of a sum graph $G = (V, E)$. (Recall that one can store variable-size numbers using at most two times as many bits when compared to Eq. (1) with Elias prefix codes [8].) Now, define

$$\textbf{storage}(G) = \min\{\textbf{storage}(\lambda, G) \mid \exists \lambda : V \to \mathbb{N} : \lambda \text{ labels } G\}.$$

Then, for an arbitrary graph $G' = (V', E')$ one could define

$$\sigma_{\textbf{store}}(G') = \min\{\textbf{storage}(G) \mid \exists s \in \mathbb{N} : G = G' + \overline{K_s} \text{ is a sum graph}\}.$$

For instance, the construction of Ellingham can be used to state: For an n-vertex tree T, Ellingham's construction leads to $\sigma_{\textbf{store}}(T) \in \mathcal{O}(n^2)$. This should be compared to a standard representation of trees by adjacency lists that obviously needs $\mathcal{O}(n \log(n))$ space. However, our results prove that also with sum label representations, this upper bound can be obtained. In our construction, it is crucial that we also consider labellings that do not lead to a minimum sum number. This is also a difference concerning the definition of spum.

As we are mostly interested in upper-bounding $\sigma_{\textbf{store}}(G')$ in this paper, we are in fact mostly discussing

$$\sigma_{\textbf{store}}^{\max}(G') = \min\{\textbf{storage}^{\max}(G) \mid \exists s \in \mathbb{N} : G = G' + \overline{K_s} \text{ is a sum graph}\},$$

where for a sum graph $G = (V, E)$,

$$\textbf{storage}^{\max}(G) = \min\{\textbf{storage}^{\max}(\lambda, G) \mid \exists \lambda : V \to \mathbb{N} : \lambda \text{ labels } G\},$$

with

$$\textbf{storage}^{\max}(\lambda, G) = |V| \cdot \max_{v \in V} \lceil \log_2(\lambda(v)) \rceil = |V| \cdot \lceil \log_2(\max \lambda(V)) \rceil .$$

By Eq. (1), $\sigma_{\textbf{store}}(G') \le \sigma_{\textbf{store}}^{\max}(G')$.

We are now ready to formulate the main result of this paper.

Theorem 1. *Let G' be a graph on n vertices and m edges with minimum degree at least one. Then, $\sigma_{store}^{max}(G') \in \mathcal{O}(m \cdot \log(n))$. More specifically,*

$$\sigma_{store}^{max}(G') \le 9m(\log_2(n) + 1)$$

for general graphs and

$$\sigma_{store}^{max}(G') \le 3m(2 \log_2(n) + \log_2(12d)) < 3dn(2 \log_2(n) + \log_2(12d))$$

for d-degenerate graphs. Furthermore, the corresponding sum labellings can be computed in polynomial time.

In particular, this means that only $\mathcal{O}(n \log(n))$ bits are necessary to store trees with sum labellings, as they are 1-degenerate graphs. A similar result holds for planar graphs, as they are 5-degenerate. We also show that these bounds are optimal for storing graphs, up to constant factors. We also relate to the literature on adjacency labelling schemes, see, e.g., [19,32], or, more recently [4,7].

3   Labelling a Disjoint Collection of Edges

This section should be treated as an introductory exercise on sum labelling, and has no bearing on our main result. A reader familiar with sum labelling schemes may skip to the next section.

It is known that trees have sum number 1; according to a remark following Theorem 5.1 in [9], this result translates to forests. However, the label sizes may grow exponentially in these constructions. As a warm-up and to explain the difficulties encountered when designing sum labellings, we are going to present three constructions that label a disjoint collection of edges, or, more mathematically speaking, how to label a 1-regular graph M_n on n vertices.

Exponential Solution. If you have n vertices (hence $n/2$ edges), label the first edge with $2 - 3$, the second one starts with the sum of the labels of the previous edge, i.e., in the beginning, this is 5, we continue with the successor 6, then we add up the previous two labels, continue with the successor, etc. ...

This can be brought into the following sum labelling scheme $\lambda : \mathbb{N} \to \mathbb{N}$ that is meant to work for any 1-regular graph:

$$\lambda(n) = \begin{cases} 2 & \text{if } n = 1 \\ \lambda(n-1) + 1 & \text{if } n \text{ is even} \\ \lambda(n-2) + \lambda(n-1) & \text{if } n \text{ is odd and } n > 1 \end{cases} \quad (2)$$

Lemma 1. *For the labelling defined in Eq. (2), we find that $\lambda(n) \in \Theta(\sqrt{2}^n)$.*

Linear Solution. Assume that n is an even number. Consider the following sum labelling scheme for 1-regular graphs on n vertices. We group endpoint labels of each edge together by parentheses.

$$(n, 2n - 1), (n + 1, 2n - 2), \ldots, \left(\frac{3n}{2} - 1, \frac{3n}{2}\right).$$

All edge labels sum up to $3n - 1$ (which is the isolate), and even the sum of the two smallest labels, i.e., $n + (n + 1) = 2n + 1$, is smaller than $3n - 1$ but bigger than any other label in the graph. As each label is in $\Theta(n)$, the overall space requirement of this labelling scheme is $\Theta(n \log(n))$. Moreover, $\sigma(M_n) = 1$.

The Union of Many Identical Components. The previous consideration was quite special to 1-regular graphs. We are now developing an argument that can be generalised towards a certain type of graph operation. One can think of M_n as being the disjoint graph union of $n/2$ times M_2. For simplicity of the exposition, assume $n/2 = 2^d$ in the following. Label the vertices $v_{1,1}, v_{2,1}, v_{1,2}, v_{2,2}, \ldots, v_{1,2^d}, v_{2,2^d}$ of M_n (with edges between $v_{1,j}$ and $v_{2,j}$) as follows, for $j = 1, \ldots, 2^d$:

$$\lambda(v_{1,j}) = 1 + 8 \cdot (j - 1) + 2^{4+d} \cdot (2^d - j)$$
$$\lambda(v_{2,j}) = 2 + 8 \cdot (2^d - j) + 2^{5+d} \cdot (j - 1)$$

For instance, for $d = 2$, we get $\lambda(v_{1,1}) = 1+8\cdot0+64\cdot3$, $\lambda(v_{2,1}) = 2+8\cdot3+64\cdot0$, so that the connecting edge is testified by the isolate label $3+8\cdot3+64\cdot3 = 219 = (11011011)_2$. Also, $\lambda(v_{1,2}) = 1+8\cdot1+64\cdot2$, $\lambda(v_{2,2}) = 2+8\cdot2+64\cdot1$, adding up again to 219. Likewise, $\lambda(v_{1,3}) = 1+8\cdot2+64\cdot1$, $\lambda(v_{2,3}) = 2+8\cdot1+64\cdot2$, and finally $\lambda(v_{1,4}) = 1+8\cdot3+64\cdot0$ and $\lambda(v_{2,4}) = 2+8\cdot0+64\cdot3$. By construction, all numbers need at most $2d + 4$ bits for labelling 2^{d+1} vertices. Hence, the overall space requirement for storing M_n is again $\mathscr{O}(n\log(n))$ bits.

Notice that the zero bit introduced in the third and sixth binary position in the example is important to avoid that any labels add up to another valid label but the ones of the edge endpoints. This technique can be easily generalised to obtain the following result:

Lemma 2. *Let G be any fixed graph. Then, the n-fold disjoint graph union G_n of G with itself obeys $\sigma_{store}(G_n) \in \mathscr{O}(n\log(n))$. Moreover, $\sigma(G_n) \leq \sigma(G)$.* □

4   Storing Graphs Using Sum Labelling

Alternative Notions. One of our motivations to return to sum labellings was the idea that one can use them to efficiently store graphs. This idea was already expressed in [23]. There, they consider the notion of the *range* of a sum graph G that is realizing $\sigma(G')$, which happens to coincide with the notion called spum later. But following this motivation (to store graphs), let us define the *range* of a labelling λ of a sum graph $G = (V, E)$ as the difference between $\max\lambda(V)$ and $\min\lambda(V)$. The idea behind is that it would suffice to store the numbers $\lambda(v) - \min\lambda(V)$ for all vertices $v \in V$, plus the value of $\min\lambda(V)$ once, instead of storing all values $\lambda(v)$, which could help us save some bits. The following lemma tells us that this variation in our considerations (which could also lead to variations of the our definition of σ_{store} and related notions) is not essential for our current considerations, as we mostly neglect constant factors. In particular, we might consider $|V| \cdot \lceil\log_2(\max\lambda(V) - \min\lambda(V))\rceil + \lceil\log_2(\min\lambda(V))\rceil$ as a more appropriate definition of the maximum estimate of the storage requirements of a sum graph $G = (V, E)$ with respect to a sum labelling λ.

Lemma 3. *Let λ be a sum labelling of a non-empty sum graph $G = (V, E)$, and let* $\mathrm{range}(\lambda(V)) = \max\lambda(V) - \min\lambda(V)$. *Then,*

$$\mathrm{range}(\lambda(V)) > \min\lambda(V);$$
$$2 \cdot \mathrm{range}(\lambda(V)) > \max\lambda(V).$$

Hence, we know that $\max\lambda(V) \in \Theta(\mathrm{range}(\lambda(V)))$.

What is the main purpose of a graph database? Clearly, one has to access the graphs. A basic operation would be to answer the query if there is an edge between two vertices. Now, if $\max\lambda(V)$ of a sum graph is polynomial in the number $n = |V|$ of its vertices, we can answer this query in time $\mathscr{O}(\log(n))$, a property also discussed as *adjacency labelling scheme* by Peleg [32]. Namely, assuming the polynomial bound on the size of the labels, we would need time

$\mathcal{O}(\log(n))$ to add the two labels of the vertices, and we also need time $\mathcal{O}(\log(n))$ to search for the sum in the ordered list of numbers, using binary search, because there are only $\mathcal{O}(n^2)$ many numbers needed to describe a graph. If $\max \lambda(V)$ would be super-polynomial, then the additional time $\mathcal{O}(\log(\max \lambda(V)))$ would be quite expensive, which probably makes the idea of storing large graphs as sum graphs in databases unattractive. This motivates in particular also considering $\max \lambda(V)$ of the labelling λ of a sum graph.

Lower Bounds. How many bits are really necessary to store graphs? We will discuss lower and upper bounds in the following, starting with a lower bound.

Lemma 4. *Given an n-vertex graph G, $\sigma_{store}^{max}(G), \sigma_{store}(G) \in \Omega(n \log n)$.*

This proves that, up to constants, a sum labelling that uses $\mathcal{O}(n \log(n))$ bits only is storage-optimal. This gives one of the motivations underlying the discussions in the next section. Moreover, $\mathcal{O}(n \log(n))$ is also the space requirement that is needed for storing sparse graphs in traditional graph storing methods. More precisely, just for writing down the names of the vertices, $\Omega(n \log n)$ bits are needed, as can be seen by a calculation similar to Lemma 4. But we have already seen examples like the complete graphs or the complete bipartite graphs, where we also find sum labellings that use only $\mathcal{O}(n \log(n))$ bits for storing them. Labellings using $\mathcal{O}(n \log(n))$ bits will be shown for graphs of fixed degeneracy in Theorem 3.

Upper Bounds. Here, we start our discussion on upper bounds for the storage requirement of storing graphs with sum labellings. First, we briefly discuss the number of isolates in this respect. Based on some probabilistic arguments, it is known that the number of isolates is about the number of edges of the graph to be encoded [14,31] for nearly all graphs.

Remark 1. As there are $2^{\Theta(n^2)}$ many graphs on n vertices, we cannot hope for a sum labelling scheme that uses only $n^{2-\varepsilon}$ many isolates and only polynomial-size labels and hence a polynomial range, because we need at least $\Omega(n^2)$ many bits just to write down n-vertex graphs. As an aside, allowing for n^2 many isolates also means always allowing exclusive labellings.   □

Conversely, assuming that we can sum-label each n-vertex graph with m edges with labels of polynomial size, then we can upper-bound σ_{store} by $\mathcal{O}(m \log(n))$. By our discussions from Lemma 4 and Remark 1, we cannot hope for anything substantially better. Can we reach this bound? Unfortunately, this seems to be an open question that we will answer to some extent below in our main result. In [23], it was shown that each n-vertex graph without isolates can be represented by a sum labelling that uses number not bigger than 4^n. In other words, one would need at most $2n$ bits to represent each vertex of an n-vertex graph. This also shows that sum graphs have a *constrained 1-labelling scheme* as defined in [19]. Hitherto, it was unknown how to sum-label arbitrary graphs with polynomial-size labels. As our main result, we are going solve this problem affirmatively, with nice consequences for d-degenerate graphs.

5   A Novel Algorithm for Sum Labelling

We will now prove our main result, showing that sum labellings can be used to store graphs (without isolated vertices) as efficiently as traditional methods can do. Notice that the two theorems shown in this section (Theorems 2, 3) imply Theorem 1.

Theorem 2. *Every n-vertex graph G of minimum degree at least one can be turned into a sum graph H by adding at most m isolates to G, such that H admits a sum labelling scheme λ satisfying*

$$\lambda(v) \leq 4 \cdot n^3 \qquad\qquad \forall\, v \in V(G); \qquad\qquad (3)$$

$$\lambda(v) \leq 8 \cdot n^3 \qquad\qquad \forall\, v \in V(H). \qquad\qquad (4)$$

Our sum labelling is an exclusive labelling, computable in polynomial time.

Proof. Note that Eq. (3) implies Eq. (4), as isolate labels are sums of labels of $V(G)$. So we will focus on showing Eq. (3) in this proof. Let the vertices of G be $\{v_1, v_2, \ldots, v_n\}$. Let G_i be the induced subgraph on the first i vertices of G, i.e.,

$$V(G_i) = \{v_1, v_2, \ldots, v_i\}.$$

For each G_i $(2 \leq i \leq n)$, we will show that there is a sum graph H_i which can be obtained by adding $r_i \leq \binom{i}{2}$ isolates to G_i (since G_i has at most $\binom{i}{2}$ edges), satisfying $\lambda(v) \leq 4 \cdot i^3$ for each $v \in V(G_i)$. Moreover, all vertices of G_i will carry labels that equal 1 modulo 4, and all isolates in H_i will carry labels that equal 2 modulo 4. This modulo condition ensures that our labelling is external. Our proof is by induction on i, yielding the claimed polynomial-time algorithm.

Although the statement of the theorem makes sense only from $n \geq 2$ onwards to meet the minimum-degree requirement, it is convenient for our inductive proof to start with $i = 1$:

Base Case $(i = 1)$: We set $\lambda(v_1) = 1$. Notice that $\lambda(v_1) = 1^3$. Set $r_1 = 0$.

Induction Hypothesis: There is a sum graph H_i for G_i such that H_i has r_i isolates (in other words, $H_i = G_i \cup \{\mathsf{iso}_1, \mathsf{iso}_2, \ldots, \mathsf{iso}_{r_i}\}$), where $r_i \leq \binom{i}{2}$, and $\lambda(v) \leq 4 \cdot i^3$ for each $v \in V(G_i)$. Moreover, all vertices of G_i carry labels that equal 1 modulo 4, and all isolates in H_i carry labels that equal 2 modulo 4.

Induction Step: We add the vertex v_{i+1} to the graph H_i and connect it to its neighbours in G_i. Suppose v_{i+1} has t_i neighbours $\{v_{j_1}, v_{j_2}, \ldots, v_{j_{t_i}}\}$ in G_i. Then add t_i isolates $\{\mathsf{iso}_{r+1}, \mathsf{iso}_{r+2}, \ldots, \mathsf{iso}_{r_i+t_i}\}$ to H_i, giving the graph H_{i+1}. Thus,

$$H_{i+1} = G_{i+1} \cup \{\mathsf{iso}_1, \mathsf{iso}_2, \ldots, \mathsf{iso}_{r_i+t_i}\}.$$

We define $r_{i+1} = r_i + t_i$. Next, we set the labels for the newly added vertices. If λ is not a valid sum labelling for H_{i+1}, then we will change their λ-values of the newly added vertices. We will show that their λ-values only need to be changed less than i^3 times until we reach a valid sum labelling for H_{i+1}.

$$\lambda(v_{i+1}) = 5; \qquad\qquad (5)$$

$$\lambda(\mathsf{iso}_{r_i+k}) = \lambda(v_{i+1}) + \lambda(v_{j_k}) \quad \forall\, k \in \{1, 2, \ldots, t_i\}. \qquad\qquad (6)$$

Claim. λ is a valid sum labelling of H_{i+1} if and only if it has none of the following *violations*.

(i) A violating pair: an ordered set of two vertices (u, w) from G_i such that $\lambda(u) = \lambda(w)$.
(ii) A violating triple: an ordered set of three vertices (u, w, y) such that $\lambda(u) < \lambda(w) < \lambda(y)$ and $\lambda(u) + \lambda(w) = \lambda(y)$ and $(u, w) \notin E(H_{i+1})$.

Observations. Notice that it could happen that some of the 'new' isolates in H_{i+1} carry labels that are already labels of isolates from H_i. In that case, we implicitly merge these isolates, which automatically avoids violating pairs among them. Then, the number t_i is decreased accordingly. Also, the modulo 4 arithmetics prevent vertices from G_i and the isolates to pair up as a violating pair. (∗)

Proof (of the claim). It is easy to see that if H_{i+1} has any of the above violations, then λ is not a valid sum labelling of H_{i+1}. Now we will prove the other direction: if λ is not a valid sum labelling of H_{i+1}, then it either has a violating pair or a violating triple.

Note that H_{i+1} has $i + r_i + t_i + 1 = (i + 1) + r_{i+1}$ many vertices, each with its corresponding λ-value. If two of the vertices have the same λ-value, then it is a type (i) violation, and we are done. So, we assume that all the λ-values are distinct. Given these $(i + 1) + r_{i+1}$ distinct numbers, we construct their corresponding sum graph H'_{i+1} on $(i + 1) + r_{i+1}$ vertices using the sum labelling property.

Both H_{i+1} and H'_{i+1} have the same set of vertices and the same labelling scheme λ. However, since λ is a valid labelling scheme for H'_{i+1} but not for H_{i+1}, they cannot have the same set of edges. Furthermore, H_{i+1} is a subgraph of H'_{i+1}. This is because every edge $e = (u, w)$ of H_{i+1} is either an edge that was also present in H_i (in which case there is a vertex labelled $u + w$ in H_{i+1} and H'_{i+1}, since H_i is a sum graph by the induction hypothesis), or it is one of the t_i new edges added (in which case one of the t_i new isolates $\{\mathsf{iso}_{r_i+1}, \mathsf{iso}_{r_i+2}, \dots, \mathsf{iso}_{r_{i+1}}\}$ is labelled $u + w$ by Eq. (6)).

Due to (∗), the only way for the edge sets of H_{i+1} and H'_{i+1} to differ is if there is an edge $e = (u, w)$ such that $e \in E(H'_{i+1})$ and $e \notin E(H_{i+1})$. This means there are three vertices $\{u, w, y\}$ in H'_{i+1} (and so also in H_{i+1}) such that $\lambda(u) + \lambda(w) = \lambda(y)$, a type (ii) violation.   ◇

Now, if H_{i+1} is a sum graph with the labelling scheme derived from Eq. (5) and Eq. (6), then we are done. Otherwise, we will (slightly) modify these labels to obtain a new labelling scheme, as follows.

$$\lambda(v_{i+1}) \leftarrow \lambda(v_{i+1}) + 4; \tag{7}$$
$$\lambda(\mathsf{iso}_{r_i+k}) \leftarrow \lambda(\mathsf{iso}_{r_i+k}) + 4. \tag{8}$$

We again check if with these new labels, H_{i+1} is a sum graph. If not, we increment these values by 4 again. We keep doing this until H_{i+1} becomes a sum graph.

The crucial point to note is that each time we increment by 4, at least one of the violations disappears, never to occur again.

To fully understand this last sentence, we need to refine our analysis of potential conflicts that might occur when running our algorithm. Namely, following up on the proof of the previous claim, consider three vertices $\{u, w, y\}$ in H_{i+1} such that (erroneously) $\lambda(u) + \lambda(w) = \lambda(y)$ in the labelling λ of H_{i+1}. First observe that not all vertices from $\{u, w, y\}$ can be isolates, as all isolates carry labels that are 2 modulo 4. As we know that λ, restricted to the vertices of H_i, turns H_i into a sum graph, not all of the vertices $\{u, w, y\}$ belong to H_i. If y is one of the isolates of H_i, then its labelling will not change when updating λ according to Eq. (8). As one of the vertices u, w does not belong to H_i, we have, w.l.o.g., $u \in V(H_i)$ and $w = v_{i+1}$, because if w would be among the isolates, the sum of the labels of u and w would equal 0 modulo 4, but all isolates carry labels that are 2 modulo 4. This means that out of the three labels of u, w, y, exactly one will change according to Eq. (7) and as it will also be the only one that might increase in further modifications, a violation will never re-appear in the triple $\{u, w, y\}$. Assume now that y is one of the new isolates, say, $y = \mathsf{iso}_{r_i+1}$. If exactly one of the two other vertices, say, u, already belongs to H_i, then the other one, w, must be v_{i+1}. As $\lambda(u) + \lambda(w) = \lambda(y) = \lambda(\mathsf{iso}_{r_i+1})$, we must have $u = v_{j_1}$, as we have no violating pairs. However, this means that the edge $\{u, w\}$ belongs both to H_{i+1} and to H'_{i+1}, contradicting our assumption. Therefore, if y is one of the new isolates, then both u and w must belong to H_i. This means that the labellings of u and of w will never change by the re-labellings described in Eqs. (7) and (8), while the labelling of y will only (further) increase, so that indeed a violation will never re-appear in the triple $\{u, w, y\}$.

How often might we have to update a labelling when moving from H_i to a valid sum graph H_{i+1}? Our previous analysis shows that the following two scenarios could be encountered for a violating triple $\{u, w, y\}$:

- y is an isolate of H_i and exactly one of $\{u, w\}$ belongs to $V(H_i)$. There are $i \cdot r_i$ many cases when this might occur.
- y is a new isolate and $\{u, w\} \subseteq V(H_i)$. There are $t_i \cdot \binom{i}{2} = t_i \cdot i(i-1)/2$ many possibilities for this situation.

Recall that r_i isolates are contained in the sum graph H_i and $t_i = r_{i+1} - r_i$ isolates are newly added to yield H_{i+1}. Our analysis shows that after at most $s_i = i \cdot r_i + t_i \cdot i(i-1)/2$ many steps, a valid sum labelling of H_{i+1} was found. By observing that r_i cannot be bigger than the number $\binom{i}{2} = i(i-1)/2$ of hypothetical edges in H_i, and t_i is upper-bounded by the number i of vertices in H_i, we can furthermore estimate:

$$s_i \leq i \cdot i(i-1)/2 + i \cdot i(i-1)/2 = i^3 - i^2 .$$

By induction hypothesis, we know that for each of the i vertices v in H_i, we have $\lambda(v) \leq i^3$. As H_i contains only i vertices that are labelled with numbers that are equal to 1 modulo 4, within at most $i^3 - i^2$ increment steps, we will find a label for v_{i+1} that is no bigger than $4 \cdot (i^3 - i^2) + 1 \leq (\sqrt[3]{4}(i+1))^3$, basically using the pigeon hole principle. As all labels of isolates are sums of labels of vertices from G_i, their sizes are upper-bounded by $4i^3 + 4(i-1)^3 < 8 \cdot i^3$.   □

Proof (of Theorem 1.1). Theorem 2 gives an upper bound of $(n+m)(\log(8n^3))$ on the total number of bits needed to store H. Since every vertex in the graph G has degree at least one, we have $n \leq 2m$. Substituting, we get an upper bound of $3m(\log(8n^3)) \leq 3m(3\log n + 3) = 9m(\log n + 1)$, as required by Theorem 1. □

As we always start with setting the label of the first vertex to one, the obtained labelling uses the number one as a label. Notice that this is related to the (to the best of our knowledge, still open) question if every graph G (without isolates) can be embedded into a sum graph H with $\sigma(G)$ many isolates such that there is a sum labelling λ of H such that $1 \in \lambda(V(H))$, see [28] (also, the authors of [21] study a relaxation of this question).

We will now look into a specific class of sparse graphs, namely, into d-degenerate graphs. Recall that a graph $G = (V, E)$ is d-degenerate if its vertices can be ordered like $V = \{v_1, v_2, \ldots, v_n\}$ such that, considering the graph G_i induced by the vertex set $V_i = \{v_1, \ldots, v_i\}$, v_i has degree at most d in G_i. We will call such an ordering a *d-degenerate vertex ordering*. Similar to Theorem 2, we can show the following result.

Theorem 3. *Every d-degenerate n-vertex graph G of minimum degree at least one can be made a sum graph H by adding at most m isolates to G, such that H admits a sum labelling scheme λ satisfying*

$$\lambda(v) \leq 6d \cdot n^2 \qquad\qquad \forall v \in V(G); \qquad\qquad (9)$$
$$\lambda(v) \leq 12d \cdot n^2 \qquad\qquad \forall v \in V(H). \qquad\qquad (10)$$

This sum labelling is an exclusive labelling, computable in polynomial time.

Proof (of Theorem 1.2). Theorem 3 gives an upper bound of $(n+m)(\log(12dn^2))$ on the total number of bits needed to store H. Since every vertex in the graph G has degree at least one, we have $n \leq 2m$. Substituting, we get an upper bound of $3m(\log(12dn^2)) \leq 3m(2\log_2(n) + \log_2(12d))$, as required by Theorem 1.   □

Remark 2. As planar graphs are 5-degenerate [26], this sum labelling needs labels with $2\log_2(n) + \mathcal{O}(1)$ bits for storing planar graphs, improving on previous published bounds for implicit representations of planar graphs [5,13,19,20,30,35,36], except the very last proposal [4], also see [7]. However, our approach generalises to sum graphs of arbitrary fixed degeneracy, which is unclear for other approaches from the literature on adjacency labelling schemes.

On the other hand, in adjacency labelling, the labels of two vertices alone are enough to decide whether the vertices are adjacent or not; for sum labelling, one needs to additionally check the labels of all the other vertices. Hence, sum labelling is not a type of adjacency labelling.

While our approach needs only $\mathcal{O}(n\log(n))$ bits to store trees, it is unknown if this can be achieved by a labelling that uses one isolate only; compare with [9].

6   Discussions

It is an interesting question how bad the labelling produced by our algorithm could get if it comes to determining the exclusive sum number of a graph. To give another example, when labelling the complete bipartite graph $K_{|P|,|Q|}$, with its vertex set V split into two independent sets P, Q, the ordering that first lists P and then Q will actually produce the optimal exclusive sum labelling as suggested in [27,34]. Also by presenting the vertices of P and Q alternatingly to our algorithm, one can produce a labelling that realizes the exclusive sum number $|P| + |Q| - 1$ of $K_{|P|,|Q|}$, but the range will then be nearly twice as big.

This brings us to the following interesting question: Is there always a vertex ordering such that our algorithm yields an optimal exclusive sum labelling?

Proposition 1. *There exists a family of graphs (G_n) such that, if our algorithm is presented with a certain ordering of $V(G_n)$, where $|V(G_n)| = n \geq 3$, then it will produce a labelling λ_n matching $\epsilon(G_n)$, but if presented with a different ordering, it will yield a labelling λ'_n requiring $|E(G_n)|$ many isolates. The ratio between the number of isolates produced by λ'_n and $\epsilon(G_n)$ grows beyond any limit.*

Namely, the family of paths on n vertices gives such a graph family. The labelling that is optimal with respect to the exclusive sum number is different from the one proposed in [27,34].

Moreover, the following computational complexity questions are of interest, in particular, if one wants to apply sum labellings for storing real-world graphs. Are there polynomial time algorithms for (any of) the following questions, given a graph G without isolates as input?

- Find the sum number $\sigma(G)$, and output a sum labelling.
- Find the exclusive sum number $\epsilon(G)$, and output an exclusive sum labelling.
- Output a sum labelling minimizing the range of the labels.
- Output a sum labelling minimizing the storage needs $\sigma_{\text{store}}^{\max}(G)$ or $\sigma_{\text{store}}(G)$.

In particular, if a question of the suggested form would be NP-hard, it would be interesting to know if there are good heuristics that order the vertices of a graph in a way that our algorithm produces a provable approximation to the best graph parameter value. As the proof of Proposition 1 shows, for instance the strategy behind the proof of Theorem 3 would actually produce a worst-case labelling in a sense, i.e., even labellings that have some good properties can be really bad with respect to another criterion. If it comes to giving an NP-hardness proof for any of these questions, one of the difficulties is that the graph parameters related to sum labelling have a non-local flavour in the sense that local modifications of a graph could have tremendous effect on the graph parameters. It seems important to further study different typical graph operations with respect to these parameters. Here, more results like Lemma 2 are needed [22].

There are hundreds of variants of graph labellings [11], some of which can be also used to store graphs (e.g., integral sum labellings [15], mod sum labellings [16,40,41,43], product labellings [3]), leading to further questions as discussed above for sum labellings, also bridging to adjacency labellings [19,32].

Acknowledgments. The authors are grateful to the organisers of GRAPHMASTERS 2020 [12] for providing the virtual environment that initiated this research.

References

1. Angles, R.: A comparison of current graph database models. In: 2012 IEEE 28th International Conference on Data Engineering Workshops, pp. 171–177. IEEE (2012)
2. Angles, R., Gutierrez, C.: Survey of graph database models. ACM Comput. Surv. (CSUR) **40**(1), 1–39 (2008)
3. Bergstrand, D., Hodges, K., Jennings, G., Kuklinski, L., Wiener, J., Harary, F.: Product graphs are sum graphs. Math. Mag. **65**(4), 262–264 (1992)
4. Bonamy, M., Gavoille, C., Pilipczuk, M.: Shorter labeling schemes for planar graphs. In: Chawla, S. (ed.) Proceedings of the 2020 ACM-SIAM Symposium on Discrete Algorithms, SODA, pp. 446–462. SIAM (2020)
5. Bonichon, N., Gavoille, C., Hanusse, N., Poulalhon, D., Schaeffer, G.: Planar graphs, via well-orderly maps and trees. Graph. Combin. **22**(2), 185–202 (2006)
6. Dominguez-Sal, D., Urbón-Bayes, P., Giménez-Vañó, A., Gómez-Villamor, S., Martínez-Bazán, N., Larriba-Pey, J.L.: Survey of Graph Database Performance on the HPC Scalable Graph Analysis Benchmark. In: Shen, H.T., et al. (eds.) WAIM 2010. LNCS, vol. 6185, pp. 37–48. Springer, Heidelberg (2010). https://doi.org/10.1007/978-3-642-16720-1_4. http://graphanalysis.org/index.html
7. Dujmovic, V., Esperet, L., Gavoille, C., Joret, G., Micek, P., Morin, P.: Adjacency labelling for planar graphs (and beyond). In: 61st IEEE Annual Symposium on Foundations of Computer Science, FOCS, pp. 577–588. IEEE (2020)
8. Elias, P.: Universal codeword sets and representations of the integers. IEEE Trans. Inf. Theory **21**(2), 194–203 (1975)
9. Ellingham, M.N.: Sum graphs from trees. Ars Combin. **35**, 335–349 (1993)
10. Fernau, H., Ryan, J.F., Sugeng, K.A.: A sum labelling for the generalised friendship graph. Discret. Math. **308**, 734–740 (2008)
11. Gallian, J.A.: A dynamic survey of graph labeling, version 23. Electron. J. Combin. DS **6** (2020). https://www.combinatorics.org/ojs/index.php/eljc/article/view/DS6/pdf
12. Gąsieniec, L., Klasing, R., Radzik, T.: Combinatorial Algorithms: 31st International Workshop, IWOCA 2020, Bordeaux, France, June 8–10, 2020, Proceedings, vol. 12126. Springer, Heidelberg (2020). https://doi.org/10.1007/978-3-030-48966-3
13. Gavoille, C., Labourel, A.: Shorter implicit representation for planar graphs and bounded treewidth graphs. In: Arge, L., Hoffmann, M., Welzl, E. (eds.) ESA 2007. LNCS, vol. 4698, pp. 582–593. Springer, Heidelberg (2007). https://doi.org/10.1007/978-3-540-75520-3_52
14. Gould, R.J., Rödl, V.: Bounds on the number of isolated vertices in sum graphs. In: Alavi, Y., Chartrand, G., Ollermann, O.R., Schwenk, A.J. (eds.) Graph Theory, Combinatorics, and Applications, 1988. Two Volume Set, pp. 553–562. Wiley (1991)
15. Harary, F.: Sum graphs over all the integers. Discret. Math. **124**(1–3), 99–105 (1994)
16. Harary, F.: Sum graphs and difference graphs. Congr. Numer. **72**, 101–108 (1990)
17. Hartsfield, N., Smyth, W.F.: A family of sparse graphs of large sum number. Discret. Math. **141**(1–3), 163–171 (1995)

18. Jouili, S., Vansteenberghe, V.: An empirical comparison of graph databases. In: 2013 International Conference on Social Computing, pp. 708–715. IEEE (2013)

19. Kannan, S., Naor, M., Rudich, S.: Implicit representation of graphs. SIAM J. Discret. Math. **5**(4), 596–603 (1992)

20. Keeler, K., Westbrook, J.R.: Short encodings of planar graphs and maps. Discret. Appl. Math. **58**(3), 239–252 (1995)

21. Konečný, M., Kučera, S., Novotná, J., Pekárek, J., Šimsa, Š, Töpfer, M.: Minimal sum labeling of graphs. J. Discrete Algorithms **52–53**, 29–37 (2018)

22. Korman, A., Peleg, D., Rodeh, Y.: Constructing labeling schemes through universal matrices. In: Asano, T. (ed.) ISAAC 2006. LNCS, vol. 4288, pp. 409–418. Springer, Heidelberg (2006). https://doi.org/10.1007/11940128_42

23. Kratochvíl, J., Miller, M., Nguyen, H.M.: Sum graph labels - an upper bound and related problems. In: 12th Australasian Workshop on Combinatorial Algorithms, AWOCA, pp. 126–131. Institut Teknologi Bandung, Indonesia (2001)

24. Kumar Kaliyar, R.: Graph databases: a survey. In: International Conference on Computing, Communication & Automation, pp. 785–790. IEEE (2015)

25. Lancichinetti, A., Fortunato, S., Radicchi, F.: Benchmark graphs for testing community detection algorithms. Phys. Rev. E **78**(4), 046110 (2008)

26. Lick, D.R., White, A.T.: k-degenerate graphs. Can. J. Math. **22**(5), 1082–1096 (1970)

27. Miller, M., Patel, D., Ryan, J., Sugeng, K.A., Slamin, Tuga, M.: Exclusive sum labeling of graphs. J. Comb. Math. Comb. Comput. **55**, 137–148 (2005)

28. Miller, M., Ryan, J., Smith, W.F.: The sum number of the cocktail party graph. Bull. Inst. Combin. Appl. **22**, 79–90 (1998)

29. Miller, M., Ryan, J.F., Ryjácek, Z.: Characterisation of graphs with exclusive sum labelling. Electron. Notes Discrete Math. **60**, 83–90 (2017)

30. Munro, J.I., Raman, V.: Succinct representation of balanced parentheses and static trees. SIAM J. Comput. **31**(3), 762–776 (2001)

31. Nagamochi, H., Miller, M., Slamin: On the number of isolates in graph labeling. Discret. Math. **243**, 175–185 (2001)

32. Peleg, D.: Proximity-preserving labeling schemes. J. Graph Theory **33**(3), 167–176 (2000)

33. Pyatkin, A.V.: New formula for the sum number for the complete bipartite graphs. Discret. Math. **239**(1–3), 155–160 (2001)

34. Ryan, J.: Exclusive sum labeling of graphs: a survey. AKCE Int. J. Graphs Comb. **6**(1), 113–136 (2009)

35. Schnyder, W.: Planar graphs and poset dimension. Order **5**, 323–343 (1989)

36. Schnyder, W.: Embedding planar graphs on the grid. In: Johnson, D.S. (ed.) Proceedings of the First Annual ACM-SIAM Symposium on Discrete Algorithms, SODA, pp. 138–148. SIAM (1990)

37. Singla, S., Tiwari, A., Tripathi, A.: Some results on the spum and the integral spum of graphs. Discrete Math. **344**(5), 112311 (2021)

38. Slamet, S., Sugeng, K.A., Miller, M.: Sum graph based access structure in a secret sharing scheme. J. Prime Res. Math. **2**, 113–119 (2006)

39. Smyth, W.F.: Sum graphs of small sum number. Colloq. Math. Soc. János Bolyai **60**, 669–678 (1991)

40. Sutton, M.: Summable graph labellings and their applications. Ph.D. thesis, Department of Computer Science, University of Newcastle, Australia (2000)

41. Sutton, M., Miller, M.: Mod sum graph labelling of $H_{m,n}$ and K_n. Aust. J. Combin. **20**, 233–240 (1999)

42. Sutton, M., Miller, M.: On the sum number of wheels. Discret. Math. **232**, 185–188 (2001)
43. Sutton, M., Miller, M., Ryan, J., Slamin: Connected graphs which are not mod sum graphs. Discret. Math. **195**(1), 287–293 (1999)
44. Tuga, M., Miller, M.: Delta-optimum exclusive sum labeling of certain graphs with radius one. In: Akiyama, J., Baskoro, E.T., Kano, M. (eds.) IJCCGGT 2003. LNCS, vol. 3330, pp. 216–225. Springer, Heidelberg (2005). https://doi.org/10.1007/978-3-540-30540-8_23
45. Wang, Y., Liu, B.: The sum number and integral sum number of complete bipartite graphs. Discret. Math. **239**(1–3), 69–82 (2001)

On Minimizing Regular Expressions
Without Kleene Star

Hermann Gruber[1], Markus Holzer[2(✉)], and Simon Wolfsteiner[3]

[1] Knowledgepark GmbH, Leonrodstr. 68, 80636 Munich, Germany
hermann.gruber@kpark.de
[2] Institut für Informatik, Universität Giessen, Arndtstr. 2, 35392 Giessen, Germany
holzer@informatik.uni-giessen.de
[3] Institut für Diskrete Mathematik und Geometrie, TU Wien,
Wiedner Hauptstr. 8–10, 1040 Vienna, Austria

Abstract. Finite languages lie at the heart of literally every regular expression. Therefore, we investigate the approximation complexity of minimizing regular expressions without Kleene star, or, equivalently, regular expressions describing finite languages. On the side of approximation hardness, given such an expression of size s, we prove that it is impossible to approximate the minimum size required by an equivalent regular expression within a factor of $O\left(\frac{s}{(\log s)^\delta}\right)$ if the running time is bounded by a quasipolynomial function depending on δ, for every $\delta > 1$, unless the exponential time hypothesis (ETH) fails. For approximation ratio $O(s^{1-\delta})$, we prove an exponential-time lower bound depending on δ, assuming ETH. The lower bounds apply to alphabets of constant size. On the algorithmic side, we show that the problem can be approximated in polynomial time within $O(\frac{s \log \log s}{\log s})$, with s being the size of the given regular expression. For constant alphabet size, the bound improves to $O(\frac{s}{\log s})$. Finally, we devise a family of superpolynomial approximation algorithms with approximation ratios matching the lower bounds, while the running times are just above the lower bounds excluded by the exponential time hypothesis.

1 Introduction

Regular expressions are used in many applications and it is well known that for each regular expression, there is a finite automaton that defines the same language and *vice versa*. Automata are very well suited for programming tasks and immediately translate to efficient data structures. On the other hand, regular expressions are well suited for human users and therefore are often used as interfaces to specify certain patterns or languages.

Regarding performance optimization, putting effort into the internal representation inside the regex engine is of course a natural choice. On the other hand, most of the time, developers use existing APIs but are not willing, or able, to change the source code of these. Thus, sometimes practitioners, as well as theory researchers, see a need for optimizing the input regular expressions, as witnessed

© Springer Nature Switzerland AG 2021
E. Bampis and A. Pagourtzis (Eds.): FCT 2021, LNCS 12867, pp. 245–258, 2021.
https://doi.org/10.1007/978-3-030-86593-1_17

by questions in pertinent Q&A forums.[1] More often than not, the regular expressions under consideration are in fact without Kleene star, that is, they describe only finite languages. Moreover, recently the descriptional complexity of finite languages attracted new attention because of its close connection to well-known measures for the complexity of formal proofs in first-order predicate logic [6].

The problem of minimizing regular expressions accepting infinite languages is PSPACE-complete, and even attaining a sublinear approximation ratio is already equally hard [10]. When restricting to finite languages, there is of course the classical reduction from 3-SAT to the equivalence problem for regular expressions without star [18, Theorem 2.3]. It is sometimes overlooked that, unlike the case of infinite languages, the classical reduction *does not imply* hardness of the corresponding minimization problem. In fact, no lower bounds for minimizing regular expressions without Kleene star were known prior to the present work at all.[2] Also, there are hardness results for minimizing acyclic nondeterministic finite automata [3,11], and also for minimizing acyclic context-free grammars [15]—but nothing thus far for regular expressions without star. In this work, we fill this gap by proving tight lower and upper approximability bounds.

As a byproduct of our proofs, we also substantially improve the inapproximability bound for minimizing nondeterministic finite automata in the case of finite languages, and give the first nontrivial approximation guarantee. The results are summarized in Table 1.

Recent years have seen a renewed interest in the analysis of computational problems, among others, on formal languages, since more fine-grained hardness results can be achieved based on the exponential time hypothesis (ETH) than with more traditional proofs based on the assumption $P \neq NP$ [1,2,5,8,23,24]. Namely, ETH posits that there is no algorithm that decides 3-SAT formulae with n variables in time $2^{o(n)}$, and is just one among other strong hypotheses that were used during the last decade to perform fine-grained complexity studies; for a short survey on results obtained by some of these hypotheses, we refer to [25].

We contribute a fine-grained analysis of approximability and inapproximability for minimizing regular expressions without Kleene star. On the side of approximation hardness, given such an expression of size s, we prove that it is impossible to approximate the minimum size required by an equivalent regular expression within a factor of $O\left(\frac{s}{(\log s)^\delta}\right)$ if the running time is bounded by a

---

[1] See for example the following questions drawn from various sites: (i) P. Krauss: Minimal regular expression that matches a given set of words, URL: https://cs. stackexchange.com/q/72344, Accessed: 2021-01-02, (ii) J. Mason: A released perl with trie-based regexps! URL: http://taint.org/2006/07/07/184022a.html, Accessed: 2020-07-21, (iii) pdanese (StackOverflow username): Speed up millions of regex replacements in Python 3, URL: https://stackoverflow.com/q/42742810, Accessed: 2021-01-02, (iv) P. Scheibe: RegEx performance: Alternation vs Trie, URL: https:// stackoverflow.com/q/56177330, Accessed: 2021-01-02, and (v) Ch. Xu: Minimizing size of regular expression for finite sets, URL: https://cstheory.stackexchange.com/ q/16860, Accessed: 2021-01-02.

[2] See, e.g., item (v) of the previous footnote.

Table 1. Coarse-grained overview of known and new results for minimization problems. For better comparability, approximability is understood to be in polynomial time, and hardness results are under classical assumptions such as $P \neq NP$.

	General	Unary languages	Finite languages
DFA	Exactly solvable in P [17]		
NFA	PSPACE-complete [22], *not* approximable within $o(n)$ [10], trivially approximable within $O(n)$	coNP-hard [22], *not* approximable within $o(n)$ [10,11], trivially approximable within $O(n)$	DP-hard [11], *not* approximable within $\frac{\sqrt{n}}{2^{(\log n)^{7/8-\varepsilon}}}$ [3,12], trivially approximable within $O(n)$
			Not approximable within $n^{1-\varepsilon}$ (Corollary 16), approximable within $\frac{n}{\log n}$ for fixed alphabet (Theorem 18)
RE			coNP-hard (Corollary 7), *not* approximable within $n^{1-\varepsilon}$ (Corollary 7), approximable within $\frac{n \log \log n}{\log n}$ (Theorem 11)

quasipolynomial function depending on δ, for every $\delta > 1$, unless the ETH fails. For approximation ratio $O(s^{1-\delta})$, we prove an exponential-time lower bound depending on δ, assuming ETH. These lower bounds apply to alphabets of constant size. On the algorithmic side, we show that the problem can be approximated in polynomial time within $O(\frac{s \log \log s}{\log s})$, where s is the size of the given regular expression. For constant alphabet size, the bound improves to $O(\frac{s}{\log s})$. Finally, we devise a family of superpolynomial approximation algorithms that attain the performance ratios of the lower bounds, while their running times are just above those excluded by the ETH. For instance, we attain an approximation ratio of $O\left(\frac{s}{(\log s)^{\delta}}\right)$ in time $2^{O((\log s)^{\delta})}$ for $\delta > 1$, and a ratio of $s^{1-\delta}$ in time $2^{O(s^{\delta})}$ for $\delta > 0$. These running times nicely fit with the excluded running times of $2^{o((\log s)^{\delta})}$ and of $2^{o(s^{\delta})}$, respectively, for these approximation ratios.

This paper is organized as follows: in the next section, we define the basic notions relevant to this paper. Section 3 covers approximation hardness results for various runtime regimes based on the ETH. Then in Sect. 4, these negative results are complemented with approximation algorithms that neatly attain these lower bounds. In Sect. 5, we transfer some of these results to the minimization problem for nondeterministic finite automata. To conclude this work, we indicate possible directions for further research in the last section. Due to space constraints, some of the proofs are omitted.

2 Preliminaries

We assume that the reader is familiar with the basic notions of formal language theory as contained in [17]. In particular, let Σ be an *alphabet* and Σ^* the *set* *of all words over the alphabet* Σ including the *empty word* ε. The *length of a word* w is denoted by $|w|$, where $|\varepsilon| = 0$, and the total number of occurrences of the alphabet symbol a in w is denoted by $|w|_a$. In this paper, we mainly deal with finite languages. The *order* of a finite language L is the length of a longest word belonging to L. A finite language $L \subseteq \Sigma^*$ is called *homogeneous* if all words in the language have the same length. We say that a homogeneous language $L \subseteq \Sigma^n$ is *full* if L is equal to Σ^n. For languages $L_1, L_2 \subseteq \Sigma^*$, the *left quotient of L_1 and L_2* is defined as $L_1^{-1}L_2 = \{\, v \in \Sigma^* \mid$ there is some $w \in L_1$ such that $wv \in L_2 \,\}$. If L_1 is a singleton, i.e., $L_1 = \{w\}$, for some word $w \in \Sigma^*$, we omit braces, that is, we write $w^{-1}L_2$ instead of $\{w\}^{-1}L_2$. The set $w^{-1}L_2$ is also called the *derivative of L_2 w.r.t. the word w*. In order to fix the notation, we briefly recall the definition of regular expressions and the languages described by them.

The *regular expressions* over an alphabet Σ are defined inductively in the usual way:[3] $\emptyset$, ε, and every letter $a \in \Sigma$ is a regular expression; and when E and F are regular expressions, then $(E + F)$, $(E \cdot F)$, and $(E)^*$ are also regular expressions. The language defined by a regular expression E, denoted by $L(E)$, is defined as follows: $L(\emptyset) = \emptyset$, $L(\varepsilon) = \{\varepsilon\}$, $L(a) = \{a\}$, $L(E + F) = L(E) \cup L(F)$, $L(E \cdot F) = L(E) \cdot L(F)$, and $L(E^*) = L(E)^*$. The *alphabetic width* or *size* of a regular expression E over an alphabet Σ, denoted by $\mathsf{awidth}(E)$, is defined as the total number of occurrences of letters of Σ in E. For a regular language L, we define its alphabetic width, $\mathsf{awidth}(L)$, as the minimum alphabetic width among all regular expressions describing L.

We are interested in regular expression minimization w.r.t. its alphabetic width (or, equivalently, its size). An algorithm that returns near-optimal solutions is called an *approximation algorithm*. Assume that we are working on a minimization problem in which each potential solution has a positive cost and that we wish to find a near-minimal solution. We say that an approximation algorithm for the problem has a *performance guarantee* of $\rho(n)$ if for any input of size n, the cost C of the solution produced by the approximation algorithm is *within a factor of* $\rho(n)$ of the cost C^* of a minimal solution: $\frac{C}{C^*} \leq \rho(n)$. If the approximation algorithm is running in polynomial time, we speak of a *polynomial-time approximation algorithm*. For most of our hardness results, we assume the exponential time hypothesis (ETH) introduced in [19].

Exponential Time Hypothesis. There is a positive constant c such that the satisfiability of a formula in 3-CNF with n variables and m clauses cannot be decided in time $2^{cn}(m + n)^{O(1)}$.

In particular, using the Sparsification Lemma [19], the ETH implies that there is no algorithm running in time $2^{o(m)}$ that decides satisfiability of a 3-SAT formula with m clauses. This is of course a much stronger assumption than $\mathsf{P} \neq \mathsf{NP}$. For more background on the topic, see, e.g., the survey [20].

---

[3] For convenience, parentheses in regular expressions are sometimes omitted and concatenation is sometimes simply written as juxtaposition. The priority of operators is specified in the usual fashion: concatenation is performed before union, and star before both concatenation and union.

3   Inapproximability

In this section, we will show that, for a given regular expression without Kleene star, the minimum size required by an equivalent regular expression cannot be approximated within a certain factor if the running time is within certain bounds, assuming the ETH. We start off with an estimate of the required regular expression size for a language which we shall use as gadget.

Lemma 1. *Let* $P_r = \{ xy \in \{0,1\}^* \mid |x| = |y| = r \text{ and } x = y^R \}$ *denote the language of all binary palindromes of length* $2r$. *Then* $2^r \leq \mathsf{awidth}(P_r) \leq 2^{r+2}-4$.

The upper bound is in fact tight, yet proving this takes a lot more effort [14]. Notwithstanding, the simple lower bound above suffices for the purpose of the present work.

It was shown in [13] that taking the quotient of a regular language can cause at most a quadratic blow-up in required regular expression size. *Vice versa*, the alphabetic width of a language can be lower-bounded by the order of the square root of the alphabetic width of any of its quotients. For our reduction, we need a tighter relationship. This is possible if we resort to special cases. Let us consider homogeneous languages and expressions in more detail. First, we need a simple observation that turns out to be very useful in the forthcoming considerations.

Lemma 2. *Let* $L \subseteq \Sigma^n$ *be a homogeneous language. If* E *is a regular expression describing* L, *then any subexpression of* E *describes a homogeneous language, too.*

Now we are ready to consider the descriptional complexity of quotients of homogeneous languages in detail.

Lemma 3. *Let* $L \subseteq \Sigma^n$ *be a homogeneous language. Then* $\mathsf{awidth}(w^{-1}L) \leq \mathsf{awidth}(L)$, *for any word* $w \in \Sigma^*$.

We build upon the classical coNP-completeness proof of the inequality problem for regular expressions without star given in [18, Theorem 2.3]. We recall the reduction to make this paper more self-contained.

Theorem 4. *Let* φ *be a formula in 3-DNF with* n *variables and* m *clauses. Then a regular expression* ζ *can be computed in time* $O(m \cdot n)$ *such that the language* $Z = L(\zeta)$ *is homogeneous and* Z *is full if and only if* φ *is a tautology.*

Proof. Let $\varphi = \bigvee_{i=1}^m c_i$ be a formula in 3-DNF. We can assume without loss of generality that no clause c_i contains both x_j and $\overline{x}_j$ as a literal. For each clause c_i, let $\zeta_i = \zeta_{i1}\zeta_{i2}\cdots\zeta_{in}$, where

$$\zeta_{ij} = \begin{cases} (0+1) & \text{if both } x_j \text{ and } \overline{x}_j \text{ are not literals in } c_i, \\ 0 & \text{if } \overline{x}_j \text{ is a literal in } c_i, \\ 1 & \text{if } x_j \text{ is a literal in } c_i. \end{cases}$$

Let $\zeta = \zeta_1 + \zeta_2 + \cdots + \zeta_m$. Clearly, $Z = L(\zeta) \subseteq \{0,1\}^n$. Let w in $\{0,1\}^n$. Then w is in Z if and only if w satisfies some clause c_i. Thus $Z = \{0,1\}^n$ if and only if φ is a tautology. This completes the reduction.   □

Now if we wanted to apply the reduction from Theorem 4 to the minimization problem for regular expressions, the trouble is that we cannot predict the minimum required regular expression size for $Z = L(\zeta)$ in case it is not full. To make this happen, we use a similar trick as recently used in [15] for the analogous case of context-free grammars. In the following lemma, we embed the language P_r of all binary palindromes of length $2r$ together with the language $Z = L(\zeta)$ (as defined in Theorem 4) into a more complex language Y. Depending on whether or not Z is full, the alphabetic width of Y is at most linear or at least quadratic, respectively, in m. Recall that m refers to the number of clauses in the given 3-DNF formula φ.

Lemma 5. *Let φ be a formula in 3-DNF with n variables and m clauses and let ζ be the regular expression constructed in Theorem 4. Furthermore, let*

$$Y = Z \cdot \{0,1\}^{2r} \cup \{0,1\}^n \cdot P_r,$$

where $Z = L(\zeta)$ and P_r, for $r \leq m$, is defined as in Lemma 1. Then $\mathsf{awidth}(Y) = O(m)$ *if Z is full, and* $\mathsf{awidth}(Y) = \Omega(2^r)$ *if Z is not full.*

The above lemma can serve as a gap introducing reduction. For example, if we take $r = 2\log m$, then $\Omega(2^r)$ is in $\Omega(m^2)$. Now we are in the position to state our first inapproximability result.

Theorem 6. *Let E be a regular expression without Kleene star of size s, and let δ be a constant such that $0 < \delta \leq \frac{1}{2}$. Then no deterministic $2^{o(s^\delta)}$-time algorithm can approximate* $\mathsf{awidth}(L(E))$ *within a factor of $o(s^{1-\delta})$, unless ETH fails.*

Proof. We give a reduction from the 3-DNF tautology problem as in Lemma 5. That is, given a formula φ in 3-DNF with n variables and m clauses, we construct a regular expression that generates the language $Y = Z \cdot \{0,1\}^{2r} \cup \{0,1\}^n \cdot P_r$. The sets P_r and Z are defined as in Lemma 1 and Theorem 4, respectively. Here, the set Y features some carefully chosen parameter r, which will be fixed later on. For now, we only assume $2\log m \leq r \leq m$.

Next, we need to show that the reduction is correct in the sense that if Z is full, then $\mathsf{awidth}(Y)$ is asymptotically strictly smaller than in the case where it is not full. By Lemma 5, it follows that $\mathsf{awidth}(Y) = O(m)$ if Z is full and $\mathsf{awidth}(Y) = \Omega(2^r)$, otherwise. Thus, the reduction is correct, since we have assumed that $r \geq 2\log m$, and consequently $2^r = \omega(m)$.

It is easy to see that the running time of the reduction is linear in the size of the constructed regular expression describing Y. Now we estimate the size of that regular expression. Recall from Theorem 4 that the regular expression ζ has size $O(m \cdot n)$. Because formula φ is a 3-DNF, we have $m \geq n/3$, and so the size of ζ is in $O(m^2)$. The set $\{0,1\}^{n+2r}$ admits a regular expression of size $O(m+n) = O(m)$; and $\mathsf{awidth}(P_r) = \Theta(2^r)$ by Lemma 1. Since we have assumed that $r \geq 2\log m$, the order of magnitude of the constructed regular expression is $s = \Theta(2^r)$.

Now we need to fix the parameter r in our reduction; let us pick $r = \frac{1}{\delta} \cdot \log m$. Recall that the statement of the theorem requires $\frac{1}{\delta} \geq 2$, thus we have $r \geq 2 \log m$. So this is a valid choice for the parameter r—in the sense that the reduction remains correct.

Towards a contradiction with the ETH, assume that there is an algorithm A_δ approximating the alphabetic width within $o\left(s^{1-\delta}\right)$ running in time $2^{o\left(s^\delta\right)}$. Then A_δ could be used to decide whether Z is full as follows: the putative approximation algorithm A_δ returns a cost C that is at most $o(s^{1-\delta})$ times the optimal cost C^*, that is, $C = o(s^{1-\delta}) \cdot C^* = o(s^{1-\delta}) \cdot \mathsf{awidth}(Y)$.

On the one hand, if Z is full, then $\mathsf{awidth}(Y) = O(m)$ by Lemma 5. In this case, the hypothetical approximation algorithm A_δ returns a cost C with $C = o(m \cdot s^{1-\delta}) = o\left(m \cdot \Theta\left(m^{\frac{1}{\delta}}\right)^{1-\delta}\right) = o\left(m^{\frac{1}{\delta}(1-\delta)+1}\right) = o\left(m^{\frac{1}{\delta}}\right) = o\left(2^r\right)$. In the second step of the above calculation, we used the fact that $s = \Theta(2^r) = \Theta(m^{\frac{1}{\delta}})$ and in the last step, we used the fact that we chose r as $r = \frac{1}{\delta} \cdot \log m$, which is equivalent to $2^r = m^{\frac{1}{\delta}}$.

On the other hand, in case Z is not full, then Lemma 5 states that $\mathsf{awidth}(Y) = \Omega\left(2^r\right)$. Using the constants implied by the O-notation, the size returned by algorithm A_δ could thus be used to decide, for large enough m, whether Z is full, and thus by Theorem 4 whether the 3-DNF formula φ is a tautology.

It remains to show that the running time of A_δ in terms of m is in $2^{o(m)}$, which contradicts the ETH. Recall again that $s = \Theta(m^{\frac{1}{\delta}})$; we thus can express the running time of the algorithm A_δ in terms of m, namely, $2^{o\left(s^\delta\right)} = 2^{o\left(\Theta(m^{\frac{1}{\delta}})^\delta\right)} = 2^{o\left((c \cdot m^{\frac{1}{\delta}})^\delta\right)} = 2^{o(m)}$, for some constant c, which yields the desired contradiction. □

Assuming ETH, the above proof also implies that the problem cannot be solved exactly in time $2^{o(\sqrt{s})}$. The inapproximability result can be stated more simply when using the classical hardness assumption $\mathsf{P} \neq \mathsf{NP}$:

Corollary 7. *Let E be a regular expression without Kleene star of size s, and let δ be a constant with $0 < \delta < 1$. Then no deterministic polynomial-time algorithm can approximate $\mathsf{awidth}(L(E))$ within a factor of $s^{1-\delta}$, unless $\mathsf{P} = \mathsf{NP}$.*

Proof. The reduction in Theorem 6 is from a coNP-complete problem and runs in polynomial time for every choice of $\delta \geq \frac{1}{2}$. Observe that it suffices to show approximation hardness for $\delta \leq \frac{1}{2}$, since the weaker hardness result for $\delta > \frac{1}{2}$ is then implied. □

Again, assuming ETH, we can change the parameter r in the reduction in Theorem 6 to trade a sharper inapproximability ratio against a weaker lower bound on the running time.

Theorem 8. *Let E be a regular expression without Kleene star of size s, and let δ be a constant with $\delta > 1$. Then no deterministic $2^{o(\log s)^\delta}$-time algorithm can approximate $\mathsf{awidth}(L(E))$ within a factor of $o\left(s/(\log s)^\delta\right)$, unless ETH fails.*

4   Approximability

From the previous section, we know that there are severe limits on what we can expect from efficient approximation algorithms. In this section, we present different approximation algorithms for minimizing regular expressions describing finite languages. Each of them introduces a new algorithmic hook, some of which might be useful in implementations. We start off with an algorithm that requires the input to be specified non-succinctly as a list of words. In case the alphabet size is sufficiently large, listing simply all words is enough; otherwise we construct a deterministic finite automaton and further distinguish on the number of states. This leads to the following result.

Theorem 9. *Let L be a finite language given as a list of words, with s being the sum of the word lengths. Then* awidth(L) *can be approximated in deterministic polynomial time within a factor of $O(\frac{s}{\sqrt{\log s}})$.*

Recall that the minimal deterministic finite automaton can be exponentially larger than regular expressions in the worst case, also for finite languages [21]. Also, the conversion from deterministic finite automata to regular expressions is only quasipolynomial in the worst case. These facts of course affect the performance guarantee. Nevertheless, we believe that the scheme from the proof of Theorem 9 is worth a look, since the minimal deterministic finite automaton may eliminate a lot of redundancy in practice. Furthermore, the algorithm works equally if we are able to construct a nondeterministic finite automaton which is smaller than the minimal deterministic finite automaton. To this end, some recently proposed effective heuristics for size reduction of nondeterministic automata could be used [4].

Admittedly, regular expressions are exponentially more succinct than a list of words and our inapproximability results crucially rely on that. So, we now turn to the second approximation algorithm. It makes use of the fact that if a given regular expression E describes very short words only, then it is not too difficult to produce a regular expression that is noticeably more succinct than E. In that case, the algorithm builds a trie, which then can be converted into an equivalent regular expression of size linear in the trie.

For the purpose of this paper, a *trie* (also known as *prefix tree*) is simply a tree-shaped deterministic finite automaton with the following properties:

1. The edges are directed away from the root, i.e., towards the leaves.
2. The root is the start state.
3. All leaves are accepting states.
4. Each edge is labelled with a single alphabet symbol.

The last condition is needed if we want to bound the size of an equivalent regular expression in terms of the nodes in the trie. The following lemma seems to be folklore; the observation is used, e.g., in [16].

Lemma 10. *Let T be a trie with n nodes accepting L. Then an equivalent regular expression of alphabetic width at most $n - 1$ can be constructed in deterministic polynomial time from T.*

Now we have collected all tools for an approximation algorithm that works with regular expressions as input, which even comes with an improved approximation ratio.

Theorem 11. *Let E be a regular expression without Kleene star of alphabetic width s. Then* $\mathsf{awidth}(L(E))$ *can be approximated in deterministic polynomial time within a factor of* $O\left(\frac{s \log \log s}{\log s}\right)$.

Proof. We again start with a case distinction by alphabet size.

1. The size of the alphabet used in L is at most $\log s$. We further distinguish the cases in which the order of $L(E)$, i.e., the length of a longest word in $L(E)$, is less than $\frac{\log s}{\log \log s}$ or not. The order of $L(E)$ can be easily computed recursively, in polynomial time, by traversing the syntax tree of E. We consider two subcases:

   (a) The order of $L(E)$ is less than $\frac{\log s}{\log \log s}$. We enumerate the words in $L(E)$, e.g., by performing a membership test for each word of length less than $\frac{\log s}{\log \log s}$. Then we use a standard algorithm to construct a trie for $L(E)$. The worst case for the size of T is when L contains all words of length less than $\frac{\log s}{\log \log s}$. Then T is a full $(\log s)$-ary trie of height $\frac{\log s}{\log \log s}$. All nodes are accepting, giving a one-to-one correspondence between the number of nodes in T and the number of words in $L(T)$. That is, the number of nodes in T is equal to $\sum_{i=0}^{\frac{\log s}{\log \log s}-1}(\log s)^i = O\left(\frac{(\log s)^{\frac{\log s}{\log \log s}}}{\log s}\right)$.

   Using the fact that $(\log s)^{\frac{\log s}{\log \log s}} = s$, this is in $O\left(\frac{s}{\log s}\right)$ and we can construct an equivalent regular expression of that size in deterministic polynomial time by virtue of Lemma 10.

   (b) The order of $L(E)$ is at least $\frac{\log s}{\log \log s}$. We make use of the observation that the order of $L(E)$, i.e., the length of a longest word in $L(E)$, is a lower bound on the required regular expression size, as observed, e.g., in [7, Proposition 6]. That is, the optimal solution is at least of size $\frac{\log s}{\log \log s}$ and thus the regular expression E given as input is already a feasible solution that is at most $\frac{s \log \log s}{\log s}$ times larger than the optimal solution.

2. The size of the alphabet used in L is greater than $\log s$. The size of the alphabet used in L is likewise a lower bound on the required regular expression size and, similarly to the previous case, the input is a feasible solution that is at most $\frac{s}{\log s}$ times greater than the optimal solution size.

This proves the stated claim.   □

For alphabets of constant size, the performance ratio can be slightly improved—by a factor of $\log \log s$.

Theorem 12. *Let E be a regular expression without Kleene star of alphabetic width s over a fixed k-ary alphabet. Then* $\mathsf{awidth}(L(E))$ *can be approximated in deterministic polynomial time within a factor of* $O\left(\frac{s}{\log s}\right)$.

A better performance ratio can be achieved if we allow a superpolynomial running time.

Theorem 13. *Let E be a regular expression without Kleene star of alphabetic width s, and let $f(s)$ be a time constructible[4] function with $f(s) = \Omega(\log s)$. Then* $\mathsf{awidth}(L(E))$ *can be approximated in deterministic time* $2^{O(f(s))}$ *within a factor of* $O\left(\frac{s \log f(s)}{f(s)}\right)$.

Proof. First, as in Theorem 11, we make a case distinction by alphabet size, and then distinguish by the order of the language. Recall that the order of the language can be computed in polynomial time in a recursive manner on the syntax tree of E. The main new ingredient of this proof is a brute-force search for an optimal solution, powered by a context-free grammar that efficiently generates the search space of candidate regular expressions. So, we again start with distinguishing by alphabet size:

1. The size k of the alphabet used in $L(E)$ is at most $f(s)$. Again, we consider two subcases:
   (a) The order of $L(E)$ is less than $f(s)/\log f(s)$. We make use of the fact that there is a context-free grammar generating all regular expressions describing finite languages over the alphabet used in E. Such a grammar can be used to enumerate all regular expressions of size less than $f(s)/\log f(s)$ with polynomial delay [9]. For finite languages, there is an efficient grammar generating at most $O(f(s))^{f(s)/\log f(s)}$ of these candidates[5] in total [16, Prop. 8.3]. Observe that $O(f(s))^{f(s)/\log f(s)} = 2^{O(f(s))}$. For each enumerated candidate regular expression C and each word w of length less than $f(s)/\log f(s)$, we test whether $w \in L(E)$, and if so, we verify that $w \in L(C)$. If C passes all these tests, we can safely conclude that $L(E) \subseteq L(C)$. To verify whether $L(C) \subseteq L(E)$, we enumerate the words in $L(E)$ and build a trie T that accepts the language. Notice that the trie has at most $f(s)^{O(f(s)/\log f(s))} = 2^{O(f(s))}$ nodes. Since T is a deterministic finite automaton, it can be easily complemented, and we can apply the usual product construction—with the position automaton of C—to check whether $L(C) \cap \Sigma^* \backslash L(T) = \emptyset$. In this way, each candidate regular expression can be tested with a running time bounded by a polynomial in $2^{O(f(s))}$. Recall that the total number of candidates is in $2^{O(f(s))}$, and that the candidates can be enumerated

---

[4] We say that a function $f(n)$ is *time constructible* if there exists an $f(n)$ time-bounded multitape Turing machine M such that for each n there exists some input on which M actually makes $f(n)$ moves [17].

[5] The grammar in [16, Proposition 8.3] does not generate all valid regular expressions, but incorporates some performance tweaks. These tweaks perfectly fit our purpose: while the grammar does not generate all feasible solutions, it still generates at least one optimal solution. More precisely, given a finite language L with $\mathsf{awidth}(L) = k$, the context-free grammar is guaranteed to enumerate a regular expression of alphabetic width k for it.

with polynomial delay. We conclude that in this case, if $L(E)$ admits an equivalent regular expression of size at most $f(s)/\log f(s)$, an *optimal* solution can be found by exhaustive search with a running time bounded by $2^{(O(f(s)))^{O(1)}} \cdot 2^{(O(f(s)))^{O(1)}} = 2^{O(f(s))}$. The performance ratio is $\frac{s \log f(s)}{f(s)}$.

(b) The order of $L(E)$ is at least $f(s)/\log f(s)$. Again, the order of $L(E)$ is a lower bound on required regular expression size. Thus the regular expression E given as input is already a feasible solution with performance ratio $\frac{s \log f(s)}{f(s)}$.

2. The size of the alphabet used in $L(E)$ is greater than $f(s)$. Similarly, the size of the used alphabet is a lower bound on the required regular expression size. Thus, the regular expression E given as input is a feasible solution, whose performance ratio is $\frac{s}{f(s)}$ in this case.

This completes the proof. □

Observe that although the statement of Theorem 13 specializes to Theorem 11 if we set $f(s) = \log s$, the two proofs nevertheless use different algorithms. Both approaches will have their own merits and their own tradeoffs between running time and performance guarantees when put to practice. Again, in case the alphabet size is bounded, we can slightly improve the performance guarantee of the previous theorem—compare with Theorem 12.

Theorem 14. *Let E be a regular expression without Kleene star of alphabetic width s over a fixed k-ary alphabet, and let $f(s)$ be a time constructible function with $f(s) = \Omega(\log s)$. Then $\mathsf{awidth}(L(E))$ can be approximated in deterministic time $2^{O(f(s))}$ within a factor of $O\left(\frac{s}{f(s)}\right)$.*

To compare this with our inapproximability results, we pick $f(s) = s^\delta$, for some $\delta \leq \frac{1}{2}$, to obtain an approximation ratio of $s^{1-\delta}$ in time $2^{O(s^\delta)}$. Here, Theorem 6 rules out an approximation ratio of $o(s^{1-\delta})$ within a running time of $2^{o(s^\delta)}$. Another pick is $f(s) = (\log s)^\delta$, for some $\delta > 1$, yielding an approximation ratio $O\left(\frac{s}{(\log s)^\delta}\right)$ in time $2^{O(\log s)^\delta}$. In contrast, Theorem 8 rules out an approximation ratio of $o\left(\frac{s}{(\log s)^\delta}\right)$ in time $2^{o(\log s)^\delta}$. In both cases, the upper bound asymptotically matches the obtained lower bounds, and thus there remains little room for improvements, unless the ETH fails.

5   Minimizing Nondeterministic Finite Automata

In this section, we show that several of our results apply *mutatis mutandis* to the problem of minimizing acyclic nondeterministic finite automata, i.e., those accepting finite languages.

Theorem 15. *Let A be an s-state acyclic nondeterministic finite automaton, and let δ be a constant such that $0 < \delta \le \frac{1}{2}$. Then no deterministic $2^{o(s^\delta)}$-time algorithm can approximate the nondeterministic state complexity of $L(A)$ within a factor of $O(s^{1-\delta})$, unless ETH fails.*

Corollary 16. *Let A be an s-state acyclic nondeterministic finite automaton, and let δ be a constant with $0 < \delta < 1$. Then no deterministic polynomial-time algorithm can approximate the nondeterministic state complexity of $L(A)$ within a factor of $O(s^{1-\delta})$, unless $P = NP$.* □

The quasipolynomial-time inapproximability result carries over as well:

Theorem 17. *Let A be an s-state acyclic nondeterministic finite automaton, and let δ be a constant with $\delta > 1$. Then no deterministic $2^{o((\log s)^\delta)}$-time algorithm can approximate the nondeterministic state complexity of $L(A)$ within a factor of $o(s/(\log s)^\delta)$, unless ETH fails.* □

Regarding positive approximability results, we cannot use the entire toolkit that we have developed for regular expressions. For instance, the size of the used alphabet does not bound the number of states needed. Also, even for binary alphabets, the number of nondeterministic s-state finite automata is in $2^{\Omega(s^2)}$, which renders the enumeration of automata with few states less feasible. At least, the polynomial-time approximation for bounded alphabet size carries over:

Theorem 18. *Let A be an s-state acyclic nondeterministic finite automaton over a fixed k-ary alphabet. Then the nondeterministic state complexity of $L(A)$ can be approximated in deterministic polynomial time within a factor of $O\left(\frac{s}{\log s}\right)$.*

6 Conclusion

We conclude by indicating some possible directions for further research. First, we would like to continue with investigating inapproximability bounds within polynomial time based on the strong exponential time hypothesis (SETH). Further topics are exact exponential-time algorithms and parameterized complexity. In addition to the natural parameter of desired solution size, the order of the finite language and the alphabet size seem to be natural choices.

Given the practical relevance of the problem we investigated, we think that implementing some of the ideas from the above approximation algorithms is worth a try. Also, POSIX regular expressions restricted to finite languages are a more complex model than the one we investigated, but a more practical one as well. Although we would rather not expect better approximability bounds in that model, we suspect that character classes and other mechanisms can offer practical hooks for reducing the size of regular expressions.

Acknowledgments. We would like to thank Michael Wehar for some discussion, and the anonymous reviewers for their valuable comments.

References

1. Abboud, A., Backurs, A., Williams, V.V.: If the current clique algorithms are optimal, so is Valiant's parlser. SIAM J. Comput. **47**(6), 2527–2555 (2015)
2. Bringmann, K., Grønlund, A., Larsen, K.G.: A dichotomy for regular expression membership testing. In: Proceedings of the 58th Annual IEEE Symposium on Foundations of Computer Science, pp. 307–318. IEEE, Berkeley, October 2017
3. Chalermsook, P., Heydrich, S., Holm, E., Karrenbauer, A.: Nearly tight approximability results for minimum biclique cover and partition. In: Schulz, A.S., Wagner, D. (eds.) ESA 2014. LNCS, vol. 8737, pp. 235–246. Springer, Heidelberg (2014). https://doi.org/10.1007/978-3-662-44777-2_20
4. Clemente, L., Mayr, R.: Efficient reduction of nondeterministic automata with application to language inclusion testing. Log. Methods Comput. Sci. **15**(1) (2019)
5. de Oliveira Oliveira, M., Wehar, M.: On the fine grained complexity of finite automata non-emptiness of intersection. In: Jonoska, N., Savchuk, D. (eds.) DLT 2020. LNCS, vol. 12086, pp. 69–82. Springer, Cham (2020). https://doi.org/10.1007/978-3-030-48516-0_6
6. Eberhard, S., Hetzl, St.: On the compressibility of finite languages and formal proofs. Inform. Comput. **259**, 191–213 (2018)
7. Ellul, K., Krawetz, B., Shallit, J., Wang, M.: Regular expressions: new results and open problems. J. Autom. Lang. Comb. **10**(4), 407–437 (2005)
8. Fernau, H., Krebs, A.: Problems on finite automata and the exponential time hypothesis. Algorithms **10**(1), 24 (2017)
9. Florêncio, Ch.C., Daenen, J., Ramon, J., Van den Bussche, J., Van Dyck, D.: Naive infinite enumeration of context-free languages in incremental polynomial time. J. Univ. Comput. Sci. **21**(7), 891–911 (2015)
10. Gramlich, G., Schnitger, G.: Minimizing NFA's and regular expressions. J. Comput. Syst. Sci. **73**(6), 908–923 (2007)
11. Gruber, H., Holzer, M.: Computational complexity of NFA minimization for finite and unary languages. In: Preproceedings of the 1st International Conference on Language and Automata Theory and Applications, Technical Report 35/07, pp. 261–272. Research Group on Mathematical Linguistics, Universitat Rovira i Virgili, Tarragona, March 2007
12. Gruber, H., Holzer, M.: Inapproximability of nondeterministic state and transition complexity assuming P $\neq$ NP. In: Harju, T., Karhumäki, J., Lepistö, A. (eds.) DLT 2007. LNCS, vol. 4588, pp. 205–216. Springer, Heidelberg (2007). https://doi.org/10.1007/978-3-540-73208-2_21
13. Gruber, H., Holzer, M.: Language operations with regular expressions of polynomial size. Theoret. Comput. Sci. **410**(35), 3281–3289 (2009)
14. Gruber, H., Holzer, M.: Optimal regular expressions for palindromes of given length. In: Bonchi, F., Puglisi, S.J. (eds.) Proceedings of the 46th International Symposium on Mathematical Foundations of Computer Science, Leibniz International Proceedings in Informatics. Schloss Dagstuhl-Leibniz-Zentrum für Informatik, Dagstuhl (2021, accepted)
15. Gruber, H., Holzer, M., Wolfsteiner, S.: On minimal grammar problems for finite languages. In: Hoshi, M., Seki, S. (eds.) DLT 2018. LNCS, vol. 11088, pp. 342–353. Springer, Cham (2018). https://doi.org/10.1007/978-3-319-98654-8_28
16. Gruber, H., Lee, J., Shallit, J.: Enumerating regular expressions and their languages. arXiv:1204.4982 [cs.FL], April 2012

17. Hopcroft, J.E., Ullman, J.D.: Introduction to Automata Theory, Languages and Computation. Addison-Wesley, Boston (1979)
18. Hunt III, H.B.: On the time and tape complexity of languages I. In: Proceedings of the 5th Annual ACM Symposium on Theory of Computing, pp. 10–19. ACM, Austin, April-May 1973
19. Impagliazzo, R., Paturi, R., Zane, F.: Which problems have strongly exponential complexity? J. Comput. Syst. Sci. **63**(4), 512–530 (2001)
20. Lokshtanov, D., Marx, D., Saurabh, S.: Lower bounds based on the exponential time hypothesis. Bull. Eur. Assoc. Theor. Comput. Sci. **105**, 41–72 (2011)
21. Mandl, R.: Precise bounds associated with the subset construction on various classes of nondeterministic finite automata. In: Proceedings of the 7th Princeton Conference on Information and System Sciences, pp. 263–267, March 1973
22. Meyer, A.R., Stockmeyer, L. J.: The equivalence problem for regular expressions with squaring requires exponential time. In: Proceedings of the 13th Annual Symposium on Switching and Automata Theory, pp. 125–129. IEEE Society Press, October 1972
23. Mráz, F., Průša, D., Wehar, M.: Two-dimensional pattern matching against basic picture languages. In: Hospodár, M., Jirásková, G. (eds.) CIAA 2019. LNCS, vol. 11601, pp. 209–221. Springer, Cham (2019). https://doi.org/10.1007/978-3-030-23679-3_17
24. Wehar, M.: Hardness results for intersection non-emptiness. In: Esparza, J., Fraigniaud, P., Husfeldt, T., Koutsoupias, E. (eds.) ICALP 2014, Part II. LNCS, vol. 8573, pp. 354–362. Springer, Heidelberg (2014). https://doi.org/10.1007/978-3-662-43951-7_30
25. Williams, V.V.: On some fine-grained questions in algorithms and complexity. In: Sirakov, B., de Souza, P.N., Viana, M. (eds.) Proceedings of the International Congress of Mathematicians, pp. 3447–3487. World Scientific, Rio de Janeiro, April 2018

Computational Complexity of Computing a Quasi-Proper Equilibrium

Kristoffer Arnsfelt Hansen[1]([✉]) [iD] and Troels Bjerre Lund[2]

[1] Aarhus University, Aarhus, Denmark
arnsfelt@cs.au.dk
[2] IT-University of Copenhagen, Copenhagen, Denmark
trbj@itu.dk

Abstract. We study the computational complexity of computing or approximating a quasi-proper equilibrium for a given finite extensive form game of perfect recall. We show that the task of computing a symbolic quasi-proper equilibrium is PPAD-complete for two-player games. For the case of zero-sum games we obtain a polynomial time algorithm based on Linear Programming. For general n-player games we show that computing an approximation of a quasi-proper equilibrium is FIXP_a-complete. Towards our results for two-player games we devise a new perturbation of the strategy space of an extensive form game which in particular gives a new proof of existence of quasi-proper equilibria for general n-player games.

1 Introduction

A large amount of research has gone into defining [1,13,18,22,23] and computing [5,7,8,10,17,25] various refinements of Nash equilibria [19]. The motivation for introducing these refinements has been to eliminate undesirable equilibria, e.g., those relying on playing dominated strategies.

The quasi-proper equilibrium, introduced by van Damme [1], is one of the more refined solution concepts for extensive form games. Any quasi-proper equilibrium is quasi-perfect, and therefore also sequential, and also trembling hand perfect in the associated normal form game. Beyond being a further refinement of the quasi-perfect equilibrium [1], it is also conceptually related in that it addresses a deficiency of the direct translation of a normal form solution concept to extensive form games. One of the most well known refinements is Selten's trembling hand perfect equilibrium, originally defined [22] for normal form games, and the solution concept is usually referred to as normal-form perfect. This can be translated to extensive form games, by applying the trembling hand definition to each information set of each player, which yields what is now known as extensive-form perfect equilibria [23]. However, this translation introduces undesirable properties, first pointed out by Mertens [16]. Specifically, Mertens

The first author is supported by the Independent Research Fund Denmark under grant no. 9040-00433B.

© Springer Nature Switzerland AG 2021
E. Bampis and A. Pagourtzis (Eds.): FCT 2021, LNCS 12867, pp. 259–271, 2021.
https://doi.org/10.1007/978-3-030-86593-1_18

presents a certain two-player voting game where all extensive-form perfect equilibria have weakly dominated strategies in their support. That is, extensive-form perfection is in general inconsistent with *admissibility*. Mertens argues that quasi-perfection is conceptually superior to Selten's notion of extensive-form perfection, as it avoids the cause of the problem in Mertens' example. It achieves this with a subtle modification of the definition of extensive-form perfect equilibria, which in effect allows each player to plan as if they themselves were unable to make any future mistakes. Further discussion of quasi-perfection can be found in the survey of Hillas and Kohlberg [11].

One of the most restrictive equilibrium refinements of normal-form games is that of Myerson's normal-form proper equilibrum [18], which is a refinement of Selten's normal-form perfect equilibrium. Myerson's definition can similarly be translated to extensive form, again by applying the definition to each information set of each player, which yields the extensive-form proper equilibria. Not surprisingly, all extensive-form proper equilibria are also extensive-form perfect. Unfortunately, this also means that Merten's critique applies equally well to extensive-form proper equilibria. Again, the definition can be subtely modified to sidestep Merten's example, which then gives the definition for quasi-proper-equilibria [1]. It is exactly this solution concept that is the focus of this paper.

1.1   Contributions

The main novel idea of the paper is a new perturbation of the strategy space of an extensive form game of perfect recall, in which a Nash equilibrium is an ε-quasi-proper equilibrium of the original game. This construction works for *any* number of players and in particular directly gives a new proof of existence of quasi-proper equilibria for general n-player games.

From a computational perspective we can, in the important case of two-player games, exploit the new pertubation in conjunction with the sequence form of extensive form games [12] to compute a symbolic quasi-proper equilibrium by solving a Linear Complementarity Problem. This immediately implies PPAD-membership for the task of computing a symbolic quasi-proper equilibrium. For the case of zero-sum games a quasi-proper equilibrium can be computed by solving just a Linear program which in turn gives a polynomial time algorithm.

For games with more than two players there is, from the viewpoint of computational complexity, no particular advantage in working with the sequence form. Instead we work directly with behavior strategies and go via so-called δ-almost ε-quasi-proper equilibrium, which is a relaxation of ε-quasi-proper equilibrium. We show $FIXP_a$-membership for the task of computing an approximation of a quasi-proper equilibrium. We leave the question of FIXP-membership as an open problem similarly to previous results about computing Nash equilibrium refinements in games with more than two players [2, 3, 8].

Since we work with refinements of Nash equilibrium, PPAD-hardness for two-player games and $FIXP_a$-hardness for n-player games, with $n \geq 3$ follow directly. This combined with our membership results for PPAD and $FIXP_a$ implies PPAD-completeness and $FIXP_a$-completeness, respectively.

1.2   Relation to Previous Work

Any strategic form game may be written as an extensive form game of comparable size, and any quasi-proper equilibrium of the extensive form representation is
a proper equilibrium of the strategic form game. Hence our results fully generalize previous results for computing [24] or approximating [8] a proper equilibrium.
The generalization is surprisingly clean, in the sense that if a bimatrix game is
translated into an extensive form game, the strategy constraints introduced in
this paper will end up being identical to those defined in [24] for the given bimatrix game. This is surprising, since a lot of details have to align for this structure
to survive through a translation to a different game model. Likewise, if a strategic form game with more than two players is translated into an extensive form
game, the fixed point problem we construct in this paper is identical to that for
strategic form games [8].

The quasi-proper equilibria are a subset of the quasi-perfect equilibria, so
our positive computational results also generalize the previous results for quasi-
perfect equilibria [17]. Again, the generalization is clean; if all choices in the game
are binary, then quasi-perfect and quasi-proper coincide, and the constraints
introduced in this paper work out to be exactly the same as those for computing
a quasi-perfect equilibrium. The present paper thus manages to cleanly generalize
two different constructions in two different game models.

2   Preliminaries

2.1   Extensive Form Games

A game Γ in *extensive form* of imperfect information with n players is given as
follows. The structure of Γ is determined by a finite tree T. For a non-leaf node
v, let $S(v)$ denote the set of immediate successor nodes. Let Z denote the set
of leaf nodes of T. In a leaf-node $z \in Z$, player i receives utility $u_i(z)$. Non-leaf
nodes are either chance-nodes or decision-nodes belonging to one of the players.
To every chance node v is associated a probability distribution on $S(v)$. The
set P_i of decision-nodes for Player i is partitioned into information sets. Let H_i
denote the information sets of Player i. To every decision node v is associated a
set of $|S(v)|$ actions and these label the edges between v and $S(v)$. Every decision
node belonging to a given information set h shares the same set C_h of actions.
Define $m_h = |C_h|$ to be the number of actions of every decision node of h. The
game Γ is of *perfect recall* if every node v belonging to an information set h of
Player i share the same sequence of actions and information sets of Player i that
are observed on the path from the root of T to v. We shall only be concerned
with games of perfect recall [14].

A local strategy for Player i at information set $h \in H_i$ is a probability
distribution b_{ih} on C_h assigning a behavior probability to each action in C_h and
in turn induces a probability distribution on $S(v)$ for every $v \in h$. A local strategy
b_{ih} for every information set $h \in H_i$ defines a behavior strategy b_i for Player i.
The behavior strategy b_i is fully mixed if $b_{ih}(c) > 0$ for every $h \in H_i$ and every

$c \in C_h$. Given a local strategy b'_{ih} denote by b_i/b'_{ih} the result of replacing b_{ih} by b'_{ih}. In particular if $c \in C_h$ we let b_i/c prescribe action c with probability 1 in h. For another behavior strategy b'_i and an information set h for Player i we let $b_i/_h b'_i$ denote the behavior strategy that chooses actions according to b_i until h is reached after which actions are chosen according to b'_i. We shall also write $b_i/_h b'_i/c = b_i/_h(b'_i/c)$. A behavior strategy profile $b = (b_1, \ldots, b_n)$ consists of a behavior strategy for each player. Let B be the set of all behavior strategy profiles of Γ. We let $b_{-i} = (b_1, \ldots, b_{i-1}, b_{i+1}, \ldots, b_n)$ and $(b_{-i}; b'_i) = b/b'_i = (b_1, \ldots, b_{i-1}, b'_i, b_{i+1}, \ldots, b_n)$. Furthermore, for simplicity of notation, we define $b/_h b'_i = b/(b_i/_h b'_i)$, and $b/_h b'_i/c = b/(b_i/_h b'_i/c)$.

A behavior strategy profile $b = (b_1, \ldots, b_n)$ gives together with the probability distributions of chance-nodes a probability distribution on the set of paths from the root-node to a leaf-node of T. We let $\rho_b(v)$ be the probability that v is reached by this path and for an information set h we let $\rho_b(h) = \sum_{v \in h} \rho_b(v)$ be the total probability of reaching a node of h. Note that we define $\rho_b(v)$ for all nodes v of T. When $\rho_b(h) > 0$ we let $\rho_b(v \mid h)$ be the conditional probability that node v is reached given that h is reached. The *realization weight* $\rho_{b_i}(h)$ for Player i of an information set $h \in H_i$ is the product of behavior probabilities given by b_i on any path from the root to h. Note that this is well-defined due to the assumption of perfect recall.

Given a behavior strategy profile $b = (b_1, \ldots, b_n)$, the payoff to Player i is $U_i(b) = \sum_{z \in Z} u_i(z)\rho_b(z)$. When $\rho_b(h) > 0$ the conditional payoff to Player i given that h is reached is then $U_{ih}(b) = \sum_{z \in Z} u_i(z)\rho_b(z \mid h)$.

Realization weights are also defined on actions, to correspond to Player i's weight assigned to the given action:

$$\forall h \in H_i, c \in C_h : \quad \rho_{b_i}(c) = \rho_{b_i}(h)b_i(c) \tag{1}$$

We note that the realization weight of an information set is equal to that of the most recent action by the same player, or is equal to 1, if no such action exists.

A realization plan for Player i is a strategy specified by its realization weights for that player. As shown by Koller et al. [12], the set of valid realization weights for Player i can be expressed by the following set of linear constraints

$$\forall h \in H_i : \quad \rho_{b_i}(h) = \sum_{c \in C_h} \rho_{b_i}(c) \quad \wedge \quad \forall c \in C_h : \rho_{b_i}(c) \geq 0 \tag{2}$$

in the variables $\rho_{b_i}(c)$ letting $\rho_{b_i}(h)$ refer to the realization weight of the most recent action of Player i before reaching information set h or to the constant 1 if h is the first time Player i moves. This formulation is known as the sequence form [12], and has the advantage that for two-player games, the utility of each player is bilinear, i.e., linear in the realization weights of each player. As shown by Koller et al. this allows the equilibria to be characterized by the solutions to a Linear Complementarity Problem for general sum games, and as solutions to a Linear Program for zero-sum games. We will build on this insight for computing quasi-proper equilibria of two-player games.

Given a behavior strategy for a player, the corresponding realization plan can be derived by multiplying the behavior probability of an action with the realization weight of its information set. However, it is not always the case that the reverse is possible. The behavior probability of an action is the ratio of the realization weight of an action to the realization weight of its information set, but if any of the preceeding actions by the player have probability 0, the ratio works out to $\frac{0}{0}$. In the present paper, the restriction on the strategy space ensures that no realization weight is zero, until we have retrieved the behavior probabilities.

A strategy profile b is a Nash equilibrium if for every i and every behavior strategy profile b_i' of Player i we have $U_i(b) \geq U_i(b/b_i')$. Our object of study is quasi-proper equilibrium defined by van Damme [1] refining the Nash equilibrium. We first introduce a convenient notation for quantities used in the definition. Let b be a behavior strategy profile, h an information set of Player i such that $\rho_b(h) > 0$, and $c \in C_h$. We then define

$$K_i^{h,c}(b) = \max_{b_i'} U_{ih}(b/_h b_i'/c) \ . \tag{3}$$

When b_i' is a pure behavior strategy we say that $b/_h b_i'$ is a h-local *purification* of b. We note that $U_{ih}(b/_h b_i'/c)$ always assumes its maximum for a pure behavior strategy b_i'.

Definition 1 (Quasi-proper equilibrium). *Given $\varepsilon > 0$, a behavior strategy profile b is an ε-quasi-proper equilibrium if b is fully mixed and satisfies for every i, every information set h of Player i, and every $c, c' \in C_h$, that $b_{ih}(c) \leq \varepsilon b_{ih}(c')$ whenever $K_i^{h,c}(b) < K_i^{h,c'}(b)$.*

A behavior strategy profile b is a quasi-proper equilibrium if and only if it is a limit point of a sequence of ε-quasi-proper equilibria with $\varepsilon \to^+ 0$.

We shall also consider a relaxation of quasi-proper equilibrium in analogy to relaxations of other equilibrium refinements due to Etessami [2].

Definition 2. *Given $\varepsilon > 0$ and $\delta > 0$, a behavior strategy profile b is a δ-almost ε-quasi-proper equilibrium if b is fully mixed and satisfies for every Player i, every information set h of Player i, and every $c, c' \in C_h$ that $b_{ih}(c) \leq \varepsilon b_{ih}(c')$ whenever $K_i^{h,c}(b) + \delta \leq K_i^{h,c'}(b)$.*

2.2   Strategic Form Games

A game Γ in *strategic form* with n players is given as follows. Player i has a set S_i of *pure strategies*. To a pure strategy profile $a = (a_1, \ldots, a_n)$ Player i is given utility $u_i(a)$. A mixed strategy x_i for Player i is a probability distribution on S_i. We identify a pure strategy with the mixed strategy that selects the pure strategy with probability 1. A strategy profile $x = (x_1, \ldots, x_n)$ consists of a mixed strategy for each player. To a strategy profile x Player i is given utility $U_i(x) = \sum_{a \sim x} u_i(a) \prod_j x_j(a_j)$. A strategy profile x is fully mixed if $x_i(a_i) > 0$ for all i and all $a_i \in S_i$. We let $x_{-i} = (x_1, \ldots, x_{i-1}, x_{i+1}, \ldots, x_n)$. Given a strategy x_i' for Player i we define $(x_{-i}; x_i') = x/x_i' = (x_1, \ldots, x_{i-1}, x_i', x_{i+1}, \ldots, x_n)$.

A strategy profile x is a Nash equilibrium if for every i and every strategy x_i' of Player i we have $U_i(x/x_i') \leq U_i(x)$. Myerson defined the notion of proper equilibrium [18] refining the Nash equilibrium.

Definition 3 (Proper equilibrium). *Given $\varepsilon > 0$, a strategy profile x is an ε-proper equilibrium if x is fully mixed and satisfies for every i and every $c, c' \in S_i$ that $x_i(c) \leq \varepsilon x_i(c')$ whenever $U_i(x_{-i}; c) < U_i(x_{-i}; c')$.*

A strategy profile x is a proper equilibrium if and only if it is a limit point of a sequence of ε-proper equilibria with $\varepsilon \to^+ 0$.

For proper equilibrium we also consider a relaxation as suggested by Etessami [2].

Definition 4. *Given $\varepsilon > 0$ and $\delta > 0$, a strategy profile x is a δ-almost ε-proper equilibrium if x is fully mixed and satisfies for every i and every $c, c' \in S_i$ that $x_i(c) \leq \varepsilon x_i(c')$ whenever $U_i(x_{-i}; c) + \delta \leq U_i(x_{-i}; c')$.*

2.3 Complexity Classes

We give here only a brief description of the classes PPAD and FIXP and refer to Papadimitriou [20] and Etessami and Yannakakis [4] for detailed definitions and discussion of the two classes.

PPAD is a class of discrete total search problems, whose totality is guaranteed based on a parity argument on a directed graph. More formally PPAD is defined by a canonical complete problem ENDOFTHELINE. Here a directed graph is given implicitly by predecessor and successor circuits, and the search problem is to find a degree 1 node different from a given degree 1 node. We do not make direct use of the definition of PPAD but instead prove PPAD-membership indirectly via Lemke's algorithm [15] for solving a Linear Complementarity Problem (LCP).

FIXP is the class of real-valued total search problems that can be cast as Brouwer fixed points of functions represented by $\{+, -, *, /, \max, \min\}$-circuits computing a function mapping a convex polytope described by a set of linear inequalities to itself. The class FIXP_a is the class of *discrete* total search problems that reduce in polynomial time to *approximate* Brouwer fixed points. We will prove FIXP_a membership directly by constructing an appropriate circuit.

3 Two-Player Games

In this section, we prove that computing a single quasi-proper equilibrium of a two-player game Γ can be done in PPAD, and in the case of zero-sum games, it can be computed in P. We are using the same overall approach as has been used for computing quasi-perfect equilibria of extensive form games [17], proper equilibria of two-player games [24], and proper equilibria of poly-matrix games [8].

The main idea is to construct a new game Γ_ε, where the strategy space is slightly restricted for both players, in such a way that equilibria of the new game are ε-quasi-proper equilibria of the original game. This construction also provides a new proof of existence for quasi-proper equilibria of n-player games, since there is nothing in neither the construction nor the proof that requires the game to have only two players. However, for two players, the strategy constraints can be enforced using a symbolic infinitesimal ε, which can be part of the solution output, thereby providing a witness of the quasi-properness of the computed strategy.

We will first describe the strategy constraints. At a glance, the construction consists of fitting the strategy constraints for ε-proper equilibria [24] into the strategy constraints of each of the information sets of the sequence form [12], discussed in the preliminaries section, Eq. (2).

The constraints for ε-proper equilibria [24] restricts the strategy space of each player to be an ε-permutahedron. Before the technical description of this, we define the necessary generalization of the permutahedron. A permutahedron is traditionally over the vector $(1, \ldots, n)$, but it generalizes directly to any other set as well.

Definition 5 (Permutahedron). *Let $\alpha \in \mathbb{R}^m$ with all coordinates being distinct. A permutation $\pi \in S_m$ acts on α by permuting the coordinates of α, i.e. $(\pi(\alpha))_i = \alpha_{\pi(i)}$. We define the permutahedron $Perm(\alpha)$ over α to be the convex hull of the set $\{\pi(\alpha) \mid \pi \in S_m\}$ of the $m!$ permutations of the coordinates of α.*

A very useful description of the permutahedron is by its $2^m - 2$ facets.

Proposition 1 (Rado [21]). *Suppose $\alpha_1 > \alpha_2 > \cdots > \alpha_m$. Then*

$$Perm(\alpha) = \left\{ x \in \mathbb{R}^m \ \middle| \ \sum_{i=1}^m x_i = \sum_{i=1}^m \alpha_i \wedge \forall S \notin \{\emptyset, [m]\} : \sum_{c \in S} x_c \geq \sum_{i=1}^{|S|} \alpha_{m-i+1} \right\} .$$

As each inequality of Proposition 1 define a facet of the permutahedron, any direct formulation of the permutahedron over n elements requires $2^n - 2$ inequalities. Goemans [6] gave an asymptotically optimal extended formulation for the permutahedron, using $O(n \log n)$ additional constraints and variables. This allows a compact representation, which allows us to use ε-permutahedra [24] as building blocks for our strategy constraints.

The ε-permutahedron defined in [24] is a permutahedron over the vector $(1, \varepsilon, \varepsilon^2, \ldots, \varepsilon^{m-1})$, normalized to sum to 1. We need to generalize this, so that it can sum to any value ρ, and in a way that does not require normalization. In the following, we will abuse notation slightly, and use ρ without subscript as a real number, since it will shortly be replaced by a realization weight for each specific information set.

Definition 6 (ε-Permutahedron). *For real $\rho > 0$, integers $k \geq 0$ and $m \geq 1$, and $\varepsilon > 0$ such that $\rho \geq \varepsilon^k$, define the vector $p_\varepsilon(\rho, k, m) \in \mathbb{R}^m$ by*

$$(p_\varepsilon(\rho, k, m))_i = \begin{cases} \rho - (\varepsilon^{k+1} + \cdots + \varepsilon^{k+m-1}) & , i = 1 \\ \varepsilon^{k+i-1} & , i > 1 \end{cases} ,$$

and define the ε-permutahedron $\Pi_\varepsilon(\rho, k, m) = Perm(p_\varepsilon(\rho, k, m)) \subseteq \mathbb{R}^m$.

We shall be viewing ε as a variable. Note that, by definition, $\|p_\varepsilon(\rho, k, m)\|_1 = \rho$.

Lemma 1. *Assume $0 < \varepsilon \leq 1/3$ and $\rho \geq \varepsilon^k$, for a given integer $k \geq 0$. Then for every $1 \leq i < m$ we have $(p_\varepsilon(\rho, k, m))_i \geq (p_\varepsilon(\rho, k, m))_{i+1}/(2\varepsilon)$.*

Proof. The statement clearly holds for $i > 1$. Next we see that $(p_\varepsilon(\rho, k, m))_1 = \rho - \varepsilon^{k+1}(1 - \varepsilon^{m-1})/(1 - \varepsilon) \geq \varepsilon^k - \varepsilon^{k+1}/(1 - \varepsilon) = (1/\varepsilon - 1/(1 - \varepsilon))\varepsilon^{k+1} \geq \varepsilon^{k+1}/(2\varepsilon) = (p_\varepsilon(\rho, k, m))_2/(2\varepsilon)$. □

We are now ready to define the perturbed game Γ_ε.

Definition 7 (Strategy constraints). *For each player i, and each informa-tion set $h \in H_i$, let $k_h = \sum_{h'<h} m_{h'}$ be the sum of the sizes of the action sets at information sets visited by Player i before reaching information set h. Now, in the perturbed game Γ_ε, restrict $(\rho_{b_i}(c_1), \rho_{b_i}(c_2), \ldots, \rho_{b_i}(c_{m_h}))$ to be in $\Pi_\varepsilon(\rho_{b_i}(h), k_h, m_h)$.*

Notice that the strategy constraints for the first information set a player visits is identical to the strategy constraints for proper equilibria of bimatrix games.

The next three lemmas describe several ways we may modify coordinates of points of $\Pi_\varepsilon(\rho, k, m)$ while staying within $\Pi_\varepsilon(\rho', k, m)$ for appropriate ρ'. These are needed for the proof of our main technical result, Proposition 2, below.

Lemma 2. *Let $0 < \varepsilon < 1/3$, $\rho \geq \varepsilon^k$, and $x \in \Pi_\varepsilon(\rho, k, m)$. Suppose for distinct c and c' we have $x_c > 2\varepsilon x_{c'}$. Then there exists $\delta > 0$ such that $x + \delta(e_{c'} - e_c) \in \Pi_\varepsilon(\rho, k, m)$ (here as usual e_i denotes the i-unit vector).*

Proof. By definition of $\Pi_\varepsilon(\rho, k, m)$ we may write x as a convex combination of the corner points of $\Pi_\varepsilon(\rho, k, m)$, $x = \sum_{\pi \in S_m} w_\pi \pi(p_\varepsilon(\rho, k, m))$, where $w_\pi \geq 0$ and $\sum_{\pi \in S_m} w_\pi = 1$. There must exist a permutation π such that $w_\pi > 0$ and $\pi^{-1}(c) < \pi^{-1}(c')$, since otherwise $x_c \leq 2\varepsilon x_{c'}$ by Lemma 1. Let $\pi' \in S_m$ such that $\pi'(\pi^{-1}(c)) = c'$, $\pi'(\pi^{-1}(c')) = c$, and $\pi'(i) = \pi(i)$ when $\pi(i) \notin \{c, c'\}$. We then have that

$$x' = x + w_\pi(\pi'(p_\varepsilon(\rho, k, m)) - \pi(p_\varepsilon(\rho, k, m))) \in \Pi_\varepsilon(\rho, k, m) .$$

Note now that $\pi'(p_\varepsilon(\rho, k, m)) - \pi(p_\varepsilon(\rho, k, m))$ is equal to

$$((p_\varepsilon(\rho, k, m))_{\pi^{-1}(c)} - (p_\varepsilon(\rho, k, m))_{\pi^{-1}(c')})(e_{c'} - e_c) .$$

Since $(p_\varepsilon(\rho, k, m))_{\pi^{-1}(c)} > (p_\varepsilon(\rho, k, m))_{\pi^{-1}(c')}$, the statement follows. □

Lemma 3. *Let $x \in \Pi_\varepsilon(\rho, k, m)$ where $\rho \geq \varepsilon^k$. Then $x + \delta e_c \in \Pi(\rho + \delta, k, m)$ for any $\delta > 0$ and c.*

Proof. This follows immediately from Proposition 1 since the inequalities defin-ing the facets of $\Pi_\varepsilon(\rho, k, m)$ and $\Pi_\varepsilon(\rho + \delta, k, m)$ are exactly the same. □

Lemma 4. *Let $x \in \Pi_\varepsilon(\rho, k, m)$ where $0 < \varepsilon \leq 1/2$ and $\rho > \max(\varepsilon^k, 2m\varepsilon^{k+1})$. Let c be such that $x_c \geq x_{c'}$ for all c'. Then $x - \delta e_c \in \Pi_\varepsilon(\rho - \delta, k, m)$ for any $\delta \leq \min(\rho - \varepsilon^k, \rho/m - 2\varepsilon^{k+1})$.*

Proof. Since $\delta \leq \rho - \varepsilon^k$ we have $\rho - \delta \geq \varepsilon^k$, thereby satisfying the definition of $\Pi_\varepsilon(\rho - \delta, k, m)$. By the choice of c we have that $x_c \geq \rho/m$. Since we also have $\delta \leq \rho/m - 2\varepsilon^{k+1}$ it follows that $x_c - \delta \geq 2\varepsilon^{k+1}$. Thus $x_c - \delta \geq \varepsilon^{k+1} + \cdots + \varepsilon^{k+m-1}$. It then follows immediately from Proposition 1 that $x - \delta e_c \in \Pi_\varepsilon(\rho - \delta, k, m)$, since any inequality given by S with $c \in S$ is trivially satisfied, and any inequality with $c \notin S$ is unchanged from $\Pi_\varepsilon(\rho, k, m)$. $\qquad\square$

We are now in position to prove correctness of our approach.

Proposition 2. *Any Nash equilibrium of Γ_ε is a 2ε-quasi-proper equilibrium of Γ, for any sufficiently small $\varepsilon > 0$.*

Proof. Let b be a Nash equilibrium of Γ_ε. Consider Player i for any i, any information set $h \in H_i$, and let $c, c' \in h$ be such that $b_{ih}(c) > 2\varepsilon b_{ih}(c')$. We are then to show that $K_i^{h,c}(b) \geq K_i^{h,c'}(b)$, when $\varepsilon > 0$ is sufficiently small. Let b_i' be such that $U_{ih}(b/_h b_i'/c') = K_i^{h,c'}(b)$. We may assume that b_i' is a pure behavior strategy thereby making b/b_i' a h-local purification. Let $H_{i,c'}$ be the set of those information sets of Player i that follow after h when taking action c' in h. Similarly, let $H_{i,c}$ be the set of those information sets of Player i that follow after taking action c in h. Note that by perfect recall of Γ we have that $H_{i,c'} \cap H_{i,c} = \emptyset$. Let b_i^* be any pure behavior strategy of Player i choosing $c_h^* \in C_h$ maximizing $b_{ih}(c_h^*)$, for all $h \in H_i$. We claim that $U_{ih}(b/_h b_i^*/c) \geq K_i^{h,c'}(b)$ for all sufficiently small $\varepsilon > 0$.

Let x_i be the realization plan given by b_i, let x' be the realization plan given by $b_i/_h b_i'/c'$, and let x_i^* be the realization plan given by $b_i/_h b_i^*/c$. We shall next apply Lemma 2 to h, Lemma 3 to all $h' \in H_{i,c'}$, and Lemma 4 to all $h^* \in H_{i,c}$ assigned positive realization weight by $b_i/_h b_i^*/c$, to obtain that for all sufficiently small $\varepsilon > 0$ there is $\delta > 0$ such that $\widetilde{x}_i = x_i + \delta(x_i' - x_i^*)$ is a valid realization plan of Γ_ε.

Lemma 3 can be applied whenever $\varepsilon > 0$ is sufficiently small, whereas Lemma 2 in addition makes use of the assumption that $b_{ih}(c) > 2\varepsilon b_{ih}(c')$. To apply Lemma 4, we need to prove that the player's realization weight is sufficiently large for the relevant information sets, specifically $\rho_{h'} > \varepsilon^{k_{h'}}$ for each relevant information set h'. Since b_i^* is pure, Player i's realization weight, $\rho_{h'}$ for each information set h' in $H_{i,c}$ is either 0 or ρ_c. Since $b_{ih}(c) > 2\varepsilon b_{ih}(c')$, we have that $\rho_c > \varepsilon^{k_h + |C_h| - 1} \geq \varepsilon^{k_{h'}}$ as needed.

Thus, consider $\varepsilon > 0$ and $\delta > 0$ such that $\widetilde{x}_i$ is a valid realization plan and let $\widetilde{b}_i$ be the corresponding behavior strategy. Since b is a Nash equilibrium we have $U_i(b/\widetilde{b}_i) \leq U_i(b)$. But $U_i(b/\widetilde{b}_i) = U_i(b) + \delta(U_i(b/_h b_i'/c') - U_i(b/_h b_i^*/c))$. It follows that $\delta(U_i(b/_h b_i'/c') - U_i(b/_h b_i^*/c)) \leq 0$, and since $\delta > 0$ we have $U_i(b/_h b_i^*/c) \geq U_i(b/_h b_i'/c')$. Equivalently, $U_{ih}(b/_h b_i^*/c) \geq U_{ih}(b/_h b_i'/c')$, which was to be proved. Since i and $h \in H_i$ were arbitrary, it follows that b is a 2ε-quasi-proper equilibrium in Γ, for any sufficient small $\varepsilon > 0$. $\qquad\square$

Theorem 1. *A symbolic ε-quasi-proper equilibrium for a given two-player extensive form game with perfect recall can be computed by applying Lemke's algorithm to an LCP of polynomial size, and can be computed in* PPAD.

Proof. Given an extensive form game Γ, construct the game Γ_ε. The strategy constraints (Definition 7) are all expressed directly in terms of the realization weights of each player. Using Goemans' [6] extended formulation, the strategy constraints require only $O(\sum_h |C_h| \log |C_h|)$ additional constraints and variables, which is linearithmic in the size of the game. Furthermore, all occurrences of ε are on the right-hand side of the linear constraints. These constraints fully replace the strategy constraints of the sequence form [12]. In the sequence form, there is a single equality per information set, ensuring conservation of the realization weight. In our case, this conservation is ensured by the permutahedron constraint for each information set.

In the case of two-player games, the equilibria can be captured by an LCP of polynomial size, which can be solved using Lemke's algorithm [15], if the strategy constraints are sufficiently well behaved. Since the added strategy constraints is a collection of constraints derived from Goemans' extended formulation, the proof that the constraints are well behaved is identical to the proofs of [24, Theorem 5.1 and 5.4], which we will therefore omit here. Following the approach of [17] the solution to the LCP can be made to contain the symbolic ε, with the probabilities of the strategies being formal polynomials in the variable ε.

By Proposition 2, equilibria of Γ_ε are ε-quasi-proper equilibria of Γ. All realization weights of the computed realization plans are formal polynomials in ε. Finally, from this we may express the ε-quasi-proper equilibrium in behavior strategies, where all probabilities are rational functions in ε.   □

Having computed a symbolic ε-quasi-proper equilibrium for Γ it is easy to compute the limit for $\varepsilon \to 0$, thereby giving a quasi-proper equilibrium of Γ. It is crucial here that we first convert into behavior strategies before computing the limit. In the case of zero-sum games, the same construction can be used to construct a linear program of polynomial size, whose solution would provide quasi-proper equilibria of the given game. This is again analogous to the approach of [17] and further details are hence omitted.

Theorem 2. *A symbolic ε-quasi-proper equilibrium for a given two-player extensive form zero-sum game with perfect recall can be computed in polynomial time.*

4   Multi-player Games

In this section we argue that approximating a quasi-proper equilibrium for a finite extensive-form game Γ with $n \geq 3$ players is FIXP_a-complete. As for two-player games, by Proposition 2 an ε-quasi-proper equilibrium for Γ could be obtained by computing an equilibrium of the perturbed game Γ_ε. But for more than two players we do not know how to make efficient use of this connection.

Indeed, from the viewpoint of computational complexity there is no advantage in doing so. Our construction instead works by directly combining the approach and ideas of the proof of FIXP_a-completeness for quasi-perfect equilibrium in extensive form games by Etessami [2] and of the proof of FIXP_a-completeness for proper equilibrium in strategic form games by Hansen and Lund [8]. We explain below how these are modified and combined to obtain the result. The approach obtains FIXP_a membership, leaving FIXP-membership as an open problem. A quasi-proper equilibrium is defined as a limit point of a sequence of ε-quasi-proper equilibria, whose existence was obtained by the Kakutani fixed point theorem by Myerson [18]. This limit point operation in itself poses a challenge for FIXP membership. The use of the Kakutani fixed point theorem presents a further challenge. In fact, it is not known if the set of ε-quasi-proper equilibria can be characterized as a set of Brouwer fixed points. However, as we show below analougous to the case of proper equilibria [8], these may be approximated by δ-almost ε-quasi-proper equilibria, which in turn can be expressed as a set of Brouwer fixed points. In fact we show that the corresponding search problem is in FIXP.

To see how to adapt the result of Hansen and Lund [8] for strategic form games to the setting of extensive form games, it is helpful to compare the definitions of ε-proper equilibrium and δ-almost ε-proper equilibrium in strategic form games to the corresponding definitions of ε-quasi-proper equilibrium and δ-almost ε-proper equilibrium in extensive form games.

In a strategic form game, Player i is concerned with the payoffs $U_i(x_{-i}, c)$, which we may think of as *valuations* of all pure strategies $c \in S_i$. The relationship between these valuations in turn place constraints on the strategy x_i chosen by Player i in an ε-proper equilibrium or a δ-almost ε-proper equilibrium. In an extensive form game, Player i is in a given information set h considering the payoffs $K_i^{h,c}$, which we may similarly think of as *valuations* of all actions $c \in C_h$. The relationship between these valuations place constraints on the local strategy b_{ih} chosen by Player i in a ε-quasi-proper equilibrium or a δ-almost ε-proper equilibrium. These constraints are completely analogous to those placed on the strategies in strategic form games. This fact will allow us to adapt the constructions of Hansen and Lund by essentially just changing the way the valuations are computed. Etessami [2] observed that these may be computed using dynamic programming and gave a construction of formulas computing them.

Lemma 5 (cf. [2, Lemma 7]). *Given an extensive form game of perfect recall Γ, a player i, an information set h of Player i, and $c \in C_h$ there is a polynomial size $\{+, -, *, /, \max\}$-formula $V_i^{h,c}$ computable in polynomial time satisfying that for any fully mixed behavior strategy profile b it holds that $V_i^{h,c}(b) = K_i^{h,c}(b)$.*

With this we can now state our result for multi-player games.

Theorem 3. *Given as input a finite extensive form game of perfect recall Γ with n players and a rational $\gamma > 0$, the problem of computing a behavior strategy profile b' such that there is a quasi-proper equilibrium b of Γ with $\|b' - b\|_\infty < \gamma$ is FIXP_a-complete.*

While the adaptation of the results of Hansen and Lund [8] to extensive-form games is conceptually simple, changes must be made in all parts of the construction and technical proof. We refer to the full version of the paper [9] for further details.

References

1. van Damme, E.: A relation between perfect equilibria in extensive form games and proper equilibria in normal form games. Int. J. Game Theory **13**, 1–13 (1984). https://doi.org/10.1007/BF01769861
2. Etessami, K.: The complexity of computing a (quasi-)perfect equilibrium for an n-player extensive form game. Games Econ. Behav. **125**, 107–140 (2021). https://doi.org/10.1016/j.geb.2019.03.006
3. Etessami, K., Hansen, K.A., Miltersen, P.B., Sørensen, T.B.: The complexity of approximating a trembling hand perfect equilibrium of a multi-player game in strategic form. In: Lavi, R. (ed.) SAGT 2014. LNCS, vol. 8768, pp. 231–243. Springer, Heidelberg (2014). https://doi.org/10.1007/978-3-662-44803-8_20
4. Etessami, K., Yannakakis, M.: On the complexity of Nash equilibria and other fixed points. SIAM J. Comput. **39**(6), 2531–2597 (2010). https://doi.org/10.1137/080720826
5. Farina, G., Gatti, N.: Extensive-form perfect equilibrium computation in two-player games. In: AAAI, pp. 502–508. AAAI Press (2017)
6. Goemans, M.X.: Smallest compact formulation for the permutahedron. Math. Program. **153**(1), 5–11 (2014). https://doi.org/10.1007/s10107-014-0757-1
7. Hansen, K.A.: The real computational complexity of minmax value and equilibrium refinements in multi-player games. Theory Comput. Syst. **63**(7), 1554–1571 (2018). https://doi.org/10.1007/s00224-018-9887-9
8. Hansen, K.A., Lund, T.B.: Computational complexity of proper equilibrium. In: Proceedings of the 2018 ACM Conference on Economics and Computation, EC 2018, pp. 113–130. ACM, New York (2018). https://doi.org/10.1145/3219166.3219199
9. Hansen, K.A., Lund, T.B.: Computational complexity of computing a quasi-proper equilibrium. arXiv:2107.04300 [cs.GT] (2021)
10. Hansen, K.A., Miltersen, P.B., Sørensen, T.B.: The computational complexity of trembling hand perfection and other equilibrium refinements. In: Kontogiannis, S., Koutsoupias, E., Spirakis, P.G. (eds.) SAGT 2010. LNCS, vol. 6386, pp. 198–209. Springer, Heidelberg (2010). https://doi.org/10.1007/978-3-642-16170-4_18
11. Hillas, J., Kohlberg, E.: Foundations of strategic equilibria. In: Aumann, R.J., Hart, S. (eds.) Handbook of Game Theory, vol. 3, chap. 42, pp. 1597–1663. Elsevier Science (2002)
12. Koller, D., Megiddo, N., von Stengel, B.: Efficient computation of equilibria for extensive two-person games. Games Econom. Behav. **14**, 247–259 (1996). https://doi.org/10.1006/game.1996.0051
13. Kreps, D.M., Wilson, R.: Sequential equilibria. Econometrica **50**(4), 863–894 (1982). https://doi.org/10.2307/1912767
14. Kuhn, H.W.: Extensive games and the problem of information. In: Kuhn, H.W., Tucker, A.W. (eds.) Contributions to the Theory of Games II, pp. 193–216. Princeton University Press, Princeton (1953)

15. Lemke, C.: Bimatrix equilibrium points and mathematical programming. Manag. Sci. **11**, 681–689 (1965). https://doi.org/10.1287/mnsc.11.7.681
16. Mertens, J.F.: Two examples of strategic equilibrium. Games Econ. Behav. **8**(2), 378–388 (1995). https://doi.org/10.1016/S0899-8256(05)80007-7
17. Miltersen, P.B., Sørensen, T.B.: Computing a quasi-perfect equilibrium of a two-player game. Econ. Theor. **42**(1), 175–192 (2010). https://doi.org/10.1007/s00199-009-0440-6
18. Myerson, R.B.: Refinements of the Nash equilibrium concept. Int. J. Game Theory **15**, 133–154 (1978). https://doi.org/10.1007/BF01753236
19. Nash, J.: Non-cooperative games. Ann. Math. **2**(54), 286–295 (1951)
20. Papadimitriou, C.H.: On the complexity of the parity argument and other inefficient proofs of existence. J. Comput. Syst. Sci. **48**(3), 498–532 (1994). https://doi.org/10.1016/S0022-0000(05)80063-7
21. Rado, R.: An inequality. J. Lond. Math. Soc. **s1–27**(1), 1–6 (1952). https://doi.org/10.1112/jlms/s1-27.1.1
22. Selten, R.: Spieltheoretische behandlung eines oligopolmodells mit nachfrageträgheit. Zeitschrift für die gesamte Staatswissenshaft **12**, 301–324 (1965)
23. Selten, R.: A reexamination of the perfectness concept for equilibrium points in extensive games. Int. J. Game Theory **4**, 25–55 (1975). https://doi.org/10.1007/BF01766400
24. Sørensen, T.B.: Computing a proper equilibrium of a bimatrix game. In: Faltings, B., Leyton-Brown, K., Ipeirotis, P. (eds.) ACM Conference on Electronic Commerce, EC '12, pp. 916–928. ACM (2012). https://doi.org/10.1145/2229012.2229081
25. von Stengel, B., van den Elzen, A., Talman, D.: Computing normal form perfect equilibria for extensive two-person games. Econometrica **70**(2), 693–715 (2002). https://doi.org/10.1111/1468-0262.00300

Computational Complexity of Synchronization Under Sparse Regular Constraints

Stefan Hoffmann[(✉)]

Informatikwissenschaften, FB IV, Universität Trier, Trier, Germany
hoffmanns@informatik.uni-trier.de

Abstract. The constrained synchronization problem (CSP) asks for a synchronizing word of a given input automaton contained in a regular set of constraints. It could be viewed as a special case of synchronization of a discrete event system under supervisory control. Here, we study the computational complexity of this problem for the class of sparse regular constraint languages. We give a new characterization of sparse regular sets, which equal the bounded regular sets, and derive a full classification of the computational complexity of CSP for letter-bounded regular constraint languages, which properly contain the strictly bounded regular languages. Then, we introduce strongly self-synchronizing codes and investigate CSP for bounded languages induced by these codes. With our previous result, we deduce a full classification for these languages as well. In both cases, depending on the constraint language, our problem becomes NP-complete or polynomial time solvable.

Keywords: Automata theory · Constrained synchronization · Computational complexity · Sparse languages · Bounded languages · Strongly self-synchronizing codes

1 Introduction

A deterministic semi-automaton is *synchronizing* if it admits a reset word, i.e., a word which leads to a definite state, regardless of the starting state. This notion has a wide range of applications, from software testing, circuit synthesis, communication engineering and the like, see [13,45,47]. The famous Černý conjecture [11] states that a minimal synchronizing word, for an n state automaton, has length at most $(n-1)^2$. We refer to the mentioned survey articles for details [45,47].

Due to its importance, the notion of synchronization has undergone a range of generalizations and variations for other automata models. The paper [19] introduced the constrained synchronization problem (CSP[1]). In this problem,

---

[1] In computer science the acronym CSP is usually used for the constraint satisfaction problem [35]. However, as here we are not concerned with constrained satisfaction problems at all, no confusion should arise.

© Springer Nature Switzerland AG 2021
E. Bampis and A. Pagourtzis (Eds.): FCT 2021, LNCS 12867, pp. 272–286, 2021.
https://doi.org/10.1007/978-3-030-86593-1_19

we search for a synchronizing word coming from a specific subset of allowed input sequences. To sketch a few applications:

Reset State. In [19] one motivating example was the demand that a system, or automaton thereof, to synchronize has to first enter a "directing" mode, perform a sequence of operations, and then has to leave this operating mode and enter the "normal operating mode" again. In the most simple case, this constraint could be modelled by ab^*a, which, as it turns out [19], yields an NP-complete CSP. Even more generally, it might be possible that a system – a remotely controlled rover on a distant planet, a satellite in orbit, or a lost autonomous vehicle – is not allowed to execute all commands in every possible order, but certain commands are only allowed in certain order or after other commands have been executed. All of this imposes constraints on the possible reset sequences.

Part Orienters. Suppose parts arrive at a manufacturing site and they need to be sorted and oriented before assembly. Practical considerations favor methods which require little or no sensing, employ simple devices, and are as robust as possible. This could be achieved as follows. We put parts to be oriented on a conveyor belt which takes them to the assembly point and let the stream of the parts encounter a series of passive obstacles placed along the belt. Much research on synchronizing automata was motivated by this application [12, 17, 18, 25, 37, 38, 47] and I refer to [47] for an illustrative example. Now, furthermore, assume the passive components could not be placed at random along the belt, but have to obey some restrictions, or restrictions in what order they are allowed to happen. These could be due to the availability of components, requirements how to lay things out or physical restrictions.

Supervisory Control. The CSP could also be viewed of as supervisory control of a discrete event system (DES) that is given by an automaton and whose event sequence is modelled by a formal language [10, 42, 50]. In this framework, a DES has a set of controllable and uncontrollable events. Dependent on the event sequence that occurred so far, the supervisor is able to restrict the set of events that are possible in the next step, where, however, he can only limit the use of controllable events. So, if we want to (globally) reset a finite state DES [2] under supervisory control, this is equivalent to CSP.

Biocomputing. In [4, 5] DNA molecules have been used as both hardware and software for finite automata of nanoscaling size, see also [47]. For instance, Benenson et al. [4] produced "a 'soup of automata', that is, a solution containing 3×10^{12} identical automata per $\mu 1$. All these molecular automata can work in parallel on different inputs, thus ending up in different and unpredictable states. In contrast to an electronic computer, one cannot reset such a system by just pressing a button; instead, in order to synchronously bring each automaton to its start state, one should spice the soup with (sufficiently many copies of) a DNA molecule whose nucleotide sequences encodes a reset word" [47]. Now, it might be possible that certain sequences, or subsequences, are not possible as they might have unwanted biological side-effects, or might destroy the molecules at all.

Reduction Procedure. This example is more formal and comes from attempts to solve the Černý conjecture [47]. In [27] a special rank factorization [41] for automata was introduced from which smaller automata could be derived. Then, it was shown that the original automaton is synchronizing if and only if the reduced automaton admits a synchronizing word in a certain regular constraint language, and the reset threshold, i.e., the lengths of the shortest synchronizing word, of the original automaton could be bounded by that of the shortest one in the constraint language for the reduced automaton.

In [19], a complete analysis of the complexity landscape when the constraint language is given by small partial automata was done. It is natural to extend this result to other language classes.

In general there exist constraint languages yielding PSPACE-complete constrained problems [19]. A language is polycyclic [31], if it is recognizable by an automaton such that every strongly connected component forms a single cycle, and a language is sparse [51] if only polynomially many words of a specific length are in the language. As shown in [31] for polycyclic languages, which, as we show, equal the sparse regular languages, the problem is always in NP. This motivates investigating this class further. Also, as written in more detail in Remark 1, a subclass of these languages has a close relation to the commutative languages, and as for commutative constraint languages a trichotomy result has been established [30], tackling the sparse languages seems to be the next logical step. In fact, we show a dichotomy result for a subclass that contains the class corresponding to the commutative languages. Additionally, as has been noted in [19], the constraint language ab^*a is the smallest language, in terms of a recognizing automaton, giving an NP-complete CSP. The class of languages for which our dichotomy holds true contains this language.

Let us mention that restricting the solution space by a regular language has also been applied in other areas, for example to topological sorting [3], solving word equations [15,16], constraint programming [39], or shortest path problems [43]. The famous road coloring theorem [1,46] states that every finite strongly connected and directed aperiodic graph of uniform out-degree admits a labelling of its edges such that a synchronizing automaton results. A related problem to our problem of constrained synchronization is to restrict the possible labelling(s), and this problem was investigated in [49].

Outline and Contribution. Here, we look at the complexity landscape for sparse regular constraint languages. In Sect. 3 we introduce the sparse languages and show that the regular sparse languages are characterized by polycyclic automata introduced in [31]. A similar characterization in terms of non-deterministic automata was already given in [21, Lemma 2]. In this sense, we extend this characterization to the deterministic case. As for polycyclic constraint automata the constrained problem is always in NP, see [31, Theorem 2], we can deduce the same for sparse regular constraint languages, which equal the bounded regular languages [34].

In Sect. 4 we introduce the letter-bounded languages, a proper subset of the sparse languages, and show that for letter-bounded constraint languages, the constrained synchronization problem is either in P or NP-complete.

The difficulty why we cannot handle the general case yet lies in the fact that in the reductions, in the general case, we need auxiliary states and it is not clear how to handle them properly, i.e., how to synchronize them properly while staying inside the constraint language.

In Sect. 5 we introduce the class of strongly self-synchronizing codes. The strongly self-synchronizing codes allow us to handle these auxiliary states mentioned before. We show that for homomorphisms given by such codes, the constrained problem for the homomorphic image of a language has the same computational complexity as for the original language. This result holds in general, and hence is of independent interest. Here we apply it to the special case of bounded, or sparse, regular languages given by such codes.

Lastly, we present a bounded language giving an NP-complete constrained problem that could not be handled by our methods so far.

2   Preliminaries and Definitions

We assume the reader to have some basic knowledge in computational complexity theory and formal language theory, as contained, e.g., in [32]. For instance, we make use of regular expressions to describe languages. By Σ we denote the *alphabet*, a finite set. For a word $w \in \Sigma^*$ we denote by $|w|$ its *length*, and, for a symbol $x \in \Sigma$, we write $|w|_x$ to denote the *number of occurrences of x* in the word. We denote the empty word, i.e., the word of length zero, by ε. For $L \subseteq \Sigma^*$, we set $\mathrm{Pref}(L) = \{u \in \Sigma^* \mid \exists v \in \Sigma^* : uv \in L\}$. A word $u \in \Sigma^*$ is a *factor* (or *infix*) of $w \in \Sigma^*$ if there exist words $x, y \in \Sigma^*$ such that $w = xuy$. For $U, V \subseteq \Sigma^*$, we set $U \cdot V = UV = \{uv \mid u \in U, v \in V\}$ and $U^0 = \{\varepsilon\}$, $U^{i+1} = U^i U$, and $U^* = \bigcup_{i \geq 0} U^i$ and $U^+ = \bigcup_{i > 0} U^i$. We also make use of complexity classes like P, NP, or PSPACE. With $\leq_m^{\log}$ we denote a logspace many-one reduction. If for two problems L_1, L_2 it holds that $L_1 \leq_m^{\log} L_2$ and $L_2 \leq_m^{\log} L_1$, then we write $L_1 \equiv_m^{\log} L_2$.

A *partial deterministic finite automaton (PDFA)* is a tuple $\mathcal{A} = (\Sigma, Q, \delta, q_0, F)$, where Σ is a finite set of *input symbols*, Q is the finite *state set*, $q_0 \in Q$ the *start state*, $F \subseteq Q$ the *final state set* and $\delta \colon Q \times \Sigma \to Q$ the *partial transition function*. The *partial transition function* $\delta \colon Q \times \Sigma \to Q$ extends to words from Σ^* in the usual way. Furthermore, for $S \subseteq Q$ and $w \in \Sigma^*$, we set $\delta(S, w) = \{\delta(q, w) \mid \delta(q, w) \text{ is defined and } q \in S\}$. We call $\mathcal{A}$ *complete* if δ is defined for every $(q, a) \in Q \times \Sigma$. If $|\Sigma| = 1$, we call $\mathcal{A}$ a *unary automaton* and $L \subseteq \Sigma^*$ is also called a *unary language*. The set $L(\mathcal{A}) = \{w \in \Sigma^* \mid \delta(q_0, w) \in F\}$ denotes the language *recognized* by $\mathcal{A}$.

A *deterministic and complete semi-automaton (DCSA)* $\mathcal{A} = (\Sigma, Q, \delta)$ is a deterministic and complete finite automaton without a specified start state and with no specified set of final states. When the context is clear, we call both deterministic finite automata and semi-automata simply *automata*.

A complete automaton $\mathcal{A}$ is called *synchronizing* if there exists a word $w \in \Sigma^*$ with $|\delta(Q, w)| = 1$. In this case, we call w a *synchronizing word* for $\mathcal{A}$. We call a state $q \in Q$ with $\delta(Q, w) = \{q\}$ for some $w \in \Sigma^*$ a *synchronizing state*. For a semi-automaton (or PDFA) with state set Q and transition function $\delta : Q \times \Sigma \to Q$, a state q is called a *sink state*, if for all $x \in \Sigma$ we have $\delta(q, x) = q$. Note that, if a synchronizing automaton has a sink state, then the synchronizing state is unique and must equal the sink state.

In [19] the *constrained synchronization problem (CSP)* was defined for a fixed PDFA $\mathcal{B} = (\Sigma, P, \mu, p_0, F)$.

Decision Problem 1: [19] $L(\mathcal{B})$-CONSTR-SYNC
Input: DCSA $\mathcal{A} = (\Sigma, Q, \delta)$.
Question: Is there a synchronizing word $w \in \Sigma^*$ for $\mathcal{A}$ with $w \in L(\mathcal{B})$?

The automaton $\mathcal{B}$ will be called the *constraint automaton*. If an automaton $\mathcal{A}$ is a yes-instance of $L(\mathcal{B})$-CONSTR-SYNC we call $\mathcal{A}$ *synchronizing with respect to $\mathcal{B}$*. Occasionally, we do not specify $\mathcal{B}$ and rather talk about L-CONSTR-SYNC. For example, for the unconstrained case, we have Σ^*-CONSTR-SYNC $\in$ P [11,47].

In our NP-hardness reduction, we will need the following problem from [31].

Decision Problem 2: DISJOINTSETTRANSPORTER
Input: DCSA $\mathcal{A} = (\Sigma, Q, \delta)$ and disjoint $S, T \subseteq Q$.
Question: Is there a word $w \in \Sigma^*$ such that $\delta(S, w) \subseteq T$?

Theorem 1. *For unary deterministic and complete input semi-automata the problem* DISJOINTSETTRANSPORTER *is* NP*-complete.*

A PDFA $\mathcal{A} = (\Sigma, Q, \delta, q_0, F)$ is called *polycyclic*, if for each $q \in Q$ there exists $u \in \Sigma^*$ such that $\{w \in \Sigma^* \mid \delta(q, w) = q\} \subseteq u^*$. A PDFA is polycyclic if and only if every strongly connected component consists of a single cycle [31, Proposition 3], where each transition in the cycle is labelled by precisely one letter. Formally, for each strongly connected component $S \subseteq Q$ and $q \in S$, we have[2] $|\{x : x \in \Sigma \text{ and } \delta(q, x) \text{ is defined and in } S\}| \leqslant 1$ (note that in the special case $|S| = 1$, the aforementioned set might be empty if the single state in S has no self-loops). A precursor of this characterization of polycyclic automata in a special case was given in [20] under the term *linear cycle automata*.

The following slight generalization of [19, Theorem 27] will be needed.

Proposition 2. *Let $\varphi : \Sigma^* \to \Gamma^*$ be a homomorphism. Then, for each regular $L \subseteq \Sigma^*$, we have $\varphi(L)$-CONSTR-SYNC $\leqslant_m^{\log} L$-CONSTR-SYNC.*

3   Sparse and Bounded Regular Languages

Here, in Theorem 5, we establish that for constraint languages from the class of sparse regular languages, which equals the class of the bounded regular languages [34], the constrained problem is always in NP.

---

[2] In [31], I made an error in my formalization by writing $|\{\delta(q, x) : x \in \Sigma, \delta(q, x) \text{ is defined }\} \cap S| \leqslant 1$.

A language $L \subseteq \Sigma^*$ is *sparse*, if there exists $c \geqslant 0$ such that, for every $n \geqslant 0$, we have $L \cap \Sigma^n \in O(n^c)$. Sparse languages were introduced into computational complexity theory by Berman & Hartmanis [6]. Later, it was established by Mahaney that if there exists a sparse NP-complete set (under polynomial-time many-one reductions), then P = NP [36]. For a survey on the relevance of sparse sets in computational complexity theory, see [28].

A language $L \subseteq \Sigma^*$ is called *bounded*, if there exist $w_1, \ldots, w_k \in \Sigma^*$ such that $L \subseteq w_1^* \ldots w_k^*$. Bounded languages were introduced by Ginsburg & Spanier [23].

We will need the following representation of the bounded regular languages.

Theorem 3 ([24]). *A language $L \subseteq w_1^* \cdots w_k^*$ is regular if and only if it is a finite union of languages of the form $L_1 \cdots L_k$, where each $L_i \subseteq w_i^*$ is regular.*

It is known that the class of sparse regular languages equals the class of bounded regular languages [34], or see [40,51], where the bounded languages are not mentioned but the equivalence is implied by their results and Theorem 3. The next results links this class to the polycyclic PDFAs.

Proposition 4. *Let $L \subseteq \Sigma^*$ be regular. Then, L is sparse if and only if it is recognizable by a polycyclic PDFA.*

In [31, Theorem 2] it was shown that for polycyclic constraint languages, the constrained problem is always in NP. So, we can deduce the next result.

Theorem 5. *If $L \subseteq \Sigma^*$ is sparse and regular, then L-CONSTR-SYNC $\in$ NP.*

We will need the following closure property stated in [51, Theorem 3.8] of the sparse regular languages.

Proposition 6. *The class of sparse regular languages is closed under homomorphisms.*

Note that the connection of the polycyclic languages to the sparse or bounded languages was not noted in [31]. However, a condition characterizing the sparse regular languages in terms of forbidden patterns was given in [40], and it was remarked that "a minimal deterministic automaton recognises a sparse language if and only if it does not contain two cycles reachable from one another". This is quite close to our characterization.

4   Letter-Bounded Constraint Languages

Fix a constraint automaton $\mathcal{B} = (\Sigma, P, \mu, p_0, F)$. Let $a_1, \ldots, a_k \in \Sigma$ be a sequence of (not necessarily distinct) letters. In this section, we assume $L(\mathcal{B}) \subseteq a_1^* \cdots a_k^*$. A language which fulfills the above condition is called *letter-bounded*. Note that the language ab^*a given in the introduction as an example is letter-bounded. In fact, it is the language with the smallest recognizing automaton yielding an NP-complete constrained problem [19].

A language such that the a_i are pairwise distinct, i.e., $a_i \neq a_j$ for $i \neq j$, is called *strictly bounded*. The class of strictly bounded languages has been extensively studied [9,14,22–24,29], where in [22–24] no name was introduced for them and in [29] they were called strongly bounded. The class of letter-bounded languages properly contains the strictly bounded languages.

Remark 1. Let $\Sigma = \{b_1, \ldots, b_r\}$ be an alphabet of size r. Then, the mappings

$$\Phi(L) = L \cap b_1^* \cdots b_r^* \text{ and } \mathrm{perm}(L) = \{w \in \Sigma^* \mid \exists u \in L \ \forall a \in \Sigma : |u|_a = |w|_a\}$$

for $L \subseteq \Sigma^*$ are mutually inverse and inclusion preserving between the languages in $b_1^* \cdots b_r^*$ and the commutative languages in Σ^*, where a language $L \subseteq \Sigma^*$ is commutative if $\mathrm{perm}(L) = L$. Furthermore, for strictly bounded languages of the form $B_1 \cdots B_r \subseteq b_1^* \cdots b_r^*$ with $B_j \subseteq \{b_j\}^*$, $j \in \{1, \ldots, r\}$, we have $\mathrm{perm}(B_1 \cdots B_r) = B_1 \sqcup \cdots \sqcup B_r$, where $U \sqcup V = \{u_1 v_1 \cdots u_n v_n \mid u_i, v_i \in \Sigma^*, u_1 \cdots u_n \in U, v_1 \cdots v_n \in V\}$ for $U, V \subseteq \Sigma^*$. Hence, $\mathrm{perm}(L)$ is regularity-preserving for strictly bounded languages. More specifically, the above correspondence between the two language classes is regularity-preserving in both directions. For commutative constraint languages, a classification of the complexity landscape has been achieved [30]. By the close relationship between commutative and certain strictly bounded languages, it is natural to tackle this language class next. However, as shown in [30], for commutative constraint languages, we can realize PSPACE-complete problems, but, by Theorem 5, for strictly bounded languages, the constrained problem is always in NP. However, by the above relations, Theorem 3 for languages in $b_1^* \cdots b_r^*$ is equivalent to [30, Theorem 5], a representation result for commutative regular languages.

Our first result says, intuitively, that if in $A_1 \cdots A_k$ with A_j unary and regular, if no infinite unary language A_j over $\{a_j\}$ lies between non-empty unary languages over a distinct letter[3] than a_j, then $(A_1 \cdots A_k)$-CONSTR-SYNC is in P.

Proposition 7. *Let $A_j \subseteq \{a_j\}^*$ be unary regular languages for $j \in \{1, \ldots, k\}$. Set $L = A_1 \cdot \ldots \cdot A_k$. If for all $j \in \{1, \ldots, k\}$, A_j infinite implies that $A_i \subseteq \{a_j\}^*$ for all $i < j$ or $A_i \subseteq \{a_j\}^*$ for all $i > j$ (or both), then L-CONSTR-SYNC $\in$ P.*

Now, we state a sufficient condition for NP-hardness over binary alphabets. This condition, together with Proposition 2, allows us to handle the general case in Theorem 9. Its application together with Proposition 2 shows, in some respect, that the language ab^*a is the prototypical language giving NP-hardness. We give a proof sketch of Lemma 8 at the end of this section.

Lemma 8. *Suppose $\Sigma = \{a, b\}$. Let $L(\mathcal{B}) \subseteq \Sigma^*$ be letter-bounded. Then, $L(\mathcal{B})$-CONSTR-SYNC is NP-hard if $L(\mathcal{B}) \cap \Sigma^* ab^{|P|} b^* a \Sigma^* \neq \varnothing$.*

So, finally, we can state our main theorem of this section. Recall that by Theorem 5, and as the class of bounded regular languages equals the class of sparse regular languages [34], for bounded regular constraint languages, the constrained problem is, in our case, in NP.

---

[3] Hence different from $\{\varepsilon\}$, as $\{\varepsilon\} \subseteq \{a\}^*$ for $a \in \Sigma$.

Theorem 9 (Dichotomy Theorem). *Let $a_1, \ldots, a_k \in \Sigma$ be a sequence of letters and $L \subseteq a_1^* \cdots a_k^*$ be regular. The problem L-CONSTR-SYNC is NP-complete if*

$$L \cap \left(\bigcup_{\substack{1 \leqslant j_1 < j_2 < j_3 \leqslant k \\ a_{j_2} \notin \{a_{j_1}, a_{j_3}\}}} L_{j_1, j_2, j_3} \right) \neq \varnothing$$

with $L_{j_1, j_2, j_3} = \Sigma^* a_{j_1} \Sigma^* a_{j_2}^{|P|} \Sigma^* a_{j_3} \Sigma^*$ for $1 \leqslant j_1 < j_2 < j_3 \leqslant k$ and solvable in polynomial time otherwise.

As the languages L_{j_1, j_2, j_3} are regular, we can devise a polynomial-time algorithm which checks the condition mentioned in Theorem 9.

Corollary 10. *Given a PDFA $\mathcal{B}$ and a sequence of letters $a_1, \ldots, a_k$ as input such that $L(\mathcal{B}) \subseteq a_1^* \cdots a_k^*$, the complexity of $L(\mathcal{B})$-CONSTR-SYNC is decidable in polynomial-time.*

Proof. An automaton for each L_{j_1, j_2, j_3} has size linear in $|P|$. So, by the product automaton construction [32], non-emptiness of $L(\mathcal{B})$ with each L_{j_1, j_2, j_3} could be checked in time $O(|P|^2)$. Doing this for every L_{j_1, j_2, j_3} gives a polynomial-time algorithm to check non-emptiness of the language written in Theorem 9.   □

Example 1. For the following constraint languages CSP is NP-complete: ab^*a, $aa(aaa)^* bbb^* d \cup a^* b \cup d^*$, $bbcc^* d^* \cup a$.

For the following constraint languages CSP is in P: $a^5 bd \cup cd^4$, $a^5 bd \cup cd^*$, $aa^* bbbbcd^* \cup bbbdd^* d$.

Proof (Proof Sketch for Lemma 8). We construct a reduction from an instance of DISJOINTSETTRANSPORTER[4] for unary input automata.

To demonstrate the basic idea, we only do the proof in the case $L \subseteq a^* b^* a^*$. By assumption we can deduce $a^{r_1} b^{r_2} a^{r_3} \in L(\mathcal{B})$ with $p_2 \geqslant |P|$ and $r_1, r_3 \geqslant 1$. By the pigeonhole principle, in $\mathcal{B}$, when reading the factor b^{r_2}, at least one state has to be traversed twice. Hence, we find $0 < p_2 \leqslant |P|$ such that $a^{r_1} b^{r_2 + i \cdot p_2} a^{r_3} \subseteq L(\mathcal{B})$ for each $i \geqslant 0$.

Let $\mathcal{A} = (\{c\}, Q, \delta)$ and $(\mathcal{A}, S, T)$ be an instance of DISJOINTSETTRANSPORTER. We can assume S and T are non-empty, as for $S = \varnothing$ it is solvable, and if $T = \varnothing$ we have no solution. Construct $\mathcal{A}' = (\Sigma, Q', \delta')$ by setting $Q' = S_{r_2} \cup \ldots \cup S_1 \cup Q \cup Q_1 \cup \ldots \cup Q_{p_2 - 1} \cup \{t\}$, where t is a new state, $S_i = \{s_i \mid s \in S\}$ for $i \in \{1, \ldots, r_2\}$ are pairwise disjoint copies of S and $Q_i = \{q^i \mid q \in Q\}$ are[5] also pairwise disjoint copies of Q. Note that also

---

[4] Note that the problem DISJOINTSETTRANSPORTER is over a unary alphabet, but for L-CONSTR-SYNC we have $|\Sigma| > 1$. Indeed, we need the additional letters in Σ.

[5] Observe that by the indices a correspondence between the sets is implied. The index in Q_i at the top to distinguish, for $s \in S$ and $i \in \{1, \ldots, \min\{r_2, p_2 - 1\}\}$, between $s_i \in S_i$ and $s^i \in Q_i$. Hence, for each $s \in S$ and $i \in \{1, \ldots, r_2\}$, the states s and s_i correspond to each other, and for $q \in Q$ and $i \in \{1, \ldots, p_2 - 1\}$ the states q and q^i.

$S_i \cap Q_j = \varnothing$ for $i \in \{1, \ldots, r_2\}$ and $j \in \{1, \ldots, p_2 - 1\}$. Set $S_0 = S$ as a shorthand. Choose any $\hat{s} \in S_{r_2}$, then, for $q \in Q$ and $x \in \Sigma$, the transition function is given by

$$\delta'(q, x) = \begin{cases} s_{i-1} & \text{if } x = b \text{ and } q = s_i \in S_i \text{ for some } i \in \{1, \ldots, r_2\}; \\ \hat{s} & \text{if } x = a \text{ and } q \in (Q \cup Q_1 \cup \ldots \cup Q_{p_2-1}) \setminus S; \\ s_{r_2} & \text{if } x = a \text{ and } q = s_i \in S_i \text{ for some } i \in \{0, \ldots, r_2\}; \\ t & \text{if } x = a \text{ and } q \in T; \\ q^{p_2-1} & \text{if } x = b \text{ and } q \in Q; \\ q^{i-1} & \text{if } x = b \text{ and } q = q^i \in Q_i \text{ for some } i \in \{2, \ldots, p_2 - 1\}; \\ \delta(q, c) & \text{if } x = b \text{ and } q = q^1 \in Q_1; \\ q & \text{otherwise.} \end{cases}$$

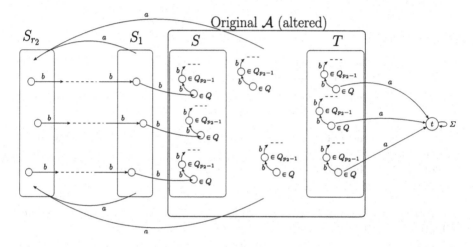

Fig. 1. The reduction from the proof sketch sketch of Lemma 8. The letter a transfers everything surjectively onto S_{r_2}, indicated by four large arrows at the top and bottom and labelled by a. The auxiliary states $Q_1, \ldots, Q_{p_2-1}$, which are meant to interpret a sequence b^{p_2} like a single symbol in the original automaton, are also only indicated inside of $\mathcal{A}$, but not fully written out.

Please see Fig. 1 for a sketch of the reduction. For the constructed automaton $\mathcal{A}'$, the following could be shown: $\exists m \geq 0 : \delta(S, c^m) \subseteq T$ if and only if $\mathcal{A}'$ has a synchronizing word in $ab^{r_2}(b^{p_2})^* a$ if and only if $\mathcal{A}'$ has a synchronizing word in $ab^* a$ if and only if $\mathcal{A}'$ has a synchronizing word in $a^* b^* a^*$.

Now, suppose $\delta(s, c^m) \subseteq T$ for some $m \geq 0$. By the above, $\mathcal{A}'$ has a synchronizing word u in $ab^{r_2}(b^{p_2})^* a$. Then, $a^{r_1-1} u a^{r_3-1} \in L(\mathcal{B})$ also synchronizes $\mathcal{A}'$.

Conversely, suppose we have a synchronizing word $w \in L$ for $\mathcal{A}'$. As $L \subseteq a^* b^* a^*$ by the above equivalences, $\delta(S, c^m) \subseteq T$ for some $m \geq 0$. □

5   Constraints from Strongly Self-synchronizing Codes

Here, we introduce strongly self-synchronizing codes and investigate L-CONSTR-SYNC for bounded constraint languages $L \subseteq w_1^* \cdots w_k^*$ where $\{w_1, \ldots, w_k\}$ is such a code.

Let $C \subseteq \Sigma^+$ be non-empty. Then, C is called a *self-synchronizing code* [7,8, 33], if $C^2 \cap \Sigma^+ C \Sigma^+ = \varnothing$. If, additionally, $C \subseteq \Sigma^n$ for some $n > 0$, then it is called[6] a *comma-free code* [26]. Every self-synchronizing code is an infix code, i.e., no proper factor of a word from C is in C [33]. A *strongly self-synchronizing code* is a self-synchronizing code $C \subseteq \Sigma^+$ such that, additionally, $(\mathrm{Pref}(C) \setminus C)C \cap \Sigma^* C \Sigma^+ = \varnothing$.

To give some intuition for the strongly self-synchronizing codes, we also present an alternative characterization, a few examples and a way to construct such codes.

Proposition 11. *A non-empty $C \subseteq \Sigma^+$ is a strongly self-synchronizing code if and only if, for all $u \in \mathrm{Pref}(C)$ and $v \in C$, if we write $uv = x_1 \cdots x_n$ with $x_i \in \Sigma$ for $i \in \{1, \ldots, n\}$, then, for all $j \in \{1, \ldots, n\}$ and $k \geq 1$ where $j + k - 1 \leq n$, we have that $x_j x_{j+1} \cdots x_{j+k-1} \in C$ implies $j = |u| + 1$ and $k = |v|$ or $j = 1$ and $k = |u|$. Intuitively, in uv only the last $|v|$ symbols form a factor in C and possibly the first $|u|$ symbols.*

When passing from letters to words by applying a homomorphism, in the reductions, we have to introduce additional states. The definition of the strongly synchronizing codes was motivated by the demand that these states also have to be synchronized, which turns out to be difficult in general.

Example 2. The code $\{aacc, bbc, bac\}$ is strongly self-synchronizing. The code $\{aab, bccc, abc\}$ is self-synchronizing, but not strongly self-synchronizing as, for example, $(a)(abc)$ contains aab or $(aa)(bccc)$ contains abc.

Remark 2 (Construction). Take any non-empty finite language $X \subseteq \Sigma^n$, $n > 0$, and a symbol $c \in \Sigma$ such that $\{c\}\Sigma^* \cap X = \varnothing$. Let $k = \max\{\ell \geq 0 \mid \exists u, v \in \Sigma^* : uc^\ell v \in X\}$. Then, $Y = c^{k+1}X$ is a strongly self-synchronizing code.

Example 3. Let $\Sigma = \{a, b, c\}$ and $C = \{ab, ba, aa\}$. Then, $\{cab, cba, caa\}$ or $\{bbab, bbaa\}$ are strongly self-synchronizing codes by Remark 2.

Our next result, which holds in general, states conditions on a homomorphism such that we not only have a reduction from the problem for the homomorphic image to our original problem, as stated in Proposition 2, but also a reduction in the other direction.

Theorem 12. *Let $\varphi : \Sigma^* \to \Gamma^*$ be a homomorphism such that $\varphi(\Sigma)$ is a strongly self-synchronizing code and $|\varphi(\Sigma)| = |\Sigma|$. Then, for each regular $L \subseteq \Sigma^*$ we have L-CONSTR-SYNC $\equiv_m^{\log} \varphi(L)$-CONSTR-SYNC.*

---

[6] In [33] this distinction is not made and self-synchronizing codes are also called comma-free codes.

Finally, we apply Theorem 12 to bounded languages such that $\{w_1, \ldots, w_k\}$ forms a strongly self-synchronizing code.

Theorem 13. *Let $L \subseteq w_1^* \cdots w_k^*$ be regular such that $\{w_1, \ldots, w_k\}$ is a strongly self-synchronizing code. Then, L-CONSTR-SYNC is either NP-complete or in P.*

Example 4. (1) $((aacc)(bbc)^*(bac))$-CONSTR-SYNC is NP-complete.
(2) $((bbc)(aacc)(bac)^* \cup (bbc)^*)$-CONSTR-SYNC is in P.

6   Conclusion and Discussion

We have looked at the constrained synchronization problem (Problem 1) – CSP for short – for letter-bounded regular constraint languages and bounded languages induced by strongly self-synchronizing codes, thereby continuing the investigation started in [19]. The complexity landscape in these cases is completely understood. Only the complexity classes P and NP-complete arise. In [31] the question was raised if we can find sparse constraint languages that give constrained problems complete for some candidate NP-intermediate complexity class. At least for the language classes investigated here this is not the case. For general sparse regular languages, it is still open if a corresponding dichotomy theorem holds, or candidate NP-intermediate problems arise. By the results obtained so far and the methods of proofs, we conjecture that in fact a dichotomy result holds true.

Let us relate our results to the previous work [31], where partial results for NP-hardness and containment in P were given. Namely, by setting $\mathrm{Fact}(L) = \{v \in \Sigma^* \mid \exists u, w \in \Sigma^* : uvw \in L\}$ and $\mathcal{B}_{p,E} = (\Sigma, P, \mu, q, E)$ for $\mathcal{B} = (\Sigma, P, \mu, p_0, F)$ with $E \subseteq P$ and $q \in P$, the following was stated.

Proposition 14 ([31]). *Suppose we find $u, v \in \Sigma^*$ such that we can write $L = uv^*U$ for some non-empty language $U \subseteq \Sigma^*$ with $u \notin \mathrm{Fact}(v^*), v \notin \mathrm{Fact}(U)$ and $\mathrm{Pref}(v^*) \cap U = \varnothing$. Then L-CONSTR-SYNC is NP-hard.*

Proposition 15 ([31]). *Let $\mathcal{B} = (\Sigma, P, \mu, p_0, F)$ be a polycyclic PDFA. If for every reachable $p \in P$ with $L(\mathcal{B}_{p,\{p\}}) \neq \{\varepsilon\}$ we have $L(\mathcal{B}_{p_0,\{p\}}) \subseteq \mathrm{Suff}(L(\mathcal{B}_{p,\{p\}}))$, then the problem $L(\mathcal{B})$-CONSTR-SYNC is solvable in polynomial time.*

Note that Proposition 14 implies that ab^*a gives an NP-complete CSP. However, in the letter-bounded case there exist constraint languages giving NP-complete problems for which this is not implied by Proposition 14, for example: ab^*ba, ab^*ab, aa^*abb^*a or $ba^*b \cup a$. Also, Proposition 15 is weaker than our Proposition 7 in the case of letter-bounded constraints. For example, it does not apply to ab^*b, every PDFA for this languages has a loop exclusively labelled by the letter b and reachable after reading the letter a from the start state, and so words along this loop cannot have a word starting with a as a suffix.

For general bounded languages, let us note the following implication of Propositions 2 and 7.

Proposition 16. *Let $u, v \in \Sigma^*$. If $L \subseteq u^* v^*$ is regular, then L-CONSTR-SYNC is solvable in polynomial time.*

Next, in Proposition 17, we give an example of a bounded regular language yielding an NP-complete synchronization problem, but for which this is not directly implied by the results we have so far.

Proposition 17. *The problem $((ab)(ba)^*(ab))$-CONSTR-SYNC is NP-complete.*

By Proposition 17, for the homomorphism $\varphi : \{a, b\}^* \to \{a, b\}^*$ given by $\varphi(a) = ab$ and $\varphi(b) = ba$ both problems ab^*a and $\varphi(ab^*a) = ab(ba)^*ab$ are NP-complete. So, this is a homomorphisms which preserves, in this concrete instance, the computational complexity. But its image $\{ab, ba\}$ is not even a self-synchronizing code. However, I do not know if this homomorphism always preserves the complexity. Similary, I do not know if the condition from Theorem 12 characterizes those homomorphisms which preserve the complexity.

In the reduction used in Lemma 8 the resulting automaton has a sink state. However, in general, for questions of synchronizability it makes a difference if we have a sink state or not, at least with respect to the Černý conjecture [11], as for automata with a sink state this conjecture holds true, even with the better bound[7] $\frac{n(n-1)}{2}$ [44, 48]. However, even in [19] certain reductions establishing PSPACE-completeness use only automata with a sink state. Hence, for hardness these automata are sufficient at least in certain instances. So, it might be interesting to know if in terms of computational complexity of the CSP, we can, without loss of generality, limit ourselves to input automata with a sink state. The methods of proof for the letter-bounded constraints show that in this case, we can actually do this, as these input automata are sufficient to establish all cases of intractability.

Lastly, let us mention the following related problem[8] one could come up with. Fix a deterministic and complete semi-automaton $\mathcal{A}$. Then, for input PDFAs $\mathcal{B}$, what is the computational complexity to determine if $\mathcal{A} = (\Sigma, Q, \delta)$ has a synchronizing word in $L(\mathcal{B})$? As the set of synchronizing words $\{w \in \Sigma^* : |\delta(Q, w)| = 1\} = \bigcup_{q \in Q} \bigcap_{q' \in Q} L((\Sigma, Q, \delta, q', \{q\}))$ is a regular language, we have to test for non-emptiness of intersection of this fixed regular language with $L(\mathcal{B})$. This could be done in NL, hence in P.

Acknowledgement. I thank anonymous reviewers of a previous version for detailed feedback. I also sincerely thank the reviewers of the current version (at least one reviewers saw both versions) for careful reading and giving valuable feedback to improve my scientific writing and pointing to two instances were I overlooked, in retrospect, two simple conclusions.

---

[7] In [44] erroneously the bound $n(n+1)/2$ was reported as being sharp, but the overall argument in fact works to yield the sharp bound $n(n-1)/2$.

[8] This was actually suggested by a reviewer of a previous version.

References

1. Adler, R., Weiss, B.: Similarity of Automorphisms of the Torus. American Mathematical Society: Memoirs of the American Mathematical Society, American Mathematical Society (1970)
2. Alves, L.V., Pena, P.N.: Synchronism recovery of discrete event systems. IFAC-PapersOnLine **53**(2), 10474–10479 (2020). 21th IFAC World Congress
3. Amarilli, A., Paperman, C.: Topological sorting with regular constraints. In: Chatzigiannakis, I., Kaklamanis, C., Marx, D., Sannella, D. (eds.) 45th International Colloquium on Automata, Languages, and Programming, ICALP 2018. LIPIcs, Prague, Czech Republic, 9–13 July 2018, vol. 107, pp. 115:1–115:14. Schloss Dagstuhl - Leibniz-Zentrum für Informatik (2018)
4. Benenson, Y., Adar, R., Paz-Elizur, T., Livneh, Z., Shapiro, E.: DNA molecule provides a computing machine with both data and fuel. Proc. Natl. Acad. Sci. U.S.A. **100**, 2191–2196 (2003)
5. Benenson, Y., Paz-Elizur, T., Adar, R., Keinan, E., Livneh, Z., Shapiro, E.: Programmable and autonomous computing machine made of biomolecules. Nature **414**, 430–434 (2001)
6. Berman, L., Hartmanis, J.: On isomorphisms and density of NP and other complete sets. SIAM J. Comput. **6**(2), 305–322 (1977)
7. Berstel, J., Perrin, D.: Theory of Codes. Pure and Applied Mathematics, vol. 117. Academic Press, Inc., Orlando, XIV, 433 (1985)
8. Berstel, J., Perrin, D., Reutenauer, C.: Codes and Automata, Encyclopedia of Mathematics and its Applications, vol. 129. Cambridge University Press (2010)
9. Blattner, M., Cremers, A.B.: Observations about bounded languages and developmental systems. Math. Syst. Theory **10**, 253–258 (1977)
10. Cassandras, C.G., Lafortune, S.: Introduction to Discrete Event Systems, 2nd edn. Springer, Heidelberg (2008)
11. Černý, J.: Poznámka k homogénnym experimentom s konečnými automatmi. Matematicko-fyzikálny časopis **14**(3), 208–216 (1964)
12. Chen, Y., Ierardi, D.: The complexity of oblivious plans for orienting and distinguishing polygonal parts. Algorithmica **14**(5), 367–397 (1995)
13. Cho, H., Jeong, S., Somenzi, F., Pixley, C.: Synchronizing sequences and symbolic traversal techniques in test generation. J. Electron. Test. **4**(1), 19–31 (1993)
14. Dassow, J., Paun, G.: On the regularity of languages generated by context-free evolutionary grammars. Discret. Appl. Math. **92**(2–3), 205–209 (1999)
15. Diekert, V.: Makanin's algorithm for solving word equations with regular constraints. Report, Fakultät Informatik, Universität Stuttgart, March 1998
16. Diekert, V., Gutiérrez, C., Hagenah, C.: The existential theory of equations with rational constraints in free groups is PSPACE-complete. Inf. Comput. **202**(2), 105–140 (2005)
17. Eppstein, D.: Reset sequences for monotonic automata. SIAM J. Comput. **19**(3), 500–510 (1990)
18. Erdmann, M.A., Mason, M.T.: An exploration of sensorless manipulation. IEEE J. Robot. Autom. **4**(4), 369–379 (1988)
19. Fernau, H., Gusev, V.V., Hoffmann, S., Holzer, M., Volkov, M.V., Wolf, P.: Computational complexity of synchronization under regular constraints. In: Rossmanith, P., Heggernes, P., Katoen, J. (eds.) 44th International Symposium on Mathematical Foundations of Computer Science, MFCS 2019. LIPIcs, Aachen, Germany, 26–30 August 2019, vol. 138, pp. 63:1–63:14. Schloss Dagstuhl - Leibniz-Zentrum für Informatik (2019)

20. Ganardi, M., Hucke, D., König, D., Lohrey, M., Mamouras, K.: Automata theory on sliding windows. In: Niedermeier, R., Vallée, B. (eds.) 35th Symposium on Theoretical Aspects of Computer Science, STACS 2018. LIPIcs, Caen, France, 28 February–3 March 2018, vol. 96, pp. 31:1–31:14. Schloss Dagstuhl - Leibniz-Zentrum für Informatik (2018)

21. Gawrychowski, P., Krieger, D., Rampersad, N., Shallit, J.O.: Finding the growth rate of a regular or context-free language in polynomial time. Int. J. Found. Comput. Sci. **21**(4), 597–618 (2010)

22. Ginsburg, S.: The Mathematical Theory of Context-free Languages. McGraw-Hill, New York (1966)

23. Ginsburg, S., Spanier, E.H.: Bounded ALGOL-like languages. Trans. Am. Math. Soc. **113**(2), 333–368 (1964)

24. Ginsburg, S., Spanier, E.H.: Bounded regular sets. Proc. Am. Math. Soc. **17**(5), 1043–1049 (1966)

25. Goldberg, K.Y.: Orienting polygonal parts without sensors. Algorithmica **10**(2–4), 210–225 (1993)

26. Golomb, S.W., Gordon, B., Welch, L.R.: Comma-free codes. Can. J. Math. **10**, 202–209 (1958)

27. Gusev, V.V.: Synchronizing automata of bounded rank. In: Moreira, N., Reis, R. (eds.) CIAA 2012. LNCS, vol. 7381, pp. 171–179. Springer, Heidelberg (2012). https://doi.org/10.1007/978-3-642-31606-7_15

28. Hartmanis, J., Mahaney, S.R.: An essay about research on sparse NP complete sets. In: Dembiński, P. (ed.) MFCS 1980. LNCS, vol. 88, pp. 40–57. Springer, Heidelberg (1980). https://doi.org/10.1007/BFb0022494

29. Herrmann, A., Kutrib, M., Malcher, A., Wendlandt, M.: Descriptional complexity of bounded regular languages. J. Autom. Lang. Comb. **22**(1–3), 93–121 (2017)

30. Hoffmann, S.: Computational complexity of synchronization under regular commutative constraints. In: Kim, D., Uma, R.N., Cai, Z., Lee, D.H. (eds.) COCOON 2020. LNCS, vol. 12273, pp. 460–471. Springer, Cham (2020). https://doi.org/10.1007/978-3-030-58150-3_37

31. Hoffmann, S.: On a class of constrained synchronization problems in NP. In: Cordasco, G., Gargano, L., Rescigno, A. (eds.) Proceedings of the 21th Italian Conference on Theoretical Computer Science, ICTCS 2020, Ischia, Italy. CEUR Workshop Proceedings, CEUR-WS.org (2020)

32. Hopcroft, J.E., Ullman, J.D.: Introduction to Automata Theory, Languages, and Computation. Addison-Wesley Publishing Company (1979)

33. Hsieh, C., Hsu, S., Shyr, H.J.: Some algebraic properties of comma-free codes. Technical report. Kyoto University Research Information Repository (KURENAI) (1989)

34. Latteux, M., Thierrin, G.: On bounded context-free languages. Elektronische Informationsverarbeitung und Kybernetik (J. Inf. Process. Cybern.) **20**(1), 3–8 (1984)

35. Lecoutre, C.: Constraint Networks: Techniques and Algorithms. Wiley, Hoboken (2009)

36. Mahaney, S.R.: Sparse complete sets of NP: solution of a conjecture of Berman and Hartmanis. J. Comput. Syst. Sci. **25**(2), 130–143 (1982)

37. Natarajan, B.K.: An algorithmic approach to the automated design of parts orienters. In: 27th Annual Symposium on Foundations of Computer Science, Toronto, Canada, 27–29 October 1986, pp. 132–142. IEEE Computer Society (1986)

38. Natarajan, B.K.: Some paradigms for the automated design of parts feeders. Int. J. Robot. Res. **8**(6), 98–109 (1989)

39. Pesant, G.: A regular language membership constraint for finite sequences of variables. In: Wallace, M. (ed.) CP 2004. LNCS, vol. 3258, pp. 482–495. Springer, Heidelberg (2004). https://doi.org/10.1007/978-3-540-30201-8_36

40. Pin, J.: Mathematical Foundations of Automata Theory (2020). https://www.irif.fr/~jep/PDF/MPRI/MPRI.pdf

41. Piziak, R., Odell, P.L.: Full rank factorization of matrices. Math. Mag. **72**(3), 193–201 (1999)

42. Ramadge, P.J., Wonham, W.M.: Supervisory control of a class of discrete event processes. SIAM J. Control Optim. **25**, 206–230 (1987)

43. Romeuf, J.: Shortest path under rational constraint. Inf. Process. Lett. **28**(5), 245–248 (1988)

44. Rystsov, I.: Reset words for commutative and solvable automata. Theor. Comput. Sci. **172**(1–2), 273–279 (1997)

45. Sandberg, S.: 1 homing and synchronizing sequences. In: Broy, M., Jonsson, B., Katoen, J.-P., Leucker, M., Pretschner, A. (eds.) Model-Based Testing of Reactive Systems. LNCS, vol. 3472, pp. 5–33. Springer, Heidelberg (2005). https://doi.org/10.1007/11498490_2

46. Trahtman, A.N.: The road coloring problem. Israel J. Math. **172**(1), 51–60 (2009). https://doi.org/10.1007/s11856-009-0062-5

47. Volkov, M.V.: Synchronizing automata and the Černý conjecture. In: Martín-Vide, C., Otto, F., Fernau, H. (eds.) LATA 2008. LNCS, vol. 5196, pp. 11–27. Springer, Heidelberg (2008). https://doi.org/10.1007/978-3-540-88282-4_4

48. Volkov, M.V.: Synchronizing automata preserving a chain of partial orders. Theor. Comput. Sci. **410**(37), 3513–3519 (2009)

49. Vorel, V., Roman, A.: Complexity of road coloring with prescribed reset words. J. Comput. Syst. Sci. **104**, 342–358 (2019)

50. Wonham, W.M., Cai, K.: Supervisory Control of Discrete-Event Systems. CCE, Springer, Cham (2019). https://doi.org/10.1007/978-3-319-77452-7

51. Yu, S.: Regular languages. In: Rozenberg, G., Salomaa, A. (eds.) Handbook of Formal Languages, pp. 41–110. Springer, Heidelberg (1997). https://doi.org/10.1007/978-3-642-59136-5_2

On Dasgupta's Hierarchical Clustering Objective and Its Relation to Other Graph Parameters

Svein Høgemo[1]($\boxtimes$), Benjamin Bergougnoux[1], Ulrik Brandes[3], Christophe Paul[2], and Jan Arne Telle[1]

[1] Department of Informatics, University of Bergen, Bergen, Norway
svein.hogemo@uib.no
[2] LIRMM, CNRS, Univ Montpellier, Montpellier, France
[3] Social Networks Lab, ETH Zürich, Zürich, Switzerland

Abstract. The minimum height of vertex and edge partition trees are well-studied graph parameters known as, for instance, vertex and edge ranking number. While they are NP-hard to determine in general, linear-time algorithms exist for trees. Motivated by a correspondence with Dasgupta's objective for hierarchical clustering we consider the total rather than maximum depth of vertices as an alternative objective for minimization. For vertex partition trees this leads to a new parameter with a natural interpretation as a measure of robustness against vertex removal.

As tools for the study of this family of parameters we show that they have similar recursive expressions and prove a binary tree rotation lemma. The new parameter is related to trivially perfect graph completion and therefore intractable like the other three are known to be. We give polynomial-time algorithms for both total-depth variants on caterpillars and on trees with a bounded number of leaf neighbors. For general trees, we obtain a 2-approximation algorithm.

1 Introduction

Clustering is a central problem in data mining and statistics. Although many objective functions have been proposed for (flat) partitions into clusters, hierarchical clustering has long been considered from the perspective of iterated merge (in agglomerative clustering) or split (in divisive clustering) operations. In 2016, Dasgupta [9] proposed an elegant objective function, hereafter referred to as DC-value, for nested partitions as a whole, and thus facilitated the study of hierarchical clustering from an optimization perspective. This work has sparked research on other objectives, algorithms, and computational complexity, and drawn significant interest from the data science community [8].

It is customary to represent the input data as an edge-weighted graph, where the weights represent closeness (in similarity clustering) or distance (in dissimilarity clustering). The bulk of work that has been done on DC-value has concentrated on assessing the performance of well-known clustering algorithms in

© Springer Nature Switzerland AG 2021
E. Bampis and A. Pagourtzis (Eds.): FCT 2021, LNCS 12867, pp. 287–300, 2021.
https://doi.org/10.1007/978-3-030-86593-1_20

Table 1. A family of graph parameters based on nested graph decompositions.

Vertex depth partition tree	Maximum (max)	Total (sum)
Edge (EPT)	Edge ranking number [18]	Dasgupta's clustering objective [9]
Vertex (VPT)	Tree-depth [23,24] vertex ranking number [12,27] minimum elimination tree height [3]	[now in this paper]

terms of this objective. In Dasgupta's original paper, a simple divisive clustering algorithm for similarity clustering, recursively splitting the input graph along an α-approximated sparsest cut, was shown to give a $O(\alpha \cdot \log n)$-approximation to DC-value. In later papers, this result was further improved upon: Charikar and Chatziafratis [4] show that this algorithm in fact achieves an $O(\alpha)$-approximation of DC-value, and complement this result with a hardness result for approximating DC-value. They also provide new approximation algorithms by way of linear and semi-definite relaxations of the problem statement. The former is also pointed out by Roy and Pokutta [26]. For dissimilarity clustering (maximizing the objective function), several algorithms achieve constant approximation ratio, including average-linkage (the most commonly used agglomerative clustering method) [8], although a semi-definite relaxation again can do a little better [5].

In a recent paper showing that Dasgupta's objective remains intractable even if the input dissimilarities are binary, i.e., when hierarchically clustering an unweighted undirected graph, Høgemo, Paul and Telle [15] initiated the study of Dasgupta's objective as a graph parameter. By the nature of Dasgupta's objective, the associated cluster trees are binary, and admit a mapping from the inner nodes to the edges of the graph such that every edge connects two vertices from different subtrees. We relate such trees to so-called edge partition trees [18], and show that minimizing Dasgupta's objective is equal to minimizing the total depth of all leaves in an edge partition tree.

If we consider the maximum depth of a leaf (the height of the tree) instead, its minimum over all edge partition trees of a graph is known as the edge ranking number of that graph [18]. The same concept applies to vertex partition trees, in which there is a one-to-one correspondence between all of its nodes (leaves and inner nodes) and the vertices of the graph such that no edge connects two vertices whose corresponding nodes are in disjoint subtrees. The minimum height of any vertex partition tree is called the vertex ranking number [12,27], but also known as tree-depth [23,24] and minimum elimination tree height [3].

The above places Dasgupta's objective, applied to unweighted graphs, into a family of graph parameters as shown in Table 1. It also suggests a new graph parameter, combining the use of vertex partition trees with the objective of minimizing the total depth of vertices. All three previously studied parameters are NP-hard to determine [15,20,25], and we show that the same holds for the new parameter. Interestingly, the proof relies on a direct correspondence with trivially perfect graph completion and thus provides one possible interpretation of the parameter in terms of intersecting communities in social networks [22]. We give an alternative interpretation in terms of robustness against network dismantling.

For both parameters based on tree height, efficient algorithms have been devised in case the input graph is a tree. For the edge ranking number, it took a decade from a polynomial-time 2-approximation [18] and an exact polynomial-time algorithm [10] to finally arrive at a linear-time algorithm [21]. Similarly, a polynomial-time algorithm for the vertex ranking number [17] was later improved to linear time [27]. No such algorithms for the input graph being a tree are known for the total-depth variants.

Our paper is organized as follows. In Sect. 2 we give formal definitions, and give a rotation lemma for general graphs to improve a given clustering tree. This allows us to show that if a clustering tree for a connected graph has an edge cut which is not minimal, or has a subtree defining a cluster that does not induce a connected subgraph, then it cannot be optimal for DC-value. In Sect. 3 we go through the 4 problems in Table 1 and prove the equivalence with the standard definitions. We also show an elegant and useful recursive formulation of each of the 4 problems. In Sect. 4 we consider the situation when the input graph is a tree. We give polynomial-time algorithms to compute the total depth variants, including DC-value, for caterpillars and more generally for trees having a bounded number of leaves in the subtree resulting from removing its leaves. We then consider the sparsest cut heuristic used by Dasgupta [9] to obtain an approximation on general graphs. When applied to trees, even to caterpillars, this does not give an optimal algorithm for DC-value. However, we show that it does give a 2-approximation on trees, which improves on an 8-approximation due to Charikar and Chatziafratis [5].

We leave as open the question if any of the two total depth variants can be solved in polynomial time on trees. On the one hand it would be very surprising if a graph parameter with such a simple formulation was NP-hard on trees. On the other hand, the graph-theoretic footing of the algorithms for the two max depth variants on trees does not seem to hold. The maximum depth variants are amenable to greedy approaches, where any vertex or edge that is reasonably balanced can be made root of the partition tree, while this is not true for the total depth variants.

2   Preliminaries

We use standard graph theoretic notation [13]. In this paper, we will often talk about several different trees: an unrooted tree which is the input to a problem, and a rooted tree which is a decompositional structure used in the problem formulation. To differentiate the two, we will denote an unrooted tree as any graph, G, while a rooted tree is denoted T. Furthermore, $V(G)$ are called the *vertices* of G, while $V(T)$ are called the *nodes* of T.

A rooted tree has the following definition; a tree (connected acyclic graph) T equipped with a special node called the *root* r, which produces an ordering on $V(T)$. Every node v in T except r has a *parent*, which is the neighbor that lies on the path from v to r. Every node in T that has v as its parent is called a *child* of v. A node with no children is called a *leaf*. Leaves are also defined on unrooted trees as vertices which have only one neighbor. The set of leaves in a tree T is denoted $L(T)$. The subtree induced by the internal vertices of T, i.e. $T\setminus L(T)$, is called the *spine-tree* of T. A *caterpillar* is a tree whose spine-tree is a path; this is the *spine* of the caterpillar.

In a rooted tree, the set of nodes on the path from v to r is called the *ancestors* of v, while the set of all nodes that include v on their paths to r is called the *descendants* of v. We denote by $T[v]$ the subtree induced by the descendants of v (naturally including v itself). As can be seen already for the paragraph above, we reserve the name *node* for the vertices in rooted trees. In unrooted trees and graphs in general we only use *vertex*; this is to avoid confusion. For a given graph G, we use $n(G)$ and $m(G)$ to denote $|V(G)|$ and $|E(G)|$, respectively, or simply n and m if clear from context. Let A be a subset of $V(G)$. Then $G[A]$ is the *induced subgraph* of G by A, i.e. the graph $(A, \{uv \in G \mid u, v \in A\})$. If B is a subset of $V(G)$ disjoint from A, then $G[A, B]$ is the bipartite subgraph of G induced by A and B, i.e. the graph $(A \cup B, \{uv \in G \mid u \in A \wedge v \in B\})$. A *cut* in a graph is a subset of the edges that, if removed, leaves the graph disconnected. If G is an unrooted tree, then every single edge uv forms a cut, and we let G_u (respectively G_v) denote the connected component of $G - uv$ containing u (respectively v). We use $[k]$ to denote the set of integers from 1 to k.

Definition 1 (Edge-partition tree, Vertex-partition tree). *Let G be a connected graph. An* edge-partition tree *(EPT for short) T of G is a rooted tree where:*

- *The leaves of T are $V(G)$ and the internal nodes of T are $E(G)$.*
- *Let r be the root of T. If $G' = G - r$ has k connected components $G'_1, \ldots, G'_k$ (note that $k \leq 2$), then r has k children $c_1, \ldots, c_k$.*
- *For all $1 \leq i \leq k$, $T[c_i]$ is an edge partition tree of G'_i.*

A vertex-partition tree *(VPT for short) T of G is a rooted tree where:*

- *The nodes of T are $V(G)$.*
- *Let r be the root of T. If $G' = G - r$ has k connected components $G'_1, \ldots, G'_k$, then r has k children $c_1, \ldots, c_k$.*
- *For all $1 \leq i \leq k$, $T[c_i]$ is a vertex partition tree of G'_i.*

The set of all edge partition trees of G is denoted $EPT(G)$ and the set of all vertex partition trees of G is denoted $VPT(G)$.

For each node x in a tree T, we denote by $ed_T(x)$ the *edge depth* of x in T, i.e. the number of tree edges on the path from the root of T to x, and by $vd_T(x)$ the *vertex depth* of x in T, equal to $ed_T(x) + 1$, i.e. the number of nodes on the path from the root of T to x. We make this distinction as the measures on

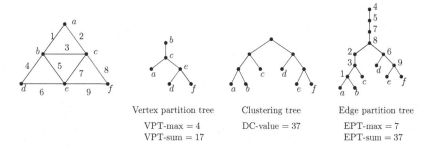

Vertex partition tree · Clustering tree · Edge partition tree

VPT-max = 4 · DC-value = 37 · EPT-max = 7

VPT-sum = 17 · EPT-sum = 37

Fig. 1. This figure shows the different types of partition trees on a small graph (the 3-sun). Vertices are marked with letters and edges with numbers. The clustering tree and the edge partition tree have the same structure. All trees are optimal for the measures defined in Sect. 3.

VPT's are defined in terms of vertex depth and vice versa for EPT's. The *vertex height* of a tree T is equal to the maximum vertex depth of the nodes in T, and the *edge height* of a tree T is equal to the maximum edge depth of the nodes in T. We generally assume that the graph G is connected; if G is disconnected, then any VPT (or EPT) is a forest, consisting of the union of VPTs (EPTs) of the components of G (Fig. 1).

A graph G is *trivially perfect* if there is a vertex partition tree T of G such that for any two vertices u, v, if u is an ancestor of v (or vice versa) in T, then uv is an edge in G. We call T a *generating tree* for G. Trivially perfect graphs are also known as *quasi-threshold graphs* or *comparability graphs of trees* (see [19]).

Definition 2 (Clustering tree). *A binary tree T is a clustering tree of G if:*

- *The leaves of T are $V(G)$. The clustering tree of K_1 is that one vertex.*
- *Let r be the root of T, with children a and b. Then $A = L(T[a])$ and $B = L(T[b])$ is a partition of $V(G)$.*
- *$T[a]$ and $T[b]$ are clustering trees of $G[A]$ and $G[B]$, respectively.*

For any node $x \in T$, we define $G[x]$ as shorthand for $G[L(T[x])]$, and for two siblings $a, b \in T$ we define $G[a, b]$ as shorthand for $G[L(T[a]), L(T[b])]$.

Definition 3 (DC-value). *The Dasgupta Clustering value of a graph G and a clustering tree T of G is*

$$\mathsf{DC\text{-}value}(G, T) = \sum_{x \in V(T) \setminus L(T)} m(G[a_x, b_x]) \cdot n(G[x])$$

where a_x and b_x are the children of x in T. The DC-value *of G,* DC-value(G), *is the minimum* DC-value *over all of its clustering trees.*

The following lemma gives a condition under which one can improve a given hierarchical clustering tree by performing either a left rotation or a right notation at some node of the tree. See Fig. 2. First off, it is easy to see that performing such a rotation maintains the property of being a clustering tree.

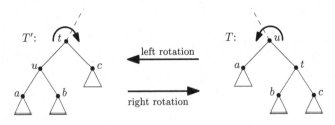

Fig. 2. Here T' is derived from T by a left tree rotation, or equivalently T is derived from T' by a right tree rotation.

Lemma 1 (Rotation Lemma). *Let G be a graph with clustering trees T and T' such that T' is the result of left rotation in T and T of a right rotation in T'. Let T and T' have nodes a, b, c, t, u as in Fig. 2. We have*

$$\text{DC-value}(T) - \text{DC-value}(T') = n(G[c]) \cdot m(G[a,b]) - n(G[a]) \cdot m(G[b,c]).$$

Proof. The DC-value of $T[u]$ is equal to

$$(n(G[a]) + n(G[b]) + n(G[c])) \cdot (m(G[a,b]) + m(G[a,c])) + (n(G[b]) + n(G[c])) \cdot$$
$$m(G[b,c]) + \text{DC-value}(T[a]) + \text{DC-value}(T[b]) + \text{DC-value}(T[c])$$

and the DC-cost of the rotated tree $T'[t]$ is equal to

$$(n(G[a]) + n(G[b]) + n(G[c])) \cdot (m(G[a,c]) + m(G[b,c])) + (n(G[a]) + n(G[b])) \cdot m(G[a,b])$$
$$+ \text{DC-value}(T[a]) + \text{DC-value}(T[b]) + \text{DC-value}(T[c])$$

See Fig. 2 for reference. By substituting the costs written above and cancelling out, we get the equality in the statement of the lemma.   □

This lemma proves useful anywhere where we would like to manipulate clustering trees. We first use it to prove an important fact about DC-value:

Theorem 1. *Let G be a connected graph, and let T be an optimal hierarchical clustering of G. Then, for any node $t \in T$, the subgraph $G[t]$ is connected.*

Proof. We assume towards a contradiction that there exists a connected graph G and an optimal hierarchical clustering T of G, with some node $t \in T$ such that the subgraph $G[t]$ is not connected. Observe that for r, the root in T, $G[r] = G$ is connected. Then there must exist a node t' such that $G[t']$ is not connected and for every ancestor $u \neq t'$ (of which there is at least one) $G[u]$ is connected. We focus on t', its parent u', its children b and c, and its sibling, a. The following claim is useful:

Claim ([9], Lemma 2). Let G be a disconnected graph. In an optimal clustering tree of G, the cut induced by the root is an empty cut.

Since T is optimal, by Claim 2 there are no edges going between the subgraphs $G[b]$ and $G[c]$ in G. Since $G[u']$ is connected, there must be at least one edge going between $G[a]$ and $G[b]$ in G. We thus have $n(G[a]) \cdot m(G[b, c]) = 0$ and $n(G[c]) \cdot m(G[a, b]) > 0$. But now, by Lemma 1, we can perform a tree rotation on T to obtain a clustering with strictly lower cost than T. This implies that T cannot be optimal after all. Thus, the theorem is true as stated.   □

Corollary 1. *Let T be an optimal clustering tree of a graph G (not necessarily connected). Then, for every internal node $t \in T$ with children u, v, the cut $E(G[u, v])$ is an inclusion-wise minimal cut in $G[t]$.*

That all optimal clustering trees have this property is hardly surprising, but is still worth pointing out. It is hard to imagine a scenario where this property would be unwanted in an application of similarity-based hierarchical clustering. Also, going forward in this paper, we will be exclusively working with this kind of clustering trees. Therefore we give it the name:

Definition 4 (Viable clustering tree). *Let T be a clustering tree of some graph. We say that T is a viable clustering tree if it has the added restriction that for every internal node $x \in T$ with children a_x, b_x, the cut induced by the partition $(L(T[a_x]), L(T[b_x]))$ is an inclusion-wise minimal cut in $G[L(T[x])]$.*

3 Four Related Problems

We define four measures on partition trees of a graph G, three of them well-known in the literature. To give a unified presentation, throughout this paper we will call them VPT-sum, VPT-max, EPT-sum and EPT-max, with no intention to replace the more well-known names. All four measures can be defined with very simple recursive formulas.

Definition 5. VPT-max(G) *is the minimum vertex height over trees $T \in VPT(G)$.*

This is arguably the most well-known of the four measures. It is known under several names, such as *tree-depth*, *vertex ranking number*, and *minimum elimination tree height*. The definition of tree-depth and minimum elimination tree height is exactly the minimum height of a vertex partition tree (an elimination tree is a vertex partition tree). The equivalence of vertex ranking number and minimum elimination tree height is shown in [11], while it is known from [24] that

$$\text{VPT-max}(G) = \min_{v \in V(G)} \left(1 + \max_{C \in cc(G-v)} \text{VPT-max}(C)\right).$$

Definition 6. EPT-max(G) *is the minimum edge height over trees $T \in EPT(G)$.*

It is known that EPT-max(G) is equivalent to the edge ranking number [18]. Statements with a ⋆ have proofs omitted due to space constraints; see full version [16] for the proofs.

Theorem 2. *For any connected graph G,*

$$\text{EPT-max}(G) = \min_{e \in E(G)} (1 + \max_{C \in cc(G-e)} \text{EPT-max}(C)).$$

Definition 7. EPT-sum(G) *is the minimum over every tree $T \in EPT(G)$ of the sum of the edge depth of all leaves in T.*

The following equivalence between EPT-sum and DC-value, and the very simple recursive formula, forms the motivation for the results we present here.

Theorem 3. *For any connected graph G,* EPT-sum$(G) = $ DC-value(G) *and*

$$\text{EPT-sum}(G) = \min_{e \in E(G)} (n(G) + \sum_{C \in cc(G-e)} \text{EPT-sum}(C)).$$

Proof. We begin proving the equivalence between EPT-sum and its recursive formulation. For $T \in EPT(G)$, we denote EPT-sum$(G,T) = \sum_{\ell \in L(T)} ed_T(\ell)$. Let T^* be an optimal EPT of G. Since G is connected, T^* has only one root r. We let $c_1, \ldots, c_k$ be the children of r. We denote by T_i^* the subtree $T^*[c_i]$ and by C_i the induced subgraph $G[L(T_i^*)]$. Observe that we have:

$$\begin{aligned}
\text{EPT-sum}(G, T^*) &= \sum_{\ell \in L(T^*)} ed_{T^*}(\ell) \\
&= n(G) + \sum_{i \in [1,k]} \sum_{\ell_i \in L(T^{*i})} ed_{T_i^*}(\ell_i) \\
&= n(G) + \sum_{i \in [1,k]} \text{EPT-sum}(C_i, T_i^*).
\end{aligned}$$

Suppose e_r is the edge of G mapped to r in T^*. As T^* is optimal, for every $i \in [1, k]$, T_i^* is an optimal EPT of C_i. We have

$$\begin{aligned}
\text{EPT-sum}(G, T^*) &= n(G) + \sum_{i \in [1,k]} \text{EPT-sum}(C_i) \\
&= n(G) + \sum_{C \in cc(G-e_r)} \text{EPT-sum}(C) \\
&= \min_{e \in E(G)} (n(G) + \sum_{C \in cc(G-e)} \text{EPT-sum}(C))
\end{aligned}$$

where the first two equalities follow from the definition of EPT's and the last from the optimality of T^*, and we conclude that the recursive formula holds.

Now, we prove the equivalence between DC-value and EPT-sum. Given a clustering tree CT of a graph G, which by Theorem 1 can be assumed to be viable, it is easy to construct an EPT T such that DC-value$(G, CT) = $ EPT-sum(G, T). For each internal node x of CT with children a_x, b_x, we replace x with a path P_x on $m(G[a_x, b_x])$ nodes, connect one end to the parent of x and the other end to the two children. Then we construct an arbitrary map between the nodes on the path P_x and the edges in $G[a_x, b_x]$. As CT is viable, $E(G[a_x, b_x])$ is an inclusion-wise minimal cut of $G[x]$ and thus T is an EPT of G. We have DC-value$(G, CT) = $ EPT-sum(G, T) because when we replace x by the path P_x, we increase the edge depth of the $n(G[x])$ leaves of the subtree rooted at x by $m(G[a_x, b_x])$. Conversely, given an EPT T of G, contracting every path with degree-two internal nodes into a single edge results in a clustering tree CT (not necessarily viable unless T is optimal) and we have DC-value$(G, CT) = $ EPT-sum(G, T) by the same argument as above. We conclude from these constructions that DC-value$(G) = $ EPT-sum(G). $\square$

Definition 8. VPT-sum(G) *is the minimum over every tree $T \in VPT(G)$ of the sum of the vertex depth of all nodes in T.*

Theorem 4. *For any connected graph G, we have*

$$\text{VPT-sum}(G) = \min_{v \in V(G)} \left(n(G) + \sum_{C \in cc(G-v)} \text{VPT-sum}(C) \right).$$

When comparing the definition of VPT-sum with the definition of trivially perfect graphs, it is not hard to see that a tree minimizing VPT-sum(G) is a generating tree of a trivially perfect supergraph of G where as few edges as possible have been added.

Theorem 5. *For any graph G, there exists a trivially perfect completion of G with at most k edges iff* VPT-sum(G) $\leq k + n(G) + m(G)$.

It is interesting that this formal relation, in addition to tree-depth, connects the class of trivially perfect graphs to another one of the four measures. Note that VPT-max (i.e. tree-depth) is also related to trivially perfect completion where the objective is to minimize the clique number of the completed graph. This parallels definitions of the related graph parameters *treewidth* and *pathwidth* as the minimum clique number of any chordal or interval supergraph, respectively. Nastos and Gao [22] have indeed proposed to determine a specific notion of community structure in social networks, referred to as familial groups, via trivially perfect editing, i.e., by applying the minimum number of edge additions and edge removals to turn the graph into a trivially perfect graph. The generating tree of a closest trivially perfect graph is then interpreted as a vertex partition tree, and thus a hierarchical decomposition into nested communities that intersect at their cores. For both familial groups and VPT-sum, an imperfect structure is transformed into an idealized one, with the difference that VPT-sum only allows for the addition of edges. Nastos and Gao [22] prefer the restriction to addition when one is "interested in seeing how individuals in a community are organized".

Viewed from the opposite perspective, another interpretation of VPT-sum is as a measure of network vulnerability under vertex removal. The capability of a network to withstand series of failures or attacks on its nodes is often assessed by observing changes in the size of the largest connected component, the reachability relation, or average distances [14]. An optimal vertex partition tree under VPT-sum represents a worst-case attack scenario in which, for all vertices simultaneously, the average number of removals in their remaining component that it takes to detach a vertex is minimized.

The problem of adding the fewest edges to make a trivially perfect graph was shown NP-hard by Yannakakis in [28] and so Theorem 5 implies the following.

Corollary 2. *Computing* VPT-sum *is NP-hard.*

4   VPT-sum and EPT-sum of Trees

In this section we consider the case when the input graph G is a tree. In this case, every minimal cut consists of one edge, and hence by Corollary 1 the optimal clustering trees are edge partition trees, i.e. the internal nodes of T are $E(G)$. This allows us to prove that the cut at any internal node t of an optimal clustering tree is an internal edge of $G[t]$, unless $G[t]$ is a star, which in turn allows us to give an algorithm for caterpillars.

Lemma 2. *Let T an optimal clustering tree of a tree G. For any internal node $t \in T$ with children u, v, if $G[t]$ is not a star, then neither u nor v are leaves in T. This implies that the edge associated with t is an internal edge of G.*

Theorem 6. *The DC-value of a caterpillar can be computed in $O(n^3)$ time.*

Proof. We view a caterpillar G as a collection of stars $(X_1, \ldots, X_p)$ that are strung together. The central vertices of the stars $(x_1, \ldots, x_p)$ form the spine of G. Thus, every internal edge $x_i x_{i+1}$ in G lies on the spine, and removing such an edge we get two sub-caterpillars, $(X_1, \ldots, X_i)$ and $(X_{i+1}, \ldots, X_p)$. For every $i, j \in [p]$ with $i \leq j$, we define $DC[i, j]$ to be the DC-value of the sub-caterpillar $(X_i, X_{i+1}, \ldots, X_j)$. Note that for a star X on n vertices we have DC-value$(X) = \binom{n+1}{2} - 1$ (one less than the n'th triangle number) as DC-value$(K_1) = 0 = \binom{2}{2} - 1$, and whichever edge we cut in a star on n vertices we end up with a single vertex and a star on $n - 1$ vertices. Therefore $DC[i, i] = \binom{n(X_i)+1}{2} - 1$ for every $i \in [p]$. From Theorem 1 and Lemma 2, we deduce the following for every $i, j \in [p]$ with $i < j$.

$$DC[i,j] = \sum_{k \in [i,j]} n(X_k) + \min_{k \in \{i, i+1, \ldots, j-1\}} DC[i,k] + DC[k+1, j]$$

Hence, to find $DC(G)$, we compute $DC[i, j]$ for every $i, j \in [p]$ with $i < j$ in order of increasing $j - i$ and return $DC[1, p]$. For the runtime, note that calculating a cell $DC[i, j]$ takes time $O(n)$ and there are $O(n^2)$ such cells in the table.   □

This dynamic programming along the spine of a caterpillar can be generalized to compute the DC-value of any tree G in time $n^{O(d_G)}$, where d_G is the number of leaves of the spine-tree G' of G. Note that for a caterpillar G we have $d_G = 2$.

In addition, we can show analogues of Theorem 1 and Lemma 2 for VPT-sum, which enables us to form a polynomial-time algorithm for the VPT-sum of caterpillars, and by the same generalization as above, for any tree G where d_G is bounded.

Theorem 7. *DC-value and VPT-sum of a tree G is found in $n^{O(d_G)}$ time.*

Lastly, we discuss the most well-studied approximation algorithm for DC-value [9], recursively partitioning the graph along a sparsest cut. A *sparsest cut* of a graph G is a partition (A, B) of $V(G)$ that minimizes the measure $\frac{m(G[A,B])}{|A| \cdot |B|}$. A sparsest cut must be a minimal cut.

For general graphs, finding a sparsest cut is NP-hard and must be approximated itself. On trees however, every minimal cut consists of one edge, and a sparsest cut is a *balanced cut*, minimizing the size of the largest component. The

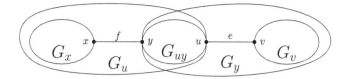

Fig. 3. Inclusion relations between the subtrees G_u, G_v, G_x, G_y and G_{uy}.

optimal cut can therefore in trees be found efficiently. We call an edge of a tree *balanced*, if it induces a balanced cut.

The results by Charikar and Chatziafratis [4] already indicate that the balanced cut algorithm gives an 8-approximation of the DC-value of trees (and other graph classes for which the sparsest cut can be found in polynomial time, like planar graphs, see [1] for more information). In the following, we prove that for trees, we can guarantee a 2-approximation. We start by showing an upper bound on the DC-value of the two subtrees resulting from removing an arbitrary edge of a tree, and follow up with a stronger bound if the removed edge is balanced.

Lemma 3. *If G is a tree and $e = uv \in E(G)$, then*
$$\text{DC-value}(G_u) + \text{DC-value}(G_v) \leq \text{DC-value}(G) - \min\{n(G_u), n(G_v)\}.$$

Lemma 4. *If G is a tree and $e = uv \in E(G)$ balanced, then* $\text{DC-value}(G_u) + \text{DC-value}(G_v) \leq \text{DC-value}(G) - \max\{n(G_u), n(G_v)\} \leq \text{DC-value}(G) - \frac{n(G)}{2}$.

Proof. The proof is by induction on the number of edges in the tree G. The single edge $e = uv$ of a K_2 induces a balanced cut and $\text{DC-value}(G_u) + \text{DC-value}(G_v) = 0 + 0 \leq 2 - 1 = \text{DC-value}(G) - \max\{n(G_u), n(G_v)\}$.

For the induction step assume that G is a tree with at least two edges and choose any balanced edge $e = uv$. Let $f = xy$ be the edge at the root of an optimal clustering of G. If $e = f$, then by definition $\text{DC-value}(G) = \text{DC-value}(G_u) + \text{DC-value}(G_v) + n$ and the lemma holds. Assume therefore that $e \neq f$ and w.l.o.g. $f \in E(G_u)$ and $e \in E(G_y)$, as in Fig. 3. We let G_{uy} denote the subgraph induced by $V(G_u) \cap V(G_y)$.

By definition, $\text{DC-value}(G_u) \leq \text{DC-value}(G_x) + \text{DC-value}(G_{uy}) + (n - n(G_v))$, and together with $\text{DC-value}(G_{uy}) + \text{DC-value}(G_v) \leq \text{DC-value}(G_y) - \eta$ where

$$\eta = \begin{cases} \max\{n(G_{uy}), n(G_v)\} & \text{if } e \text{ is balanced in } G_y \text{ (from induction hypothesis)} \\ \min\{n(G_{uy}), n(G_v)\} & \text{otherwise (from Lemma 3)} \end{cases}$$

we get

$$\begin{aligned} \text{DC-value}(G_u) &+ \text{DC-value}(G_v) \\ &\leq \underbrace{\text{DC-value}(G_x) + \text{DC-value}(G_y) + n - n(G_v) - \eta}_{=\text{DC-value}(G)}. \end{aligned}$$

It remains to show that $n(G_v) + \eta \geq \max\{n(G_u), n(G_v)\}$, which is obvious if $n(G_u) \leq n(G_v)$. So we suppose that $n(G_u) > n(G_v)$ and proceed to show that $n(G_v) + \eta \geq n(G_u)$. Since e is balanced in G, we have by definition

$$\max\{n(G_x) + n(G_{uy}), n(G_v)\} \leq \max\{n(G_x), n(G_{uy}) + n(G_v)\},$$

implying that $n(G_v) \geq n(G_x)$. It follows that

$$n(G_v) + n(G_{uy}) \geq n(G_x) + n(G_{uy}) = n(G_u).$$

See Fig. 3. We have two cases to consider. If $n(G_{uy}) \leq n(G_v)$, then $n(G_{uy}) = \min\{n(G_{uy}), n(G_v)\} \leq \eta$, implying that $n(G_v) + \eta \geq n(G_u)$. If $n(G_{uy}) > n(G_v)$, then e is balanced in G_y, so η is the maximum of $n(G_v)$ and $n(G_{uy})$ and we are done since $n(G_u) \leq n(G_v) + \eta$. □

Let a *balanced clustering* of a tree G be a clustering tree $\hat{T}$ such that for every internal node e in $\hat{T}$, the edge corresponding to e is a balanced edge in $G[e]$. As discussed earlier, a balanced clustering of a tree can be found efficiently. We now prove the guarantee of 2-approximation:

Theorem 8. *Let G be a tree, and $\hat{T}$ a balanced clustering of G. We then have* DC-value$(G, \hat{T}) \leq 2 \cdot$ DC-value(G).

Proof. The overall proof goes by strong induction. For the base case, we easily see that for every tree on at most 2 vertices, all the balanced clustering trees are actually optimal; for these trees the statement follows trivially. For the induction step, we assume that for all trees on at most some k vertices, the statement holds. We then look at a tree G on $n = k + 1$ vertices. We focus on two different clustering trees of G: T^*, which is an optimal clustering tree and has DC-value $W^* =$ DC-value(G), and $\hat{T}$, which is a balanced clustering tree and has DC-value $\hat{W}$. Our aim is now to prove that $\hat{W} \leq 2 \cdot W^*$.

We denote the root of $\hat{T}$ by $r = uv$ and its two children by c_u and c_v. By definition, $\hat{T}[c_u]$ and $\hat{T}[c_v]$ are balanced clustering trees of G_u and G_v, respectively. By our induction hypothesis, we know that DC-value$(G_u, \hat{T}[c_u]) \leq 2 \cdot$ DC-value(G_u) and respectively for c_v. By definition we have $\hat{W} =$ DC-value$(G_u, \hat{T}[c_u]) +$ DC-value$(G_v, \hat{T}[c_v]) + n$ which gives $\hat{W} \leq 2 \cdot ($DC-value$(G_u) +$ DC-value$(G_v)) + n$. By Lemma 4 we have DC-value$(G_u) +$ DC-value$(G_v) \leq$ DC-value$(G) - \frac{n}{2}$ and so $\hat{W} \leq 2 \cdot ($DC-value$(G) - \frac{n}{2}) + n = 2 \cdot W^*$ and we are done. □

On the other hand, the recursive sparsest cut algorithm will not necessarily compute the optimal value on trees. Actually, it fails already for caterpillars.

Theorem 9. *There is an infinite family of caterpillars $\{B_k \mid k \geq 3\}$ such that* DC-value$(B_k) \geq 2^k$ *and for any balanced clustering tree $\hat{T}_k$ of B_k has the property that* DC-value$(B_k, \hat{T}_k)/$DC-value$(B_k) = 1 + \Omega(1/\sqrt{\text{DC-value}(B_k)})$.

We conjecture that the actual performance of the balanced cut algorithm on trees is $1 + O(1/\log n))$, which would substantially improve the 2-approximation ratio given by Theorem 8.

Note: After we submitted this paper to *FCT 2021*, it has come to our attention that parameters equivalent to EPT-sum and VPT-sum have been studied before in a completely different context, with edge and vertex partition trees seen as a generalization of binary search trees. Relevant results include: finding the EPT-sum of a node-weighted tree is NP-hard [6]; the balanced cut-approach gives a 1.62-approximation for EPT-sum on (node-weighted) trees [7] (this surpasses the upper bound found in this paper); and VPT-sum on (node-weighted) trees admits a PTAS [2].

References

1. Abboud, A., Cohen-Addad, V., Klein, P.N.: New hardness results for planar graph problems in p and an algorithm for sparsest cut. In: Proceedings of the 52nd Annual ACM SIGACT Symposium on Theory of Computing, STOC 2020, pp. 996–1009. ACM (2020). https://doi.org/10.1145/3357713.3384310

2. Berendsohn, B.A., Kozma, L.: Splay trees on trees. CoRR, abs/2010.00931 (2020). arXiv:2010.00931

3. Bodlaender, H.L., et al.: Rankings of graphs. SIAM J. Discret. Math. 11(1), 168–181 (1998). https://doi.org/10.1137/S0895480195282550

4. Charikar, M., Chatziafratis, V.: Approximate hierarchical clustering via sparsest cut and spreading metrics. In: Annual ACM-SIAM Symposium on Discrete Algorithms (SODA), pp. 841–854 (2017)

5. Charikar, M., Chatziafratis, V., Niazadeh, R.: Hierarchical clustering better than average-linkage. In: Proceedings of the Thirtieth Annual ACM-SIAM Symposium on Discrete Algorithms, pp. 2291–2304 (2019)

6. Cicalese, F., Jacobs, T., Laber, E., Molinaro, M.: On the complexity of searching in trees and partially ordered structures. Theoret. Comput. Sci. 412(50), 6879–6896 (2011)

7. Cicalese, F., Jacobs, T., Laber, E., Molinaro, M.: Improved approximation algorithms for the average-case tree searching problem. Algorithmica 68(4), 1045–1074 (2014)

8. Cohen-Addad, V., Kanade, V., Mallmann-Trenn, F., Mathieu, C.: Hierarchical clustering: objective functions and algorithms. J. ACM 66(4), 26:1–26:42 (2019)

9. Dasgupta, S.: A cost function for similarity-based hierarchical clustering. In: Annual ACM symposium on Theory of Computing (STOC), pp. 118–127 (2016)

10. de la Torre, P., Greenlaw, R., Schäffer, A.A.: Optimal edge ranking of trees in polynomial time. Algorithmica 13(6), 592–618 (1995)

11. Deogun, J.S., Kloks, T., Kratsch, D., Müller, H.: On vertex ranking for permutation and other graphs. In: Enjalbert, P., Mayr, E.W., Wagner, K.W. (eds.) STACS 1994. LNCS, vol. 775, pp. 747–758. Springer, Heidelberg (1994). https://doi.org/10.1007/3-540-57785-8_187

12. Deogun, J.S., Kloks, T., Kratsch, D., Müller, H.: On the vertex ranking problem for trapezoid, circular-arc and other graphs. Discret. Appl. Math. 98(1), 39–63 (1999). https://doi.org/10.1016/S0166-218X(99)00179-1

13. Diestel, R.: Graph Theory, 5th edn. Springer, Heidelberg (2017). https://doi.org/10.1007/978-3-662-53622-3

14. Dong, S., Wang, H., Mostafavi, A., Gao, J.: Robust component: a robustness measure that incorporates access to critical facilities under disruptions. J. R. Soc. Interface 16(157), 20190149 (2019)

15. Høgemo, S., Paul, C., Telle, J.A.: Hierarchical clusterings of unweighted graphs. In: International Symposium on Mathematical Foundations of Computer Science (MFCS 2020), vol. 170, pp. 47:1–47:13 (2020). https://doi.org/10.4230/LIPIcs. MFCS.2020.47

16. Høgemo, S., Bergougnoux, B., Brandes, U., Paul, C., Telle, J.A.: On dasgupta's hierarchical clustering objective and its relation to other graph parameters. arXiv preprint arXiv:2105.12093 (2021)

17. Iyer, A.V., Ratliff, H.D., Vijayan, G.: Optimal node ranking of trees. Inf. Process. Lett. **28**(5), 225–229 (1988). https://doi.org/10.1016/0020-0190(88)90194-9

18. Iyer, A.V., Ratliff, H.D., Vijayan, G.: On an edge ranking problem of trees and graphs. Discret. Appl. Math. **30**(1), 43–52 (1991). https://doi.org/10.1016/0166-218X(91)90012-L

19. Jing-Ho, Y., Jer-Jeong, C., Chang, G.J.: Quasi-threshold graphs. Discret. Appl. Math. **69**(3), 247–255 (1996). https://doi.org/10.1016/0166-218X(96)00094-7

20. Lam, T.W., Yue, F.L.: Edge ranking of graphs is hard. Discret. Appl. Math. **85**(1), 71–86 (1998). https://doi.org/10.1016/S0166-218X(98)00029-8

21. Lam, T.W., Yue, F.L.: Optimal edge ranking of trees in linear time. Algorithmica **30**(1), 12–33 (2001)

22. Nastos, J., Gao, Y.: Familial groups in social networks. Soc. Netw. **35**(3), 439–450 (2013). https://doi.org/10.1016/j.socnet.2013.05.001

23. Nešetřil, J., Ossona de Mendez, P.: On low tree-depth decompositions. Graph. Combin. **31**(6), 1941–1963 (2015)

24. Nešetřil, J., Ossona de Mendez, P.: Tree-depth, subgraph coloring and homomorphism bounds. Eur. J. Combin. **27**(6), 1022–1041 (2006). https://doi.org/10.1016/j.ejc.2005.01.010

25. Pothen, A.: The complexity of optimal elimination trees. Technical report (1988)

26. Roy, A., Pokutta, S.: Hierarchical clustering via spreading metrics. In: Proceedings of the 30th International Conference on Neural Information Processing Systems, NIPS 2016, pp. 2324–2332 (2016)

27. Schäffer, A.A.: Optimal node ranking of trees in linear time. Inf. Process. Lett. **33**(2), 91–96 (1989). https://doi.org/10.1016/0020-0190(89)90161-0

28. Yannakakis, M.: Computing the minimum fill-in is NP-complete. SIAM J. Algebraic Discret. Methods **2**, 77–79 (1981)

Mengerian Temporal Graphs Revisited

Allen Ibiapina$^{(\boxtimes)}$ ⓘ and Ana Silva ⓘ

ParGO Group - Parallelism, Graphs and Optimization, Departamento de
Matemática, Universidade Federal do Ceará, Fortaleza, Brazil
allen.ibiapina@alu.ufc.br, anasilva@mat.ufc.br
https://www.ufc.br/

Abstract. A temporal graph $\mathcal{G}$ is a graph that changes with time. More
specifically, it is a pair (G, λ) where G is a graph and λ is a function
on the edges of G that describes when each edge $e \in E(G)$ is active.
Given vertices $s, t \in V(G)$, a temporal s, t-path is a path in G that
traverses edges in non-decreasing time; and if s, t are non-adjacent, then
a vertex temporal s, t-cut is a subset $S \subseteq V(G)$ whose removal destroys
all temporal s, t-paths.

It is known that Menger's Theorem does not hold on this context, i.e.,
that the maximum number of internally vertex disjoint temporal s, t-
paths is not necessarily equal to the minimum size of a vertex temporal
s, t-cut. In a seminal paper, Kempe, Kleinberg and Kumar (STOC'2000)
defined a graph G to be Mengerian if equality holds on (G, λ) for every
function λ. They then proved that, if each edge is allowed to be active
only once in (G, λ), then G is Mengerian if and only if G has no gem as
topological minor. In this paper, we generalize their result by allowing
edges to be active more than once, giving a characterization also in terms
of forbidden structures. We also provide a polynomial time recognition
algorithm.

Keywords: Temporal graphs · Menger's Theorem · Forbidden minors

1 Introduction

Temporal graphs have been the subject of a lot of interest in recent years (see
e.g. the surveys [9,11,14,15]). They appear under many distinct names (temporal networks [9], edge-scheduled networks [1], dynamic networks [17], time-
varying graphs [4], stream graphs, link streams [11], etc.), but with very little (if
any) distinction between the various models. Here, we favor the name temporal
graphs.

An example where one can apply a temporal graph is the modeling of proximity of people within a region, with each vertex representing a person, and two
people being linked by an edge at a given moment if they are close to each other.

Supported by CNPq grants 303803/2020-7 and 437841/2018-9, FUNCAP/CNPq grant
PNE-0112-00061.01.00/16, and CAPES (PhD student funding).

ⓒ Springer Nature Switzerland AG 2021
E. Bampis and A. Pagourtzis (Eds.): FCT 2021, LNCS 12867, pp. 301–313, 2021.
https://doi.org/10.1007/978-3-030-86593-1_21

This and similar ideas have been used also to model animal proximity networks, human communication, collaboration networks, travel and transportation networks, etc. We refer the reader to [9] for a plethora of applications, but just to cite a simple (and trendy) example where these structures could be used, imagine one wants to track, in the proximity network previously described, the spreading of a contagious disease. In this case it is important to be able to detect whether there was an indirect contact between two people, and of course the contact is only relevant if it occurred after one of these people got sick (see e.g. [6]). Therefore, when studying temporal graphs, it makes sense to be concerned with paths between vertices that respect the flow of time; these are called *temporal paths* and are essentially different from the traditional concept in static graphs, allowing even for multiple definitions of minimality (see e.g. [5,13,17]).

Temporal graphs are being used in practice in a variety of fields for many decades now, with the first appearances dating back to the 1980's, but only recently there seems to be a bigger effort to understand these structures from a more theoretical point of view. An issue that is often raised is whether results on the static version of a certain problem are inherited. This is not always the case, as has been shown for the temporal versions of connected components [2], Menger's Theorem [1,10], and Edmonds' Theorem [3,10].

In particular, in [10] they define and characterize a Mengerian graph as being a graph[1] for which all assignment of activity times for its edges produces a temporal graph on which Menger's Theorem holds. However, their definition does not allow for a certain edge to be active multiple times, which is the case in most of the recent studies. In this paper, we fill this gap, giving a complete characterization of Mengerian graphs. Like the characterization given in [10], ours is in terms of forbidden structures. We also provide a polynomial-time recognition algorithm for these graphs.

Related Work and Results. Given a graph G, and a time labeling $\lambda : E(G) \to \mathbb{N}\backslash\{0\}$, we call the pair (G, λ) a *temporal graph*. Also, given vertices $s, t \in V(G)$, a *temporal s,t-path* in (G, λ) is a path between s and t in G such that if edge e appears after e' in the path, then $\lambda(e') \leq \lambda(e)$. We say that two such paths are *vertex disjoint* if their internal vertices are distinct. Also, a subset $S \subset V(G)$ ($S \subseteq E(G)$) is a *vertex temporal s,t-cut* if there is no temporal s,t-path in $(G - S, \lambda')$, where λ' is equal to λ restricted to $E(G - S)$. From now on, we make an abuse of language and write simply $(G - S, \lambda)$. The analogous notions on static graphs, and on edges, can be naturally defined and the following is one of the most celebrated theorems in Graph Theory.

Theorem 1 (Menger [12]). *Let G be a graph, and $s, t \in V(G)$ be such that $st \notin E(G)$. Then, the maximum number of vertex (edge) disjoint s,t-paths in G is equal to the minimum size of a vertex (edge) cut in G.*

In [1], the author already pointed out that this might not be the case for temporal paths. Here, we present the example given later in [10]; see Fig. 1(a).

---

[1] We adopt the definition of [16], where a graph can have multiple edges incident to the same pair of vertices. This is sometimes called multigraph.

Observe that the only temporal s,t-path using sw also uses v, and since $\{w,v\}$ separates s from t, we get that there are no two vertex disjoint temporal s,t-paths. At the same time, no single vertex among u,v and w breaks all temporal s,t-paths, i.e., there is no vertex temporal s,t-cut of size 1.

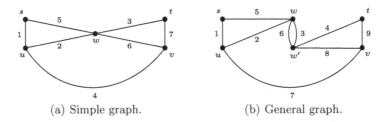

| (a) Simple graph. | (b) General graph. |

Fig. 1. Examples of graphs where the temporal version of Menger's Theorem applied to vertices does not hold.

Inspired by this, in [10] the authors define the *Mengerian graphs*, which are those graphs G for which, no matter what is the time labeling λ, we get that the maximum number of temporal s,t-paths in (G,λ) is equal to the minimum size of a vertex temporal s,t-cut in (G,λ), for every pair $s,t \in V(G)$ such that $st \notin E(G)$. They then give a very interesting characterization for when the graph is simple: a simple graph G is Mengerian if and only if G does not have the graph depicted in Fig. 1(a) (also known as the gem) as a topological minor. Despite being a very nice result, it does not allow for an edge of a simple graph to be active more than once, which is generally the case in practice. Indeed, as the graph depicted in Fig. 1(b) tells us, if an edge can be active more than once, then a graph which does not contain the gem as a topological minor might still be non-Mengerian. In this paper, we allow G to be a general graph (i.e., it can contain multiple edges incident to the same pair of vertices; this is also sometimes called multigraph), and give a similar characterization below. The formal definition of an m-topological minor is given in Sect. 2, but for now it suffices to say that, when subdividing an edge e, the multiplicity of the obtained edges is the same as e.

Theorem 2. *Let G be any graph. Then, G is a Mengerian graph if and only if G does not have one of the graphs in Fig. 2 as an m-topological minor.*

We also provide a polynomial time recognition algorithm to decide whether a given graph G is Mengerian. We use the algorithm in [8] that decides whether H is a topological minor of a given graph G in time FPT when parameterized by $|V(H)|$ to solve a number of subproblems, getting the complexity below.

Theorem 3. *Let G be a graph on n vertices. Then, one can decide whether G is a Mengerian graph in time $O(n^3 m)$, where m is the number of edges of the underlying graph of G.*

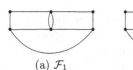 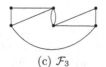 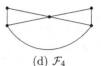

(a) $\mathcal{F}_1$ (b) $\mathcal{F}_2$ (c) $\mathcal{F}_3$ (d) $\mathcal{F}_4$

Fig. 2. Forbidden m-topological minors.

Now, we comment on other papers that attack related problems. In [1], the author proves that the temporal version of Menger's Theorem applied to edges always holds. In [13], the authors give an alternative formulation of Menger's Theorem that holds on the temporal context. There, they define the notion of *out-disjointness*, where two paths are disjoint if they do not share the same departure time for a given vertex, and the notion of *node departure time cut*, where one removes a time label from the possible departure times of a vertex. They then prove that the maximum number of out-disjoint paths between s, t is equal to the minimum size of a node departure time cut. In [1], the author also proves that deciding whether there are k vertex disjoint temporal s, t-paths is NP-complete. This was improved for fixed $k = 2$ in [10], where they also prove that deciding whether there is a vertex temporal s, t-cut of size at most k is NP-complete, for given k. Observe that the latter problem can be easily solved in time $O(|V(G)|^k)$, which raises the question about whether it can be FPT when parameterized by k. This is answered negatively in [18], where they prove that this is W[1]-hard. Finally, in [7], the authors further investigate the cut problem, giving some more strict hardness results (e.g. that the problem is hard even if G is a line graph), as well as some positive ones (e.g., that the problem is polynomial when G has bounded treewidth).

Our paper is organized as follows. We give the formal definitions in Sect. 2, and outlines of the proofs in Sect. 3 (necessity of Theorem 2), Sect. 4 (sufficiency of Theorem 2), and Sect. 5 (Theorem 3).

2   Preliminaries

A *graph* is a tuple $G = (V, E)$ together with a function that associates each $e \in E$ with two (not necessarily distinct) elements in V. We do not refer explicitly to this function, and instead we write e *has endpoints* uv to denote the fact that e is mapped to $\{u, v\}$; we also write uv to represent the pair of endpoints $\{u, v\}$. The elements in V are called *vertices*, and the elements in E, *edges*. If the graph is denoted by G, we also use $V(G), E(G)$ to denote its vertex and edge sets, respectively. A graph H such that $V(H) \subseteq V(G)$ and $E(H) \subseteq V(G)$ is called a *subgraph of* G, and we write $H \subseteq G$.

The *degree* of a vertex $v \in V$ is the number of edges having v as endpoint, and it is denoted by $d_G(v)$. The *maximum degree* of G is then the maximum among the degree of vertices of G; it is denoted by $\Delta(G)$.

An edge with equal endpoints is called a *loop*. Also, if there are exactly k edges in $E(G)$ with endpoints uv, then we say that uv has *multiplicity* k; we also call uv a *multiple edge*. A graph is *simple* if it does not have loops nor multiple edges. For further basic definitions, we refer the reader to [16].

Given a graph G, we relate to G a simple graph G' such that $V(G') = V(G)$ and $E(G')$ is obtained from $E(G)$ by removing loops and multiple edges, i.e., $E(G') = \{uv \mid$ there exists $e \in E(G)$ with endpoints uv and $u \neq v\}$. We call G' the *underlying simple graph of G*, and denote it by $U(G)$.

A *temporal graph* is a pair (G, λ) such that G is a graph and λ is a *time labeling* function that assigns to each edge e the time where e is active (i.e., $\lambda : E(G) \to \mathbb{N}\setminus\{0\}$). Given $S \subseteq E(G)$, we denote by $\max \lambda(S)$ the value $\max_{e \in S} \lambda(e)$. We say that (G, λ) has *lifetime T* if $T = \max \lambda(E(G))$. We call a value $i \in [T]$ a *timestamp*. In the literature, the model found more often uses a simple graph G and a time labeling function that allows each edge to appear more than once during the lifetime of (G, λ). Observe that this is equivalent to our model.

We refer the reader to the introduction for the definition of temporal s, t-path and of vertex temporal s, t-cut. We say that s *reaches t in (G, λ)* if there exists a temporal s, t-path in (G, λ). Given $S \subseteq V(G)$, we denote by $(G - S, \lambda)$ the temporal graph obtained by removing S from G and restricting λ to $E(G - S)$.

The maximum number of vertex disjoint temporal s, t-walks is denoted by $p_{G,\lambda}(s, t)$ (recall that by vertex disjoint, we mean internally vertex disjoint). If $s, t \in V(G)$ are non-adjacent, then $c_{G,\lambda}(s, t)$ denotes the minimum size of a vertex temporal s, t-cut. And if $s, t \in V(G)$ are adjacent, then $c_{G,\lambda}(s, t)$ denotes $|S| + |E'|$, where E' is the set of edges of G with endpoints st and S is a minimum vertex temporal s, t-cut in $(G - E', \lambda)$.

A graph G is *Mengerian* if $p_{G,\lambda}(s, t) = c_{G,\lambda}(s, t)$ for every time labeling $\lambda : E(G) \to \mathbb{N}\setminus\{0\}$, and every pair of vertices $s, t \in V(G)$. In order to present our characterization of Mengerian graphs, we first need to define a new type of graph subdivision.

Given a graph H and an edge $e \in E(H)$ with endpoints uv, a *subdivision of e* is the graph obtained from H by removing e, adding a new vertex w and edges uw and vw; and given an edge $uv \in E(U(H))$ with multiplicity k, a *multiple subdivision of uv* (or m-subdivision for short) is the graph obtained by removing every $e \in E(H)$ with endpoints uv, adding a new vertex w, and k edges with endpoints uw, as well as k edges with endpoints wv. If a graph G is obtained from H by a series of edge subdivisions, we say that G is a subdivision of H, and if G is obtained from H by a series of edge m-subdivisions, we say that G is an m-subdivision of H. Also, if G has a subgraph that is a subdivision of H, we say that H is a *topological minor* of G, writing $H \preceq G$; and if G has a subgraph that is an m-subdivision of H, we say that H is an *m-topological minor* of G, writing $H \preceq_m G$. Denote by $\mathcal{F}$ the set of graphs in Fig. 2. Observe that the relations $\preceq$ and $\preceq_m$ are distinct. For instance, the graph G obtained from $\mathcal{F}_1$ by m-subdividing the multiple edge once does not contain a subdivision of $\mathcal{F}_1$, i.e., $\mathcal{F}_1 \preceq_m G$ and $\mathcal{F}_1 \not\preceq G$. This explains the need for Theorem 3, since the algorithm to recognize topological minors [8] cannot be directly applied to m-topological minors.

A *chain* in a graph G is a path $(z_0, \cdots, z_q)$ in $U(G)$ such that $z_{i-1}z_i$ has multiplicity at least 2, for each $i \in [q]$. Note that any m-subdivision H of a graph in $\{\mathcal{F}_1, \mathcal{F}_2, \mathcal{F}_3\}$ contains a unique chain, which we call the *chain of H*.

The following lemma is used in Sects. 3 and 4.

Lemma 1. *G is Mengerian if and only if every $H \subseteq G$ is Mengerian.*

Proof. Sufficiency is trivial since $G \subseteq G$. To prove necessity, suppose that $H \subseteq G$ is non-Mengerian, and let s, t, λ be such that $p_{H,\lambda}(s,t) < c_{H,\lambda}(s,t)$. Consider the time labeling function λ' for $E(G)$ defined as follows.

$$\lambda'(e) = \begin{cases} \lambda(e) + 1 & \text{, if } e \in E(H), \\ 1 & \text{, if } e \in E(G) \backslash E(H) \text{ has endpoints } yt, \text{ and} \\ \max \lambda(E(H)) + 2 & \text{, otherwise.} \end{cases}$$

Let E' be the set of edges of H with endpoints in s, t and E'', the set of edges of G with endpoints s, t that are not in H. Also, let $H' = H - E'$ and $G' = G - E''$. We first prove that $p_{G',\lambda}(s,t) = p_{H',\lambda}(s,t)$ and $c_{G',\lambda}(s,t) = c_{H',\lambda}(s,t)$. Because $H' \subseteq G'$, we clearly have $p_{G',\lambda}(s,t) \geq p_{H',\lambda}(s,t)$ and $c_{G',\lambda}(s,t) \geq c_{H',\lambda}(s,t)$. For the other way around, suppose that P is a temporal s,t-path in G' not contained in H'. This means that P contains at least one edge $e \in E(G') \backslash E(H')$. This is a contradiction because either e is not incident to t, and hence appear after all the edges incident to t, or e is incident to t and appears before all the other edges of P. Because the set of temporal s,t-paths in G' and H' are the same, the desired equalities follow. Finally, observe that $p_{G,\lambda}(s,t) = p_{G',\lambda}(s,t) + |E'| + |E''| = p_{H',\lambda}(s,t) + |E'| + |E''| = p_{H,\lambda}(s,t) + |E''|$, while $c_{G,\lambda}(s,t) = c_{G',\lambda}(s,t) + |E'| + |E''| = c_{H',\lambda}(s,t) + |E'| + |E''| = c_{H,\lambda}(s,t) + |E''|$. It thus follows that $p_{G,\lambda}(s,t) < c_{G,\lambda}(s,t)$, and G is non-Mengerian, as we wanted to prove. □

Finally, given a simple graph G and vertices $Z = \{z_1, \cdots, z_k\} \subseteq V(G)$, the *identification of Z* is the simple graph obtained from $G - Z$ by adding a new vertex z and edges zw for every $w \in N_G(Z)$. This operation will be used in some of our proofs.

3   Outline of the Proof of Necessity of Theorem 2

We prove the contraposition, i.e., that if $F \preceq_m G$ for some $F \in \mathcal{F}$, then G is non-Mengerian. Observe that, because of Lemma 1, it suffices to prove that each $F \in \mathcal{F}$ is non-Mengerian and that if G is an m-topological subdivision of F, then G is non-Mengerian. Observe Fig. 3 to see that the graphs in $\mathcal{F}$ are non-Mengerian. For the second part, we prove that a single m-subdivision preserves the property of being non-Mengerian.

Lemma 2. *If F is non-Mengerian, and G is obtained from F by an m-subdivision, then G is non-Mengerian.*

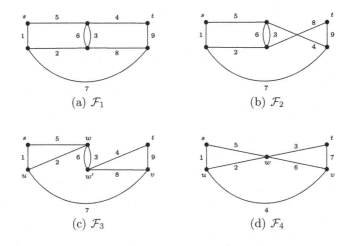

Fig. 3. Forbidden m-topological minors and a function such that $p(s,t) < c(s,t)$.

4   Outline of the Proof of Sufficiency of Theorem 2

Again because of Lemma 1, we only need to prove that if G is a minimal non-Mengerian graph, then G is an m-subdivision of some $F \in \mathcal{F}$. Before we start, we present a definition and a result from [10] that will be used in our proofs. Given a simple graph G, vertices $v, w \in V(G)$, and a positive integer d, a graph G is called (v, w, d)-separable if:

- v and w each have degree exactly d.
- Either $G - \{v, w\}$ consists of d components or $vw \in E(G)$ and $G - \{v, w\}$ has $d - 1$ components.

Lemma 3 ([10]). *Let G be a 2-connected simple graph with $\Delta(G) \geq 4$. If $\mathcal{F}_4 \npreceq G$, then G is (v, w, d)-separable for some $v, w \in V(G)$ and integer $d \geq 4$.*

The following lemma is also useful in our proof.

Lemma 4. *Let G be a minimal non-Mengerian graph, and s, t, λ be such that $p_{G,\lambda}(s,t) < c_{G,\lambda}(s,t)$. Then, G is 2-connected, $st \notin E(U(G))$ and every edge incident in s or t has multiplicity 1.*

We first investigate graphs whose underlying simple graph has maximum degree 3. It is known that if G is a simple graph of maximum degree 3, then the number of edges and the number of vertices needed to disconnect $s, t \in V(G)$ are equal. Therefore, by the result in [1] that says that the edge version of Menger's Theorem holds on temporal graphs, we get that every simple graph with maximum degree 3 is Mengerian. In other words, this case did not need to be investigated in [10]. This cannot be directly applied to our case, as indeed there are graphs in $\mathcal{F}$ that are non-Mengerian whose underlying simple graph has degree at most 3.

Lemma 5. *Let G be a minimal non-Mengerian graph such that $\Delta(U(G)) \leq 3$. Then $F \preceq_m G$ for some $F \in \{\mathcal{F}_1, \mathcal{F}_2, \mathcal{F}_3\}$.*

Proof. Let s, t, λ be such that $p_{G,\lambda}(s,t) < c_{G,\lambda}(s,t)$. We first find a useful substructure inside of G.

Claim 1. There exists a chain $(z_0, \ldots, z_q)$ in G, $q \geq 1$, such that $d_{U(G)}(z_0) = d_{U(G)}(z_q) - 3$, and $z_i \notin \{s, t\}$ for every $i \in \{0, \cdots, q\}$.

Proof (of Claim 1). Denote by $c'_{G,\lambda}(s,t)$ the minimum cardinality of a subset $S \subseteq E(G)$ such that s does not reach t in $(G - S, \lambda)$, and by $p'_{G,\lambda}(s,t)$ the maximum number of edge disjoint temporal s, t-paths. Observe that, since by Lemma 4 we get that $st \notin E(U(G))$, then it is possible to remove one endpoint of each $e \in S$ distinct from s and t that also destroys every temporal s, t-path; hence $c'_{G,\lambda}(s,t) \geq c_{G,\lambda}(s,t) > p_{G,\lambda}(s,t)$. Using the equality $c'_{G,\lambda}(s,t) = p'_{G,\lambda}(s,t)$ proved in [1], one can find at least $c_{G,\lambda}(s,t)$ edge disjoint temporal s, t-paths. Because $c_{G,\lambda}(s,t) > p_{G,\lambda}(s,t)$, at least two of these paths, say J_1, J_2, must intersect in an internal vertex. Let z_0 be the vertex in $V(J_1) \cap V(J_2)$ which is closest to s in J_1. By Lemma 4 and the fact J_1, J_2 intersect in an internal vertex, we get that $z_0 \notin \{s, t\}$. For each $i \in \{1, 2\}$, let y_i be the neighbor of z_0 in $sJ_i z_0$. Since $d_{U(G)}(z_0) \leq 3$, $z_0 \neq t$ and $y_1 \neq y_2$ by the choice of z_0, it follows that $d_{U(G)}(z_0) = 3$ and the next vertex in J_1 and J_2 also coincides, i.e., that there exists $z_1 \in (N(z_0) \cap V(J_1) \cap V(J_2)) \backslash \{y_1, y_2\}$. Now consider $(z_0, z_1, \cdots, z_q)$ be maximal such that $(z_0, e_1, z_1, \cdots, e_q, z_q)$ is a subpath in J_1 and $(z_0, e'_1, z_1, \cdots, e'_q, z_q)$ is a subpath in J_2. Since J_1 and J_2 are edge disjoint, we get that $e_i \neq e'_i$ for every $i \in [q]$, and again by Lemma 4 we have $z_i \notin \{s, t\}$ for every $i \in \{0, \cdots, q\}$. Observe that it also follows that $d_{U(G)}(z_q) = 3$.

Now let $P = (v = z_0, \cdots, z_q = w)$ be the path given by the above claim, and suppose that $d_{U(G)}(z_i) = 2$ for every $i \in [q-1]$; otherwise, it suffices to take the subpath $(z_0, \cdots, z_i)$ where $i \geq 1$ is smallest such that $d(z_i) = 3$. Let G' be obtained by the identification of $\{z_0, \cdots, z_q\}$, and let Z denote the new vertex. Notice that $U(G')$ is 2-connected and has exactly one vertex of degree at least 4, namely Z. Applying Lemma 3 we get a subdivision $H \subseteq U(G')$ of $\mathcal{F}_4$, since the other possibility would imply existence of two vertices of degree at least 4. Let $h_1, \cdots, h_5$ be the vertices of H such that h_5 corresponds to the degree-4 vertex of $\mathcal{F}_4$ (hence $h_5 = Z$), and $h_1, \cdots, h_4$ correspond to the path on 4 vertices in $\mathcal{F}_4$, in this order. By definition of subdivision, we know that for each $i \in [4]$, there exists a path P_i between h_i and Z in H corresponding to the subdivision of an edge of $\mathcal{F}_4$. Let $u_i Z$ denote the last edge in P_i, for each $i \in [4]$. Notice that each of these edges corresponds to an edge of G with endpoints $u_i z_{j_i}$ for some z_{j_i} in P. Now recall that $d_{U(G)}(z_i) = 2$ for every $i \in [q-1]$. In other words, we have $N_{U(G)}(z_i) = \{z_{i-1}, z_{i+1}\}$ for every $i \in [q-1]$, which implies that $z_{j_i} \in \{z_0, z_q\}$ for every $i \in [4]$. All the possible cases are depicted in Fig. 4, each giving rise to some $F \in \{\mathcal{F}_1, \mathcal{F}_2, \mathcal{F}_3\}$.

The following lemma finishes our proof. The case analysis is similar to the one made in [10], making use of the fact that $U(G)$ is (v, w, d)-separable. However,

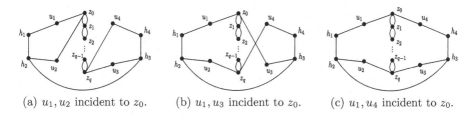

(a) u_1, u_2 incident to z_0.   (b) u_1, u_3 incident to z_0.   (c) u_1, u_4 incident to z_0.

Fig. 4. Possible adjacencies of $u_1, \cdots, u_4$ in $\{z_0, z_q\}$. The cases u_1, u_2 incident to z_q, u_1, u_3 incident to z_q, and u_1, u_4 incident to z_q are clearly analogous.

the fact that the edges between v, w and the components of $G - \{v, w\}$ might be multiple edges complicates the proof. Indeed, unlike the proof in [10], one of our cases is not reducible and this is why we needed a separate theorem to prove the polynomial-time solvability of the recognition problem. Because of space constraints, we give only a general idea of the proof.

Lemma 6. *Let G be a minimal non-Mengerian graph such that $\Delta(U(G)) \geq 4$. Then $F \preceq_m G$ for some $F \in \{\mathcal{F}_3, \mathcal{F}_4\}$.*

Proof (Outline). Consider $s, t \in V(G)$ such that $p_{G,\lambda}(s,t) < c_{G,\lambda}(s,t)$, and suppose $\mathcal{F}_4 \not\preceq G$, as otherwise we are done. It is possible to prove that $U(G)$ must be 2-connected, which by Lemma 3 gives us that $U(G)$ is (v, w, d)-decomposable, for some $v, w \in V(G)$ and $d \geq 4$. If $s \notin \{v, w\}$, let H be the component of $G - \{v, w\}$ containing s, and if $t \notin \{v, w\}$, let H' be the one containing t. The proof consists of analysing the cases where: H and H' are defined and $H = H'$; H and H' are defined and $H \neq H'$; $v \in \{s, t\}$ and $w \notin \{s, t\}$ (i.e. exactly one between H, H' is defined); and $v = s$ and $w = t$ (none is defined). For each of these cases, if the edges linking v, w to the case in point components have multiplicity 1, we fall back in the cases analyzed in [10]. And when at least one of them has multiplicity greater than 1, say vv', we find a chain containing vv' with endpoints z_0, z_q, two disjoint cycles C_1, C_2 containing z_0, z_q, respectively, and a path P between these cycles that is disjoint from the chain (see Fig. 5(a)). Because of space constraints, we give only an outline of the case where $H = H'$, which is the case that differs the most from the proof in [10].

In what follows, we say that a temporal x, y-path $J = (x = z_0, e_1, \cdots, e_p, y = z_p)$ can be *concatenated* with a temporal y, z-path $J' = (y = z'_0, e'_1, \cdots, e'_q, z = z'_q)$ if $\lambda(e_p) \leq \lambda(e'_1)$. Observe that the union of J and J' must contain a temporal x, z-path (it suffices to remove eventual vertex repetitions).

First, notice that there must exist a temporal s, t-path not contained in H as otherwise H would be a proper subgraph of G which is also non-Mengerian. So, let $\mathcal{P}$ be a maximum set of vertex disjoint temporal s, t-paths of (G, λ), and let $J_1 \in \mathcal{P}$ be such that J_1 is not contained in H. Without loss of generality, we can suppose that J_1 visits v before w. Now let G^* be the subgraph of G induced by $V(H) \cup V(J_1)$. As any set of vertex disjoint temporal s, t-paths contains at most one such path passing out of H, and by the choice of J_1, we have

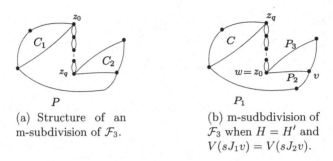

(a) Structure of an
m-subdivision of $\mathcal{F}_3$.

(b) m-sudbdivision of
$\mathcal{F}_3$ when $H = H'$ and
$V(sJ_1v) = V(sJ_2v)$.

Fig. 5. General structure of $\mathcal{F}_3$ to the left, and particular case to the right.

$p_{G,\lambda}(s,t) = p_{G^*,\lambda}(s,t)$. Also, because $G^* \subseteq G$, we get $c_{G,\lambda}(s,t) \geq c_{G^*,\lambda}(s,t)$, and by the minimality of G, we have that: (*) $c_{G,\lambda}(s,t) > c_{G^*,\lambda}(s,t)$. Then, let S be a minimum vertex temporal s,t-cut in (G^*,λ); note that S is not a cut in (G,λ) as this would contradict (*). Similarly, if there is a vertex $u \in S$ out of H, then $(S\backslash\{u\}) \cup \{v\}$ is a vertex temporal s,t-cut in (G,λ), again contradicting (*). Therefore we get that $S \subset V(H)$, and also that there must exist a temporal s,t-path J_2 in (G,λ) not passing by S. Because $H \subseteq G^*$, such temporal path is not contained in H. We first analyse the case where J_2 visits v before w.

Let v_1 be the neighbor of v in H, and w_1, the neighbor of w in H. Suppose first that $V(sJ_2v) = V(sJ_1v)$, and notice that the part of sJ_1w contained in $V(H)$ does not intersect S, as this has vertex set equal to sJ_2v_1. This implies that we cannot concatenate sJ_1w with wJ_2t, as otherwise we obtain a temporal s,t-path contained in G^* that does not intersect S. Therefore the edges of J_1 and J_2 with endpoints ww_1 must be distinct. This also implies that $S \cap V(wJ_1t) \neq \emptyset$ since $S \subseteq V(H)$ and $S \cap V(J_1) \neq \emptyset$. Since $S \cap V(J_2) = \emptyset$, we get that $V(w_1J_2t) \neq V(w_1J_1t)$. Now, let $(w = z_0, w_1 = z_2, \cdots, z_q)$ be maximal such that $\{z_0, \cdots, z_q\} \subseteq V(wJ_1t) \cap V(wJ_2t)$. By the same reason why ww_1 must have multiplicity at least two, we get that the same holds for every pair z_iz_{i+1}, $i \in [q-1]$. Also, because $V(w_1J_2t) \neq V(w_1J_1t)$ and these paths must intersect again after z_q, namely in t, there exists a cycle $C \subseteq V(w_1J_1t) \cup V(w_1J_2t)$ containing z_q of size at least 3, and disjoint from $\{z_0, \cdots, z_{q-1}\}$. Now, note that if there is no temporal s,t-path contained in H, then we would get $p_{G,\lambda}(s,t) = 1$, and $c_{G,\lambda}(s,t) = 1$ as $\{v\}$ would be a vertex temporal s,t-cut. Since this cannot occur, we know that there is at least one temporal s,t-path disjoint from J_1, i.e., $|\mathcal{P}| \geq 2$. Observe that this implies the existence of a path P_1 between v_1 and C not passing by $\{z_0, \cdots, z_q\}$. Finally, let P_2, P_3 be vertex disjoint v,w-paths, each passing by components distinct from H (recall that $G - \{v,w\}$ has at least 3 components). Observe Fig. 5(b) to see that $V(C \cup P_1 \cup P_2 \cup P_3)$ contains an m-subdivision of F_3. In the figure, only P_3 is shown to have at least one internal vertex, but that also holds for P_2, giving us the second desired cycle of size at least 3. A similar argument can be made when sJ_2v_1 can be concatenated with v_1J_1w, and when neither applies, then we get vv_1 with multiplicity at least 2 and again a similar argument can be made.

Now, suppose that J_2 visits w before v, and for each $a \in \{1,2\}$, denote by e_a^v (e_a^w) the edge in J_a with endpoints vv_1 (ww_1). First, suppose that there exists $e \in E(sJ_1v) \cap E(vJ_2t)$. Because e appears before e_1^w in J_1 and after e_2^w in J_2, we get that: (**) $\lambda(e_1^w) \geq \lambda(e) \geq \lambda(e_2^w)$. Suppose that ww_1 has multiplicity 1 and note that (**) gives us that $\lambda(e') = \lambda(e)$ for every $e' \in E(v_1J_1w_1) \cup E(w_1J_2v_1)$. This implies that sJ_2w_1, $w_1J_1v_1$ and v_1J_2t can be concatenated to obtain a temporal s,t-path, contradicting the fact that S is a vertex temporal s,t-cut in G^*. Therefore, ww_1 has multiplicity at least 2. Now, observe that if the neighbor z_1 of w_1 in $J_1\backslash\{w\}$ is equal to the neighbor of w_1 in $J_2\backslash\{w\}$, then a similar argument can be applied to conclude that the edge w_1z_1 has multiplicity at least 2, and so on. More formally, let $P = (w = z_0, w_i = z_1, \cdots, z_q)$ be maximal such that $z_0, \cdots, z_q$ appears consecutively in this order in wJ_1t and consecutively in the reverse order in sJ_2w. Applying the previous argument inductively, we get that all internal edges in P have multiplicity at least 2. Now, for each $a \in \{1,2\}$, let y_a be the neighbor of z_q in $J_a\backslash\{z_{q-1}\}$. Because w_1J_1t and sJ_2w_1 have only one common endpoint, namely w_1, we get that $y_1 \neq y_2$. Also, observe that the union of sJ_1v_1, v_1J_2t, sJ_2z_q and z_qJ_1t contains a closed walk W disjoint of P. Therefore, there exists a cycle C of length at least 3 in H intersecting P exactly in z_q. One can see that W also implies the existence of a path P' between v_1 and C disjoint from P. One can finish building an m-subdivision of $\mathcal{F}_3$ by taking a cycle containing v and w not intersecting H (which exists because $p \geq 3$). Finally, suppose that $E(sJ_1v) \cap E(vJ_2t) = \emptyset$. In particular this implies that vv_1 has multiplicity at least 2, since an edge with endpoints vv_1 must be contained in both J_1 and J_2. By taking a maximal subpath P with endpoint v common to sJ_1v and vJ_2t we can apply the same argument as before to get the desired m-subdivision of $\mathcal{F}_3$. $\quad\square$

5 Mengerian Graphs Recognition

In [8], they show that given two simple graphs G,H, one can check if H is a topological minor of G in time $O(f(|V(H)|)|V(G)|^3)$. Thus, for a finite family of graphs $\mathcal{H} = \{H_1, \ldots, H_k\}$ each of constant size, the problem of deciding whether $H \preceq G$ for some $H \in \mathcal{H}$ can be solved in polynomial time. If the same holds for m-topological minor, then Theorem 2 implies that we can recognize Mengerian graphs in polynomial time. We prove that this is indeed the case by giving an algorithm that makes use of the one presented in [8]. The following lemma is the key for reaching polynomial time.

Lemma 7. *Let G be a graph such that $F \preceq_m G$ for some $F \in \{\mathcal{F}_1, \mathcal{F}_2, \mathcal{F}_3\}$. Then there is an m-subdivision $H \subseteq G$ of a graph in $\{\mathcal{F}_1, \mathcal{F}_2, \mathcal{F}_3\}$ such that $d_{U(G)}(v) = 2$ for every $v \in V(H)$ that is an internal vertex of the chain of H.*

Proof (of Theorem 3). By Theorem 2, it suffices to decide whether $F \preceq_m G$ for some $F \in \mathcal{F}$. If G is not 2-connected, we simply apply the algorithm to each 2-connected component of G; so suppose G is 2-connected. Also, note that $\mathcal{F}_4 \preceq_m G$ if and only if $\mathcal{F}_4 \preceq G$, as $\mathcal{F}_4$ has no multiple edges; so we also suppose that $\mathcal{F}_4 \not\preceq G$, which can be tested in time $O(n^3)$ by the result in [8]. We first do a preprocessing of G as follows. For every $u \in V(G)$ such that $d_{U(G)}(u) = 2$, let $N(u) = \{v, w\}$. If either uv and uw have both multiplicity greater than 1, or both multiplicity 1, then contract u with either one of its neighbors, maintaining the multiplicity of the new edge equal to the multiplicity of uv. Observe that if we find an m-subdivision containing the new edge, then we can easily restore the edges of G to obtain the m-subdivision in G. We can then suppose that if $N(u) = \{v, w\}$ then uv has multiplicity 1, while uw has multiplicity at least 2. By Lemma 7, we get that if $F \preceq_m G$ for some $F \in \{\mathcal{F}_1, \mathcal{F}_2, \mathcal{F}_3\}$, then there exists an m-subdivision H of F in G whose chain is a single multiple edge. Observe that this implies that G' obtained from contracting this edge has $\mathcal{F}_4$ as a topological minor. A case analysis equal to the one that was made in the proof of Lemma 5 gives us that the other way around is also true, i.e., if G' has $\mathcal{F}_4$ as a topological minor, then G has an m-topological minor in $\{\mathcal{F}_1, \mathcal{F}_2, \mathcal{F}_3\}$. We therefore apply the algorithm in [8] to such a graph G' obtained by the contraction of a multiple edge, and we do this for every multiple edge of G, obtaining therefore an algorithm that runs in time $O(n^3 m)$, since the previous steps have complexity dominated by this.

6   Conclusion

We have characterized general Mengerian graphs and provided a recognition algorithm for them, thus generalizing a result given in the seminal paper [10].

Here, we allowed the graph G to have multiple edges, which can also be seen as allowing the edges of a simple graph to appear more than once in the lifetime of (G, λ). Observe that this can also be modeled by letting λ be a subset of positive integers instead, and indeed this model is usually the one applied in the literature. If we use this definition of temporal graph instead, and consider only simple graphs, then our theorem says that G is Mengerian if and only if $F \not\preceq G$ for every $F \in \{U(\mathcal{F}_1), U(\mathcal{F}_2), U(\mathcal{F}_3), \mathcal{F}_4\}$. But observe that this might be too restrictive. Indeed, if G does not have $\mathcal{F}_4$ as a topological minor, there could still be an assignment that allows edges to appear more than once and still have a Mengerian graph, as long as one takes care not to create a bad m-topological minor. A good question therefore is:

Problem 1. Let G be a simple graph. Can one find, in polynomial time, a function $\alpha : E(G) \to \mathbb{N} \setminus \{0\}$ such that G' is Mengerian, where G' is the graph obtained from G by making each $e \in E(G)$ have multiplicity $\alpha(e)$?

Observe also that, even if G is non-Mengerian, it could still happen that Menger's Theorem holds on (G, λ) for a given function λ. Up to our knowledge this has not been investigated yet.

Problem 2. Given a temporal graph (G, λ), and a pair of vertices $s, t \in V(G)$, can one decide in polynomial time whether $p_{G,\lambda}(s, t) = c_{G,\lambda}(s, t)$?

References

1. Berman, K.A.: Vulnerability of scheduled networks and a generalization of Menger's theorem. Netw.: Int. J. **28**(3), 125–134 (1996)
2. Bhadra, S., Ferreira, A.: Complexity of connected components in evolving graphs and the computation of multicast trees in dynamic networks. In: Pierre, S., Barbeau, M., Kranakis, E. (eds.) ADHOC-NOW 2003. LNCS, vol. 2865, pp. 259–270. Springer, Heidelberg (2003). https://doi.org/10.1007/978-3-540-39611-6_23
3. Campos, V., Lopes, R., Marino, A., Silva, A.: Edge-disjoint branchings in temporal graphs. In: Gąsieniec, L., Klasing, R., Radzik, T. (eds.) IWOCA 2020. LNCS, vol. 12126, pp. 112–125. Springer, Cham (2020). https://doi.org/10.1007/978-3-030-48966-3_9
4. Casteigts, A., Flocchini, P., Quattrociocchi, W., Santoro, N.: Time-varying graphs and dynamic networks. Int. J. Parallel Emergent Distrib. Syst. **27**(5), 387–408 (2012)
5. Casteigts, A., Himmel, A.-S., Molter, H., Zschoche, P.: Finding temporal paths under waiting time constraints. In: 31st International Symposium on Algorithms and Computation, ISAAC, volume 181 of LIPIcs, pp. 30:1–30:18 (2020)
6. Jessica Enright and Rowland Raymond Kao: Epidemics on dynamic networks. Epidemics **24**, 88–97 (2018)
7. Fluschnik, T., Molter, H., Niedermeier, R., Renken, M., Zschoche, P.: Temporal graph classes: a view through temporal separators. Theoret. Comput. Sci. **806**, 197–218 (2020)
8. Grohe, M., Kawarabayashi, K., Marx, D., Wollan, P.: Finding topological subgraphs is fixed-parameter tractable. In: Proceedings of the 43rd Annual ACM Symposium on Theory of Computing, pp. 479–488 (2011)
9. Holme, P.: Modern temporal network theory: a colloquium. Eur. Phys. J. B **88**(9), 1–30 (2015). https://doi.org/10.1140/epjb/e2015-60657-4
10. Kempe, D., Kleinberg, J., Kumar, A.: Connectivity and inference problems for temporal networks. J. Comput. Syst. Sci. **64**, 820–842 (2002)
11. Latapy, M., Viard, T., Magnien, C.: Stream graphs and link streams for the modeling of interactions over time. Soc. Netw. Anal. Min. **8**(1), 1–29 (2018). https://doi.org/10.1007/s13278-018-0537-7
12. Menger, K.: Zur allgemeinen kurventheorie. Fundam. Math. **10**(1), 96–115 (1927)
13. Mertzios, G.B., Michail, O., Spirakis, P.G.: Temporal network optimization subject to connectivity constraints. Algorithmica **81**(4), 1416–1449 (2019). https://doi.org/10.1007/s00453-018-0478-6
14. Michail, O.: An introduction to temporal graphs: an algorithmic perspective. Internet Math. **12**(4), 239–280 (2016)
15. Nicosia, V., Tang, J., Mascolo, C., Musolesi, M., Russo, G., Latora, V.: Graph metrics for temporal networks. In: Holme, P., Saramäki, J. (eds.) Temporal Networks. Understanding Complex Systems, pp. 15–40. Springer, Heidelberg (2013). https://doi.org/10.1007/978-3-642-36461-7_2
16. Douglas Brent West: Introduction to Graph Theory, vol. 2. Prentice Hall, Upper Saddle River (1996)
17. Bui Xuan, B., Ferreira, A., Jarry, A.: Computing shortest, fastest, and foremost journeys in dynamic networks. Int. J. Found. Comput. Sci. **14**(02), 267–285 (2003)
18. Zschoche, P., Fluschnik, T., Molter, H., Niedermeier, R.: The complexity of finding small separators in temporal graphs. J. Comput. Syst. Sci. **107**, 72–92 (2020)

Faster FPT Algorithms for Deletion to Pairs of Graph Classes

Ashwin Jacob[1]([✉]), Diptapriyo Majumdar[2], and Venkatesh Raman[1]

[1] The Institute of Mathematical Sciences, HBNI, Chennai, India
{ajacob,vraman}@imsc.res.in
[2] Royal Holloway, University of London, Egham, UK
diptapriyo.majumdar@rhul.ac.uk

Abstract. Let Π be a hereditary graph class. The problem of deletion to Π, takes as input a graph G and asks for a minimum number (or a fixed integer k) of vertices to be deleted from G so that the resulting graph belongs to Π. This is a well-studied problem in paradigms including approximation and parameterized complexity. This problem, for example, generalizes VERTEX COVER, FEEDBACK VERTEX SET, CLUSTER VERTEX DELETION, PERFECT DELETION to name a few. The study of this problem in parameterized complexity has resulted in several powerful algorithmic techniques including iterative compression and important separators.

Recently, the study of a natural extension of the problem was initiated where we are given a finite set of hereditary graph classes, and the goal is to determine whether k vertices can be deleted from a given graph, so that the connected components of the resulting graph belong to one of the given hereditary graph classes. The problem has been shown to be FPT as long as the deletion problem to each of the given hereditary graph classes is fixed-parameter tractable, and the property of being in any of the graph classes can be expressible in the counting monadic second order (CMSO) logic. While this was shown using some black box theorems, faster algorithms were shown when each of the hereditary graph classes has a finite forbidden set.

In this paper, we do a deep dive on pairs of specific graph classes (Π_1, Π_2) in which we would like the connected components of the resulting graph to belong to, and design simpler and more efficient FPT algorithms. We design two general algorithms for pairs of graph classes (possibly having infinite forbidden sets) satisfying certain conditions on their forbidden sets. These algorithms cover a number of pairs of popular graph classes.

Our algorithms make non-trivial use of the branching technique and as black box, FPT algorithms for deletion to individual graph classes.

1 Introduction

Graph modification problems are the class of problems in which the input instance is a graph, and the goal is to check if the input can be transformed into

© Springer Nature Switzerland AG 2021
E. Bampis and A. Pagourtzis (Eds.): FCT 2021, LNCS 12867, pp. 314–326, 2021.
https://doi.org/10.1007/978-3-030-86593-1_22

a graph of a specified graph class by using some specific number of "allowed" graph operations. Depending on the allowed operations, *vertex or edge deletion problems, edge editing or contraction problems* have been extensively studied in various algorithmic paradigms.

In the last two decades, graph modification problems, specifically vertex deletion problems have been extensively studied in the field of parameterized complexity. Examples of vertex deletion problems include VERTEX COVER, CLUSTER VERTEX DELETION, FEEDBACK VERTEX SET and CHORDAL DELETION SET. We know from the classical result by Lewis and Yannakakis [8] that the problem of whether removing a set of at most k vertices results in a graph satisfying a hereditary property π is NP-complete for every non-trivial property π. It is well-known that any hereditary graph class[1] can be described by a forbidden set of graphs, finite or infinite, that contains all minimal forbidden graphs in the class. It is also well-known [1] that if a hereditary graph class has a finite forbidden set, then deletion to the graph class has a simple fixed-parameter tractable (FPT) algorithm using a hitting set based approach.

Recently Jacob et al. [4], building on the work of Ganian et al. [3] for constraint satisfaction problems, introduced a natural extension of vertex deletion problems to deletion to scattered graph classes. Here we want to delete vertices from a given graph to put the connected components of the resulting graph to one of a few given graph classes. A scattered graph class $(\Pi_1, \ldots, \Pi_d)$ consists of graphs whose connected components are in one of the graph classes $\Pi_1, \ldots, \Pi_d$. The vertex deletion problem to this class cannot be solved by a hitting set based approach, even if the forbidden graphs for these classes are finite. For example, the solution could possibly be disjoint from the forbidden subgraphs present, as long as it separates them so that the forbidden subgraphs of the d classes don't belong to the same component.

Jacob et al. [4] proved that the vertex deletion problem for the scattered graph class $(\Pi_1, \ldots, \Pi_d)$ is FPT when the vertex deletion problem to each of the individual graph classes is FPT and for each graph class, the property that a graph belongs to the graph class is expressible by Counting Monadic Second Order logic. Unfortunately the running of the algorithm incurs a gargantuan constant factor (a function of k) overhead. The authors also proved that if the forbidden families corresponding to all the graph classes $\Pi_1, \ldots, \Pi_d$ are finite, then the problem is FPT with running time $2^{poly(k)} n^{\mathcal{O}(1)}$. The technique involves iterative compression and important separators.

Since the algorithms in [4] incur a huge running time and use sophisticated techniques, it is interesting to see whether we can get simpler and faster algorithms for some special cases of the problem. In this paper we do a deep dive on the vertex deletion problems to a pair of graph classes when at least one of the graph classes has an infinite forbidden family.

Our Problems, Results and Techniques: We look at specific variants of the following problem.

---

[1] A hereditary graph class is a class of graphs that is closed under induced subgraphs.

Π_1 OR Π_2 DELETION   **Parameter:** k

Input: An undirected graph $G = (V, E)$, two hereditary graph classes Π_1 and Π_2 defined by forbidden graphs $\mathcal{F}_1$ and $\mathcal{F}_2$ respectively.

Question: Is there a set $S \subseteq V(G)$ of size at most k such that every connected component of $G - S$ is in Π_1 or in Π_2?

The authors in [5] studied one such specific case where Π_1 is the class of forests (having infinite forbidden set of all cycles) and Π_2 is the class of cluster graphs, to which they gave an $\mathcal{O}^*(4^k)^2$ algorithm. Here, we describe two general algorithms covering pairs of a variety of graph classes. While the specific conditions, on the pairs of classes to be satisfied by these algorithms, are somewhat technical and are explained in the appropriate sections, we give a high level description here.

We first make the reasonable assumption that the vertex deletion problems to the graph class Π_1 and to Π_2 have FPT algorithms. As we want every connected component of the graph after removing the solution vertices to be in Π_1 or in Π_2, any pair of forbidden subgraphs $H_1 \in \mathcal{F}_1$ and $H_2 \in \mathcal{F}_2$ cannot both be in a connected component of G. Let us look at such a component C with $J_1, J_2 \subseteq V(C)$ such that $G[J_i]$ is isomorphic to H_i for $i \in \{1, 2\}$ and look at a path P between the sets J_1 and J_2. We know that the solution has to hit the set $J_1 \cup J_2 \cup P$.

The first generalization comes up from our observation that for certain pairs of graph classes, if we focus on a pair of forbidden subgraphs $H_1 \in \mathcal{F}_1$ and $H_2 \in \mathcal{F}_2$ that are "closest" to each other, then there is always a solution that does not intersect the shortest path P between them. This helps us to branch on the vertex sets of these forbidden graphs. However, note that the forbidden graphs may have unbounded sizes. We come up with a notion of *forbidden pair* (Definition 2 in Sect. 2) and show that there are pairs of graph classes that have finite number of forbidden pairs even if each of them has infinite forbidden sets. For some of them, we are able to bound the branching step to obtain the FPT algorithm. This problem variant covers a number of pairs of graph classes including (Interval, Trees) and (Chordal, Bipartite Permutation). We observe that this class of algorithms also yield good approximation algorithms for the deletion problem.

In the second general algorithm, we assume that $\mathcal{F}_1$ is finite and has a path P_i for a constant i, and the family of graphs of $\mathcal{F}_2$ that is present in the graph class Π_1 obtained after a pruning step (details in Sect. 4) is finite. Note that this restricts the length of P to i. Under these assumptions, we come up with a finite family of graphs to branch on and we complete the algorithm (at the leaf level of the branching algorithms) with the FPT algorithm for the deletion to each of the individual graph classes. The variant satisfying these conditions covers a number of pairs of graph classes including when Π_1 is the class of all cliques, and Π_2 is the class of planar or bounded treewidth graphs.

---

[2] $\mathcal{O}^*$ notation suppresses polynomial factors.

The running time of all the algorithms that we come up in this paper are substantially better than those in [4].

2   Forbidden Characterization for Π_1 OR Π_2 DELETION

We use $\Pi_{(1,2)}$ to denote the class of graphs whose connected components are in the graph classes Π_1 or Π_2. The following characterization for $\Pi_{(1,2)}$ is easy to see.

Lemma 1 (⋆). *A graph G is in the graph class $\Pi_{(1,2)}$ if and only if no connected component C of G contains H_1 and H_2 as induced graphs in C, where $(H_1, H_2) \in \mathcal{F}_1 \times \mathcal{F}_2$.*

We now define the notion of super-pruned family which gives us a minimal family of graphs which are forbidden in the graph class $\Pi_{(1,2)}$.

We say that family of graphs is *minimal* if no element of it is an induced subgraph of some other element of the family.

Definition 1 (Super-Pruned Family).
An element of super-pruned family $\mathsf{sp}(\mathcal{G}_1, \mathcal{G}_2)$ of two minimal families of graphs $\mathcal{G}_1$ and $\mathcal{G}_2$ is a graph that (i) belongs to one of the two families and (ii) has an element of the other family as induced subgraph.

The family $\mathsf{sp}(\mathcal{G}_1, \mathcal{G}_2)$ can be obtained from an enumeration of all pairs in $\mathcal{G}_1 \times \mathcal{G}_2$ and adding the supergraph if one of the graph is induced subgraph of the other. The family obtained is made minimal by removing the elements that are induced subgraphs of some other elements.

For example, let (Π_1, Π_2) be (Interval, Trees), with the forbidden families $\mathcal{F}_1 = \{$net, sun, long claw, whipping top, †-AW, ‡-AW$\} \cup \{C_i : i \geq 4\}$ (See Fig. 2 in [2]) and $\mathcal{F}_2$ as the set of all cycles. Note that all graphs C_i with $i \geq 4$ are in $\mathsf{sp}(\mathcal{F}_1, \mathcal{F}_2)$ as they occur in both $\mathcal{F}_1$ and $\mathcal{F}_2$. The remaining pairs of $\mathcal{F}_1 \times \mathcal{F}_2$ contain triangles from $\mathcal{F}_2$. If the graph from $\mathcal{F}_1$ is a net, sun, whipping top, †-AW or ‡-AW, it contains triangle as an induced subgraph. Hence these graphs are also in the family $\mathsf{sp}(\mathcal{F}_1, \mathcal{F}_2)$.

We now show that graphs in $\mathsf{sp}(\mathcal{F}_1, \mathcal{F}_2)$ are forbidden in the graph class $\Pi_{(1,2)}$.

Lemma 2 (⋆).[3] *If a graph G is in the graph class $\Pi_{(1,2)}$, then no connected component of G contains a graph in $\mathsf{sp}(\mathcal{F}_1, \mathcal{F}_2)$ as induced subgraphs.*

The family $\mathsf{sp}(\mathcal{F}_1, \mathcal{F}_2)$ does not cover all the pairs in $\mathcal{F}_1 \times \mathcal{F}_2$. We now define the following to capture the remaining pairs.

Definition 2 (Forbidden Pair). *A forbidden pair of $\mathcal{F}_1$ and $\mathcal{F}_2$ is a pair $(H_1, H_2) \in \mathcal{F}_1 \times \mathcal{F}_2$ such that both $H_1 \notin \mathsf{sp}(\mathcal{F}_1, \mathcal{F}_2)$ and $H_2 \notin \mathsf{sp}(\mathcal{F}_1, \mathcal{F}_2)$.*

---

[3] Proofs of Theorems and Lemmas marked ⋆ are moved to the full version due to lack of space.

For example, if Π_1 is the class of interval graphs and Π_2 is the class of forests, we have already shown that $\mathsf{sp}(\mathcal{F}_1, \mathcal{F}_2)$ contains all the graphs in $\mathcal{F}_1$ except long-claw. The only remaining pair is (long-claw, triangle) which is a forbidden pair.

Now we characterize $\Pi_{(1,2)}$ based on the super-pruned family and a family of forbidden pairs associated to $\mathcal{F}_1$ and $\mathcal{F}_2$. This is used in the algorithms in Sect. 3.

Lemma 3 (⋆). *The following statements are equivalent.*

- *Each connected component of G is in Π_1 or Π_2.*
- *The graph G does not contain graphs in the super-pruned family $\mathsf{sp}(\mathcal{F}_1, \mathcal{F}_2)$ as induced subgraphs. Furthermore, for each forbidden pair (H_1, H_2) of $\mathcal{F}_1$ and $\mathcal{F}_2$, the graphs H_1 and H_2 both cannot appear as induced subgraphs in a connected component of G.*

We now define a useful notion of forbidden sets for the graph class $\Pi_{(1,2)}$, and the notion of closest forbidden pairs.

Definition 3. *We call a minimal vertex subset $Q \subseteq V(G)$ as a forbidden set corresponding to the graph class $\Pi_{(1,2)}$ if $G[Q]$ is isomorphic to a graph in $\mathsf{sp}(\mathcal{F}_1, \mathcal{F}_2)$ or $G[Q]$ is connected and contains both H_1 and H_2 as induced subgraphs for some forbidden pair (H_1, H_2) of $\Pi_{(1,2)}$.*

Definition 4. *We say that a forbidden pair (H_1, H_2) is a closest forbidden pair in a graph G if there exists subsets $J_1, J_2 \subseteq V(G)$ such that $G[J_1]$ is isomorphic to H_1, $G[J_2]$ is isomorphic to H_2 and the distance between J_1 and J_2 in G is the smallest among all such pairs J_1, J_2 corresponding to all forbidden pairs of $\mathcal{F}_1$ and $\mathcal{F}_2$. We call the pair of vertex subsets (J_1, J_2) as the vertex subsets corresponding to the closest forbidden pair.*

3 Π_1 or Π_2 Deletion with a Constant Number of Forbidden Pairs

We start with the following reduction rule for Π_1 OR Π_2 DELETION whose correctness easily follows.

Reduction Rule 1. *If a connected component C of G is in Π_1 or in Π_2, then delete C from G. The new instance is $(G - V(C), k)$.*

In this section we assume that the forbidden pair family (as defined in Sect. 2) for Π_1 and Π_2 is finite. Before we define the further conditions, we give an algorithm for an example pair of graph classes that satisfy our problem conditions.

3.1 Interval or Trees

We define the problem.

INTERVAL-OR-TREE DELETION   **Parameter:** k

Input: An undirected graph $G = (V, E)$ and an integer k.

Question: Is there $S \subseteq V(G)$ of size at most k such that every connected component of $G - S$ is either an interval graph, or a tree?

We have the following forbidden subgraph characterization of interval graphs.

Lemma 4 ([7]). *A graph is an interval graph if and only if it does not contain net, sun, hole, whipping top, long-claw, †-AW, or ‡-AW as its induced subgraphs. See Fig. 2 in [2] for an illustration of the forbidden subgraphs for interval graph.*

Let $\mathcal{G}_1$ be the family of graphs in $\mathsf{sp}(\mathcal{F}_1, \mathcal{F}_2)$ of size at most 10. We now define the following branching rule where we branch on all induced subgraphs of G isomorphic to a member in $\mathcal{G}_1$. The correctness follows as $\mathcal{G}_1$ is a finite family of finite-sized graphs.

Branching Rule 1. *Suppose that (G, k) be the input instance and there exist a forbidden set $Q \subseteq V(G)$ such that $G[Q]$ is isomorphic to a member in $\mathcal{G}_1$. Then, for each $v \in V(Q)$, we delete v from G and decrease k by 1. The resulting instance is $(G - v, k - 1)$.*

From here on we assume that Branching Rule 1 is not applicable for G and so it is $\mathcal{G}_1$-free.

We now focus on connected components of G which contain both long-claw and triangle as induced subgraphs. We describe a branching rule corresponding to the closest forbidden pair.

Branching Rule 2. *Let (J^*, T^*) be the vertex subsets of a closest long-claw, triangle pair in a connected component of G, where J^* is a long-claw, and T^* is a triangle. Then for each $v \in J^* \cup T^*$, delete v and decrease k by 1, resulting in the instance $(G - v, k - 1)$.*

We now prove that Branching Rule 2 is sound. Let $P^* = \{u = x_0, x_1, \ldots, x_{d-1}, x_d = v\}$ be a shortest path of length $d_G(J^*, T^*) = d$ that witnesses a path from $u \in J^*$ to $v \in T^*$.

A *caterpillar* graph is a tree in which all the vertices are within distance 1 of a central path. In the graph G, let C be the connected component of $G - (J^* \cup T^*)$ containing the internal vertices of P^*. We have the following lemma.

Lemma 5 $(\star)$. *The graph C is a caterpillar with the central path being P^*. Furthermore, the only vertices of C adjacent to $J^* \cup T^*$ are x_1 and x_{d-1} which are only adjacent to x_0 and x_d respectively.*

We now use Lemma 5 to prove that Branching Rule 2 is sound.

Lemma 6 $(\star)$. *Branching Rule 2 is sound.*

From here on, assume that (G, k) is an instance for which Reduction Rule 1, Branching Rule 1, and Branching Rule 2 are not applicable. The following results are now easy to see.

Lemma 7 *(⋆).* *Let C be a connected component of G that has no triangle, but has a long-claw as induced subgraph. If $G[C]$ has no feedback vertex set of size k, then (G, k) is a no-instance. Otherwise, let X be a minimum feedback vertex set of $G[C]$. Then (G, k) is a yes-instance if and only if $(G - V(C), k - |X|)$ is a yes-instance.*

Lemma 8 *(⋆).* *Let C be a connected component of G that has no long-claw, but has a triangle as induced subgraph. If $G[C]$ has no interval vertex deletion set of size k, then (G, k) is a no-instance. Otherwise, let X be a minimum interval vertex deletion set of $G[C]$. Then (G, k) is a yes-instance if and only if $(G - V(C), k - |X|)$ is a yes-instance.*

We are ready to prove our main theorem statement of this section.

Theorem 1 *(⋆).* INTERVAL-OR-TREE DELETION *can be solved in $\mathcal{O}^*(10^k)$-time.*

We can also give an approximation algorithm for INTERVAL-OR-TREE DELETION.

Theorem 2 *(⋆).* INTERVAL-OR-TREE DELETION *has a 10-approximation algorithm.*

3.2   Algorithm for SPECIAL INFINITE-(Π_1, Π_2)-DELETION

Now, we show that the algorithm idea of the last section is applicable for a larger number of pairs of graph classes by identifying the properties that enabled the algorithm. We now define the variant of Π_1 OR Π_2 DELETION satisfying the following properties.

1. The vertex deletion problems for the graph classes Π_1 and Π_2 are FPT with algorithms to respective classes being $\mathcal{A}_1$ and $\mathcal{A}_2$.
2. The number of forbidden pairs of $\mathcal{F}_1$ and $\mathcal{F}_2$ is a constant.
3. All graphs in $\mathcal{F}_1$ and $\mathcal{F}_2$ are connected.
4. Let (H_1, H_2) be a closest forbidden pair in the graph G with (J_1, J_2) being the vertex subsets corresponding to the pair and P being a shortest path between J_1 and J_2. There is a subfamily $\mathcal{G}_1 \subseteq \mathsf{sp}(\mathcal{F}_1, \mathcal{F}_2)$ such that
    - $\mathcal{G}_1$ is a finite family of finite-sized (independent of the size of G) graphs and
    - in the graph G that is $\mathcal{G}_1$-free, if a forbidden set Q intersects the internal vertices of P, then $V(P) \subseteq Q$ where $V(P)$ is the vertex set of the path P.

SPECIAL INFINITE-(Π_1, Π_2)-DELETION   **Parameter:** k
Input: An undirected graph $G = (V, E)$, graph classes Π_1, Π_2 with forbidden families $\mathcal{F}_1$ and $\mathcal{F}_2$ such that conditions 1–4 are satisfied and an integer k.
Question: Is there a vertex set S of size at most k such that every connected component of $G - S$ is either in Π_1 or in Π_2?

Towards an FPT algorithm for SPECIAL INFINITE-(Π_1, Π_2)-DELETION, We define the following branching rule whose soundness is easy to see.

Branching Rule 3. *Let (G, k) be the input instance and let $Q \subseteq V(G)$ such that $G[Q]$ is isomorphic to a graph in $\mathcal{G}_1$. Then, for each $v \in V(Q)$, delete v from G and decrease k by 1. The resulting instance is $(G - v, k - 1)$.*

From here on we assume that Branching Rule 3 is not applicable for G and so G is $\mathcal{G}_1$-free.

We now focus on connected components of G which contain forbidden pairs. For $i \in \{1, 2\}$, let $\mathcal{F}_p^i$ denote the family of graphs H_i where (H_1, H_2) is a forbidden pair.

We have the following branching rule.

Branching Rule 4. *Let (J^*, T^*) be the vertex subsets of a closest forbidden pair (H_1, H_2) of $\mathcal{F}_1$ and $\mathcal{F}_2$. Then for each $v \in J^* \cup T^*$, we delete v and decrease k by 1, resulting in the instance $(G - v, k - 1)$.*

The soundness of the above branching rule comes from condition 4 where we assume that if a forbidden set Q intersects the internal vertices of a shortest path P between J^* and T^*, then $V(P) \subseteq Q$.

Lemma 9 $(\star)$. *Branching Rule 4 is sound.*

From here on, assume that (G, k) is an instance for which Reduction Rule 1, Branching Rule 3, and Branching Rule 4 are not applicable. Note that any component of G is now free of forbidden pairs. Hence it is $\mathcal{F}_p^1$-free or $\mathcal{F}_p^2$-free. The following results are now easy to see.

Lemma 10 $(\star)$. *Let C be a connected component of G that is $\mathcal{F}_p^1$-free. If $G[C]$ has no Π_1-deletion vertex set of size k, then (G, k) is a no-instance. Otherwise, let X be a minimum Π_1-deletion vertex set of $G[C]$. Then (G, k) is a yes-instance if and only if $(G - V(C), k - |X|)$ is a yes-instance.*

The proof of the following lemma is similar to that of Lemma 10.

Lemma 11. *Let C be a connected component of G that is $\mathcal{F}_p^2$-free. If $G[C]$ has no Π_2-deletion vertex set of size k, then (G, k) is a no-instance. Otherwise, let X be a minimum Π_2-deletion vertex set of $G[C]$. Then (G, k) is a yes-instance if and only if $(G - V(C), k - |X|)$ is a yes-instance.*

We are ready to prove our main theorem statement of this section. Let $f(k) = \max\{f_1(k), f_2(k)\}$ where $O^*(f_i(k))$ is the running time for the algorithm $\mathcal{A}_i$. Also let c the maximum among the size of graphs in $\mathcal{G}_1$ and $\max_{(H_1, H_2)}(|H_1| + |H_2|)$ where (H_1, H_2) is a forbidden pair.

Theorem 3 *(⋆)*. SPECIAL INFINITE-(Π_1, Π_2)-DELETION *can be solved in* $O^*(\max\{f(k), c^k\})$-*time.*

We now give an approximation algorithm for SPECIAL INFINITE-(Π_1, Π_2)-DELETION when for $i \in \{1, 2\}$, Π_i VERTEX DELETION has an approximation algorithm with approximation factor c_i.

Theorem 4 *(⋆)*. SPECIAL INFINITE-(Π_1, Π_2)-DELETION *has a d-approximation algorithm where* $d = \max(c, c_1, c_2)$.

The problem SPECIAL INFINITE-(Π_1, Π_2)-DELETION applies for a number of pairs of graphs of Π_1 OR Π_2 DELETION. We list a few below.

Corollary 1 *(⋆)*.

- PROPER INTERVAL-OR-TREE DELETION *can be solved in $O^*(7^k)$-time and has a 7-approximation algorithm.*
- CHORDAL-OR-BIPARTITE PERMUTATION DELETION *can be solved in $O^*(k^{O(k)})$-time and has a $\log^2(|OPT|)$-approximation algorithm where OPT denotes the optimal solution.*

4 Π_1 OR Π_2 DELETION When $\mathcal{F}_2$ is Infinite and P_i is Forbidden in Π_1

Here, we give an algorithm for another variant of Π_1 OR Π_2 DELETION where Π_1 has a finite forbidden set with P_i a path of length i, for a constant i, being one of them. To explain the further conditions necessary for the pair of classes to be satisfied, we define the following notion of sub-pruned family which is similar to the super-pruned family defined before.

Definition 5 (Sub-Pruned Family). *We define a family $\mathcal{F}_p$ associated with $\mathcal{F}_1 \times \mathcal{F}_2$ as follows.*

A graph H is in $\mathcal{F}_p$ if there exists a graph G such that (a) H is an induced subgraph of G and (H, G) or (G, H) is in $\mathcal{F}_1 \times \mathcal{F}_2$ and (b) there is no graph G' such that G' is an induced subgraph of H and (G', H) or (H, G') is in $\mathcal{F}_1 \times \mathcal{F}_2$.

$\mathcal{F}_p$ can be obtained from an enumeration of all pairs $(H_1, H_2) \in \mathcal{F}_1 \times \mathcal{F}_2$, and for each such pair, adding H_1 to $\mathcal{F}_p$ and removing H_2 from $\mathcal{F}_p$ (if it already exists) whenever H_1 is an induced subgraph of H_2, and adding H_2 to $\mathcal{F}_p$ and removing H_1 from $\mathcal{F}_p$ (if it already exists) if H_2 is an induced subgraph of H_1. The resulting family $\mathcal{F}_p$ is called the sub-pruned family of $\mathcal{F}_1 \times \mathcal{F}_2$ denoted by $\mathsf{SubPrune}(\mathcal{F}_1 \times \mathcal{F}_2)$.

Example. Let (Π_1, Π_2) be (Interval, Trees). Note that all graphs C_i with $i \geq 4$ are in $\mathsf{SubPrune}(\mathcal{F}_1 \times \mathcal{F}_2)$ as they occur in both $\mathcal{F}_1$ and $\mathcal{F}_2$. The remaining pairs contain triangles from $\mathcal{F}_2$. If the graph from $\mathcal{F}_1$ is a net, sun, whipping top, †-AW or ‡-AW, it contains triangle as subgraphs. Hence triangle is also added to the family $\mathsf{SubPrune}(\mathcal{F}_1 \times \mathcal{F}_2)$.

A key difference in the definitions of super-pruned family and sub-pruned family is that for the latter, we do not assume that the associated families are minimal. If for some $H \in \mathcal{F}$ for a family $\mathcal{F}$ of graphs, we have that H is an induced subgraph of some other element of $\mathcal{F}$, we call the graph H an *irrelevant* graph. In the following lemma, we prove that applying the sub-pruning operation to $\mathcal{F} \times \mathcal{F}$ helps us to remove the irrelevant graphs from $\mathcal{F}$ if $\mathcal{F}$ is not minimal.

Lemma 12 *(⋆). If $\mathcal{F}$ is a forbidden family for Π, then $\mathsf{SubPrune}(\mathcal{F} \times \mathcal{F})$ is a forbidden family for Π.*

We now define the variant of Π_1 OR Π_2 DELETION that we look at in this section. Let $\mathcal{F}' = \Pi_1 \cap \mathcal{F}_2$. We assume that

1. $\mathcal{F}_1$ is finite.
2. P_i, for some constant i, is forbidden in Π_1 where P_i is the path on i vertices.
3. there is an FPT algorithm $\mathcal{A}$ for Π_2-vertex deletion problem and
4. $\mathsf{SubPrune}(\mathcal{F}' \times \mathcal{F}')$ is known to be finite.

SPECIAL MIXED-(Π_1, Π_2)-DELETION   **Parameter:** k
Input: An undirected graph $G = (V, E)$, two graph classes Π_1 and Π_2 defined by forbidden graphs $\mathcal{F}_1$ and $\mathcal{F}_2$ respectively. Furthermore, conditions 1–4 above are satisfied.
Question: Is there $S \subseteq V(G)$ of size at most k such that every connected component of $G - S$ is either in Π_1 or in Π_2?

Before we develop the algorithm for this general version of scattered vertex deletion, we explain the algorithm for a specific example pair in the next subsection.

4.1   Clique or Planar Graphs

CLIQUE OR PLANAR VERTEX DELETION   **Parameter:** k
Input: An undirected graph $G = (V, E)$ and an integer k.
Question: Is there $S \subseteq V(G)$ of size at most k such that every connected component of $G - S$ is a clique or a planar graph?

We first show that the problem is indeed an example of SPECIAL MIXED-(Π_1, Π_2)-DELETION problem. We have $\mathcal{F}_1 = \{2K_1\}$ which is finite. Since P_3 is forbidden in cliques, condition 2 is satisfied. The condition 3 is satisfied as there is an FPT algorithm with $\mathcal{O}^*(k^{O(k)})$ running time for PLANAR VERTEX DELETION [6].

Finally we have $\mathcal{F}' = \mathcal{F}_2 \cap \Pi_1 = \{K_5, K_6, \dots, \}$ as planar graphs are K_5-free. We have $\mathsf{SubPrune}(\mathcal{F} \times \mathcal{F}') = \{K_5\}$ which being finite satisfies condition 4.

Let $\mathcal{F}'_h$ be the family of graphs that contain P_3 and K_5 as induced subgraphs. We define the family $\mathcal{F}_h = \mathsf{SubPrune}(\mathcal{F}'_h \times \mathcal{F}'_h)$.

Lemma 13. *The family $\mathcal{F}_h$ is of finite size with graphs of size at most 8.*

Proof. Let H be a graph in $\mathcal{F}_h$. Let J_1 and J_2 be vertex subsets of H such that $H[J_1]$ is isomorphic to P_3 and $H[J_2]$ is isomorphic to K_5. First, we observe that $d_H(J_1, J_2) \leq 1$. Suppose not. Then there is a path P between J_1 and J_2 of length at least 2. Let J_3 be the last three vertices of P. Note that $G[J_3]$ is isomorphic to P_3 as well. Then $H[J_2 \cup J_3] \in \mathcal{F}_h$ is also a graph that contains P_3 and K_5 as induced subgraphs. This contradicts that $H \in \mathcal{F}_h$ as it will be removed from $\mathcal{F}_h$ for the pair $(H, H[J_2 \cup J_3])$. Furthermore, note that $H[J_1 \cup J_2]$ is a connected graph which has P_3 and K_5 as induced subgraphs. Hence $H = H[J_1 \cup J_2]$ as otherwise it will be removed from $\mathcal{F}_h$ for the pair $(H, H[J_1 \cup J_2])$. This proves the lemma as $|J_1 \cup J_2| \leq 8$. □

We now describe the algorithm. The following branching rule and its soundness is easy to see.

Branching Rule 5. *Suppose the graph G of the input instance (G, k) has a graph $\hat{H} \in \mathcal{F}_h$ as its induced subgraph. Then for each $v \in V(\hat{H})$, we delete v and decrease k by 1, resulting in the instance $(G - v, k - 1)$.*

The following lemma is easy to see.

Lemma 14 ($\star$). *Let G be a graph such that Branching Rule 5 is not applicable. Let C be a component of G which contains K_5 as an induced subgraph. Then C is a clique.*

Since Reduction Rule 1 removes components of G that are cliques, we have the following corollary.

Corollary 2. *If Reduction Rule 1 and Branching Rule 5 are not applicable, then no connected component of the graph G has K_5 as induced subgraph.*

The following lemma allows us to apply the algorithm for PLANAR VERTEX DELETION in the remaining components of G to solve the problem.

Lemma 15 ($\star$). *Let (G, k) be the resulting instance after applying the reduction rules. If G has no planar vertex deletion set of size k, then (G, k) is a no-instance. Otherwise, let X be a minimum sized set such that $G - X$ is planar. Then X is also a solution (G, k).*

The main theorem now follows.

Theorem 5 ($\star$). CLIQUE OR PLANAR VERTEX DELETION *has a $\mathcal{O}^*(k^{O(k)})$ time FPT algorithm.*

4.2   Algorithm for SPECIAL MIXED-(Π_1, Π_2)-DELETION

We now give the algorithm of SPECIAL MIXED-(Π_1, Π_2)-DELETION. The ideas here can be seen as generalizations of the algorithm for CLIQUE OR PLANAR VERTEX DELETION.

Let $\mathcal{F}'_h$ denote the family of graphs H that satisfies the following.

- there exists a pair $(H_1, H_2) \in \mathcal{F}_1 \times \mathsf{SubPrune}(\mathcal{F}' \times \mathcal{F}')$ such that both H_1 and H_2 occur as induced subgraphs of H.
- Let $Q_1, Q_2 \subseteq V(H)$ such that $H[Q_1]$ is isomorphic to H_1 and $H[Q_2]$ is isomorphic to H_2. Then $d_H(Q_1, Q_2) \leq i$.

We define $\mathcal{F}_h = \mathsf{SubPrune}(\mathcal{F}'_h \times \mathcal{F}'_h)$.

We give algorithm SPECIAL MIXED-(Π_1, Π_2)-DELETION} by branching over graphs in $\mathcal{F}_h$ and later applying algorithm $\mathcal{A}$ for vertex deletion to graph class Π_2. The details are moved to the full version of the paper.

Theorem 6 $(\star)$. SPECIAL MIXED-(Π_1, Π_2)-DELETION *has an FPT algorithm with running time* $(d_1 + i + d_2)^k f(k) poly(n)$ *where* $d_1 = \max\{|F| : F \in \mathcal{F}_1\}$, $d_2 = \max\{|H| : H \in \mathcal{F}_h\}$ *and supposing algorithm* $\mathcal{A}$ *takes* $\mathcal{O}^*(f(k))$ *time.*

The problem SPECIAL MIXED-(Π_1, Π_2)-DELETION applies for a number of pairs of graphs of Π_1 OR Π_2 DELETION. We list a few below. Note that the second algorithm in the list covers a number of graph classes for Π_2. For example Π_2 can be the class of all planar graphs, or the class of all graphs with treewidth t.

Corollary 3 $(\star)$. Π_1 OR Π_2 DELETION *is FPT for the following pairs* (Π_1, Π_2) *of graph classes.*

1. Π_1 *is the class of cliques and* Π_2 *is the class of cactus graphs. The FPT algorithm has running time* $\mathcal{O}^*(26^k)$.
2. Π_1 *is the class of cliques and* Π_2 *is the class of graphs that has an FPT algorithm with* $\mathcal{O}^*(f(k))$ *running time for its deletion problem and its forbidden family* $\mathcal{F}_2$ *contains the graph* K_t *for some constant* t. *The FPT algorithm has running time* $\mathcal{O}^*((t + 1)^k f(k))$.
3. Π_1 *is the class of split graphs and* Π_2 *is the class of bipartite graphs. The FPT algorithm has running time* $\mathcal{O}^*(13^k)$.

5   Conclusion

We gave faster algorithms for some vertex deletion problems to pairs of scattered graph classes with infinite forbidden families. The existence of a polynomial kernel for all the problems studied are open. It is even open when all the scattered graph classes have finite forbidden families.

References

1. Cai, L.: Fixed-parameter tractability of graph modification problems for hereditary properties. Inf. Process. Lett. **58**(4), 171–176 (1996)
2. Cao, Y., Marx, D.: Interval deletion is fixed-parameter tractable. ACM Trans. Algorithms (TALG) **11**(3), 1–35 (2015)
3. Ganian, R., Ramanujan, M.S., Szeider, S.: Discovering archipelagos of tractability for constraint satisfaction and counting. ACM Trans. Algorithms **13**(2), 29:1-29:32 (2017)
4. Jacob, A., de Kroon, J.J., Majumdar, D., Raman, V.: Parameterized complexity of deletion to scattered graph classes. arXiv preprint arXiv:2105.04660 (2021)
5. Jacob, A., Majumdar, D., Raman, V.: Parameterized complexity of deletion to scattered graph classes. In: 15th International Symposium on Parameterized and Exact Computation (IPEC 2020). Schloss Dagstuhl-Leibniz-Zentrum für Informatik (2020)
6. Jansen, B.M., Lokshtanov, D., Saurabh, S.: A near-optimal planarization algorithm. In: Proceedings of the Twenty-Fifth Annual ACM-SIAM Symposium on Discrete Algorithms, pp. 1802–1811. SIAM (2014)
7. Lekkeikerker, C., Boland, J.: Representation of a finite graph by a set of intervals on the real line. Fundam. Math. **51**(1), 45–64 (1962)
8. Lewis, J.M., Yannakakis, M.: The node-deletion problem for hereditary properties is NP-complete. J. Comput. Syst. Sci. **20**(2), 219–230 (1980)

Fast Algorithms for the Rooted Triplet Distance Between Caterpillars

Jesper Jansson[1,2] and Wing Lik Lee[1(✉)]

[1] Department of Computing, The Hong Kong Polytechnic University,
Hung Hom, Kowloon, Hong Kong
`jesper.jansson@polyu.edu.hk`, `wing-lik.lee@connect.polyu.hk`
[2] Graduate School of Informatics, Kyoto University, Kyoto, Japan

Abstract. The *rooted triplet distance* measures the structural dissimilarity between two rooted *phylogenetic trees* (unordered trees with distinct leaf labels and no outdegree-1 nodes) having the same leaf label set. It is defined as the number of 3-subsets of the leaf label set that induce two different subtrees in the two trees. The fastest currently known algorithm for computing the rooted triplet distance was designed by Brodal *et al.* (SODA 2013). It runs in $O(n \log n)$ time, where n is the number of leaf labels in the input trees, and a long-standing open question is whether this is optimal or not. In this paper, we present two new $o(n \log n)$-time algorithms for the special case of *caterpillars* (rooted phylogenetic trees in which every node has at most one non-leaf child), thus breaking the $O(n \log n)$-time bound for a fundamental class of trees. Our first algorithm makes use of a dynamic rank-select data structure by Raman *et al.* (WADS 2001) and runs in $O(n \log n / \log \log n)$ time. Our second algorithm relies on an efficient orthogonal range counting algorithm invented by Chan and Pătraşcu (SODA 2010) and runs in $O(n\sqrt{\log n})$ time.

Keywords: Phylogenetic tree · Caterpillar · Rooted triplet distance · Dynamic rank-select data structure · Orthogonal range counting

1 Introduction

Phylogenetics is the study of evolutionary relationships between different species or groups. To describe a set of inferred evolutionary relationships, scientists commonly use a *phylogenetic tree*, which is a leaf-labeled tree where each leaf represents one entity such as a biological species. It is a diagrammatic representation of evolutionary history such that the more closely related two species are, the closer they are to each other in the tree.

Similarity measures between phylogenetic trees are frequently used in phylogenetics. For instance, analysis methods such as Bayesian inference [9] produce a set of phylogenetic trees that are the most likely to represent true evolutionary history. A consensus method is then applied to condense them into a single tree which is close to the original set based on some well-defined measure. Another

© Springer Nature Switzerland AG 2021
E. Bampis and A. Pagourtzis (Eds.): FCT 2021, LNCS 12867, pp. 327–340, 2021.
https://doi.org/10.1007/978-3-030-86593-1_23

use case of similarity measures is for evaluating new phylogenetic tree or network reconstruction methods. As explained in [15], in one evaluation method, some biomolecular sequences are evolved according to a model of evolution represented by a base tree and then the new reconstruction method generates a tree from these evolved sequences. The similarity between the base tree and the generated tree then gives an indicator of the quality of the reconstruction method.

One of the earliest similarity measures proposed that is still popular today is the Robinson–Foulds metric [19], which counts the number of *clusters* (leaf label sets of subtrees rooted at the nodes in the trees) that can only be found in one tree and not the other. It can be computed in linear time [7], but has the disadvantage that a small change in the input trees can lead to a huge change in the value. As an example, suppose the input consists of two identical caterpillars. If we were to move a single leaf from the bottom of one tree to the top, the Robinson-Foulds distance would go from zero to near its maximum possible value even though the two trees still share a lot of branching structure.

Another previously well-studied measure is the is the NNI (nearest neighbor interchange) distance [18], which counts the number of branch-swapping transformations needed to turn one input tree into the other. However, it has remained of mostly theoretical interest as it has been proved that computing the NNI distance is NP-hard [6].

The Kendall-Colijn metric [12] is computed by finding the lowest common ancestors of each leaf pair in the two input trees, and comparing the difference in distance to the root node. This measure can be computed in $O(n^2)$ time, but has the disadvantage that leaves close to the root will influence the score more than leaves further away from the root.

Finally, the rooted triplet distance [8] (defined formally in Sect. 1.1 below) counts the number of 3-subsets of the leaf label set that induce two different subtrees in the two trees. Its main advantages are that it is robust to small changes in the input [1] and that it can be computed in polynomial time (see Sect. 1.2). The rooted triplet distance has since its inception found a wide range of practical applications in phylogenetics, as well as in other fields of biology. See [13,14,16] for examples.

In this paper, we focus on fast algorithms for the rooted triplet distance.

1.1 Problem Definitions

A *rooted phylogenetic tree* (from here on called a *tree* for short) is a rooted, unordered tree where each internal node has at least two children, and the leaves are distinctly labeled by a set of leaf labels. A *caterpillar* is a tree where every node has at most one non-leaf child. To simplify the presentation below, we will identify every leaf in a tree with its unique label.

A *rooted triplet* is a tree with exactly three leaves. A *resolved triplet* is a rooted triplet where two of its leaves, say x, y, have a common parent that is not a parent of the remaining leaf z. We write $xy|z$ to denote such a resolved triplet. In contrast, an *unresolved triplet* is a rooted triplet where all three leaves have the same parent. An unresolved triplet with leaves x, y, z will be denoted by $x|y|z$. Unresolved triplets are also referred to as *fan triplets* in the literature.

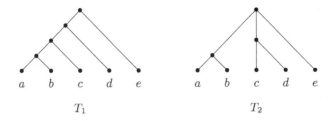

Fig. 1. An example: T_1 and T_2 are two rooted phylogenetic trees with leaf label set $\{a, b, c, d, e\}$, and T_1 is a caterpillar. Observe that, e.g., the resolved triplet $ac|d$ is consistent with T_1 but not T_2, while the unresolved triplet $a|c|e$ is consistent with T_2 but not T_1. Since $rt(T_1) = \{ab|c, ab|d, ab|e, ac|d, ac|e, ad|e, bc|d, bc|e, bd|e, cd|e\}$ and $rt(T_2) = \{ab|c, ab|d, ab|e, cd|a, a|c|e, a|d|e, cd|b, b|c|e, b|d|e, cd|e\}$, we have $d_{rt}(T_1, T_2) = 6$.

A resolved triplet $xy|z$ is said to be *consistent* with a tree T if the lowest common ancestor (LCA) of x, y in T is a proper descendant of the LCA of all three leaves in T. An unresolved triplet $x|y|z$ is said to be *consistent* with T if the LCA of all three leaves are the same in T as the LCA of any two leaves. Define $rt(T)$ to be the set of all rooted triplets that are consistent with T. If T_1, T_2 share the same set of leaf labels, define the *rooted triplet distance* between T_1 and T_2, written as $d_{rt}(T_1, T_2)$, as $\frac{1}{2}|rt(T_1) \triangle rt(T_2)|$, where $\triangle$ denotes the symmetric difference. See Fig. 1 for an example.

The rooted triplet distance problem can be stated as follows:

The Rooted Triplet Distance Problem
Input: Two rooted phylogenetic trees T_1, T_2 with the same leaf label set Λ.
Output: The rooted triplet distance $d_{rt}(T_1, T_2)$.

In the rest of the paper, we let n be the number of leaves in each input tree.

1.2 Previous Results

The rooted triplet distance was introduced in 1975 by Dobson [8]. A straightforward algorithm for computing it runs in $O(n^3)$ time. In 1996, Critchlow *et al.* [5] presented an $O(n^2)$-time algorithm for the special case of two binary trees which categorizes triplets based on their potential ancestor pairs. In 2011, Bansal *et al.* [1] gave an $O(n^2)$-time algorithm that computes the distance between two unrestricted trees by using postorder tree traversals. In 2013, Sand *et al.* [20] described an $O(n \log^2 n)$-time algorithm for binary trees using a data structure called the hierarchical decomposition tree (HDT). Brodal *et al.* [2] developed an HDT-based $O(n \log n)$-time algorithm that works for trees of arbitrary degrees. Jansson and Rajaby [11] later proposed an $O(n \log^3 n)$-time algorithm, modified from [2] by using a simpler data structure called the centroid path decomposition tree, that although slower in theory, runs faster in practice for values of n up to $4,000,000$. Subsequently, Brodal and Mampentzidis [3] designed an even more practical $O(n \log n)$-time algorithm that scales to external memory, using a

Table 1. Previous results on the rooted triplet distance problem

Year	Reference	Degree	Time complexity
1975	Dobson [8]	Arbitrary	$O(n^3)$
1996	Critchlow et al. [5]	Binary	$O(n^2)$
2011	Bansal et al. [1]	Arbitrary	$O(n^2)$
2013	Sand et al. [20]	Binary	$O(n \log^2 n)$
2013	Brodal et al. [2]	Arbitrary	$O(n \log n)$
2017	Jansson and Rajaby [11]	Arbitrary	$O(n \log^3 n)$
2017	Brodal and Mampentzidis [3]	Arbitrary	$O(n \log n)$
2019	Jansson et al. [10]	Arbitrary	$O(qn)$

further modified centroid decomposition technique. Recently, an algorithm with time complexity $O(qn)$, where at least one of the input trees has at most q internal nodes, was given by Jansson et al. [10]. See Table 1 for a summary.

1.3   New Results and Organization of Paper

We present two new algorithms for the rooted triplet distance problem restricted to caterpillars. The first algorithm is simple to implement and performs well in practice, while the second algorithm is even faster in theory, having a lower time complexity. These two algorithms are the first to achieve sub-$O(n \log n)$ time complexity for any non-trivial special cases of the rooted triplet distance problem when the number of internal nodes is unrestricted.

In Sect. 2 we give definitions, preliminary results, and summaries of data structures that will be used in the following sections. The first algorithm, presented in Sect. 3, computes the distance in $O(n \log n / \log \log n)$ time by defining a series of steps to transform one input tree to the other, and counts the number of rooted triplets that change in each step using the rank-select data structure by Raman et al. [17]. The second algorithm, presented in Sect. 4, uses the orthogonal range counting algorithm by Chan and Pǎtraşcu [4], and computes the distance in $O(n\sqrt{\log n})$ time by mapping each leaf label onto a 2-D grid and then making $O(n)$ orthogonal range counting queries on the grid. Finally, Sect. 5 summarizes our new results and lists some open problems.

2   Preliminaries

First, we describe the tree transformation steps to be used in the first algorithm. They allow us to break down the transformation of one input tree into the other into a series of steps, and compute the rooted triplet distance by adding up the changes in triplets that occur in each step.

Given two input trees on label set Λ, call one of them the *start tree* T_{start}, and the other one the *goal tree* T_{goal}. Fixing T_{goal}, define $\text{good}(T)$ to be the set of all 3-subsets of Λ that induce the same rooted triplet in T as in T_{goal}. Define $\text{bad}(T)$ to be the set of all 3-subsets of Λ that induce different triplets in T and T_{goal}. Then, given trees T, T', define $\Phi(T, T') = |\text{bad}(T) \cap \text{good}(T')| - |\text{good}(T) \cap \text{bad}(T')|$.

We may view $\Phi(T, T')$ as the change in d_{rt} with respect to T_{goal} as we transform T into T'. Φ counts the number of triplets that are turned from bad to good, and subtracts the number of triplets turning from good to bad. Rewriting $d_{rt}(T, T_{goal}) = |\text{bad}(T)|$ and $d_{rt}(T', T_{goal}) = |\text{bad}(T')|$, we see that $d_{rt}(T, T_{goal}) = \Phi(T, T') + d_{rt}(T', T_{goal})$.

If we can find a sequence of trees and the Φ-values between any two adjacent trees, then we can compute the triplet distance between any two trees in the sequence by summing up these Φ-values. This gives us a method for computing $d_{rt}(T_{start}, T_{goal})$:

Lemma 1. *Let $T_1 = T_{start}, T_2, \ldots, T_{k-1}, T_k = T_{goal}$ be a sequence of trees. Then $d_{rt}(T_{start}, T_{goal}) = \sum_{i=1}^{k-1} \Phi(T_i, T_{i+1})$.*

Next, we demonstrate that we can ignore most leaves of a tree if the changes in a transformation step are contained in some small subsets:

Lemma 2. *Given trees T_a, T_b, $u \in T_a$, and $v \in T_b$, let T_a', T_b' be subtrees of T_a, T_b rooted at u, v respectively. Obtain T_a'', T_b'' by replacing T_a', T_b' each by a single node. Then $\Phi(T_a, T_b) = \Phi(T_a', T_b')$ if T_a'' and T_b'' are isomorphic.*

The lemma holds because the condition implies that only those rooted triplets whose three leaves all lie inside T_a' and T_b' will affect $\Phi(T_a, T_b)$.

Our first algorithm makes use of a data structure based on the dynamic rank-select data structure by Raman *et al.* [17]. A query tree Q stores a multiset of integers in $[1..n]$, and supports the following operations:

- *insert(a)*: insert a value a to Q.
- *query(a)*: output the number of inserted values less than a.

The *insert, query* operations are respectively direct analogs to the *update, sum* operations in [17]. The next lemma summarizes its time complexity.

Lemma 3. *There is a data structure that supports the insert and query operations in $O(\log n / \log \log n)$ time.*

Finally, in our second algorithm, we will use an orthogonal range counting result given by Chan and Pătrașcu [4]. We restate Corollary 2.3 from [4] here:

Lemma 4. *Given n points and n axis-aligned rectangles on the grid, we can obtain the number of points inside each rectangle in $O(n\sqrt{\log n})$ total time.*

3   The First Algorithm

The main idea behind this algorithm is to treat the input caterpillars as lists of leaves, and transform one list into the other by performing insertion sort on the list while moving one group of leaves at a time. See Algorithm 1 for the pseudocode. We first describe the sequence of intermediate caterpillars used in the transformation from T_{start} to T_{goal}. We then show how $d_{rt}(T_{start}, T_{goal})$ can be computed by performing $O(n)$ insertions and queries to the rank-select query tree, which by Lemma 3 yields a total time complexity of $O(n \log n / \log \log n)$.

Algorithm 1: Rank-Select Method

Input: Caterpillars T_{start}, T_{goal} on label set Λ.
Output: $d_{rt}(T_{start}, T_{goal})$
1 Relabel T_{start}, T_{goal};
2 Build query tree Q according to input size n;
3 Parse T_{start} to obtain leaf groups $G_1, \ldots, G_m$;
4 **foreach** G_i **do**
5 $\quad$ Perform insertions and queries to Q to get values $|A_i|, |B_i|, \ldots$;
6 $\quad$ Compute $\Phi(T_{i-1}, T_i)$ using values $|A_i|, \ldots$;
7 **end**
8 Compute and output $\sum \Phi(T_{i-1}, T_i)$;

Define a mapping $\Lambda \to \mathbb{Z}$ such that the labels of each leaf in T_{goal} is mapped to the distance from the leave to the internal node that is the farthest from the root. Then, apply this mapping to both T_{goal} and T_{start}. See Fig. 2 for an example. Note that the leaves will no longer be distinctly labeled and that all leaves in T_{goal} having the same parent will receive the same label.

Define a *leaf group* of a tree T as a maximal multiset of identical leaf labels in which each element corresponds to a distinct leaf in T and the leaves corresponding to its elements all have the same parent in T. In Fig. 2, the multisets $\{5\}, \{2,2\}, \{1,1\}$ are leaf groups in T_{start}, while $\{2\}$ and $\{4,4\}$ are not. Two leaf groups G_1, G_2 are said to be *connected* if their leaves are siblings in T.

Let U, V be multisets of Λ. Define $U \prec V$ if for all $u \in U, v \in V$, $u < v$ by their value. Write $U \simeq V$ if $u = v$, and $U \preceq V$ if $u \leq v$, for all $u \in U, v \in V$.

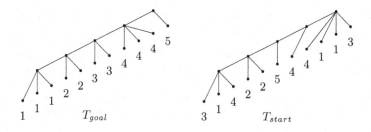

Fig. 2. In this example, after relabeling the leaves according to T_{goal}, the leaf groups in T_{start} are $\{3\}, \{1\}, \{4\}, \{2,2\}, \{5\}, \{4\}, \{4\}, \{1,1\}, \{3\}$.

3.1 Algorithm Description

To apply the insertion sort strategy, represent trees as lists of leaf groups. Given any T, define the *ordering* of T to be a list of leaf groups in T, subject to the following additional rules:

- The ordering of the leaf groups follows a post order traversal.
- A set of connected leaf groups appear in the list in ascending order, so that if $G_1, G_2, \ldots, G_k$ are connected groups listed in this order, then $G_1 \preceq G_2 \preceq \cdots \preceq G_k$.

In Fig. 2, the ordering of T_{goal} is $(\{1,1,1\}, \{2,2\}, \{3,3\}, \{4,4,4\}, \{5\})$, and the ordering of T_{start} is $(\{1\}, \{3\}, \{4\}, \{2,2\}, \{5\}, \{4\}, \{1,1\}, \{3\}, \{4\})$.

Suppose T_{start} has m leaf groups, so that its ordering is $(G_1, G_2, \ldots, G_m)$. We define a sequence of trees $T_1 = T_{start}, T_2, \ldots, T_m = T_{goal}$, where each T_i is the tree with ordering $(G_{\sigma_i(1)}, \ldots, G_{\sigma_i(i)}, G_{i+1}, \ldots, G_m)$, σ_i being a permutation of $\{1, \ldots, i\}$, and:

- $G_{\sigma_i(1)} \preceq G_{\sigma_i(2)} \preceq \cdots \preceq G_{\sigma_i(i)}$.
- Leaf groups $G_{\sigma_i(i)}$ and G_{i+1} are not connected.
- Leaf groups in $\{G_{i+1}, \ldots, G_m\}$ are connected if and only if they are also connected in T_{start}.
- Adjacent leaf groups $G_{\sigma_i(j)}, G_{\sigma_i(j+1)}$ are connected if and only if $G_{\sigma_i(j)} \simeq G_{\sigma_i(j+1)}$.

By Lemma 1, $d_{rt}(T_{start}, T_{goal}) = \sum_{i=1}^{m-1} \Phi(T_i, T_{i+1})$.

Consider leaf group G_i in T_i. G_i may be connected with some $G_{i+1}, \ldots, G_{i+j}$. By Lemma 2, to compute $\Phi(T_i, T_{i+1})$ it suffices to consider the subtree of T_i containing $G_1, \ldots, G_{i+j}$. Separate the leaves in this subtree minus G_i into four (possibly empty) subsets, A_i, B_i, C_i, D_i, so that $G_{i+1}, \ldots, G_{i+j} \subset D_i$, $B_i \simeq G_i$, and $A_i \prec G_i \prec C_i$.

To transform T_i into T_{i+1}, we may think of detaching G_i and attaching it to the parent of B_i, or, if $B_i = \varnothing$, attaching it to a new internal node at the appropriate position. Figure 3 illustrates this process.

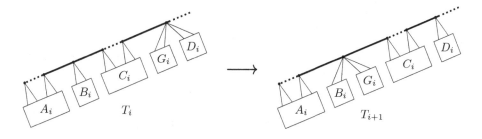

Fig. 3. Moving G_i in step i

Table 2. Listing for each type, the induced triplets in T_i, T_{i+1}, its effects on Φ, and its counts. Lowercase letters represent leaves in the corresponding subsets, so that $a_i \in A_i, b_i \in B_i, \ldots$, and repeated labels represent distinct leaves. Here, E_i, F_i are defined as sets of leaves $\{c_i, d_i\}$ satisfying the given constraints.

Type	T_i	T_{i+1}	Effect on Φ	Count
$\{g_i, a_i, a_i\}$	$a_i a_i \lvert g_i$	$a_i a_i \lvert g_i$	None	–
$\{g_i, a_i, b_i\}$	$a_i b_i \lvert g_i$	$a_i \lvert b_i \lvert g_i$	Increased	$\lvert G_i \rvert \lvert A_i \rvert \lvert B_i \rvert$
$\{g_i, a_i, c_i\}$	$a_i c_i \lvert g_i$	$a_i g_i \lvert c_i$	Increased	$\lvert G_i \rvert \lvert A_i \rvert \lvert C_i \rvert$
$\{g_i, a_i, d_i\}$	$a_i \lvert g_i \lvert d_i$	$a_i g_i \lvert d_i$	Increased	$\lvert G_i \rvert \lvert A_i \rvert \lvert D_i \rvert$
$\{g_i, b_i, b_i\}$	$b_i b_i \lvert g_i$	$b_i \lvert b_i \lvert g_i$	Increased	$\lvert G_i \rvert \binom{\lvert B_i \rvert}{2}$
$\{g_i, b_i, c_i\}$	$b_i c_i \lvert g_i$	$b_i g_i \lvert c_i$	Increased	$\lvert G_i \rvert \lvert B_i \rvert \lvert C_i \rvert$
$\{g_i, b_i, d_i\}$	$b_i \lvert g_i \lvert d_i$	$b_i g_i \lvert d_i$	Increased	$\lvert G_i \rvert \lvert B_i \rvert \lvert D_i \rvert$
$\{g_i, c_i, c_i\}$	$c_i c_i \lvert g_i$	$g_i c_i \lvert c_i$ $g_i \lvert c_i \lvert c_i$	Increased	$\lvert G_i \rvert \binom{\lvert C_i \rvert}{2}$
$\{g_i, c_i, d_i\}$	$c_i \lvert g_i \lvert d_i$	$g_i c_i \lvert d_i$	Decreased $(c_i = d_i)$	$\lvert G_i \rvert \lvert E_i \rvert$
			None $(c_i > d_i)$	–
			Increased $(c_i < d_i)$	$\lvert G_i \rvert \lvert F_i \rvert$
$\{g_i, d_i, d_i\}$	$g_i \lvert d_i \lvert d_i$	$g_i \lvert d_i \lvert d_i$	None	–
$\{g_i, g_i, a_i\}$	$a_i \lvert g_i \lvert g_i$	$a_i \lvert g_i \lvert g_i$	None	–
$\{g_i, g_i, b_i\}$	$b_i \lvert g_i \lvert g_i$	$b_i \lvert g_i \lvert g_i$	None	–
$\{g_i, g_i, c_i\}$	$c_i \lvert g_i \lvert g_i$	$g_i g_i \lvert c_i$	Increased	$\binom{\lvert G_i \rvert}{2} \lvert C_i \rvert$
$\{g_i, g_i, d_i\}$	$g_i \lvert g_i \lvert d_i$	$g_i g_i \lvert d_i$	Increased	$\binom{\lvert G_i \rvert}{2} \lvert D_i \rvert$

3.2 Computing Φ

We now show how each Φ can be computed by making $O(1)$ queries to the query tree Q. Fill the query tree following the ordering of T_{start}, where for each leaf group G of size k and value a, we perform $insert(a)$ k times. By making $O(1)$ queries at different states of Q, we can get the number of leaves within a range of values in any continuous range of leaf groups.

By Lemma 2, we only need to consider triplets where each of its leaves are in one of the subsets A_i, B_i, C_i, G_i, D_i. Categorize these triplets based on where each of its leaves are located. We can immediately disregard most triplet types: any triplet types not containing at least one leaf from G_i are unchanged, so are any triplet types where all three of its leaves are in the same subset. For the remaining types, we list their effect on Φ and counts in Table 2.

The values $\lvert A_i \rvert, \lvert B_i \rvert, \lvert C_i \rvert$ can be found by making $O(1)$ queries to Q for the number of leaves in $G_1, \ldots, G_{i-1}$ that are less than, equal to, or greater than g_i, respectively. $\lvert G_i \rvert$ can be directly read from each leaf group.

The values $|D|, |E|, |F|$ can be found using a similar approach. Let $\{G_i, \dots, G_{i+j}\}$ be a maximal set of connected leaf groups. $|D|$-values can be computed using formulas $|D_i| = \sum_{k=1}^{j} |G_{i+k}|$ and $|D_{i+k}| = |D_{i+k-1}| - |G_{i+k}|$. For the $|E|$-values, first make $O(j)$ queries to Q to obtain the number of all possible $\{c, d\}$ leaf pairs, where $c = d, c \in \{G_1, \dots, G_{i-1}\}$, and $d \in \{G_i, \dots, G_{i+j}\}$. Subtract $|B_i||G_i|$ from this total to get $|E_i|$, and then subtract $|B_{i+j+1}||G_{i+j+1}|$ from each $|E_{i+j}|$ to obtain $|E_{i+j+1}|$. Modifying the query ranges, the same method can be used to find the $|F|$-values. The precomputing steps require a number of queries proportional to the number of leaf groups in the maximal set, therefore adds $O(1)$ amortized number of queries per leaf group.

For the time complexity of the algorithm, preprocessing involves building a label map, applying it to T_{start}, and retrieving the list of leaf groups. Each one of these tasks can be done in $O(n)$ time. Then, $O(n)$ insertion and query operations are needed to compute all Φ-values. By Lemma 3, this proves:

Theorem 1. *Given two caterpillars on the same leaf label set of size n, Algorithm 1 computes the rooted triplet distance between them in $O(n \log n / \log \log n)$ time.*

4   The Second Algorithm

Our second method maps each leaf label onto a 2-D integer grid of size $n \times n$, according to their positions in the two input caterpillars. This is done so that by making $O(n)$ queries, we retrieve the total number of *good triplets*, triplets that are consistent with both input trees, and thus the rooted triplet distance. See Algorithm 2 for the pseudocode.

Algorithm 2: Orthogonal Range Counting Method

Input: Caterpillars T_1, T_2 on leaf label set Λ.
Output: $d_{rt}(T_1, T_2)$.
1 **foreach** $\ell \in \Lambda$ **do**
2 | Compute and store the point $f(\ell)$;
3 **end**
4 **foreach** *point* **p** *in* Im f **do**
5 | Perform range counting queries to get the values A, B, C, D;
6 | Compute the number of good triplets rooted at **p**;
7 **end**
8 Compute and output $d_{rt}(T_1, T_2)$ using Lemma 5;

4.1   Mapping Leaves to the Grid

First, we define the mapping of leaves into the grid. Index the internal nodes of each input caterpillar T_1, T_2 in ascending order from the bottom to top, so that the lowest internal node is labeled 1, its parent is labeled 2, etc.

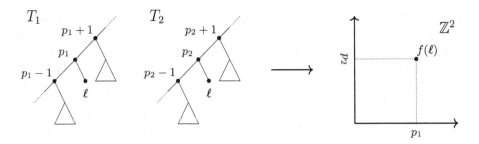

Fig. 4. f maps each leaf to a point in $\mathbb{Z}^2$ according to its positions in T_1 and T_2.

Suppose T_1 has h_1 internal nodes, and T_2 has h_2 internal nodes. Let $S = [1, h_1] \times [1, h_2]$ be a subset of $\mathbb{Z}^2$, and define a mapping $f : \Lambda \to S$, which maps each leaf label to a point in the grid according to the indices of its parent nodes in the two input trees. If for leaf ℓ, its parents in T_1, T_2 are indexed p_1, p_2 respectively, then ℓ is mapped to the point (p_1, p_2). See Fig. 4 for an illustration.

Next, define mapping f' from the set of good triplets to S as follows. For each good triplet τ, if the root node of the induced subtree in T_1, T_2 is indexed p_1, p_2 respectively, then $f'(\tau) = (p_1, p_2)$. Summing up the number of good triplets mapped to each point in $\mathrm{Im}\, f'$, we get the total number of good triplets.

4.2 Counting Good Triplets

Consider any point $\mathbf{p} = (p_1, p_2)$. Write $A_{\mathbf{p}}, B_{\mathbf{p}}, C_{\mathbf{p}}, D_{\mathbf{p}}$ for the number of leaves mapped to the regions $\{p_1\} \times \{p_2\}, \{p_1\} \times [1, p_2-1], [1, p_1-1] \times \{p_2\}, [1, p_1-1] \times [1, p_2-1]$ respectively. See Fig. 5. These values will be used to count the number of good triplets mapped to $\mathbf{p}$. For clarity of presentation, the $\mathbf{p}$-subscripts will be omitted below.

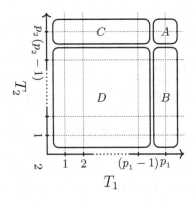

Fig. 5. Divide $[1, p_1] \times [1, p_2]$ into four regions, and define A, B, C, D as the number of leaves mapped to each region.

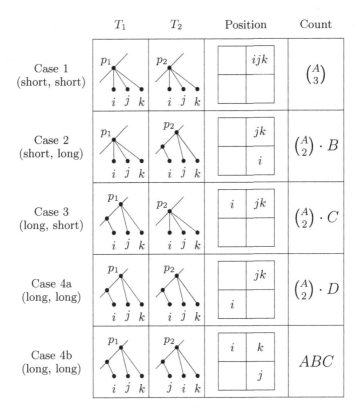

Fig. 6. The cases for counting unresolved triplets rooted at point (p_1, p_2). The Position column shows the partition of $[1, p_1] \times [1, p_2]$ mirroring Fig. 5, so that the top right quadrant corresponds to A, etc.

For every resolved triplet, one of its leaves must be a child of node p_1 in T_1, and p_2 in T_2, therefore it is mapped to the area A. The other two leaves are more related to each other than to the first leaf, which happens only if both leaves are descendants of an internal node in T_1 with index less than p_1, and similarly in T_2. This means these leaves are mapped to the area D. To count the number of good resolved triplets mapped to $\mathbf{p}$, we choose one leaf mapped to A, then choose two leaves mapped to D. Therefore the number of such triplets is $A \cdot \binom{D}{2}$.

For unresolved triplets, we proceed as follows. An unresolved triplet may either be *short*, where all three leaves share the same parent node, or it may be *long*, where only two leaves share a parent node, and the third leaf is located lower in the tree. Counting these triplets splits into four main cases. See Fig. 6.

- Case 1: The triplet is short in both T_1, T_2. This means that each leaf in such a triplet must share the same parents, namely, nodes p_1, p_2 in T_1, T_2 respectively. Therefore, these leaves are all mapped to the point $\mathbf{p}$. The number of such triplets is thus $\binom{A}{3}$.

- Case 2: The triplet is short in T_1, and long in T_2. The number of such triplets is $\binom{A}{2} \cdot B$.
- Case 3: The triplet is long in T_1, and short in T_2. The number of such triplets is $\binom{A}{2} \cdot C$.
- Case 4: The triplet is long in both T_1, T_2. In this triplet, the leaf that is in the lower position of T_1, T_2 may or may not be the same leaf. Thus, two subcases arise:
  - Case 4a: They are the same leaf. The number of such triplets is $\binom{A}{2} \cdot D$.
  - Case 4b: They are different leaves. The number of such triplets is ABC.

Repeating the above for each point to calculate the total number of good triplets subsequently gives us the rooted triplet distance:

Lemma 5.

$$d_{rt}(T_1, T_2) = \binom{n}{3} - \sum_{\mathbf{p} \in \operatorname{Im} f'} \left(A \cdot \binom{D}{2} + \binom{A}{3} + \binom{A}{2}(B + C + D) + ABC \right).$$

The mapping f can be built in $O(n)$ time. Then, for each point $\mathbf{p} \in \operatorname{Im} f'$, we make four range counting queries to find the values A, B, C, D. Afterwards, apply Lemma 5 to obtain $d_{rt}(T_1, T_2)$.

If $f'(\tau) = \mathbf{p}$, then there is at least one leaf $\ell \in \tau$ where $f(\ell) = \mathbf{p}$ also. Therefore, we have $\operatorname{Im} f' \subseteq \operatorname{Im} f$. Since $|\operatorname{Im} f| \leq n$, applying Lemma 4 yields:

Theorem 2. *Given two caterpillars on the same leaf label set of size n, Algorithm 2 computes the rooted triplet distance between them in $O(n\sqrt{\log n})$ time.*

5 Conclusion

The only known lower bound on the time complexity of computing the rooted triplet distance is the trivial one of $\Omega(n)$, which holds because any algorithm has to look at all of its input at least once. Thus, there is a gap between the known upper and lower bounds, and to close this gap is a major open problem. In this paper, we have presented two algorithms that go below the $O(n \log n)$-time upper bound for a certain special class of inputs, namely caterpillars. Although this doesn't solve the open problem, it makes some partial progress. Whether or not the techniques developed here can be extended to more general (non-caterpillar) inputs remains to be seen, but we believe our findings open up an interesting new research direction and that they show there is hope for an $o(n \log n)$-time algorithm for the general case.

Initial experiments on the first algorithm show promising practical performance. A C++ implementation of the algorithm, running on a computer with AMD Ryzen 7 2700X, 16 GB RAM, Arch Linux with kernel version 5.10.16, and g++ compiler version 10.2.0, was able to process inputs of size $n = 1,000,000$ in 3.5 s, using 69 MB of memory. The outcome of the experimental results will be reported in the full version of this paper.

We conclude with some open questions:

- Can the two algorithms be extended to work on a larger class of input trees?
- Our first algorithm uses a dynamic rank-select data structure that provides additional operations such as *delete* and *select* that are not needed by our algorithm. Is it possible to design a simpler and faster data structure that still fits our purposes?
- Is a practical implementation of the second algorithm possible? We note that certain steps such as the mapping and query steps can be parallelized easily.

Acknowledgment. JJ was partially funded by RGC/GRF project 15217019.

References

1. Bansal, M.S., Dong, J., Fernández-Baca, D.: Comparing and aggregating partially resolved trees. Theoret. Comput. Sci. **412**(48), 6634–6652 (2011)
2. Brodal, G.S., Fagerberg, R., Mailund, T., Pedersen, C.N.S., Sand, A.: Efficient algorithms for computing the triplet and quartet distance between trees of arbitrary degree. In: Proceedings of the Twenty-Fourth Annual ACM-SIAM Symposium on Discrete Algorithms (SODA 2013), pp. 1814–1832. SIAM (2013)
3. Brodal, G.S., Mampentzidis, K.: Cache oblivious algorithms for computing the triplet distance between trees. In: 25th Annual European Symposium on Algorithms (ESA 2017). Schloss Dagstuhl-Leibniz-Zentrum fuer Informatik (2017)
4. Chan, T.M., Pǎtraşcu, M.: Counting inversions, offline orthogonal range counting, and related problems. In: Proceedings of the Twenty-First Annual ACM-SIAM Symposium on Discrete Algorithms (SODA 2010), pp. 161–173. SIAM (2010)
5. Critchlow, D.E., Pearl, D.K., Qian, C.: The triples distance for rooted bifurcating phylogenetic trees. Syst. Biol. **45**(3), 323–334 (1996)
6. Dasgupta, B., He, X., Jiang, T., Li, M., Tromp, J.: On computing the nearest neighbor interchange distance. In: Proceedings of the DIMACS Workshop on Discrete Problems with Medical Applications, DIMACS Series in Discrete Mathematics and Theoretical Computer Science, vol. 55, pp. 125–143. American Mathematical Soc. (2000)
7. Day, W.H.E.: Optimal algorithms for comparing trees with labeled leaves. J. Classif. **2**(1), 7–28 (1985)
8. Dobson, A.J.: Comparing the shapes of trees. In: Street, A.P., Wallis, W.D. (eds.) Combinatorial Mathematics III. LNM, vol. 452, pp. 95–100. Springer, Heidelberg (1975). https://doi.org/10.1007/BFb0069548
9. Huelsenbeck, J.P., Nielsen, R., Ronquist, F., Bollback, J.P.: Bayesian inference of phylogeny and its impact on evolutionary biology. Science **294**(5550), 2310–2314 (2001)
10. Jansson, J., Mampentzidis, K., T.P, S.: Building a small and informative phylogenetic supertree. In: 19th International Workshop on Algorithms in Bioinformatics (WABI 2019). Schloss Dagstuhl-Leibniz-Zentrum fuer Informatik (2019)
11. Jansson, J., Rajaby, R.: A more practical algorithm for the rooted triplet distance. J. Comput. Biol. **24**(2), 106–126 (2017)
12. Kendall, M., Colijn, C.: Mapping phylogenetic trees to reveal distinct patterns of evolution. Mol. Biol. Evol. **33**(10), 2735–2743 (2016)

13. Liao, W., et al.: Alignment-free transcriptomic and metatranscriptomic comparison using sequencing signatures with variable length Markov chains. Sci. Rep. **6**(1), 1–15 (2016)
14. Moreno-Dominguez, D., Anwander, A., Knösche, T.R.: A hierarchical method for whole-brain connectivity-based parcellation. Hum. Brain Mapp. **35**(10), 5000–5025 (2014)
15. Nakhleh, L., Sun, J., Warnow, T., Linder, C.R., Moret, B.M.E., Tholse, A.: Towards the development of computational tools for evaluating phylogenetic network reconstruction methods. In: Biocomputing 2003, pp. 315–326. World Scientific (2002)
16. Page, R.D.M., Cruickshank, R., Johnson, K.P.: Louse (Insecta: Phthiraptera) mitochondrial 12S rRNA secondary structure is highly variable. Insect Mol. Biol. **11**(4), 361–369 (2002)
17. Raman, R., Raman, V., Rao, S.S.: Succinct dynamic data structures. In: Dehne, F., Sack, J.-R., Tamassia, R. (eds.) WADS 2001. LNCS, vol. 2125, pp. 426–437. Springer, Heidelberg (2001). https://doi.org/10.1007/3-540-44634-6_39
18. Robinson, D.F.: Comparison of labeled trees with valency three. J. Combin. Theory Ser. B **11**(2), 105–119 (1971)
19. Robinson, D.F., Foulds, L.R.: Comparison of phylogenetic trees. Math. Biosci. **53**(1–2), 131–147 (1981)
20. Sand, A., Brodal, G.S., Fagerberg, R., Pedersen, C.N.S., Mailund, T.: A practical $O(n \log^2 n)$ time algorithm for computing the triplet distance on binary trees. BMC Bioinform. **14**, S18 (2013). https://doi.org/10.1186/1471-2105-14-S2-S18

Deciding Top-Down Determinism of Regular Tree Languages

Peter Leupold$^{(\boxtimes)}$ and Sebastian Maneth

Faculty of Informatics, Universität Bremen, Bremen, Germany
{leupold,maneth}@uni-bremen.de

Abstract. It is well known that for a regular tree language it is decidable whether or not it can be recognized by a deterministic top-down tree automaton (DTA). However, the computational complexity of this problem has not been studied. We show that for a given deterministic bottom-up tree automaton it can be decided in quadratic time whether or not its language can be recognized by a DTA. Since there are finite tree languages that cannot be recognized by DTAs, we also consider finite unions of DTAs and show that also here, definability within deterministic bottom-up tree automata is decidable in quadratic time.

Keywords: Deterministic top-down tree automata · Definability · Decision problems

1 Introduction

Unlike for strings, where left-to-right and right-to-left deterministic automata recognize the same class of languages, this is not the case for deterministic tree automata: deterministic top-down tree automata (DTA) only recognize a strict subset of the regular tree languages. The most notorious example of a tree language that cannot be recognized by DTA is the language $\{f(a,b), f(b,a)\}$. Nevertheless, DTA bear some advantages over their bottom-up counterpart: they can be implemented more efficiently, because a tree is typically represented top-down and identified by its root node (also, a DTA may reject a given tree earlier than a bottom-up tree automaton). For (deterministic) tree transducers it has recently be shown that given a bottom-up one (with slight restrictions) one can decide whether or not there exists an equivalent top-down one [10].

Several properties have been defined that characterize DTA within the regular tree languages. Viragh [18] proves that the regular, "path-closed" tree languages are exactly the ones that are recognized by DTA. He proves this via the construction of what he calls the powerset automaton for the path-closure of a regular language. Gécseg and Steinby use a very similar method in their textbook [5]. Another approach is Nivat and Podelksi's homogeneous closure [16]. Also here the constructed tree automaton has as state set the powerset of the original state set. In neither case an exact running time has been investigated.

© Springer Nature Switzerland AG 2021
E. Bampis and A. Pagourtzis (Eds.): FCT 2021, LNCS 12867, pp. 341–353, 2021.
https://doi.org/10.1007/978-3-030-86593-1_24

In our approach, starting from a given deterministic bottom-up tree automaton, we first construct an equivalent minimal automaton. This takes quadratic time, following well known methods. We then lift the "subtree exchange property" of Nivat and Podelski [16] to such an automaton; essentially it means, that if certain transitions are present, e.g., $f(q_1, q_2) \rightarrow q$ and $f(q_2, q_1) \rightarrow q$, then also other transitions *must* be present (here, also $f(q_i, q_i) \rightarrow q$ for $i = 1, 2$). This property characterizes the DTA languages and can be decided in linear time. Finally, if the decision procedure is affirmative, we show how to construct an equivalent deterministic top-down tree automaton. The construction replaces so called "conflux groups" (e.g., the four transitions from above), one at a time by introducing new states. Care has to be taken, because the removal of one conflux group may introduce new copies of other conflux groups. However, after all original conflux groups are eliminated, the removal of newly introduced conflux groups does not cause new conflux groups to be introduced. We then generalize our results to *finite unions of deterministic top-down tree languages*. We show that they are characterized by minimal bottom-up tree automata where a finite number of "violations" to the above exchange property are present. This finiteness test can be achieved in linear time.

For unranked trees several classes of deterministic top-down tree languages have been considered. For all of them, the decision whether a given unranked regular tree language belongs to one of these classes takes exponential time [6, 11,13]. This is in sharp contrast to our results. The reason is that the unranked automata use regular expressions in their rules, and that inclusion needs to be tested for these expressions.

2 Preliminaries

Trees. For a *ranked alphabet* Σ we denote by Σ_k the set of all symbols which have rank k. Let $X = \{x_1, \dots\}$ be a set of constants called *variables*; for an integer n we denote by X_n the set $\{x_1, \dots, x_n\}$ of n variables. The set $T(\Sigma, X)$ of *trees over the ranked alphabet* Σ and the set X of variables is the smallest set defined by:

- $\Sigma_0 \subseteq T(\Sigma, X)$,
- $X \subseteq T(\Sigma, X)$, and
- if $k \geq 1$, $f \in \Sigma_k$ and $t_1, \dots, t_k \in T(\Sigma, X)$, then $f(t_1, \dots, t_k) \in T(\Sigma, X)$.

We denote by $T(\Sigma)$ the set of trees in $T(\Sigma, X)$ which do not contain variables. For a tree $t = f(t_1, \dots, t_k) \in T(\Sigma, X)$ we define its *set of nodes* as

$$N(t) := \{\epsilon\} \cup \{iu \mid i \in \{1, \dots, k\}, u \in N(t_i)\}.$$

Here ϵ denotes the root node. Let $t \in T(\Sigma, X_n)$ and $t_1, \dots, t_n \in T(\Sigma, X)$. Then $t[x_1 \leftarrow t_1, \dots, x_n \leftarrow t_n]$ denotes the tree obtained from t by replacing each occurrence of x_i by t_i.

Tree Automata and Transducers. A *(bottom-up) tree automaton* (BA) is a tuple $A = (Q, \Sigma, Q_f, \delta)$ where Q is a finite set of states, $Q_f \subseteq Q$ is a set of final states, and δ is a set of transition rules of the following form:

$$f(q_1, \ldots, q_k) \to q,$$

where $k \geq 0$, $f \in \Sigma_k$, and q, q_1, $\ldots$, $q_k \in Q$.

A tree automaton is *deterministic* (DBA) if there are no two rules with the same left-hand side. By $A(t)$ we denote the unique state that is reached in a deterministic bottom-up tree automaton by processing the tree t. For a bottom-up tree automaton A, by A_q we denote the same automaton just with q as the single final state, that is $Q_f = \{q\}$.

A *top-down tree automaton* (TA) is a tuple $A = (Q, \Sigma, I, \delta)$ where Q is a set of states, $I \subseteq Q$ is a set of initial states, and δ is a set of transition rules of the following form:

$$q(f) \to f(q_1, \ldots, q_k),$$

where $k \geq 0$, $f \in \Sigma_k$, and q, q_1, $\ldots$, $q_k \in Q$. A top-down tree automaton (Q, Σ, I, δ) is *deterministic* (DTA) if there is one initial state and there are no two rules with the same left-hand side.

A run of a BA on a tree t is a mapping $\beta : N(t) \to Q$ which fulfills the following properties: for all nodes $u \in N(t)$, if u of rank k has label f, $\beta(u) = q$ and for all $i \in \{1, \ldots, k\}$ we have $\beta(ui) = q_i$, then $f(q_1, \ldots, q_k) \to q$ is a transition in δ. We denote the transition that corresponds to $\beta(u)$ by $\tau(\beta(u))$. Sometimes we will view β as a tree and refer to nodes $\beta(u)$; here we mean a relabeling of t where every node u is labeled by $\beta(u)$.

For a bottom-up tree automaton A the run β recognizes the tree t if $\beta(\epsilon) \in Q_f$. A tree is recognized by A if there exists an accepting run for it. The language recognized by the automaton denoted by $L(A)$ is the set of all trees which are recognized. A tree language is *regular*, if it is recognized by some bottom-up tree automaton.

For an BA A its *corresponding* TA $c(A)$ is obtained by reading A's transitions from right to left and taking A's final states as initial states. In the same way for a TA its *corresponding* BA is defined. The language of a TA B is defined as $L(c(B))$.

Syntactic Congruence. A tree $C \in T(\Sigma, X_1)$ is called a *context*, if it contains exactly one occurrence of the variable x_1. Because there is only one fixed variable, we write $C[t]$ instead of $C[x_1 \leftarrow t]$. We denote by $\mathcal{C}(\Sigma)$ the set of all contexts. For a given tree language L we define the syntactic congruence $\equiv_L$ on $T(\Sigma)$ by: $s \equiv_L t$ if for all contexts $C \in \mathcal{C}(\Sigma)$ we have $C[s] \in L$ iff $C[t] \in L$.

In the case of string languages the Myhill-Nerode-Theorem states that a language is regular if and only if its syntactic congruence is of finite index [14,15]. An analogous result exists for tree languages and was long regarded as folklore; Kozen explains its history and provides a rigorous proof [8].

A concept closely related to the syntactic congruence is the *minimal deterministic bottom-up automaton* (MDBA). It is defined as follows: Let Q be the finite set of equivalence classes of $\equiv_L$ for a language L minus the unique equivalence class $C_\perp$ of all trees t for which there does not exist any context C such that $C[t] \in L$. We denote by $[t]$ the equivalence class of a tree t and define the transition function δ by: $\delta(f([t_1], \ldots, [t_k])) = [f(t_1, \ldots, t_k)]$ for all $t_1, \ldots, t_k \in T(\Sigma) \backslash C_\perp$ and $[f(t_1, \ldots, t_k)] \neq C_\perp$. With $Q_f = \{[u] \mid u \in L\}$ the DBA $M_L := (Q, \Sigma, Q_f, \delta)$ recognizes the tree language L. So the states of the MDBA for a language correspond to the equivalence classes of the syntactic congruence [3].

Proposition 1. *Let $L \subseteq T(\Sigma)$ and $M = (Q, \Sigma, Q_f, \delta)$ be the corresponding MDBA. Then the following properties hold.*

(i) For all $q \in Q$ the language $L(M_q)$ is not empty,
(ii) every transition in δ is useful, i.e., it is used in some accepting run,
(iii) for all $t \in T(\Sigma)$ we have $|\{q \in Q \mid t \in L(M_q)\}| \leq 1$.

(i) holds because the syntactic congruence does not have empty classes. For every tree, which is not in the class $C_\perp$ there exists a context C such that $C[x_1 \leftarrow t] \in L$ by the definition of the equivalence classes, which proves (ii). M_L's determinism has (iii) as a direct consequence.

Subtree Exchange Property.   The class of all languages that are recognized by DTAs is defined via these automata. However, there are several other characterizations by different means. An early one that later became known as the *path-closed* languages was provided by Viragh [18]. The path language $\pi(t)$ of a tree t, is defined inductively by:

– if $t \in \Sigma_0$, then $\pi(t) = t$
– if $t = f(t_1, \ldots, t_k)$, then $\pi(t) = \bigcup_{i=1}^{i=k} \{f i w \mid w \in \pi(t_i)\}$

For a tree language L the path language of L is defined as $\pi(L) = \bigcup_{t \in L} \pi(t)$, the path closure of L is defined as $pc(L) = \{t \mid \pi(t) \subseteq \pi(L)\}$. A tree language is path-closed if $pc(L) = L$. Viragh proved that the regular, path-closed tree languages are exactly the ones that are recognized by deterministic top-down automata. Nivat and Podelski argued that in these languages it must be possible to exchange certain subtrees [16]. We will extensively use this so-called exchange property in a formulation by Martens et al. [12].

Definition 2. A regular tree language L fulfills the *exchange property* if, for every $t \in L$ and every node $u \in N(t)$, if $t[u \leftarrow f(t_1, \ldots, t_k)] \in L$ and also $t[u \leftarrow f(s_1, \ldots, s_k)] \in L$, then $t[u \leftarrow f(t_1, \ldots, t_{i-1}, s_i, t_{i+1}, \ldots, t_k)] \in L$ for each $i = 1, \ldots, k$.

From the references cited above we obtain the following statement.

Proposition 3. *A regular tree language fulfills the exchange property if and only if it is recognized by a deterministic top-down tree automaton.*

3   Decidability of Top-Down Determinism

It is well-known that it is decidable for a regular tree language whether or not it is top-down deterministic. Viragh proved this via the construction of what he calls the *powerset automaton* for the path-closure of a regular language; the language is deterministic top-down, if it is equal to the language of the powerset automaton [18]. Gécseg and Steinby used a very similar method in their textbook [5].

Another approach can be taken via an application of Nivat and Podelksi's homogeneous closure [16]. A tree language is *homogeneous* if, for every $t \in L$ and every node $u \in N(t)$, if $t[u \leftarrow f(t_1, t_2)] \in L$, $t[u \leftarrow f(s_1, t_2)] \in L$ and also $t[u \leftarrow f(t_1, s_2)] \in L$, then $t[u \leftarrow f(s_1, s_2)] \in L$. The smallest homogeneous set containing a tree language is its *homogeneous closure*. One could construct the automaton for the language's homogeneous closure. The original language is deterministic top-down, if it is equal to its homogeneous closure.

In both approaches the automaton of the respective closure has as state set the powerset of the original state set. Thus already computing this automaton takes an exponential amount of time and even space. The second step is in both cases the decision of the equivalence of two non-deterministic automata, which is EXPTIME-complete in the size of these automata (Corollary 1.7.9 in [3]). In neither case the exact running time has been investigated. Also the approach of Cristau et al. [4] for unranked trees follows similar lines and does not have a better runtime.

We present a new method for deciding whether a regular tree language is top-down deterministic which runs in polynomial time.

In corresponding BAs and TAs, non-determinism in one direction corresponds to different transitions converging to the same right-hand side in the other direction. We now formalize this phenomenon.

Definition 4. Let A be a deterministic, minimal bottom-up tree automaton. A pair of distinct transitions $f(q_{1,1}, \ldots, q_{1,k}) \to q$ and $f(q_{2,1}, \ldots, q_{2,k}) \to q$ is called a *conflux*. A maximal set of transitions, which pairwise form confluxes (on the same input symbol f and with same right-hand side q), is called a *conflux group*.

The subtree exchange property from Definition 2 essentially states that trees that appear in the same positions can be interchanged. For states in a TA an analogous property would say that these must be exchangeable on the right-hand sides of rules; but this is not the case, because despite its determinism the runs for distinct occurrences of the same subtree can be distinct. However, when we look at the minimal deterministic bottom-up automaton for a deterministic top-down tree language, then we can establish a kind of exchange property for its states.

Lemma 5. *Let L be a deterministic top-down tree language and let M be the minimal deterministic bottom-up automaton recognizing it. If M has a conflux* of the transitions $f(q_{1,1}, \ldots, q_{1,k}) \to q$ and $f(q_{2,1}, \ldots, q_{2,k}) \to q$, then all the transitions from the set

$$\{f(q_{i_1,1}, \ldots, q_{i_k,k}) \to q \mid i_1, \ldots, i_k \in \{1, 2\}\}$$

are also present in M.

Proof. Let $t_{i,1}, \ldots t_{i,k}$ be trees such that $t_{i,j} \in L(M_{q_{i,j}})$ for all $j \in \{1, \ldots, k\}$ and $i \in \{1, 2\}$. Such trees exist by Proposition 1 (i). It follows from Proposition 1 (ii) that there exists a context $C \in \mathcal{C}(\Sigma)$ such that $C[f(t_{i,1}, \ldots t_{i,k})] \in L$ for $i \in \{1, 2\}$. Because the transitions $f(q_{1,1}, \ldots, q_{1,k}) \to q$ and $f(q_{2,1}, \ldots, q_{2,k}) \to q$ are distinct there exists a $j \in \{1, \ldots, k\}$ such that $q_{1,j} \neq q_{2,j}$. By Proposition 1 (iii) this implies that $t_{1,j} \neq t_{2,j}$.

The tree $t = C[f(t_{1,1}, \ldots, t_{1,j-1}, t_{2,j}, t_{1,j+1}, \ldots, t_{1,k})]$ must be in L by Proposition 3, because L is deterministic top-down. Thus M must apply a transition of the form $f(q_{1,1}, \ldots, q_{1,j-1}, q_{2,j}, q_{1,j+1}, \ldots, q_{1,n}) \to p$ for some state p distinct from q at the node v where f occurs.

The two corresponding subtrees $\hat{t}_1 = t/v$ and $\hat{t}_2 = \hat{t}_1[vj \leftarrow t_{2,j}]$ rooted in v are not syntactically equivalent, because M's states correspond to the equivalence classes of the syntactic congruence. Thus there is some context C such that $C[\hat{t}_1] \in L$ but $C[\hat{t}_2] \notin L$. If there is no such context, then there is one such that $C[\hat{t}_1] \notin L$ but $C[\hat{t}_2] \in L$, because otherwise the two trees would be syntactically equivalent; without loss of generality we treat only the former case.

Because L is a deterministic top-down tree language, by the exchange property from Proposition 3 the tree $C[\hat{t}_2]$ should be in L if $C[\hat{t}_1]$ is, since one is obtained from the other by exchanging $t_{1,j}$ for $t_{2,j}$ or the other way around, while the context C remains equal. This shows that no context distinguishing the trees $\hat{t}_1$ and $\hat{t}_2$ can exist, and thus p must actually be equal to q. Absolutely symmetrically we can show that also $f(q_{2,1}, \ldots, q_{1,j}, \ldots, q_{2,k}) \to q$ must be present in M. The same argument applies to each one of the k positions in the conflux, which proves the statement. □

Lemma 5 provides us with a necessary condition for a language to be deterministic top-down. We introduce the notion of violation for the case where the conditions of the lemma are not met.

Definition 6. Let M be a minimal deterministic bottom-up tree automaton. If there is a pair of transitions $f(q_{1,1}, \ldots, q_{1,k}) \to q$ and $f(q_{2,1}, \ldots, q_{2,k}) \to q$ in M which constitute a conflux, but not all the transitions from the set

$$\{f(q_{i_1,1}, \ldots, q_{i_k,k}) \to q \mid i_1, \ldots, i_k \in \{1, 2\}\}$$

are also present in M, then we say that this conflux constitutes a *violation*.

The transitions that form part of violations, which read the same symbol and result in the same state on the right-hand side form the corresponding *violating group*. For such a transition $f(q_{1,1}, \ldots, q_{1,k}) \to q$ its violating group is $\{x(p_1, \ldots, p_k) \to p \in \delta \mid x = f \text{ and } p = q)\}$.

As the symbol, which is read, and the resulting state uniquely identify each violating group, each transition of a violation belongs to exactly one group.

Now in the terminology of Definition 6 the statement of Proposition 5 says that the MDBA for a deterministic top-down tree language cannot contain any violation. Now we show that the absence of violations in the minimal automaton necessarily means that the language is top-down deterministic.

Lemma 7. *If the minimal deterministic bottom-up automaton M for a language L contains no violation, then L is top-down deterministic.*

Proof. If M does not contain any conflux, then its corresponding TA is deterministic and the statement holds. Otherwise we construct an equivalent automaton without confluxes. The first step in this construction is the elimination of one arbitrary conflux group.

So let the set $\{f(q_{i,1}, \ldots, q_{i,k}) \to q \mid i \in \{1, \ldots, \ell\}\}$ be the conflux group, which we choose to eliminate, where ℓ is the number of transitions in this group. We construct a new automaton without this conflux group that recognizes the same language.

– Its set of states is $Q \cup \{p_j \mid j \in \{1, \ldots, k\}\}$ with one new state for each position on the left-hand sides of the transitions of the conflux.
– We remove all the transitions of the conflux group.
– Instead we add the single *substitute transition* $f(p_1, \ldots, p_k) \to q$.
– Then for every transition $\lambda \to q_{i,j}$ that has one of the states $q_{i,j}$ on its right hand side we add the transition $\lambda \to p_j$ that has p_j instead, while the left-hand side is identical; we call these copies *adapter transitions*.

Let $M' = (Q \cup \{p_j \mid j \in \{1, \ldots, k\}\}, \Sigma, Q_f, \delta')$ be the resulting automaton, where δ' is obtained from δ by removing the conflux transitions and adding the substitution and adapter transitions as described. In what follows we will call the components of M the *original* ones.

The idea behind this construction is the following: M' essentially does the same runs as M. Only when M applies a transition of the eliminated conflux group M' applies the substitute transition instead. In order to be able to do this, M' must guess in the previous steps that instead of the original transitions applied by M it should use the corresponding adapter transitions.

Claim 1. *The tree automaton M' recognizes the same language as M.*

The accepting runs of M and M' are in one-to-one correspondence. This is proved in detail in the extended version available online [9].

We have seen how to eliminate one conflux group. Unfortunately, this elimination does not necessarily reduce the number of conflux groups. If the state q', which is on the right-hand side of all rules of a different conflux group, appears as one of the $q_{i,j}$ in the eliminated conflux group, then a copy of the entire group with q' is made in the adapter transitions. Note that also the newly introduced $f(p_1, \ldots, p_k) \to q$ could be a transition with one of the $q_{i,j}$ on its right hand side if this $q_{i,j}$ is equal to q, see also Example 9. In this case, however, the conflux group is not copied, because its transitions are removed before the adapter transitions are introduced.

So the number of conflux groups can stay the same and even increase. Nonetheless we start by removing all the original conflux groups in the way described.

Claim 2. *After all the original conflux groups are removed, further removals always decrease the number of conflux groups.*

New conflux groups can only be added in the step where the adapter transitions are introduced, because the substitute transition obviously creates no new conflux group. So all non-original conflux groups consist of adapter transitions and thus do not have states from the original Q on their right-hand sides. But these new states never occur on the left-hand side of any transition except their corresponding substitute transitions, which cannot form part of any conflux. Consequently they are never copied for new adapter transitions. Therefore only original conflux groups can be copied and Claim 2 holds.

Summarizing, we do one elimination step for each original conflux group. After this a number of copies of original conflux groups can have appeared. During their elimination their number decreases by one in every step. Therefore this process terminates and we obtain a bottom-up automaton, which does not have any confluxes and is equivalent to the original MDBA. It is not deterministic anymore, but now its corresponding TA is, because only confluxes result in nondeterministic choices in the reversal.   □

We illustrate the construction in the proof of Lemma 7 with two examples.

Example 8. Consider the language

$$L = \{f(a, f(a, b)), f(a, f(b, a)), f(a, f(a, a)), f(a, f(b, b))\}.$$

It is deterministic top-down, but its MDBA contains a conflux, which is not a violation. Its transitions are $q_0(a) \to q_a$, $q_0(b) \to q_b$, $f(q_a, q_b) \to q$, $f(q_b, q_a) \to q$, $f(q_a, q_a) \to q$, $f(q_b, q_b) \to q$, and $f(q_a, q) \to q_f$. The four transitions with q on the right-hand side constitute a conflux but not a violation.

Applying the construction we introduce the new states p_1 and p_2. The four transitions of the conflux are replaced by the substitute transition $f(p_1, p_2) \to q$. Further the adapter transitions $q_0(a) \to p_1$, $q_0(b) \to p_1$, $q_0(a) \to p_2$, and $q_0(b) \to p_2$ are added. The resulting automaton has the same number of transitions and two additional states. In this case q_b could be deleted, because it can only be read by the transitions of the conflux; in general original states do not become obsolete as shown by q_a. The recognized language is the same, but on leaves labeled a there is the non-deterministic choice of going into state q_a, p_1, or p_2, similarly for leaves labeled by b.   □

Example 9. An interesting case for the construction in the proof of Lemma 7 is the occurrence of a state on both the left-hand and the right-hand side of a transition of the conflux. Let $f(q', q) \to q$ be such a transition. When it (along with the other ones) is removed, it is replaced by $f(p_1, p_2) \to q$. Then also $f(p_1, p_2) \to p_2$ is added. If we did the latter step before adding $f(p_1, p_2) \to q$, then this recursiveness would be lost. So the order in which transitions are added and removed is essential.   □

Together, Lemmas 7 and 5 provide us with a characterization of the deterministic top-down tree languages.

Theorem 10. *A regular tree language is top-down deterministic if and only if its minimal deterministic bottom-up automaton contains no violations.*

This provides us with a new method to decide whether a regular tree language L given as a deterministic bottom-up automaton is top-down deterministic:

(i) Compute the minimal deterministic bottom-up tree automaton M for L.
(ii) Find all confluxes in M's set of transitions.
(iii) For each conflux check whether it constitutes a violation.

Step (i) can be computed in quadratic time. Carrasco et al. [2] showed in detail how to minimize a deterministic bottom-up automaton within this time bound. Minimization algorithms were already known early on, but their runtime was not analyzed in detail [1,5].

Both Steps (ii) and (iii) are purely syntactical analyses of the set of transitions. To optimize the runtime we can group the strings describing transitions into classes $T(q_f) = \{(q_1, \ldots, q_k) \mid f(q_1, \ldots, q_k) \to q \in \delta\}$ for all states q and all node labels f in linear time in the style of bucket sort. The different transitions of a possible conflux group are all in the same class which, on the other hand, is not longer than the total description of the automaton. Thus linearly many transitions need to be compared in order to determine whether there is a conflux and whether it constitutes a violation. Also this takes an amount of time at most quadratic in the size of the input.

Theorem 11. *For a regular tree language given as a DBA it is decidable in quadratic time whether it is also deterministic top-down.*

4   Finite Unions of Deterministic Top-Down Tree Languages

In the preceding section we have provided a new characterization of the class of top-down deterministic tree languages. One deficiency of this class is that it is not closed under basic operations such as set-theoretic union. Moreover, even simple finite languages such as $\{f(a, b), f(b, a)\}$ are not included in this class. To remedy these deficiencies, we consider finite unions of top-down deterministic tree languages. They contain many common examples for non-top-down deterministic tree languages, but still are characterized by deterministic top-down tree automata. We denote this class by $\mathcal{FU\text{-}DT}$.

In Sect. 3 we have seen that violations in the minimal deterministic bottom-up automaton can be used to decide whether a language is deterministic top-down. Among the automata with violations, some recognize languages that are still in $\mathcal{FU}$-$\mathcal{DT}$ while other ones recognize languages outside this class. We now explore how an analysis of the occurring violations can be used to determine to which one of the classes a given language belongs.

To this end we use a context-free grammar $G(M)$ to analyze where and how a given MDBA M uses the transitions of its violations. This *violation grammar*

- has M's state set plus a new start symbol S as its set of non-terminals.
- The terminals are [,] and one distinct *violation symbol* for each of the violating groups of M.
- For every transition $f(q_1, \ldots, q_k) \to q$ from a violating group ν we add the production $q \to \nu[q_1 \cdots q_k]$;
- for all transitions that are not from any violating group we add the production $q \to q_1 \cdots q_k$. Hence for initial states q_0 there are rules $q_0 \to \epsilon$.
- Finally, there is the transition $S \to [q_f]$ for each final state q_f of M.

For a run β of M we call its *corresponding string* $\theta(\beta)$ the terminal string that is generated by G by using the productions corresponding to the transitions used in β in the corresponding order. The *violation tree* of β is obtained from $\theta(\beta)$ as follows: All brackets [ ] without any other non-terminals between them are removed from $\theta(\beta)$. A root note is added and the bracket structure is seen as a tree. For example, a string $[\eta_1[\,]\eta_2[\eta_1[\,]\,]\,]$ results in the tree $\epsilon(\eta_1, \eta_2(\eta_1))$. Nodes with symbols of violations are called *violation nodes*.

Lemma 12. *Let M be an MDBA and let n be the number of transitions that form part of violations and are applied in the run β of M on a tree t. Then the corresponding string $\theta(\beta)$ for this run has length $3n + 2$.*

Proof. The unique production for the start state adds two terminals, namely [and]. The only other productions that generate terminals are the ones corresponding to transitions that form part of violations. Each one adds three terminals one of which is a violation symbol. □

So the language $L(G(M))$ is finite if and only if there is some number n such that every accepting run of M uses at most n times transitions that form part of some violation.

Lemma 13. *Let M be an MDBA. If M's violation grammar $G(M)$ generates an infinite language, then L is not in $\mathcal{FU}$-$\mathcal{DT}$.*

Essentially, every violation symbol represents a choice that cannot be made in a top-down deterministic way. All of these choices are pairwise independent in the sense that for each one a new DTA is necessary. So if there is no bound on their number, no finite union can be found. See [9] for more details. If the violation grammar's language is not infinite as in Lemma 13, then we can construct a family of DTAs that demonstrate that the given language is in $\mathcal{FU}$-$\mathcal{DT}$.

Lemma 14. *Let M be an MDBA. If M's violation grammar produces a finite language, then L is a finite union of deterministic top-down tree languages.*

Proof. We first treat the case where $L(G(M))$ is a singleton set. If the violation tree contains at most one violation node per group, then we decompose the MDBA M as follows: for every possible combination of transitions from the violating groups that contains exactly one transition from each group we make one automaton that contains exclusively these transitions from the respective violating groups. In addition it contains all the other transitions that do not belong to any violating group. The total number of automata is $\prod\{|\eta| \mid \eta$ is a violating group in $M\}$.

These automata do not contain violations any more, because all the existing ones have been removed and no new transitions have been added. Thus their corresponding TAs are deterministic top-down automata or can be transformed as in the proof of Lemma 7 by eliminating all confluxes. Finally, let K be the union of the languages of all the new automata. $K \subseteq L$, because every run in one of the new automata can be done by exactly the same transitions in M; on the other hand, also for every run of M there is one new automaton that contains all the transitions that are used, and thus $L \subseteq K$ and consequently $L = K$. So we have decomposed L into a union of deterministic top-down tree languages.

From the proof of Lemma 13 we can see that for every pair (L_1, L_2) of these languages there is a pair of trees that show that $L_1 \cup L_2$ can never be part of a deterministic top-down subset of L. Thus there cannot be any decomposition with fewer components.

We only sketch how to generalize this construction to several occurrences of the same violating group in the string and then to $L(G(M))$ consisting of several strings. If some violating group ν appears several times in the string s, at each occurrence of ν a different transition from ν could be used in a run of M. So instead of choosing one fixed transition from the group, we independently choose one for each occurrence and with it its position in the tree; we index the transition with the position of the occurrence in the violation tree. When the new automaton applies one of these transitions, it remembers its position and verifies it, while moving up in the input tree. Similarly, occurrences of ν in distinct strings can be distinguished. See [9] for more details.   □

The number of automata introduced in the proof of Lemma 14 is exponential in the number of nodes in the violation grammar's output language. This might seem bad at first sight; however, from the proof of Lemma 13 we can see that for a single tree in the output language this number cannot be improved.

Theorem 15. *For a regular tree language given as a DBA M it is decidable in quadratic time whether or not it belongs to the class $\mathcal{FU}$-$\mathcal{DT}$.*

Proof. We proceed as follows:

(i) Construct the minimal deterministic bottom-up automaton M' for L.
(ii) Detect all violations in M'.

(iii) Construct the violation grammar for M'.

(iv) Decide whether the grammar's language is finite.

Steps (i) and (ii) are just as in the procedure following Theorem 10. The construction of the violation grammar has been described above. Now the question of Step (iv) is equivalent to our decision problem by Lemmas 13 and 14. For this decision we first eliminate all deleting rules from the grammar, which can be done in linear time [7]. With this reduced grammar the finiteness of the language can be decided essentially by detecting cycles in the transition graph. This can be done in time linear in the number of edges and nodes of the graph by detecting the *strongly connected components (SCC)* [17]. If in any SCC a rule is used that produces more than one non-terminal, then the grammar's language is infinite, otherwise it is not. Also this check and therefore the entire Step (iv) can be done in linear time. See [9] for more details   □

Acknowledgment. We are grateful to Wim Martens, Helmut Seidl, Magnus Steinby, and Martin Lange for pointing us to some of the literature.

References

1. Brainerd, W.S.: The minimalization of tree automata. Inf. Control **13**(5), 484–491 (1968)
2. Carrasco, R.C., Daciuk, J., Forcada, M.L.: An implementation of deterministic tree automata minimization. In: Holub, J., Žd'árek, J. (eds.) CIAA 2007. LNCS, vol. 4783, pp. 122–129. Springer, Heidelberg (2007). https://doi.org/10.1007/978-3-540-76336-9_13
3. Comon, H., et al.: Tree automata techniques and applications (2007). http://www.grappa.univ-lille3.fr/tata. Accessed 12 October 2007
4. Cristau, J., Löding, C., Thomas, W.: Deterministic automata on unranked trees. In: Liśkiewicz, M., Reischuk, R. (eds.) FCT 2005. LNCS, vol. 3623, pp. 68–79. Springer, Heidelberg (2005). https://doi.org/10.1007/11537311_7
5. Gécseg, F., Steinby, M.: Tree automata. Akadéniai Kiadó, Budapest (1984)
6. Gelade, W., Idziaszek, T., Martens, W., Neven, F., Paredaens, J.: Simplifying XML schema: single-type approximations of regular tree languages. J. Comput. Syst. Sci. **79**(6), 910–936 (2013)
7. Harrison, M.A., Yehudai, A.: Eliminating null rules in linear time. Comput. J. **24**(2), 156–161 (1981)
8. Kozen, D.: On the Myhill-Nerode theorem theorem for trees. Bull. EATCS **47**, 170–173 (1992)
9. Leupold, P., Maneth, S.: Deciding top-down determinism of regular tree languages. ArXiv e-prints 2107.03174 (2021). https://arxiv.org/abs/2107.03174
10. Maneth, S., Seidl, H.: When is a bottom-up deterministic tree translation top-down deterministic? In: ICALP, pp. 134:1–134:18 (2020)
11. Martens, W.: Static analysis of XML transformation and schema languages. Ph.D. thesis, Hasselt University (2006)
12. Martens, W., Neven, F., Schwentick, T.: Deterministic top-down tree automata: past, present, and future. In: Logic and Automata: History and Perspectives, pp. 505–530 (2008)

13. Martens, W., Neven, F., Schwentick, T., Bex, G.J.: Expressiveness and complexity of XML schema. ACM Trans. Database Syst. **31**(3), 770–813 (2006)
14. Myhill, J.: Finite automata and the representation of events. Technical report 57–264, WADC (1957)
15. Nerode, A.: Linear automaton transformations. Proc. AMS **9**, 541–544 (1958)
16. Nivat, M., Podelski, A.: Minimal ascending and descending tree automata. SIAM J. Comput. **26**(1), 39–58 (1997). https://doi.org/10.1137/S0097539789164078
17. Tarjan, R.E.: Depth-first search and linear graph algorithms. SIAM J. Comput. **1**(2), 146–160 (1972)
18. Virágh, J.: Deterministic ascending tree automata I. Acta Cyb. **5**(1), 33–42 (1980)

Propositional Gossip Protocols

Joseph Livesey and Dominik Wojtczak$^{(\boxtimes)}$ (iD)

University of Liverpool, Liverpool, UK
{joseph livesey,d.wojtczak}@liverpool.ac.uk

Abstract. Gossip protocols are programs that can be used by a group of n agents to synchronise what they know. Namely, assuming each agent holds a secret, the goal of a protocol is to reach a situation in which all agents know all secrets. Distributed epistemic gossip protocols use epistemic formulas in the component programs for the agents. In this paper, we solve open problems regarding one of the simplest classes of such gossip protocols: propositional gossip protocols, in which whether an agent can make a call depends only on his currently known set of secrets. Specifically, we show that all correct propositional gossip protocols, i.e., the ones that always terminate in a situation where all agents know all secrets, require the underlying undirected communication graph to be complete and at least $2n - 3$ calls to be made. We also show that checking correctness of a given propositional gossip protocol is a co-NP-complete problem. Finally we report on implementing such a check with model checker nuXmv.

1 Introduction

Gossip protocols have the goal of spreading information through a network via point-to-point communications (which we refer to as calls). Each agent holds initially a secret and the aim is to arrive at a situation in which all agents know each other secrets. During each call the caller and callee exchange all secrets that they know at that point. Such protocols were successfully used in a number of domains, for instance communication networks [18], computation of aggregate information [22], and data replication [24]. For a more recent account see [21] and [23]. One of the early results established by a number of authors in the seventies, e.g., [25], is that for n agents $2n - 4$ calls are necessary and sufficient when every agent can communicate with any other agent. When such a communication graph is not complete, $2n - 3$ calls may be needed [11] but are sufficient for any connected communication graph [17]. However, all such protocols considered in these papers were centralised.

In [10] a dynamic epistemic logic was introduced in which gossip protocols could be expressed as formulas. These protocols rely on agents' knowledge and are distributed, so they are *distributed epistemic gossip protocols*. This also means that they can be seen as special cases of knowledge-based programs introduced in [15].

© Springer Nature Switzerland AG 2021
E. Bampis and A. Pagourtzis (Eds.): FCT 2021, LNCS 12867, pp. 354–370, 2021.
https://doi.org/10.1007/978-3-030-86593-1_25

In [2] a simpler modal logic was introduced that is sufficient to define these protocols and to reason about their correctness. This logic is interesting in its own rights and was subsequently studied in a number of papers. In this paper, we are going to focus on its simplest propositional fragment.

Propositional gossip protocols are a particular type of epistemic gossip protocol in which all guards are propositional. This means that calls being made by each agent are dependent only on the secrets that the agent have had access to. Clearly, this can lead to states where multiple calls are possible at the same time. Then a scheduler would decide which call takes priority. Throughout this paper, we assume that the scheduler is demonic and it picks the order of calls in a way such that the protocol fails or to maximise the number of calls made before termination. In other words, we study these gossip protocols in their worst-case scenario.

In [9], many challenging open problems about general as well as propositional gossip protocols were listed. In the paper we manage to resolve some of them. In particular, we solve its Problem 5, which asks for a characterisation of the class of graphs for which propositional protocols exist. In Sect. 3 we show that, when we ignore the direction of edges in the communication graph, the only class possible is complete graphs. In order to show that we need to establish many interesting properties of computations of such protocols. We also partially resolve its Problem 6, which asks to show that a gossip protocol needs at least $2n - 3$ calls to be correct. We prove this is indeed the case at least for the class of propositional gossip protocols. Note that, unlike what was shown in [11], this lower bound holds even if the communication graph is a complete graph. Finally, we also partially address Problem 7, which asks for the precise computational complexity of checking the correctness of such protocols. It is known that for gossip protocols without nesting of modalities this problem is in coNP^{NP} [4]. We improve this to co-NP-completeness for propositional ones.

Related Work. Much work has been done on general epistemic gossip protocols. The various types of calls used in [10] and [2] were presented in a uniform framework in [3], where in total 18 types of communication were considered and compared w.r.t. their epistemic strength. In [5], and its full version [8], the decidability of the semantics of the gossiping logic and truth was established for its limited fragment (namely, without nesting of modalities). Building upon these results it was proved in [5] that the distributed gossip protocols, the guards of which are defined in this logic, are implementable, that their partial correctness is decidable, and in [7] that termination and two forms of fair termination of these protocols are decidable, as well. Building on that, [29] showed decidability of the full logic for various variants of the gossiping model. Further, in [4] the computational complexity of this fragment was studied and in [6] an extension with the common knowledge operator was considered and analogous decidability results were established there.

Despite how simple this modal logic seems to be, there remain natural open problems about it and the gossip protocols defined using it. These problems were discussed at length in [9], where partial results were also presented that be build upon in this paper. Some of these open problems were subsequently tackled in [29], but propositional protocols were not studied there and questions regarding them were left open.

Centralised gossip protocols were studied in [19] and [20]. These had the goal to achieve higher-order shared knowledge. This was investigated further in [13], where optimal protocols for various versions of such a generalised gossip problem were given. These protocols depend on various parameters, such as the type of the underlying graph or communication. Additionally, different gossip problems which contained some negative goals, for example that certain agents must not know certain secrets, were studied. Such problems were further studied in [14] with temporal constraints, i.e., a given call has to (or can only) be made within a given time interval.

The number of calls needed to reach the desired all expert situation in the distributed but synchronous setting was studied in [26]. In the synchronous setting, agents are notified if a call was made, but may not necessary know which agents were involved. In this paper we study the more complex fully distributed asynchronous setting, where agents are not aware of the calls they do not participate in. In [27,28] the expected time of termination of several gossip protocols for complete graphs was studied.

Dynamic distributed gossip protocols were studied in [30], in which the calls allow the agents to transmit the links as well as share secrets. These protocols were characterised in terms of the class of graphs for which they terminate. Various dynamic gossip protocols were proposed and analysed in [31]. In [16] these protocols were analysed by embedding them in a network programming language NetKAT [1].

Structure of the Paper. We first introduce the logic, originally defined in [2], and then move on to tackle some open problems for the propositional gossip protocols. The first aim is to look at the communication graph required for the existence of a correct propositional gossip protocol (Sect. 3) and then a lower bound on the number of calls needed by such a protocol (Sect. 4). We then move on to looking at the complexity of the natural decision problems for these protocols (Sect. 5), before touching on some computational attempts to see how quickly a computer can determine the correctness of a given propositional gossip protocol. Due to the space limit some details of the proofs are omitted and will be published in a journal version of this paper later.

2   Gossiping Logic

We recall here the framework of [2], which we restrict to the propositional setting. We assume a fixed set A of $n \geq 3$ *agents* and stipulate that each agent holds exactly one *secret*, and that there exists a bijection between the set of agents and the set of secrets. We denote by S the set of all secrets.

The propositional language $\mathcal{L}_p$ is defined by the following grammar:

$$\phi ::= F_a S \mid \neg \phi \mid \phi \wedge \phi,$$

where $S \in \mathsf{S}$ and $a \in \mathsf{A}$. We will distinguish the following sublanguage $\mathcal{L}_p^a$, where $a \in \mathsf{A}$ is a fixed agents, which disallow all F_b operators for $b \neq a$.

So $F_a S$ is an atomic formula, which we read as 'agent a is familiar with the secret S'. Note that in [2], a compound formula $K_a \phi$, i.e., 'agent a knows the formula ϕ is true', was used. Dropping $K_a \phi$ from the logic simplifies greatly its semantics and the execution of a gossip protocol, while it is still capable of describing a rich class of protocols. Below we shall freely use other Boolean connectives that can be defined using $\neg$ and $\wedge$ in a standard way. We shall use the following formula

$$Exp_i \equiv \bigwedge_{S \in \mathsf{S}} F_i S,$$

that denotes the fact that agent i is an **expert**, i.e., he is familiar with all the secrets.

Each **call**, written as ab or a, b, concerns two different agents, the **caller**, a, and the **callee**, b. After the call the caller and the callee learn each others secrets. Calls are denoted by c, d. Abusing notation we write $a \in \mathsf{c}$ to denote that agent a is one of the two agents involved in the call c. We refer to any such call an a-**call** (b-call for agent b, etc.).

In what follows we focus on call sequences. Unless explicitly stated each call sequence is assumed to be finite. The empty sequence is denoted by ϵ. We use $\mathbf{c}$ to denote a call sequence and $\mathbf{C}$ to denote the set of all finite call sequences. Given call sequences $\mathbf{c}$ and $\mathbf{d}$ and a call c we denote by $\mathbf{c}.\mathsf{c}$ the outcome of adding c at the end of the sequence $\mathbf{c}$ and by $\mathbf{c}.\mathbf{d}$ the outcome of appending the sequences $\mathbf{c}$ and $\mathbf{d}$. We say that $\mathbf{c}'$ is an **extension** of a call sequence $\mathbf{c}$ if for some call sequence $\mathbf{d}$ we have $\mathbf{c}' = \mathbf{c}.\mathbf{d}$.

To describe what secrets the agents are familiar with, we use the concept of a **gossip situation**. It is a sequence $\mathsf{s} = (Q_a)_{a \in \mathsf{A}}$, where $\{A\} \subseteq Q_a \subseteq \mathsf{S}$ for each agent a. Intuitively, Q_a is the set of secrets a is familiar with in the gossip situation s. The **initial gossip situation** is the one in which each Q_a equals $\{A\}$ and is denoted by root. It reflects the fact that initially each agent is familiar only with his own secret. Note that an agent a is an expert in a gossip situation s iff $Q_a = \mathsf{S}$.

Each call transforms the current gossip situation by modifying the sets of secrets the agents involved in the call are familiar with as follows. Consider a gossip situation $\mathsf{s} := (Q_d)_{d \in \mathsf{A}}$ and a call ab.
Then

$$ab(\mathsf{s}) := (Q_d')_{d \in \mathsf{A}},$$

where $Q_a' = Q_b' = Q_a \cup Q_b$, and for $c \notin \{a, b\}$, $Q_c' = Q_c$.

So the effect of a call is that the caller and the callee share the secrets they are familiar with.

The result of applying a call sequence to a gossip situation s is defined inductively as follows:

$$\epsilon(\mathsf{s}) := \mathsf{s}, \quad (\mathbf{c}.\mathsf{c})(\mathsf{s}) := \mathbf{c}(\mathsf{c}(\mathsf{s})).$$

Example 1. We will use the following concise notation for gossip situations. Sets of secrets will be written down as lists. e.g., the set $\{A, B, C\}$ will be written as ABC. Gossip situations will be written down as lists of lists of secrets separated by a comma. e.g., if there are three agents, a, b and c, then root $= A, B, C$ and the gossip situation $(\{A, B\}, \{A, B\}, \{C\})$ will be written as AB, AB, C.

Let $\mathsf{A} = \{a, b, c\}$. Consider the call sequence $ac.cb.ac$. It generates the following successive gossip situations starting from root:

$$A, B, C \xrightarrow{ac} AC, B, AC \xrightarrow{cb} AC, ABC, ABC \xrightarrow{ac} ABC, ABC, ABC.$$

Hence $(ac.cb.ac)(\mathsf{root}) = (ABC, ABC, ABC)$.   □

Definition 2. *Consider a call sequence* $\mathbf{c} \in \mathbf{C}$. *We define the satisfaction relation* $\models$ *inductively as follows:*

$$\mathbf{c} \models F_a S \text{ iff } S \in \mathbf{c}(\mathsf{root})_a,$$
$$\mathbf{c} \models \neg\phi \text{ iff } \mathbf{c} \not\models \phi,$$
$$\mathbf{c} \models \phi_1 \wedge \phi_2 \text{ iff } \mathbf{c} \models \phi_1 \text{ and } \mathbf{c} \models \phi_2.$$

So a formula $F_a S$ is true after the call sequence $\mathbf{c}$ whenever secret S belongs to the set of secrets agent a is familiar with in the situation generated by the call sequence $\mathbf{c}$ applied to the initial situation root. Hence $\mathbf{c} \models Exp_a$ iff agent a is an expert in $\mathbf{c}(\mathsf{root})$.

By a **propositional component program**, in short a **program**, for an agent a we mean a statement of the form

$$*[\![]_{j=1}^{m} \psi_j \to \mathsf{c}_j],$$

where $m \geq 0$ and each $\psi_j \to \mathsf{c}_j$ is such that

– a is the caller in the call c_j,
– $\psi_j \in \mathcal{L}_p^a$.

We call each such construct $\psi \to \mathsf{c}$ a **rule** and refer in this context to ψ as a **guard**.

Intuitively, $*$ denotes a repeated execution of the rules, one at a time, where each time non-deterministically a rule is selected whose guard is true.

We assume that in each gossip protocol the agents are the nodes of a directed graph (digraph) and that each call ab is allowed only if $a \to b$ is an edge in this digraph. A minimal digraph that satisfies this assumption is uniquely determined by the syntax of the protocol and we call this digraph the **communication graph** of a given protocol. Given that the aim of each gossip protocol is that all agents become experts it is natural to consider connected communication graphs only. On the other hand, the **underlying undirected communication graph** of a given protocol is the undirected graph we obtain when all directed edges in the communication graph are replaced with undirected ones connecting the same nodes.

Consider a propositional gossip protocol, P, that is a parallel composition of the propositional component programs $*[[]_{j=1}^{m_a} \psi_j^a \rightarrow c_j^a]$, one for each agent $a \in \mathsf{A}$.

The **computation tree** of P is a directed tree defined inductively as follows. Its nodes are call sequences and its root is the empty call sequence ϵ. Further, if $\mathbf{c}$ is a node and for some rule $\psi_j^a \rightarrow c_j^a$ we have $\mathbf{c} \models \psi_j^a$, then $\mathbf{c}.c_j^a$ is a node that is a direct descendant of $\mathbf{c}$. Intuitively, the arc from $\mathbf{c}$ to $\mathbf{c}.c_j^a$ records the effect of the execution of the rule $\psi_j^a \rightarrow c_j^a$ performed after the call sequence $\mathbf{c}$ took place.

By a **computation** of a gossip protocol P we mean a maximal rooted path in its computation tree. In what follows we identify each computation with the unique call sequence it generates. Any prefix of such a call sequence is called a **prefix of** P. We say that the gossip protocol P is **partially correct** if for all leaves $\mathbf{c}$ of the computation tree of P, and all agents a, we have $\mathbf{c} \models Exp_a$, i.e., if each agent is an expert in the gossip situation $\mathbf{c}(\mathsf{root})$.

We say furthermore that P **terminates** if all its computations are finite and say that P **is correct** if it is partially correct and terminates.

In [10] the following correct propositional gossip protocol, called *Learn New Secrets* (LNS in short), for complete digraphs was proposed.

Example 3 (LNS protocol). The following program is used by agent i:

$$*[[]_{j \in \mathsf{A}} \neg F_i J \rightarrow ij].$$

Informally, agent i calls agent j if agent i is not familiar with j's secret.   □

We now define a new propositional protocol whose communication graph is not complete. First of all, agents will only be able to call agents with a higher "index", which for instance can be his phone number or name, with the corresponding total order ($>$) on A. Second, just like in the LNS protocol, agents can only call another agent if they do not know their secret. Finally, we require that an agent can make a call to another agent only if he already knows all the secrets of agents with the index value lower than the agent to be called. We will call this protocol **Learn Next Secret (LXS)** and its formal definition is as follows.

Example 4 (LXS protocol). The following program is used by agent i:

$$*[[]_{\{j \in \mathsf{A}|j>i\}} \neg F_i J \wedge \bigwedge_{\{k \in \mathsf{A}|k<j\}} F_i K \rightarrow ij].$$

□

Note that although the communication graph of LXS protocol is not complete, its underlying undirected communication graph is, which we show is always the case for correct propositional protocols in the next section.

3 Required Communication Graph

We now show that for natural classes of connected graphs no correct propositional gossip protocol exists. We first show that by carefully removing some of the calls in a prefix of P one can get another prefix of P.

Lemma 5 (Call Removal). *Consider a propositional gossip protocol P. Let* **c.d** *be a prefix of P such that* **c.d** $\not\models F_a B$. *Let* **d$'$** *be* **d** *where all calls that involve an agent familiar with B are removed, then* **c.d$'$** *is also a prefix of P and, moreover,* $(\mathbf{c.d})(\text{root})_a = (\mathbf{c.d'})(\text{root})_a$.

Proof. It suffices to show that we can remove the last such call in **d**, because that clearly preserves the **c.d$'$** $\not\models F_a B$ property and then we can simply repeat this procedure until no such calls are left in **d**.

Let $\mathbf{d} = \mathbf{d}_1.cd.\mathsf{c}_1.\mathsf{c}_2 \ldots \mathsf{c}_k$, where cd is the last call that involves an agent that is familiar with B, i.e., $\mathbf{c.d}_1 \models F_c B \vee F_d B$. Straight from the definition of the outcome of the cd call, for all agents $x \notin \{c, d\}$, $\mathbf{c.d}_1(\text{root})_x = (\mathbf{c.d}_1.cd)(\text{root})_x$. At the same time, agents c, d cannot be involved in any of the calls $\mathsf{c}_1, \ldots, \mathsf{c}_k$. Therefore, we also have for all agents $x \notin \{c, d\}$, $(\mathbf{c.d}_1.\mathsf{c}_1)(\text{root})_x = (\mathbf{c.d}_1.cd.\mathsf{c}_1)(\text{root})_x$, and by induction $(\mathbf{c.d}_1.\mathsf{c}_1 \ldots \mathsf{c}_i)(\text{root})_x = (\mathbf{c.d}_1.cd.\mathsf{c}_1 \ldots \mathsf{c}_i)(\text{root})_x$ for all $i \leq k$. Note that $a \notin \{c, d\}$, because $\mathbf{c.d} \not\models F_a B$, so $(\mathbf{c.d})(\text{root})_a = (\mathbf{c.d'})(\text{root})_a$ holds as desired.

Now consider the guard ϕ_i associated with the call c_i where $i \leq k$. By assumption on P, ϕ_i is a propositional formula built out of the atomic formulas of the form $F_x S$ where $x \notin \{c, d\}$ is the agent making the call c_i. We already showed that $(\mathbf{c.d}_1.\mathsf{c}_1 \ldots \mathsf{c}_{i-1})(\text{root})_x = (\mathbf{c.d}_1.cd.\mathsf{c}_1 \ldots \mathsf{c}_{i-1})(\text{root})_x$, so the truth of these atomic formulas is not affected by the removal of the call cd from **d**. This shows that we have $\mathbf{c.d}_1.cd.\mathsf{c}_1 \ldots \mathsf{c}_{i-1} \models \phi_i$ iff $\mathbf{c.d}_1.\mathsf{c}_1 \ldots \mathsf{c}_{i-1} \models \phi_i$ and so c_i can also be made by the protocol P after $\mathbf{c.d}_1.\mathsf{c}_1 \ldots \mathsf{c}_{i-1}$.   □

We establish now what the correctness of a propositional protocol implies for the order of calls.

Lemma 6 (Initiation). *Consider any call sequence* **c** *which is a prefix of a computation of a correct propositional gossip protocol P such that* **c** $\models F_a B$. *There does not exist a call sequence* **d** *such that* **c.d.ab** *is a prefix of P. (In other words, agent a will never call agent b if agent a already knows B).*

Proof. Suppose such a call sequence **d** exists. If $\mathbf{c.d}(\text{root})_b \subseteq \mathbf{c.d}(\text{root})_a$, then we have $\mathbf{c.d}(\text{root})_a = \mathbf{c.d.}ab(\text{root})_a$ and so the guard ϕ of the call ab is still true after ab is made. As a result, ab could be repeated indefinitely after **c.d**; a contradiction with the assumption that P is terminating.

Otherwise, there exists a secret, X, such that $\mathbf{c.d} \models F_b X \wedge \neg F_a X$ before the call ab takes place. Let us now remove all calls from **c.d** that involve agents that are familiar with the secret X, which results in a call sequence $\mathbf{c'.d'}$. Lemma 5 then implies that $\mathbf{c'.d'}$ is also a prefix of P and $\mathbf{c.d}(\text{root})_a = \mathbf{c'.d'}(\text{root})_a$, so the call ab can still be made after $\mathbf{c'.d'}$. At the same time, $\mathbf{c'.d'}(\text{root})_b \subsetneq \mathbf{c.d}(\text{root})_b$, because agent b is no longer familiar with secret X and possibly other secrets as well.

If there is still a secret Y left such that $\mathbf{c}'.\mathbf{d}' \models F_b Y \wedge \neg F_a Y$ then we again remove all calls from $\mathbf{c}'.\mathbf{d}'$ that involve agents that are familiar with the secret Y, which results in a call sequence $\mathbf{c}''.\mathbf{d}''$. We keep repeating this process until we reach a call sequence $\mathbf{c}^*.\mathbf{d}^*$ such that $\mathbf{c}^*.\mathbf{d}^*(\mathsf{root})_b \subseteq \mathbf{c}^*.\mathbf{d}^*(\mathsf{root})_a$ and $\mathbf{c}^*.\mathbf{d}^*.ab$ is a prefix of P, because $\mathbf{c}.\mathbf{d}(\mathsf{root})_a = \mathbf{c}^*.\mathbf{d}^*(\mathsf{root})_a$. Just like before, we arrive to a contradiction, because the call ab can now be repeated indefinitely. □

Already these two lemmas allow us to show non-existence of a correct propositional gossip protocol for a wide range of natural communication graph classes. The first graph class that we consider is the star graph, i.e., when communication is only possible via a single central agent. This was already shown in [9], however our proof is much more simplistic.

Theorem 7. *Suppose that the underlying undirected communication graph forms a star graph with at least 3 agents. No correct propositional protocol exists.*

Proof. Suppose such a protocol P exists. From Lemma 6 each agent, apart from the central agent, is involved in at most one call, as otherwise the protocol will not be correct. Therefore, the non-central agent involved in the first call will not have any further calls, and so will never become an expert. □

We now proceed to show that no correct propositional protocol exists when the underlying undirected communication graph is not complete. Note that if there are only two agents then this statement is trivial. In the case of three agents, a undirected connected graph with a missing link is a star graph so the statement follows from Theorem 7. The proof of this statement in the general case is quite complex, so we break it down into several lemmas. In all these lemmas, we make the assumption that a correct protocol P exists where there is no link between two agents denoted by a and b. Theorem 11 will later show how this assumption leads to a contradiction.

Lemma 8. *There exists a computation of P such that agent b learns A by receiving a call from another agent.*

Proof. We prove this by contradiction and so assume instead that in all computations agent b learns A by calling another agent.

Let us pick a computation where b knows the greatest number of secrets before learning A. In other words, if $\mathbf{c}.bc$ is a prefix of P where b learns A during the call bc, we require the size of $\mathbf{c}(\mathsf{root})_b$ to be the largest possible. This is well-defined as this value is an integer between 1 and $|A|$ and agent b has learn to A in every computation as P is correct.

We know that $\mathbf{c} \models \neg F_b C$, because otherwise bc would not be possible due to Lemma 6. We can then remove all calls of agents familiar with C in $\mathbf{c}$ and obtain $\mathbf{c}'$. Due to Lemma 5, $\mathbf{c}'.bc$ is still a prefix of P. Note that c does not know A (nor any other secret apart from his own for that matter) after $\mathbf{c}'$, because all his calls were removed. At the same time we know that $\mathbf{c}(\mathsf{root})_b = \mathbf{c}'(\mathsf{root})_b$. If P

is indeed correct, then it has to be possible to extend $\mathbf{c'}.bc$ to a prefix $\mathbf{c'}.bc.\mathbf{d}.bd$ of P, for some $\mathbf{d}$ and d, such that b finally learns A during bd. (Note that due to our original assumption, it cannot be db.) But then $\mathbf{c}(\mathrm{root})_b$ is smaller than $\mathbf{c'}.bc.\mathbf{d}(\mathrm{root})_b$, because the latter includes at least one more secret (namely C); this is a contradiction with the pick of the prefix $\mathbf{c}.bc$ of P as the one where b knows the most number of secrets before learning A. □

Lemma 9. *For any call sequence $\mathbf{c}$ without a-calls and any agent $c \in \mathsf{A}\backslash\{a,b\}$, if $\mathbf{c}.ca$ is a prefix of P such that $\mathbf{c} \models F_cB$ (i.e., a learns B from c), then $\mathbf{c}.\mathbf{d}.bc$ is also a prefix of P for some call sequence $\mathbf{d}$.*

Proof. Let us pick any prefix of P $\mathbf{c}.ca$ where $\mathbf{c}$ is without a-calls, such that $\mathbf{c} \models F_cB$. Note that after $\mathbf{c}.ca$, all agents that know A (agents a and c, only) also know B. As in every call all secrets are exchanged, any extension of $\mathbf{c}.ca$ would also have this property.

Due to Lemma 6, no agent that knows A would call b after $\mathbf{c}.ca$, because he already knows B. Moreover, b will never call a (missing link) and let us assume she does not call c either. Then, as b must learn A eventually, she must call a different agent. From here the proof follows similarly as in Lemma 8, but with an initial call sequence $\mathbf{c}.ca$.

Let us pick a computation that starts with $\mathbf{c}.ca$ where b knows the greatest number of secrets before learning A. In other words, if $\mathbf{c}.ca.\mathbf{d}.bd$ is a prefix of P where b learns A during the call bd for some $d \in \mathsf{A}\backslash\{a,b,c\}$, we require the size of $(\mathbf{c}.ca.\mathbf{d})(\mathrm{root})_b$ to be the largest possible. This is well-defined as this value is an integer between 1 and $|\mathsf{A}|$.

We know that $\mathbf{c}.ca.\mathbf{d} \models \neg F_bD$, because otherwise bd would not be possible due to Lemma 6. We can then remove all calls of agents familiar with D in $\mathbf{d}$ and obtain $\mathbf{d'}$. Due to Lemma 5, $\mathbf{c}.ca.\mathbf{d}(\mathrm{root})_b = \mathbf{c}.ca.\mathbf{d'}(\mathrm{root})_b$, so $\mathbf{c}.ca.\mathbf{d'}.bd$ is also a prefix of P.

Note that d does not know A after $\mathbf{c}.ca.\mathbf{d'}$, because $\mathbf{c}$ and $\mathbf{d'}$ have no calls involving agents familiar with A. Hence $\mathbf{c}.ca.\mathbf{d'}.bd \models \neg F_bA$. So if P is indeed correct, then it has to be possible to extend $\mathbf{c}.ca.\mathbf{d'}$ to a prefix $\mathbf{c}.ca.\mathbf{d'}.bd.\mathbf{e}.be$ of P, for some call sequence $\mathbf{e}$ and agent e, such that b finally learns A during be. (Note that it cannot be eb, because no agent familiar with A would call b after $\mathbf{c}.ca$.) But then $\mathbf{c}.ca.\mathbf{d}(\mathrm{root})_b$ is smaller than $\mathbf{c}.ca.\mathbf{d'}.bd.\mathbf{e}(\mathrm{root})_b$, because the latter includes at least one more secret (namely D); this is a contradiction with the pick of the prefix $\mathbf{c}.ca.\mathbf{d}.bd$ of P as the one where b knows the most number of secrets before learning A. In conclusion, it must be possible for b to call c after $\mathbf{c}.ca$.

We have shown so far that $\mathbf{c}.ca.\mathbf{d}.bc$ has to be a prefix of P for some call sequence $\mathbf{d}$. Note that $\mathbf{c}.ca.\mathbf{d} \not\models F_bC$ due to Lemma 6. We can then remove all calls of agents familiar with C from the suffix $ca.\mathbf{d}$ of $\mathbf{c}.ca.\mathbf{d}$, to obtain $\mathbf{c}.\mathbf{d'}$. Then due to Lemma 5, $\mathbf{c}.\mathbf{d'}.bc$ is also a prefix of P. □

Using very similar techniques we can show the following.

Lemma 10. *For any call sequence* **c** *without a-calls and any agent* $c \in \mathsf{A}\backslash\{a,b\}$, *if* **c**.*ca is a prefix of* P *such that* **c** $\models F_c B$, *and* d *is the last agent in a call with* c *before call ca takes place, then* **c**.*ca*.**d**.*da is also a prefix of* P *for some* **d**.

We now have all the ingredients needed to prove the main result of this section.

Theorem 11. *Suppose that the underlying undirected communication graph is not complete. No correct propositional protocol exists.*

Proof. Suppose such a correct propositional protocol P exists and there are two agents, say a and b, which cannot call each other.

From Lemma 8 there exists an agent $c \in \mathsf{A}\backslash\{a,b\}$ and a prefix **c**.*ca* of P such that **c** $\models F_c B$. We know that **c**, $\not\models F_c A$, because otherwise *ca* would not be possible due to Lemma 6. We can then remove all calls of agents familiar with A in **c** and obtain **c**′. Due to Lemma 5, **c**(root)$_c$ = **c**′(root)$_c$, so **c**′.*ca* is also a prefix of P, with a not yet having been in a call until *ca* takes place.

Since **c**′.*ca* is a prefix of P where **c**′ has no a-calls and **c**′ $\models F_c B$ then from Lemma 9 we get that **c**′.**d**.*bc* is also a prefix of P for some **d**. Therefore, **c**′ cannot have *bc* nor *cb* call due to Lemma 6.

Note that there has to be at least one c-call in **c**′, because **c**′ $\models F_c B$. We already excluded *bc* and *cb*. It cannot be *ac* nor *ca* either as *ca* takes place after **c**′. Therefore, the last agent to be in a c-call in **c**′ is some $d \in \mathsf{A}\backslash\{a,b,c\}$. From Lemma 10 we get that **c**′.*ca*.**e**.*da* must also be a prefix of P for some **e**. Note that also **c**′ $\models F_d B$, because when the call between d and c takes place, c already has to know B.

We know that **c**′.*ca*.**e** $\models \neg F_d A$, because otherwise *da* would not be possible due to Lemma 6. We can then remove all calls of agents familiar with A in **e** and obtain **e**′. Due to Lemma 5, **c**′.*ca*.**e**(root)$_d$ = **c**′.*ca*.**e**′(root)$_d$, so **c**′.*ca*.**e**′.*da* is also a prefix of P. As **e** has had all calls to agents familiar with A removed, **e**′ contains no c-calls. Hence, **c**′(root)$_c$ = **c**′.**e**′(root)$_c$, and so **c**′.**e**′.*ca* is also a prefix of P. (As **e**′ contains no call to or from agents familiar with A, **e**′ can now occur before *ca*, because none of the calls can involve c or a.) As *ca* does not change the set of secrets known by d, and *da* does not change the set of secrets known by c, we get that both **c**′.**e**′.*ca*.*da* and **c**′.**e**′.*da*.*ca* are also prefixes of P.

As **c**′.**e**′.*ca* is prefix of P, **c**′.**e**′ does not have any a-calls and **c**′.**e**′ $\models F_c B$ then from Lemma 9 we get that **c**′.**e**′.**f**.*bc* is also a prefix of P for some **f**.

We know that **c**′.**e**′.**f** $\models \neg F_b C$, because otherwise *bc* would not be possible due to Lemma 6. We can then remove all calls of agents familiar with C in **f** and obtain **f**′. Due to Lemma 5, **c**′.**e**′.**f**(root)$_b$ = **c**′.**e**′.**f**′(root)$_b$, so **c**′.**e**′.**f**′.*bc* is also a prefix of P.

Note that **c**′.**e**′.**f**′.*bc*(root)$_d$ = **c**′.**e**′(root)$_d$, because all calls of agents familiar with C were removed from **f**′ and d is familiar with C already after **c**′. Hence, **c**′.**e**′.**f**′.*bc*.*da* is also a prefix of P, because *da* can take place immediately after **c**′.**e**′.

Now due to Lemma 9 we get that **c**′.**e**′.**f**′.*bc*.*da*.**g**.*bd* is also a prefix of P, because **c**′.**e**′.**f**′.*bc* does not have any a-calls and **c**′.**e**′.**f**′.*bc* $\models F_d B$, because

$c' \models F_d B$. We now get a contradiction with Lemma 6, because in this prefix b calls d even though b already knows D after the bc call in $c'.e'.f'.bc.da.g.bd$. □

4   Minimal Number of Calls

In this section we establish a lower bound on the number of calls needed for a propositional protocol to terminate in a state were all agents are experts. First, we start with one very useful observation.

Lemma 12 (Conversation). *For a protocol on n agents to correctly terminate in m calls, every agent must be involved in a call after at most $m - n + 2$ calls. Furthermore, after $m - n + p$ calls, each secret must be known by at least p agents.*

Proof. For an agent a and its secret A, each call can increase the number of agents that know A by at most 1. If a has not yet been involved in any calls, then the only agent which knows A is a. If after $m - n + 2$ calls, a has not yet been involved in a call, then a is the only agent which knows A. However, only $n - 2$ calls remain for $n - 1$ agents to learn A, which is impossible.

Similarly, if after $m - n + p$ calls, fewer than p know A, then $n - p$ calls remain for at least $n - p + 1$ agents to learn A. Again this is impossible as at most 1 agent can learn A in each call. □

We are now ready to partially resolve Problem 6 from [9] for the special case of propositional protocols.

Theorem 13. *No correct proposition protocol on n agents exists with fewer than $2n - 3$ calls.*

Proof. Let us assume a correct propositional protocol P exists which always terminates after at most $2n - 4$ calls.

First, Lemma 5 in [9] shows that every gossip protocol has a computation that starts with the same agent being involved in its first two calls. By relabelling the names of the agents, we can assume that we have a call between a and b, followed by a call between b and c. W.l.o.g., we can assume that these two calls are $ab.bc$, because the resulting gossip situation is always (AB, ABC, ABC) for a, b, c, respectively, and all other agents know just their own secret.

We claim that there must be a prefix of P of the form $ab.bc.c.ac$ where c does not involve agents familiar with C. First of all, a has to learn C eventually. From Lemma 6, we know that after $ab.bc$ agent c will not call a, because he already knows A. Furthermore, due to Lemma 6, no agent that will learn C later will initiate a call with a, because he will learn A at the same time as C. So the only option left is that a learns C by calling another agent. Let $ab.bc.d.ad$ be such a prefix of P where a learns C from d. Clearly d cannot be b due to Lemma 6 and if d is c we are done. Thanks to Lemma 5 we can remove all calls in d that involve agents familiar with C and get a new prefix $ab.bc.d'.ad$ of P after which a still does not know C. Therefore, there has to be an extension $ab.bc.d'.ad.e.ae$ of this prefix after which a learns C. If e is c we are done. Otherwise, we again remove all calls in **e** that involve agents familiar with C. We continue this process until finally a calls c. This has to happen eventually, because a can call each agent at most once and a has to learn C at some point.

Now, consider any prefix $ab.bc.\mathbf{c}.ac$ of P where **c** does not involve agents familiar with C, or any agents familiar with a secret which a is not familiar with after ac. This can be done by repeated use of Lemma 5 on **c** as necessary.

The length of **c** can be between 0 and $n-4$, by Lemma 12. If more calls were in **c** then after $n-1$ calls only two agents would know C, leaving at most $n-3$ calls for $n-2$ agents to learn C. This implies not every agent can be involved in **c**.

Let the number of calls in **c** be p. At most p agents (not including a) can be involved in **c**. Hence, after ac, at most $p+3$ agents have been involved in a call, after $p+3$ calls. This leaves $n-p-5$ calls for all remaining agents to have been involved in a call, with at least $n-p-3$ agents still to have a call.

From here, if we take every call either directly from or directly to this connected component which we shall now refer to a CC. These calls must be made and be available without any calls between two agents not in CC. Hence, at most 1 extra agent can be involved in a call for each of these call. Therefore, we can say that after $q+3$ calls, we have $n-q-5$ calls remaining before all must have been in a call with at least $n-q-3$ agents still to have a call. We shall now refer to all agents yet to be in a call when there are no calls left directly to or from CC as NCC. The next call then must be between two NCC agents (if any such agents exist).

If NCC is empty, then either the last agent called was called after n calls and hence we are done by Lemma 12, or ac was the final call, in which case after n calls only three agents know C after n calls, and we are also done by Lemma 12.

Assume then that NCC is non-empty. We know that the next call cannot be directly involved with CC, hence the call must be between two agents in NCC, say a' and b'. If a' or b' can now make a call to CC, then we have added 2 agents and their secrets to CC, whilst having 2 extra calls, so we are in the same situation and can repeat. If we ever end up with 2 or fewer agents in NCC then by Lemma 12 this can no longer be completed in $2n-4$ calls, as at least $n-2$ calls will have now taken place.

Again, due to Lemma 5 in [9] it must be possible for the next two calls between agents in NCC to involve the same agent (denoted by $\bar{b}$). If this was not the case then the protocol would either terminate, or these agents would be communicating with CC, as NCC is initially totally disconnected. Let $\bar{a}$ and $\bar{c}$ be the other agents involved. By relabelling the names of the agents, we can assume that these two calls are $\bar{a}\bar{b}.\bar{b}\bar{c}$. We now repeat the process as for CC, noting that $\bar{a}$ must call $\bar{c}$ eventually. This includes another r calls involving at most r agents and their secrets. By repeating this argument we get the desired result. $\square$

We can even strengthen our lower bound further in the case of 4 agents.

Theorem 14. *No correct proposition protocol for 4 agents exists with fewer than 6 calls.*

5   Decision Problems for Propositional Protocols

We now move on to analysing the computation complexity of important decision problems for propositional gossip protocols such as checking termination, partial correctness and correctness. We say a protocol terminates if all computations are finite, i.e., there is no way for the scheduler to force the protocol to make an infinite number of calls. We first establish the necessary and sufficient condition for a propositional protocol to terminate.

Lemma 15. *A protocol will not terminate if and only if a call is made without the caller gaining new information within the first $|A|^2$ calls.*

Proof. ($\Leftarrow$) Consider a propositional gossip protocol P. Let $\mathbf{c}.ab$ be a prefix of P such that $\mathbf{c}(\text{root})_a = \mathbf{c}.ab(\text{root})_a$. Hence there exists a rule $\psi \to ab$ for agent a such that $\mathbf{c} \models \psi$. But we clearly have then $\mathbf{c}.ab \models \psi$, because ψ only depends on the secrets agent a knows and they do not change after ab is made. Therefore, call ab can be performed again after $\mathbf{c}.ab$. It is easy to see that $\mathbf{c}(\text{root})_a = \mathbf{c}.(ab)^k(\text{root})_a$ and that $\mathbf{c}.(ab)^k$ is a prefix of P for any k, so P may not terminate.

($\Rightarrow$) Consider any infinite computation $\mathbf{c}$ of P. Along this computation, the current gossip situation can change at most $|A|^2$ times, because each agent can be familiar with at most A secrets. So within the first $|A|^2$ calls of $\mathbf{c}$ there has to be a call after which the caller does not learn any new secrets. $\square$

As we will see, the previous lemma suffices to establish that all the decision problems studied in this section are in co-NP. To show them to be co-NP-hard, we show three different but similar reductions from the well-known 3-SAT problem.

Theorem 16. *The problem of checking if a given propositional gossip protocol P terminates is co-NP-complete.*

Proof (sketch). First, we show the problem to be in co-NP. Due to Lemma 15, to show *non-termination*, it suffices to guess a call sequence $\mathbf{c}$ of length $|A|^2$ and a rule $\psi \to ab$ of P such that $\mathbf{c} \models \psi$ and $\mathbf{c}(\text{root})_a = \mathbf{c}.ab(\text{root})_a$. All of that is of polynomial size and can be checked in polynomial time.

To show the problem is co-NP-hard we will create a polynomial time reduction from the 3-SAT problem, such that termination's NO instances match with 3-SAT's YES instances. The basic idea of this is to have an agent which will become an expert iff the original problem is satisfiable. Once this agent becomes an expert, the scheduler can make that call indefinitely.

One final agent, f, is created. Three agents are created for each variable, one for true (true agent), one for false (false agent), and one to decide the truth assignment to this variable (trigger agent). One agent is created for each variable for each clause, to pass on information for the option above not chosen to the final agent (loser agents). One agent is created for each variable for each clause, to pass on information for the option above chosen to relevant clauses (winner agents). One agent is created for each variable for each clause, to pass on information for the option chosen to the final agent (pass agents). One agent is created for each clause.

The protocol is now set up in such a way that the scheduler has very few options, and each agent has very few calls. For each trigger agent, t, while this agent only knows its own secret T, t wants to call agent tv (variable true) or agent tf (variable false) and the scheduler will choose which. This is essentially the same as determining if the variable is true or false. Whichever is called first, we will take the opposite. Whichever is called first will know only its only secret, and T. At this stage the agent knows it has lost, as it does not know its negation's secret. It still calls the same agents as if it has won, however, these become losers agents when they realise they do not know the extra secret. These now call f and terminate.

t now makes a second call to the winner, and terminates. The winner now knows secrets TV and F. It calls the winner agents in turn. Firstly, these agents call its unique pass agent, which then passes the secrets on to f. This is to ensure that f can become an expert even if a variable is not used to satisfy any clauses. There is one winner agent for every clause. A call is made to the particular clause agent if it will satisfy the clause. This would be easy to do in the set up, as we know what is needed to satisfy each clause. Once a clause is satisfied, it will initiate a call with f. This call is only made once. If we assume the scheduler wants the protocol to not terminate, we just need 3 functions, one for each variable (and accompanying secrets), however, as there are only 7 permutations, we can include all of these to ensure the clause calls f in any circumstance.

Now, f will learn all secrets, apart from clause secrets in any scenario, but will only learn a clause secret if that clause is satisfied. Therefore, f becomes an expert iff all clauses are satisfied. f only makes a call if it becomes an expert, and trivially once this call is made the protocol will never terminate, as it can be repeated indefinitely. It is easy to see that this whole construction can be done in polynomial time and size.   □

Using similar techniques we can show the other two problems are co-NP-complete.

Theorem 17. *The problem of checking if a given propositional gossip protocol P is partially correct is co-NP-complete.*

Theorem 18. *The problem of checking if a given propositional gossip protocol P is correct is co-NP-complete.*

Experimental Evaluation. With the knowledge that the correctness check for a propositional gossip protocol is co-NP-complete, we ran experiments using NUXMV [12] in order to see for how many agents a computer can solve this problem in a reasonable amount of time. The experiments were run on an OMEN by HP Laptop PC - 15-ax000na (ENERGY STAR), with Intel® Core™ i5-6300HQ (2.3 GHz, up to 3.2 GHz, 6 MB cache, 4 cores) microprocessor and 8 GB DDR4-2133 SDRAM (2×4 GB) memory.

Experiments were carried out on LNS protocol, which would return a positive result, and LNS with a single link missing between two agents, which would return a negative result. We simulate the behaviour of the LNS gossip protocol with several optimization as NUXMV processes.

For the correct LNS protocol on 3 and 4 agents the results were almost instant, however on 5 agents results took 4 min, rising to over an hour and a half by 6 agents. At the same time, running the program on the incorrect LNS (when one edge was removed) on 6 agents gave a result in 9 min. This suggests that on large protocols simply running a model checker on the direct encoding of the gossip protocol is not a practical algorithm for checking its correctness.

6   Conclusions

In this paper we solved several open problems about propositional gossip problems proposed in [9], but many interesting questions remain open. One is to further increase the lower bound on the minimal number of calls needed by a correct propositional protocol. No linear upper bound is known at the moment (the $2n - 3$ upper bound from [9] applies to general gossip protocols only). Another is to study simulation and bisimulation between such protocols as proposed in [9]. Finally, finding a practical correctness checking algorithm for propositional protocols would be a challenge as we established its co-NP-hardness.

Acknowledgments. We would like to thank the anonymous reviewers whose comments helped to improve this paper. The first author was supported by EPSRC NPIF PhD studentship. The second author was supported by EPSRC grant EP/P020909/1.

References

1. Anderson, C.J., et al.: NetKAT: semantic foundations for networks. In: The 41st Annual ACM SIGPLAN-SIGACT Symposium on Principles of Programming Languages, POPL 2014, pp. 113–126. ACM (2014)
2. Apt, K.R., Grossi, D., van der Hoek, W.: Epistemic protocols for distributed gossiping. In: Proceedings of the 15th Conference on Theoretical Aspects of Rationality and Knowledge (TARK 2015). EPTCS, vol. 215, pp. 51–66 (2016)
3. Apt, K.R., Grossi, D., van der Hoek, W.: When are two gossips the same? In: Barthe, G., Sutcliffe, G., Veanes, M. (eds.) LPAR-22. 22nd International Conference on Logic for Programming, Artificial Intelligence and Reasoning. EPiC Series in Computing, vol. 57, pp. 36–55. EasyChair (2018)
4. Apt, K.R., Kopczyński, E., Wojtczak, D.: On the computational complexity of gossip protocols. In: Proceedings of the Twenty-Sixth International Joint Conference on Artificial Intelligence, IJCAI 2017, pp. 765–771 (2017)
5. Apt, K.R., Wojtczak, D.: On decidability of a logic of gossips. In: Michael, L., Kakas, A. (eds.) JELIA 2016. LNCS (LNAI), vol. 10021, pp. 18–33. Springer, Cham (2016). https://doi.org/10.1007/978-3-319-48758-8_2
6. Apt, K.R., Wojtczak, D.: Common knowledge in a logic of gossips. In: Proceedings of the 16th Conference on Theoretical Aspects of Rationality and Knowledge (TARK 2017). EPTCS, vol. 251, pp. 10–27 (2017)

7. Apt, K.R., Wojtczak, D.: Decidability of fair termination of gossip protocols. In: Proceedings of the 21st International Conference on Logic for Programming, Artificial Intelligence and Reasoning (LPAR 21). Kalpa Publications in Computing, vol. 1, pp. 73–85 (2017)

8. Apt, K.R., Wojtczak, D.: Verification of distributed epistemic gossip protocols. J. Artif. Intell. Res. (JAIR) **62**, 101–132 (2018)

9. Apt, K.R., Wojtczak, D.: Open problems in a logic of gossips. In: Proceedings Seventeenth Conference on Theoretical Aspects of Rationality and Knowledge (TARK 2019). EPTCS, vol. 297, pp. 1–18 (2019)

10. Attamah, M., Van Ditmarsch, H., Grossi, D., van der Hoek, W.: Knowledge and gossip. In: Proceedings of ECAI 2014, pp. 21–26. IOS Press (2014)

11. Bumby, R.T.: A problem with telephones. SIAM J. Algebraic Discrete Methods **2**(1), 13–18 (1981)

12. Cavada, R., et al.: The nuXmv symbolic model checker. In: Biere, A., Bloem, R. (eds.) CAV 2014. LNCS, vol. 8559, pp. 334–342. Springer, Cham (2014). https://doi.org/10.1007/978-3-319-08867-9_22

13. Cooper, M.C., Herzig, A., Maffre, F., Maris, F., Régnier, P.: Simple epistemic planning: generalised gossiping. In: Proceedings of ECAI 2016. Frontiers in Artificial Intelligence and Applications, vol. 285, pp. 1563–1564. IOS Press (2016)

14. Cooper, M.C., Herzig, A., Maris, F., Vianey, J.: Temporal epistemic gossip problems. In: Slavkovik, M. (ed.) EUMAS 2018. LNCS (LNAI), vol. 11450, pp. 1–14. Springer, Cham (2019). https://doi.org/10.1007/978-3-030-14174-5_1

15. Fagin, R., Halpern, J.Y., Moses, Y., Vardi, M.Y.: Knowledge-based programs. Distrib. Comput. **10**(4), 199–225 (1997)

16. Gattinger, M., Wagemaker, J.: Towards an analysis of dynamic gossip in Netkat. In: Desharnais, J., Guttmann, W., Joosten, S. (eds.) RAMiCS 2018. LNCS, vol. 11194, pp. 280–297. Springer, Cham (2018). https://doi.org/10.1007/978-3-030-02149-8_17

17. Harary, F., Schwenk, A.J.: The communication problem on graphs and digraphs (1974)

18. Hedetniemi, S.M., Hedetniemi, S.T., Liestman, A.L.: A survey of gossiping and broadcasting in communication networks. Networks **18**(4), 319–349 (1988)

19. Herzig, A., Maffre, F.: How to share knowledge by gossiping. In: Rovatsos, M., Vouros, G., Julian, V. (eds.) EUMAS/AT -2015. LNCS (LNAI), vol. 9571, pp. 249–263. Springer, Cham (2016). https://doi.org/10.1007/978-3-319-33509-4_20

20. Herzig, A., Maffre, F.: How to share knowledge by gossiping. AI Commun. **30**(1), 1–17 (2017)

21. Hromkovič, J., Klasing, R., Pelc, A., Ruzicka, P., Unger, W.: Dissemination of Information in Communication Networks - Broadcasting, Gossiping, Leader Election, and Fault-Tolerance. Texts in Theoretical Computer Science. An EATCS Series, Springer, Heidelbergt (2005). https://doi.org/10.1007/b137871

22. Kempe, D., Dobra, A., Gehrke, J.: Gossip-based computation of aggregate information. In: Proceedings of the 44th Annual IEEE Symposium on Foundations of Computer Science, FOCS 2003, pp. 482–491. IEEE (2003)

23. Kermarrec, A., van Steen, M.: Gossiping in distributed systems. Oper. Syst. Rev. **41**(5), 2–7 (2007)

24. Ladin, R., Liskov, B., Shrira, L., Ghemawat, S.: Providing high availability using lazy replication. ACM Trans. Comput. Syst. (TOCS) **10**(4), 360–391 (1992)

25. Tijdeman, R.: On a telephone problem. Nieuw Arch. voor Wiskunde **3**(XIX), 188–192 (1971)

26. van Ditmarsch, H., Grossi, D., Herzig, A., van der Hoek, W., Kuijer, L.B.: Parameters for epistemic gossip problems. In: Proceedings of the 12th Conference on Logic and the Foundations of Game and Decision Theory (LOFT 2016) (2016)

27. van Ditmarsch, H., Kokkinis, I.: The expected duration of sequential gossiping. In: Belardinelli, F., Argente, E. (eds.) EUMAS/AT -2017. LNCS (LNAI), vol. 10767, pp. 131–146. Springer, Cham (2018). https://doi.org/10.1007/978-3-030-01713-2_10

28. van Ditmarsch, H., Kokkinis, I., Stockmarr, A.: Reachability and expectation in gossiping. In: An, B., Bazzan, A., Leite, J., Villata, S., van der Torre, L. (eds.) PRIMA 2017. LNCS (LNAI), vol. 10621, pp. 93–109. Springer, Cham (2017). https://doi.org/10.1007/978-3-319-69131-2_6

29. van Ditmarsch, H., van Der Hoek, W., Kuijer, L.B.: The logic of gossiping. Artif. Intell. **286**, 103306 (2020)

30. van Ditmarsch, H., van Eijck, J., Pardo, P., Ramezanian, R., Schwarzentruber, F.: Epistemic protocols for dynamic gossip. J. Appl. Log. **20**(C), 1–31 (2017)

31. van Ditmarsch, H., van Eijck, J., Pardo, P., Ramezanian, R., Schwarzentruber, F.: Dynamic gossip. Bull. Iran. Math. Soc. **45**, 1–28 (2018)

Complexity of Word Problems
for HNN-Extensions

Markus Lohrey[(⊠)]

Universität Siegen, Siegen, Germany
`lohrey@eti.uni-siegen.de`

Abstract. The computational complexity of the word problem in HNN-extension of groups is studied. HNN-extension is a fundamental construction in combinatorial group theory. It is shown that the word problem for an ascending HNN-extension of a group H is logspace reducible to the so-called compressed word problem for H. The main result of the paper states that the word problem for an HNN-extension of a hyperbolic group H with cyclic associated subgroups can be solved in polynomial time. This result can be easily extended to fundamental groups of graphs of groups with hyperbolic vertex groups and cyclic edge groups.

Keywords: Word problems · HNN-extensions · Hyperbolic groups

1 Introduction

The study of computational problems in group theory goes back more than 100 years. In a seminal paper from 1911, Dehn posed three decision problems [8]: The *word problem*, the *conjugacy problem*, and the *isomorphism problem*. In this paper, we mainly deal with the word problem: It is defined for a finitely generated group G. This means that there exists a finite subset $\Sigma \subseteq G$ such that every element of G can be written as a finite product of elements from Σ. This allows to represent elements of G by finite words over the alphabet Σ. For the word problem, the input consists of such a finite word $w \in \Sigma^*$ and the goal is to check whether w represents the identity element of G.

In general the word problem is undecidable. By a classical result of Boone [5] and Novikov [28], there exist finitely presented groups (finitely generated groups that can be defined by finitely many equations) with an undecidable word problem; see [32] for an excellent exposition. On the positive side, there are many classes of groups with decidable word problems. In his paper from 1912 [9], Dehn presented an algorithm that solves the word problem for fundamental groups of orientable closed 2-dimensional manifolds. This result was extended to one-relator groups (finitely generated groups that can be defined by a single equation) by Dehn's student Magnus [22]. Other important classes of groups with a decidable word problem are:

- automatic groups [11] (including important classes like braid groups [1], Coxeter groups [4], right-angled Artin groups [7], hyperbolic groups [13]),

E. Bampis and A. Pagourtzis (Eds.): FCT 2021, LNCS 12867, pp. 371–384, 2021.
https://doi.org/10.1007/978-3-030-86593-1_26

- finitely generated linear groups, i.e., finitely generated groups that can be faithfully represented by matrices over some field [29] (including polycyclic groups and nilpotent groups), and
- finitely generated metabelian groups (they can be embedded in direct products of linear groups [34]).

With the rise of computational complexity theory in the 1960's, also the computational complexity of group theoretic problems moved into the focus of research. From the very beginning, this field attracted researchers from mathematics as well as computer science. It turned out that for many interesting classes of groups the word problem admits quite efficient algorithms. Lipton and Zalcstein [18] and Simon [31] proved that deterministic logarithmic space (and hence polynomial time) suffices to solve the word problem for a linear group. For automatic groups, the word problem can be solved in quadratic time [11], and for the subclass of hyperbolic groups the word problem can be solved in linear time [16] and belongs to the complexity class LogCFL [19]. For one-relator groups in general, only a non-elementary algorithm is known for the word problem, but for important subclasses polynomial time algorithms are known, see [23,27] for recent progress.

The complexity of the word problem is also preserved by several important group theoretic constructions, e.g. graph products (which generalize free products and direct products) [10] and wreath products [33]. Two other important constructions in group theory are *HNN-extensions* and *amalgamated free products*. A theorem of Seifert and van Kampen links these constructions to algebraic topology. Moreover, HNN-extensions are used in all modern proofs for the undecidability of the word problem in finitely presented groups. For a base group H with two isomorphic subgroups A and B and an isomorphism $\varphi \colon A \to B$, the corresponding HNN-extension is the group

$$G = \langle H, t \mid t^{-1}at = \varphi(a)\,(a \in A)\rangle. \tag{1}$$

Intuitively, it is obtained by adjoing to H a new generator t (the *stable letter*) in such a way that conjugation of A by t realizes φ. The subgroups A and B are also called the *associated subgroups*. If H has a decidable word problem, A and B are finitely generated subgroups of H, and the subgroup membership problems for A and B are decidable, then also the word problem for G in (1) is decidable via *Britton reduction* [6] (iterated application of rewriting steps $t^{-1}at \to \varphi(a)$ and $tbt^{-1} = \varphi^{-1}(b)$ for $a \in A$ and $b \in B$). For the special case where $A = B$ and φ is the identity, it is shown in [33] that the word problem for the HNN-extension G in (1) is NC^1-reducible to the following problems: (i) the word problem for H, (ii) the word problem for the free group of rank two, and (iii) the subgroup membership problem for A. On the other hand, it is not clear whether this result can be extended to arbitrary HNN-extensions (even if we allow polynomial time Turing reductions instead of NC^1-reductions). A concrete open problem is the complexity of the word problem for an HNN-extension $\langle F, t \mid t^{-1}at = \varphi(a)\,(a \in A)\rangle$ of a free group F with finitely generated associated subgroups A and B. The word problem for a free group is known to be in logspace (it is a linear group) [18] and the subgroup membership problem for finitely generated subgroups of a free group can be solved in polynomial time [2]. The problem with Britton reduction in the group $\langle F, t \mid t^{-1}at = \varphi(a)\,(a \in A)\rangle$ is that every Britton reduction step may increase the length of the word by a constant multiplicative factor. This may lead to words of exponential length. One might try to solve this problem by representing the exponentially long words by so-called straight-line programs (context-free grammars that produce a single word). This idea works for the word problems of automorphism groups and certain group extensions [20, Section 4.2]. But it is not clear whether the words that arise from Britton reduction can be compressed down to polynomial size using straight-line programs. The problem arises from the fact that both A and B might be proper subgroups of H. On the other hand, if one of the associated subgroups A and B coincides with the base group H (G is then called an *ascending HNN-extension*) then one can show that the word problem for G is logspace-reducible to the so-called *compressed word problem* for H (Theorem 4). The latter problem asks whether a given straight-line program that produces a word over the generators of H evaluates to the group identity of H. The compressed word problem is known to be solvable in polynomial time for nilpotent groups, virtually special groups, and hyperbolic groups. For every linear group one still has a randomized polynomial time algorithm for the compressed word problem; see [20] for details.

Our main result deals with HNN-extensions, where the associated subgroups A and B are allowed to be proper subgroups of the base group H but are cyclic (i.e., generated by a single element) and *undistored* in H (the latter is defined in Sect. 4). We show that in this situation the word problem for G is polynomial time Turing-reducible to the *compressed power problem* for H (Theorem 7). In the compressed power problem for H, the input consists of two elements $p, q \in H$, where p is given explicitly as a word over a generating set and q is given in compressed form by a straight-line program over a generating set. The question is whether there exists an integer $z \in \mathbb{Z}$ such that $p^z = q$ in H. Moreover, in the positive case we also want to compute such a z.

Our main application of Theorem 7 concerns hyperbolic groups. We show that the compressed power problem for a hyperbolic group can be solved in polynomial time (Theorem 6). For this, we make use of the well-known fact that cyclic subgroups of hyperbolic groups are undistorted. As a consequence of Theorems 6 and 7, the word problem for an HNN-extension of a hyperbolic group with cyclic associated subgroups can be solved in polynomial time (Corollary 1). One should remark that HNN-extensions of hyperbolic groups with cyclic associated subgroups are in general not even automatic; a well-known example is the Baumslag-Solitar group $\mathsf{BS}(1, 2) = \langle a, t \mid t^{-1}at = a^2 \rangle$ [11, Section 7.4].

Corollary 1 can be generalized to fundamental groups of graphs of groups (which generalize HNN-extensions and amalgamated free products) with hyperbolic vertex groups and cyclic edge groups, see the full version [21]. For the special case where all vertex groups are free, the existence of a polynomial time algorithm for the word problem has been stated in [35, Remark 5.11] without proof. For a fundamental group of a graph of groups, where all vertex groups are copies of $\mathbb{Z}$, the word problem can be even solved in logspace [36].

2   Groups

For real numbers $a \leq b$ we denote with $[a, b] = \{r \in \mathbb{R} \mid a \leq r \leq b\}$ the closed interval from a to b. For $k, \ell \in \mathbb{N}$ we write $[k : \ell]$ for $\{i \in \mathbb{N} \mid k \leq i \leq \ell\}$. We use standard notations for words (over some alphabet Σ). As usual, the empty word is denoted with ε. Given a word $w = a_1 a_2 \cdots a_n$ (where $a_1, a_2, \ldots, a_n \in \Sigma$) and numbers $i, j \in \mathbb{N}$ with $1 \leq i \leq j$ we define $w[i : j] = a_i a_{i+1} \cdots a_{\min\{j, n\}}$.

For a group G and a subset $\Sigma \subseteq G$, we denote with $\langle \Sigma \rangle$ the subgroup of G generated by Σ. It is the smallest subgroup of G containing Σ. If $G = \langle \Sigma \rangle$ then Σ is a *generating set* for G. The group G is *finitely generated (f.g.)* if it has a finite generating set. We mostly consider f.g. groups in this paper.

Assume that $G = \langle \Sigma \rangle$ and let $\Sigma^{-1} = \{a^{-1} \mid a \in \Sigma\}$. For a word $w = a_1 \cdots a_n$ with $a_i \in \Sigma \cup \Sigma^{-1}$ we define the word $w^{-1} = a_n^{-1} \cdots a_1^{-1}$. This defines an involution on the free monoid $(\Sigma \cup \Sigma^{-1})^*$. We obtain a surjective monoid homomorphism $\pi \colon (\Sigma \cup \Sigma^{-1})^* \to G$ that preserves the involution: $\pi(w^{-1}) = \pi(w)^{-1}$. We also say that the word w represents the group element $\pi(w)$. For words $u, v \in (\Sigma \cup \Sigma^{-1})^*$ we say that $u = v$ in G if $\pi(u) = \pi(v)$. For $g \in G$ one defines $|g|_\Sigma = \min\{|w| : w \in \pi^{-1}(g)\}$ as the length of a shortest word over $\Sigma \cup \Sigma^{-1}$ representing g. If Σ is clear, we also write $|g|$ for $|g|_\Sigma$. If $\Sigma = \Sigma^{-1}$ then Σ is a finite *symmetric* generating set for G.

We will describe groups by presentations. In general, if H is a group and $R \subseteq H$ is a set of so-called *relators*, then we denote with $\langle H \mid R \rangle$ the quotient group H/N_R, where N_R is the smallest normal subgroup of H with $R \subseteq N_R$. Formally, we have $N_R = \langle\langle \{hrh^{-1} \mid h \in H, r \in R\} \rangle\rangle$. For group elements $g_i, h_i \in H$ $(i \in I)$ we also write $\langle H \mid g_i = h_i \ (i \in I) \rangle$ for the group $\langle H \mid \{g_i h_i^{-1} \mid i \in I\} \rangle$.

In most cases, one takes a free group for the group H from the previous paragraph. Fix a set Σ and let $\Sigma^{-1} = \{a^{-1} \mid a \in \Sigma\}$ be a set of formal inverses of the elements in Σ with $\Sigma \cap \Sigma^{-1} = \emptyset$. A word $w \in (\Sigma \cup \Sigma^{-1})^*$ is called *freely reduced* if it neither contains a factor aa^{-1} nor $a^{-1}a$ for $a \in \Sigma$. For every word $w \in (\Sigma \cup \Sigma^{-1})^*$ there is a unique freely reduced word that is obtained from w by deleting factors aa^{-1} and $a^{-1}a$ ($a \in \Sigma$) as long as possible. The *free group* generated by Σ consists of all freely reduced words together with the multiplication defined by $u \cdot v = \mathrm{nf}(uv)$ for u, v freely reduced. For a set $R \subseteq F(\Sigma)$ of relators we also write $\langle \Sigma \mid R \rangle$ for the group $\langle F(\Sigma) \mid R \rangle$. Every group G that is generated by Σ can be written as $\langle \Sigma \mid R \rangle$ for some $R \subseteq F(\Sigma)$. A group $\langle \Sigma \mid R \rangle$ with Σ and R finite is called *finitely presented*, and the pair (Σ, R) is a *presentation* for the group $\langle \Sigma \mid R \rangle$. Given two groups $G_1 = \langle \Sigma_1 \mid R_1 \rangle$ and $G_2 = \langle \Sigma_2 \mid R_2 \rangle$, where w.l.o.g. $\Sigma_1 \cap \Sigma_2 = \emptyset$, we define their *free product* $G_1 * G_2 = \langle \Sigma_1 \cup \Sigma_2 \mid R_1 \cup R_2 \rangle$.

Consider a group G with the finite symmetric generating set Σ. The *word problem for G w.r.t. Σ* is the following decision problem:

input: a word $w \in \Sigma^*$.

question: does $w = 1$ hold in G?

It is well known that if Σ' is another finite symmetric generating set for G, then the word problem for G w.r.t. Σ' is logspace many-one reducible to the word problem for G w.r.t. Σ. This justifies one to speak just of the word problem for the group G.

HNN-extensions. HNN-extension is an extremely important operation for constructing groups that arises in all parts of combinatorial group theory. Take a group H and a generator $t \notin H$, from which we obtain the free product $H * \langle t \rangle \cong H * \mathbb{Z}$. Assume that $A, B \leq H$ are two isomorphic subgroups of H and let $\varphi \colon A \to B$ be an isomorphism. Then, the group $\langle H * \langle t \rangle \mid t^{-1}at = \varphi(a)\ (a \in A) \rangle$ is called the *HNN-extension of A with associated subgroups A and B* (usually, the isomorphism φ is not mentioned explicitly). The above HNN-extension is usually written as $\langle H, t \mid t^{-1}at = \varphi(a)\ (a \in A) \rangle$. Britton [6] proved the following fundamental result for HNN-extensions. Let us fix a finite symmetric generating set Σ for H.

Theorem 1 (Britton's lemma [6]). *Let $G = \langle H, t \mid t^{-1}at = \varphi(a)\ (a \in A) \rangle$ be an HNN-extension. If a word $w \in (\Sigma \cup \{t, t^{-1}\})^*$ represents the identity of G then w contains a factor of the form $t^{-1}ut$ (resp., tut^{-1}), where $u \in \Sigma^*$ represents an element of A (resp., B).*

A subword of the form $t^{-1}ut$ (resp., tut^{-1}), where $u \in \Sigma^*$ represents an element of A (resp., B) is also called a *pin*.

A simple corollary of Britton's lemma is that H is a subgroup of the HNN-extension $\langle H, t \mid t^{-1}at = \varphi(a)\ (a \in A) \rangle$. Britton's lemma can be also used to solve the word problem for an HNN-extension $\langle H, t \mid t^{-1}at = \varphi(a)\ (a \in A) \rangle$. For this we need several assumptions:

- The word problem for H is decidable.
- There is an algorithm that decides whether a given word $u \in \Sigma^*$ represents an element of A (resp., B).
- Given a word $u \in \Sigma^*$ that represents an element $a \in A$ (resp., $b \in B$), one can compute a word $v \in \Sigma^*$ that represents the element $\varphi(a)$ (resp., $\varphi^{-1}(b)$). Let us denote this word v with $\varphi(u)$ (resp., $\varphi^{-1}(u)$).

Then, given a word $w \in (\Sigma \cup \{t, t^{-1}\})^*$ one replaces pins $t^{-1}ut$ (resp., tut^{-1}) by $\varphi(u)$ (resp., $\varphi^{-1}(u)$) in any order, until no more pins occur. If the final word does not belong to Σ^* then we have $w \neq 1$ in the HNN-extension. If the final word belongs to Σ^* then one uses the algorithm for the word problem of H to check whether it represents the group identity. This algorithm is known as *Britton reduction*.

An HNN-extension $G = \langle H, t \mid t^{-1}at = \varphi(a)\ (a \in A) \rangle$ with $\varphi \colon A \to B$ is called *ascending* if $A = H$ (it is also called the *mapping torus* of φ). Note that we do not require $B = H$. Ascending HNN-extensions play an important role in many group theoretical results. For instance, Bieri and Strebel [3] proved that if N is a normal subgroup of a finitely presented group G such that $G/N \cong \mathbb{Z}$ then G is an ascending HNN-extension of a finitely generated group or contains a free subgroup of rank two.

Hyperbolic Groups. Let G be a f.g. group with the finite symmetric generating set Σ. The *Cayley-graph* of G (with respect to Σ) is the undirected graph $\Gamma =$

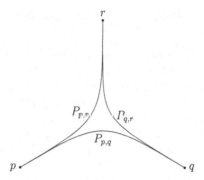

Fig. 1. The shape of a geodesic triangle in a hyperbolic group

$\Gamma(G)$ with node set G and all edges (g, ga) for $g \in G$ and $a \in \Sigma$. We view Γ as a geodesic metric space, where every edge (g, ga) is identified with a unit-length interval. It is convenient to label the directed edge from g to ga with the generator a. The distance between two points p, q is denoted with $d_\Gamma(p, q)$. Note that $|g|_\Sigma = d_\Gamma(1, g)$ for $g \in G$. For $r \geq 0$, let $\mathcal{B}_r(1) = \{g \in G \mid d_\Gamma(1, g) \leq r\}$.

Paths can be defined in a very general way for metric spaces, but we only need paths that are induced by words over Σ. Given a word $w \in \Sigma^*$ of length n, one obtains a unique path $P[w]: [0, n] \to \Gamma$, which is a continuous mapping from the real interval $[0, n]$ to Γ. It maps the subinterval $[i, i+1] \subseteq [0, n]$ isometrically onto the edge (g_i, g_{i+1}) of Γ, where g_i (resp., g_{i+1}) is the group element represented by the word $w[1 : i]$ (resp., $w[1 : i+1]$). The path $P[w]$ starts in $1 = g_0$ and ends in g_n (the group element represented by w). We also say that $P[w]$ is the unique path that starts in 1 and is labelled with the word w. More generally, for $g \in G$ we denote with $g \cdot P[w]$ the path that starts in g and is labelled with w. When writing $u \cdot P[w]$ for a word $u \in \Sigma^*$, we mean the path $g \cdot P[w]$, where g is the group element represented by u.

Let $\lambda, \zeta > 0$, $\epsilon \geq 0$ be real constants. A path $P\,colon[0, n] \to \Gamma$ of the above form is *geodesic* if $d_\Gamma(P(0), P(n)) = n$; it is a (λ, ϵ)-*quasigeodesic* if for all points $p = P(a)$ and $q = P(b)$ we have $|a - b| \leq \lambda \cdot d_\Gamma(p, q) + \epsilon$; and it is ζ-*local* (λ, ϵ)-*quasigeodesic* if for all points $p = P(a)$ and $q = P(b)$ with $|a - b| \leq \zeta$ we have $|a - b| \leq \lambda \cdot d_\Gamma(p, q) + \epsilon$.

A word $w \in \Sigma^*$ is *geodesic* if the path $P[w]$ is geodesic, which means that there is no shorter word representing the same group element from G. Similarly, we define the notion of (ζ-local) (λ, ϵ)-quasigeodesic words. A word $w \in \Sigma^*$ is *shortlex reduced* if it is the length-lexicographically smallest word that represents the same group element as w. For this, we have to fix an arbitrary linear order on Σ. Note that if $u = xy$ is shortlex reduced then x and y are shortlex reduced too. For a word $u \in \Sigma^*$ we denote with $\text{shlex}(u)$ the unique shortlex reduced word that represents the same group element as u (the underlying group G will be always clear from the context).

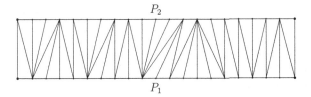

Fig. 2. Paths that asynchronously K-fellow travel

A *geodesic triangle* consists of three points $p, q, r \in G$ and geodesic paths $P_1 = P_{p,q}$, $P_2 = P_{p,r}$, $P_3 = P_{q,r}$ (the three sides of the triangle), where $P_{x,y}$ is a geodesic path from x to y. We call a geodesic triangle δ-*slim* for $\delta \geq 0$, if for all $i \in \{1, 2, 3\}$, every point on P_i has distance at most δ from a point on $P_j \cup P_k$, where $\{j, k\} = \{1, 2, 3\} \setminus \{i\}$. The group G is called δ-*hyperbolic*, if every geodesic triangle is δ-slim. Finally, G is hyperbolic, if it is δ-hyperbolic for some $\delta \geq 0$. Figure 1 shows the shape of a geodesic triangle in a hyperbolic group. Finitely generated free groups are for instance 0-hyperbolic. The property of being hyperbolic is independent of the chosen generating set Σ. The word problem for every hyperbolic group can be decided in real time [16].

Fix a δ-hyperbolic group G with the finite symmetric generating set Σ for the rest of the section, and let Γ be the corresponding geodesic metric space. Consider two paths $P_1 \colon [0, n_1] \to \Gamma$, $P_2 \colon [0, n_2] \to \Gamma$ and let $K \in \mathbb{R}$, $K \geq 0$. The paths P_1 and P_2 *asynchronously K-fellow travel* if there exist two continuous non-decreasing mappings $\varphi_1 \colon [0, 1] \to [0, n_1]$ and $\varphi_2 \colon [0, 1] \to [0, n_2]$ such that $\varphi_1(0) = \varphi_2(0) = 0$, $\varphi_1(1) = n_1$, $\varphi_2(1) = n_2$ and for all $0 \leq t \leq 1$, $d_\Gamma(P_1(\varphi_1(t)), P_2(\varphi_2(t))) \leq K$. Intuitively, this means that one can travel along the paths P_1 and P_2 asynchronously with variable speeds such that at any time instant the current points have distance at most K. If P_1 and P_2 asynchronously K-fellow travel, then by slightly increasing K one obtains a subset $E \subseteq [0 : n_1] \times [0 : n_2]$ with (i) $(0, 0), (n_1, n_2) \in E$, $d_\Gamma(P_1(i), P_2(j)) \leq K$ for all $(i, j) \in E$ and (iii) if $(i, j) \in E \setminus \{(n_1, n_2)\}$ then $(i+1, j) \in E$ or $(i, j+1) \in E$. We write $P_1 \approx_K P_2$ in this case. Intuitively, this means that a ladder graph as shown in Fig. 2 exists, where the edges connecting the horizontal P_1- and P_2-labelled paths represent paths of length $\leq K$ that connect elements from G.

Lemma 1 (c.f. [25, Lemma 1]). *Let P_1 and P_2 be (λ, ϵ)-quasigeodesic paths in Γ and assume that P_i starts in g_i, ends in h_i, and $d_\Gamma(g_1, g_2), d_\Gamma(h_1, h_2) \leq h$. Then there is a constant $K = K(\delta, \lambda, \epsilon, h) \geq h$ such that $P_1 \approx_K P_2$.*

2.1 Compressed Words and the Compressed Word Problem

Straight-line programs offer succinct representations of long words that contain many repeated substrings. We here review the basics, referring to [20] for a more in-depth introduction.

Fix a finite alphabet Σ. A *straight-line program* $\mathcal{G}$ (SLP for short) is a context-free grammar that generates exactly one word $\mathrm{val}(\mathcal{G}) \in \Sigma^*$. More formally, an SLP over Σ is a triple $\mathcal{G} = (V, S, \rho)$ where

- V is a finite set of *variables*, disjoint from Σ,
- $S \in V$ is the *start variable*, and
- $\rho \colon V \to (V \cup \Sigma)^*$ is the *right-hand side mapping*, which is acyclic in the sense that the binary relation $\{(A, B) \in V \times V \mid B \text{ appears in } \rho(A)\}$ is acyclic.

We define the *size* $|\mathcal{G}|$ of $\mathcal{G}$ as $\sum_{A \in V} |\rho(A)|$. The *evaluation* function $\mathrm{val} = \mathrm{val}_{\mathcal{G}} \colon (V \cup \Sigma)^* \to \Sigma^*$ is the unique homomorphism between free monoids such that (i) $\mathrm{val}(a) = a$ for $a \in \Sigma$, and (ii) $\mathrm{val}(A) = \mathrm{val}(\rho(A))$ for $A \in V$. We finally take $\mathrm{val}(\mathcal{G}) = \mathrm{val}(S)$. We call $\mathrm{val}(\mathcal{G})$ the word defined by the SLP $\mathcal{G}$.

Example 1. Let $\Sigma = \{a, b\}$ and fix $n \geq 0$. We define $\mathcal{G}_n = (\{A_0, \ldots, A_n\}, A_n, \rho)$, where $\rho(A_0) = ab$ and $\rho(A_{i+1}) = A_i A_i$ for $0 \leq i \leq n - 1$. It is an SLP of size $2(n + 1)$. We have $\mathrm{val}(A_0) = ab$ and more generally $\mathrm{val}(A_i) = (ab)^{2^i}$. Thus $\mathrm{val}(\mathcal{G}_n) = \mathrm{val}(A_n) = (ab)^{2^n}$.

The SLP $\mathcal{G} = (V, S, \rho)$ is *trivial* if S is the only variable and $\rho(S) = \varepsilon = \mathrm{val}(\mathcal{G})$. An SLP is in *Chomsky normal form* if it is either trivial or all right-hand sides $\rho(A)$ are of the form $a \in \Sigma$ or BC with $B, C \in V$. There is a linear-time algorithm that transforms a given SLP $\mathcal{G}$ into an SLP $\mathcal{G}'$ in Chomsky normal such that $\mathrm{val}(\mathcal{G}) = \mathrm{val}(\mathcal{G}')$; see [20, Proposition 3.8].

The following theorem is the technical main result from [17]:

Theorem 2 (c.f. [17]). *Let G be a hyperbolic group with the finite symmetric generating set Σ. Given an SLP $\mathcal{G}$ over Σ one can compute in polynomial time an SLP $\mathcal{H}$ over Σ such that $\mathrm{val}(\mathcal{H}) = \mathrm{shlex}(\mathrm{val}(\mathcal{G}))$.*

If G is a f.g. group with the finite and symmetric generating set Σ, then we define the *compressed word problem* of G as the following problem:

input: an SLP $\mathcal{G}$ over Σ.
question: does $\mathrm{val}(\mathcal{G})$ represent the group identity of G?

An immediate consequence of Theorem 2 is the following result:

Theorem 3 (c.f. [17]). *The compressed word problem for a hyperbolic group can be solved in polynomial time.*

The compressed word problem turns out to be useful for the solution of the word problem for an ascending HNN-extension:

Theorem 4. *Let H be a finitely generated group. The word problem for an ascending HNN-extension $G = \langle H, t \mid t^{-1}at = \varphi(a) \ (a \in H)\rangle$ is logspace-reducible to the compressed word problem for H.*

The proof is similar to corresponding results for automorphism groups and semi-direct products [20, Section 4.2]; see the full version [21] for details.

We will also need a generalization of straight-line programs, known as composition systems [14, Definition 8.1.2] (in [20] they are called *cut straight-line programs*). A *composition system* over Σ is a tuple $\mathcal{G} = (V, S, \rho)$, with V and S as for an SLP, and where we also allow, as right-hand sides for ρ, expressions of the form $B[i : j]$, with $B \in V$ and $i, j \in \mathbb{N}$, $1 \leq i \leq j$. The numbers i and j are stored in binary encoding. We again require ρ to be acyclic. When $\rho(A) = B[i : j]$ we define $\mathrm{val}(A) = \mathrm{val}(B)[i : j]$. We define the *size* $|\mathcal{G}|$ of the composition system $\mathcal{G}$ as the total number of occurrences of symbols from $V \cup \Sigma$ in all right-hand sides. Hence, a right-hand $B[i : j]$ contributes 1 to the size, and we ignore the numbers i, j. Adding the bit lengths of the numbers i and j to the size $|\mathcal{G}|$ would only lead to a polynomial blow-up for $|\mathcal{G}|$. To see this, first normalize the composition system so that all right-hand sides have the form a, BC or $B[i : j]$ with $a \in \Sigma$ and $B, C \in V$; analogously to the Chomsky normal form of SLPs this can be achieved in polynomial time. If n is the number of variables of the resulting composition system, then every variable produces a string of length at most 2^n. Hence, we can assume that all numbers i, j that appear in a right-hand side $B[i : j]$ are of bit length $\mathcal{O}(n)$.

We can now state an important result of Hagenah; see [14, Algorithmus 8.1.4] as well as [20, Theorem 3.14].

Theorem 5. *There is a polynomial-time algorithm that, given a composition system $\mathcal{G}$, computes an SLP $\mathcal{G}'$ such that $\mathrm{val}(\mathcal{G}) = \mathrm{val}(\mathcal{G}')$.*

It will be convenient to allow in composition systems also more complex right-hand sides. For instance $(ABC)[i : j]D$ would first concatenate the strings produced from A, B, and C. From the resulting string the substring from position i to position j is cut out and this substring is concatenated with the string produced by D.

3   The Compressed Power Problem

In the next section we want to study the word problem in HNN-extensions with cyclic associated subgroups. For this, the following computational problem turns out to be important. Let G be a f.g. group with the finite symmetric generating set Σ. We define the *compressed power problem* for G as the following problem:

input: a word $w \in \Sigma^*$ and an SLP $\mathcal{G}$ over Σ.
output: the binary coding of an integer $z \in \mathbb{Z}$ such that $w^z = \mathrm{val}(\mathcal{G})$ in G if such an integer exists, and *no* otherwise.

Theorem 6. *For every hyperbolic group G, the compressed power problem can be solved in polynomial time.*

Proof. Fix the word $w \in \Sigma^*$ and the SLP $\mathcal{G} = (V, \rho, S)$ over Σ, w.l.o.g. in Chomsky normal form. We have to check whether the equation

$$w^z = \mathrm{val}(\mathcal{G}) \tag{2}$$

has a solution in G, and compute in the positive case a solution $z \in \mathbb{Z}$. Let g be the group element represented by w.

In a hyperbolic group G the order of torsion elements is bounded by a fixed constant that only depends on G, see also the proof of [26, Theorem 6.7]. This allows to check in polynomial time whether g has finite order in G. If g has finite order, say d, then it remains to check for all $0 \le i \le d-1$ whether $w^i = \mathrm{val}(\mathcal{G})$ in G, which can be done in polynomial time by Theorem 3. This solves the case where g has finite order in G.

Now assume that g has infinite order in G. Then (2) has at most one solution. By considering also the equation $(w^{-1})^z = \mathrm{val}(\mathcal{G})$, it suffices to search for a solution $z \in \mathbb{N}$. We can also assume that w is shortlex-reduced. Using techniques from [12] one can further ensure that for every $n \in \mathbb{N}$, w^n is (λ, ϵ)-quasigeodesic for fixed constants λ and ϵ that only depend on the group G; see [21] for details. Finally, by Theorem 2 we can also assume that the word $\mathrm{val}(\mathcal{G})$ (and hence every word $\mathrm{val}(X)$ for X a variable of $\mathcal{G}$) is shortlex-reduced. Hence, if $w^z = \mathrm{val}(\mathcal{G})$ for some $z \in \mathbb{N}$, then by Lemma 1 we have $P[w^z] \approx_K P[\mathrm{val}(\mathcal{G})]$ for a fixed constant K that only depends on G. We proceed in two steps.

Step 1. We compute in polynomial time for all variables $X \in V$ of the SLP $\mathcal{G}$, all group elements $a, b \in \mathcal{B}_K(1)$ (there are only constantly many), and all factors w' of w a bit $\beta[X, a, b, w'] \in \{0, 1\}$ which is set to 1 if and only if (i) $\mathrm{val}(X) = aw'b$ in G and (ii) $P[\mathrm{val}(X)] \approx_K a \cdot P[w']$.

We compute these bits $\beta[X, a, b, w']$ in a bottom-up process where we begin with variables X such that $\rho(X)$ is a terminal symbol and end with the start variable S. So, let us start with a variable X such that $\rho(X) = c \in \Sigma$ and let a, b, w' as above. Then we have to check whether $c = aw'b$ in G and $P[c] \approx_K a \cdot P[w']$. The former can be checked in linear time (it is an instance of the word problem) and the latter can be done in polynomial time as well: we have to check whether the path $a \cdot P[w']$ splits into two parts, where all vertices in the first (resp., second) part belong to $\mathcal{B}_K(1)$ (resp., $\mathcal{B}_K(c)$).

Let us now consider a variable X with $\rho(X) = YZ$ such that all bits $\beta[Y, a, b, w']$ and $\beta[Z, a, b, w']$ have been computed. Let us fix $a, b \in \mathcal{B}_K(1)$ and a factor w' of w. We have $\beta[X, a, b, w'] = 1$ if and only if there exists a factorization $w' = w'_1 w'_2$ and $c \in \mathcal{B}_K(1)$ such that $\beta[Y, a, c, w'_1] = 1$ and $\beta[Z, c^{-1}, b, w'_2] = 1$. This allows us to compute $\beta[X, a, b, w']$ in polynomial time.

Step 2. We compute in polynomial time for all variables $X \in V$, all group elements $a, b \in \mathcal{B}_K(1)$, all proper suffixes w_2 of w, and all proper prefixes w_1 of w the unique number $z = z[X, a, b, w_2, w_1] \in \mathbb{N}$ (if it exists) such that (i) $\mathrm{val}(X) = aw_2 w^z w_1 b$ in G and (ii) $P[\mathrm{val}(X)] \approx_K a \cdot P[w_2 w^z w_1]$. If such an integer z does not exist we set $z[X, a, b, w_2, w_1] = \infty$. Note that the integers $z[X, a, b, w_2, w_1]$ are unique since w represents a group element of infinite order. We represent $z[X, a, b, w_2, w_1]$ in binary encoding. As in step 1, the computation of the numbers $z[X, a, b, w_2, w_1]$ begins with variables X such that $\rho(X)$ is a terminal symbol and ends with the start variable S; see [21] for details. The bits $\beta[X, a, b, w']$ from step 1 are needed in the computation. Finally, $z[S, 1, 1, \varepsilon, \varepsilon]$ is the unique solution of Eq. (2) if $z[S, 1, 1, \varepsilon, \varepsilon] < \infty$. $\square$

4   HNN-extensions with Cyclic Associated Subgroups

Let H be a f.g. group and fix a generating set Σ for H. We say that a cyclic subgroup $\langle g \rangle \leq H$ is *undistorted* in H if there exists a constant δ such that for every $h \in \langle g \rangle$ there exists $z \in \mathbb{Z}$ with $h = g^z$ and $|z| \leq \delta \cdot |h|_\Sigma$ (this definition does not depend on the choice of Σ).[1] This is clearly the case if $\langle g \rangle$ is finite.

Note that if $g, h \in H$ are elements of the same order then the group $\langle H, t \mid t^{-1}gt = h \rangle$ is the HNN-extension $\langle H, t \mid t^{-1}at = \varphi(a) \ (a \in \langle g \rangle) \rangle$, where $\varphi \colon \langle g \rangle \to \langle h \rangle$ is the isomorphism with $\varphi(g^z) = h^z$ for all $z \in \mathbb{Z}$. In the following theorem we consider a slight extension of the word problem for such an HNN-extension $G = \langle H, t \mid t^{-1}gt = h \rangle$ which we call the *semi-compressed word problem* for G. In this problem the input is a sequence $\mathcal{G}_0 t^{\epsilon_1} \mathcal{G}_1 t^{\epsilon_2} \mathcal{G}_2 \cdots t^{\epsilon_n} \mathcal{G}_n$ where every $\mathcal{G}_i$ $(0 \leq i \leq n)$ is an SLP (or a composition system) over the alphabet Σ and $\epsilon_i \in \{-1, 1\}$ for $1 \leq i \leq n$. The question is whether $\mathrm{val}(\mathcal{G}_0) t^{\epsilon_1} \mathrm{val}(\mathcal{G}_1) t^{\epsilon_2} \mathrm{val}(\mathcal{G}_2) \cdots t^{\epsilon_n} \mathrm{val}(\mathcal{G}_n) = 1$ in G.

Theorem 7. *Let H be a fixed f.g. group and let $g, h \in H$ be elements with the same order in H (so that the cyclic subgroups $\langle g \rangle$ and $\langle h \rangle$ are isomorphic) such that $\langle g \rangle$ and $\langle h \rangle$ are undistorted. Then the semi-compressed word problem for the HNN-extension $\langle H, t \mid t^{-1}gt = h \rangle$ is polynomial-time Turing-reducible to the compressed power problem for H.*

Proof. The case where $\langle g \rangle$ and $\langle h \rangle$ are both finite is easy. In this case, by the main result of [15], even the compressed word problem for $\langle H, t \mid t^{-1}gt = h \rangle$ is polynomial time Turing-reducible to the compressed word problem for H, which is a special case of the compressed power problem.

Let us now assume that $\langle g \rangle$ and $\langle h \rangle$ are infinite. Fix a symmetric finite generating set Σ for H. Let $W = \mathcal{G}_0 t^{\epsilon_1} \mathcal{G}_1 t^{\epsilon_2} \mathcal{G}_2 \cdots t^{\epsilon_n} \mathcal{G}_n$ be an input for the semi-compressed word problem for $\langle H, t \mid t^{-1}gt = h \rangle$, where $\mathcal{G}_i$ is a composition system over Σ for $0 \leq i \leq n$ and $\epsilon_i \in \{-1, 1\}$ for $1 \leq i \leq n$. Basically, we do Britton reduction in any order on the word $\mathrm{val}(\mathcal{G}_0) t^{\epsilon_1} \mathrm{val}(\mathcal{G}_1) t^{\epsilon_2} \mathrm{val}(\mathcal{G}_2) \cdots t^{\epsilon_n} \mathrm{val}(\mathcal{G}_n)$. The number of Britton reduction steps is bounded by $n/2$. After the i-th step we have a sequence $U = \mathcal{H}_0 t^{\zeta_1} \mathcal{H}_1 t^{\zeta_2} \mathcal{H}_2 \cdots t^{\zeta_m} \mathcal{H}_m$ where $m \leq n$, $\mathcal{H}_i = (V_i, S_i, \rho_i)$ is a composition system over Σ, and $\zeta_i \in \{-1, 1\}$. Let $u_i = \mathrm{val}(\mathcal{H}_i)$, $s_i = |\mathcal{H}_i|$ and define $s(U) = m + \sum_{i=0}^m s_i$, which is a measure for the encoding length of U. We then search for an $1 \leq i \leq m-1$ such that one of the following two cases holds:

(i) $\zeta_i = -1$, $\zeta_{i+1} = 1$ and there is an $\ell \in \mathbb{Z}$ such that $u_i = g^\ell$ in H.
(ii) $\zeta_i = 1$, $\zeta_{i+1} = -1$ and there is an $\ell \in \mathbb{Z}$ such that $u_i = h^\ell$ in H.

Using oracle access to the compressed power problem for H we can check in polynomial time whether one of these cases holds and compute the corresponding integer ℓ. We then replace the subsequence $\mathcal{H}_{i-1} t^{\zeta_i} \mathcal{H}_i t^{\zeta_{i+1}} \mathcal{H}_{i+1}$ by a composition system $\mathcal{H}_i'$ where $\mathrm{val}(\mathcal{H}_i')$ is $u_{i-1} h^\ell u_{i+1}$ in case (i) and $u_{i-1} g^\ell u_{i+1}$ in case (ii). Let U' be the resulting sequence. It remains to bound $s(U')$. For this we have to

---

[1] The concept of undistorted subgroups is defined for arbitrary finitely generated subgroups but we will need it only for the cyclic case.

bound the size of the composition system $\mathcal{H}'_i$. Assume that $\zeta_i = -1$, $\zeta_{i+1} = 1$, and $u_i = g^\ell$ in H (the case where $\zeta_i = 1$, $\zeta_{i+1} = -1$ and $u_i = h^\ell$ in H is analogous). It suffices to show that h^ℓ can be produced by a composition system $\mathcal{H}''_i$ of size $s_i + O(1)$. Then we can easily bound the size of $\mathcal{H}'_i$ by $s_{i-1} + s_i + s_{i+1} + O(1)$, which yields $s(U') \leq s(U) + O(1)$. This shows that every sequence V that occurs during the Britton reduction satisfies $S(V) \leq S(W) + O(n)$ (recall that W is the initial sequence and that the number of Britton reductions is bounded by $n/2$).

Fix the constant δ such that for every $g' \in \langle g \rangle$ the unique (since g has infinite order) $z \in \mathbb{Z}$ with $g' = g^z$ satisfies $|z| \leq \delta \cdot |g'|_\Sigma$. Hence, we have $|\ell| \leq \delta \cdot |u_i|$. W.l.o.g. we can assume that $\delta \in \mathbb{N}$. The variables of $\mathcal{H}''_i$ are the variables of $\mathcal{H}_i$ plus two new variables A_h and S'_i. Define a morphism η by $\eta(a) = A_h$ for all $a \in \Sigma$ and $\eta(A) = A$ for every variable A of $\mathcal{H}_i$. We define the right-hand side mapping ρ''_i of $\mathcal{H}''_i$ by: $\rho''_i(A_h) = h$ if $\ell \geq 0$ and $\rho''_i(A_h) = h^{-1}$ if $\ell < 0$, $\rho''_i(S'_i) = (S^\delta_i)[1 : |\ell| \cdot |h|]$ and $\rho''_i(A) = \eta(\rho_i(A))$ for all variables A of $\mathcal{H}_i$. Note that S^δ_i derives to $h^{\delta \cdot |u_i|}$ if $\ell \geq 0$ and to $h^{-\delta \cdot |u_i|}$ if $\ell < 0$. Since $|\ell| \leq \delta \cdot |u_i|$, $(S^\delta_i)[1 : |\ell| \cdot |h|]$ derives to h^ℓ. The start variable of $\mathcal{H}''_i$ is S'_i. The size of $\mathcal{H}''_i$ is $s_i + |h| + \delta = s_i + O(1)$, since $|h|$ and δ are constants. $\square$

A subgroup of a hyperbolic group is undistorted if and only if it is quasiconvex [24, Lemma 1.6]. That cyclic subgroups in hyperbolic groups are quasiconvex was shown by Gromov [13, Corollary 8.1.D]. Hence, infinite cyclic subgroups of a hyperbolic group are undistorted. Together with Theorems 6 and 7 we get:

Corollary 1. *Let H be a hyperbolic group and let $g, h \in H$ have the same order. Then the word problem for $\langle H, t \mid t^{-1}gt = h \rangle$ can be solved in polynomial time.*

5   Future Work

There is no hope to generalize Corollary 1 to the case of finitely generated associated subgroups (there exists a finitely generated subgroup A of a hyperbolic group G such that the membership problem for A is undecidable [30]). On the other hand, it is known that the membership problem for quasiconvex subgroups of hyperbolic groups is decidable. What is the complexity of the word problem for an HNN-extension of a hyperbolic group H with finitely generated quasiconvex associated subgroups? Even for the case where H is free (where all subgroups are quasiconvex) the existence of a polynomial time algorithm is not clear.

The best known complexity bound for the word problem of a hyperbolic group is LogCFL, which is contained in the circuit complexity class NC^2. This leads to the question whether the complexity bound in Corollary 1 can be improved to NC. Also the complexity of the compressed word problem for an HNN-extension of a hyperbolic group H with cyclic associated subgroups is open (even in the case where the base group H is free). Recall that the compressed word problem for a hyperbolic group can be solved in polynomial time [17].

Acknowledgments. This work is supported by the DFG project LO748/12-1.

References

1. Artin, E.: Theorie der Zöpfe. Abh. Math. Semin. Univ. Hambg. **4**(1), 47–72 (1925)
2. Avenhaus, J., Madlener, K.: The Nielsen reduction and P-complete problems in free groups. Theoret. Comput. Sci. **32**(1–2), 61–76 (1984)
3. Bieri, R., Strebel, R.: Almost finitely presented soluble groups. Commentarii Mathematici Helvetici **53**, 258–278 (1978)
4. Björner, A., Brenti, F.: Combinatorics of Coxeter Groups. Graduate Texts in Mathematics, vol. 231. Springer, New York (2005). https://doi.org/10.1007/3-540-27596-7
5. Boone, W.W.: The word problem. Ann. Math. Second Series **70**, 207–265 (1959)
6. Britton, J.L.: The word problem. Ann. Math. **77**(1), 16–32 (1963)
7. Charney, R.: An introduction to right-angled Artin groups. Geom. Dedicata. **125**, 141–158 (2007). https://doi.org/10.1007/s10711-007-9148-6
8. Dehn, M.: Über unendliche diskontinuierliche Gruppen. Math. Ann. **71**, 116–144 (1911)
9. Dehn, M.: Transformation der Kurven auf zweiseitigen Flächen. Math. Ann. **72**, 413–421 (1912)
10. Diekert, V., Kausch, J.: Logspace computations in graph products. J. Symb. Comput. **75**, 94–109 (2016)
11. Epstein, D.B.A., Cannon, J.W., Holt, D.F., Levy, S.V.F., Paterson, M.S., Thurston, W.P.: Word Processing in Groups. Jones and Bartlett (1992)
12. Epstein, D.B.A., Holt, D.F.: The linearity of the conjugacy problem in word-hyperbolic groups. Internat. J. Algebra Comput. **16**(2), 287–306 (2006)
13. Gromov, M.: Hyperbolic groups. In: Gersten, S.M. (ed.) Essays in Group Theory. Mathematical Sciences Research Institute Publications, vol. 8, pp. 75–263. Springer, Heidelberg (1987). https://doi.org/10.1007/978-1-4613-9586-7_3
14. Hagenah, C.: Gleichungen mit regulären Randbedingungen über freien Gruppen. Ph.D. thesis, University of Stuttgart (2000)
15. Haubold, N., Lohrey, M.: Compressed word problems in HNN-extensions and amalgamated products. Theory Comput. Syst. **49**(2), 283–305 (2011). https://doi.org/10.1007/s00224-010-9295-2
16. Holt, D.: Word-hyperbolic groups have real-time word problem. Internat. J. Algebra Comput. **10**, 221–228 (2000)
17. Holt, D.F., Lohrey, M., Schleimer, S.: Compressed decision problems in hyperbolic groups. In: 36th International Symposium on Theoretical Aspects of Computer Science, STACS 2019, Berlin, Germany, 13–16 March 2019, LIPIcs, vol. 126, pp. 37:1–37:16. Schloss Dagstuhl - Leibniz-Zentrum für Informatik (2019). http://www.dagstuhl.de/dagpub/978-3-95977-100-9
18. Lipton, R.J., Zalcstein, Y.: Word problems solvable in logspace. J. ACM **24**(3), 522–526 (1977)
19. Lohrey, M.: Decidability and complexity in automatic monoids. Int. J. Found. Comput. Sci. **16**(4), 707–722 (2005)
20. Lohrey, M.: The Compressed Word Problem for Groups. Springer Briefs in Mathematics, Springer, Heidelberg (2014). https://doi.org/10.1007/978-1-4939-0748-9
21. Lohrey, M.: Complexity of word problems for HNN-extensions. CoRR abs/2107.01630 (2021). https://arxiv.org/abs/2107.01630
22. Magnus, W.: Das Identitätsproblem für Gruppen mit einer definierenden Relation. Math. Ann. **106**(1), 295–307 (1932). https://doi.org/10.1007/BF01455888

23. Mattes, C., Weiß, A.: Parallel algorithms for power circuits and the word problem of the Baumslag group. CoRR abs/2102.09921 (2021). https://arxiv.org/abs/2102.09921

24. Minasyan, A.: On products of quasiconvex subgroups in hyperbolic groups. Int. J. Algebra Comput. **14**(2), 173–195 (2004)

25. Myasnikov, A., Nikolaev, A.: Verbal subgroups of hyperbolic groups have infinite width. J. Lond. Math. Soc. **90**(2), 573–591 (2014)

26. Myasnikov, A., Nikolaev, A., Ushakov, A.: Knapsack problems in groups. Math. Comput. **84**, 987–1016 (2015)

27. Myasnikov, A., Ushakov, A., Won, D.W.: The word problem in the Baumslag group with a non-elementary Dehn function is polynomial time decidable. J. Algebra **345**(1), 324–342 (2011)

28. Novikov, P.S.: On the algorithmic unsolvability of the word problem in group theory. Am. Math. Soc. Transl. II. Ser. **9**, 1–122 (1958)

29. Rabin, M.O.: Computable algebra, general theory and theory of computable fields. Trans. Am. Math. Soc. **95**, 341–360 (1960)

30. Rips, E.: Subgroups of small cancellation groups. Bull. Lond. Math. Soc. **14**, 45–47 (1982)

31. Simon, H.U.: Word problems for groups and contextfree recognition. In: Proceedings of Fundamentals of Computation Theory, FCT 1979, pp. 417–422. Akademie-Verlag (1979)

32. Stillwell, J.: Classical Topology and Combinatorial Group Theory, 2nd edn. Springer, Heidelberg (1995). https://doi.org/10.1007/978-1-4612-4372-4

33. Waack, S.: The parallel complexity of some constructions in combinatorial group theory. J. Inf. Process. Cybern. EIK **26**, 265–281 (1990)

34. Wehrfritz, B.A.F.: On finitely generated soluble linear groups. Math. Z. **170**, 155–167 (1980)

35. Weiß, A.: On the complexity of conjugacy in amalgamated products and HNN extensions. Ph.D. thesis, University of Stuttgart (2015)

36. Weiß, A.: A logspace solution to the word and conjugacy problem of generalized Baumslag-Solitar groups. In: Algebra and Computer Science. Contemporary Mathematics, vol. 677. American Mathematical Society (2016)

On Finding Separators in Temporal Split and Permutation Graphs

Nicolas Maack[1], Hendrik Molter[1,2]($\boxtimes$) [iD], Rolf Niedermeier[1] [iD], and Malte Renken[1] [iD]

[1] Algorithmics and Computational Complexity, TU Berlin, Berlin, Germany
nicolas.km.maack@campus.tu-berlin.de,
{rolf.niedermeier,m.renken}@tu-berlin.de
[2] Department of Industrial Engineering and Management,
Ben-Gurion University of the Negev, Beer-Sheva, Israel
molterh@post.bgu.ac.il

Abstract. Disconnecting two vertices s and z in a graph by removing a minimum number of vertices is a fundamental problem in algorithmic graph theory. This (s,z)-Separation problem is well-known to be polynomial solvable and serves as an important primitive in many applications related to network connectivity.

We study the NP-hard Temporal (s,z)-Separation problem on temporal graphs, which are graphs with fixed vertex sets but edge sets that change over discrete time steps. We tackle this problem by restricting the layers (i.e., graphs characterized by edges that are present at a certain point in time) to specific graph classes.

We restrict the layers of the temporal graphs to be either all split graphs or all permutation graphs (both being perfect graph classes) and provide both intractability and tractability results. In particular, we show that in general Temporal (s,z)-Separation remains NP-hard both on temporal split and temporal permutation graphs, but we also spot promising islands of fixed-parameter tractability particularly based on parameterizations that measure the amount of "change over time".

Keywords: Temporal graphs · Connectivity problems · Special graph classes · NP-hardness · Fixed-parameter tractability

1 Introduction

Finding a smallest set of vertices whose deletion disconnects two designated vertices—the separation problem—is a fundamental problem in algorithmic graph theory. The problem, which is a backbone of numerous applications related to network connectivity, is well-known to be polynomial-time solvable in (static)

Based on the Bachelor thesis of N. Maack. H. Molter was supported by the DFG, project MATE (NI 369/17), and by the ISF, grant No. 1070/20. Main part of this work was done while H. Molter was affiliated with TU Berlin. M. Renken was supported by the DFG, project MATE (NI 369/17).

© Springer Nature Switzerland AG 2021
E. Bampis and A. Pagourtzis (Eds.): FCT 2021, LNCS 12867, pp. 385–398, 2021.
https://doi.org/10.1007/978-3-030-86593-1_27

graphs. Driven by the need of understanding and mastering dynamically changing network structures, in recent years the study of temporal graphs—graphs with a fixed vertex set but edge sets that may change over discrete time steps—has enjoyed an enormous growth. One of the earliest systematic studies on temporal graphs dealt with the separation problem [15], where it turned out to be NP-hard. This motivates the study of parameterized complexity aspects as well as of the complexity behavior on special temporal graph classes [12,21]. Continuing and extending this line of research, we provide a first in-depth study on temporal versions of split and permutation graphs, two classes of perfect graphs on which many generally NP-hard problems become polynomial-time solvable [5,13]. We present both intractability as well as (fixed-parameter) tractability results.

Formally, a *temporal graph* is an ordered triple $\mathcal{G} = (V, \mathcal{E}, \tau)$, where V denotes the set of vertices, $\mathcal{E} \subseteq \binom{V}{2} \times \{1, 2, ..., \tau\}$ the set of time-edges where $(\{v, w\}, t) \in \mathcal{E}$ represents an edge between vertices v and w available at time t, and $\tau \in \mathbb{N}$ is the maximum time label. We can think of it as a series of τ static graphs, called *layers*. The graph containing the union of the edges of all layers is called the *underlying graph* of $\mathcal{G}$.

Recently, connectivity and path-related problems have been extensively studied on temporal graphs [2,6–10,12,18,21]. In the temporal setting, paths, walks, and reachability are defined in a time-respecting way [15]: A *temporal (s, z)-walk* (or *temporal walk*) of length ℓ from vertex $s = v_0$ to vertex $z = v_k$ in a temporal graph $\mathcal{G}$ is a sequence $P = ((v_{i-1}, v_i, t_i))_{i=1}^{\ell}$ such that $(\{v_{i-1}, v_i\}, t_i)$ is a time-edge of $\mathcal{G}$ for all $i \in \{1, 2, ..., \ell\}$ and $t_i \leq t_{i+1}$ for all $i \in \{1, 2, ..., \ell-1\}$.[1] A temporal walk is a temporal *path* if it visits every vertex at most once.

We study the TEMPORAL (s, z)-SEPARATION problem. Here, a *temporal (s, z)-separator* is a set of vertices (not containing s and z) whose removal destroys all temporal paths from s to z.

TEMPORAL (s, z)-SEPARATION

Input: A temporal graph $\mathcal{G} = (V, \mathcal{E}, \tau)$, two distinct vertices $s, z \in V$, and $k \in \mathbb{N}$.

Question: Does $\mathcal{G}$ admit a temporal (s, z)-separator of size at most k?

TEMPORAL (s, z)-SEPARATION is NP-complete [15] and W[1]-hard when parameterized by the separator size k [21]. On the positive side, one can verify a solution in $\mathcal{O}(|\mathcal{G}|)$ time (see e.g. Bui-Xuan et al. [6]). Zschoche et al. [21] investigated the differences between the computational complexity of finding temporal separators that remove (non-strict) temporal paths vs. *strict* temporal paths. Fluschnik et al. [12] studied the impact of restrictions on the layers or the underlying graph on the computational complexity of TEMPORAL (s, z)-SEPARATION and found that it remains NP-complete even under severe restrictions. In particular, the problem stays NP-complete and W[1]-hard when parameterized by the separator size even if every layer contains only one edge and for several restrictions of the underlying graph [12]. They further investigated the case

---

[1] Such walks are also called "non-strict", whereas "strict" walks require $t_i < t_{i+1}$. We focus on non-strict walks in this work.

where every layer of the temporal graph is a unit interval graph and obtained fixed-parameter tractability for TEMPORAL (s, z)-SEPARATION when parameterized by the lifetime τ and the so-called "shuffle number", a parameter that measures how much the relative order of the intervals changes over time. This result initiated research on the amount of "change over time" measured by a parameter that is tailored to the graph class into which all layers fall.

In our work, we follow this paradigm of restricting each layer to a certain graph class and measuring the amount of change over time with parameters that are tailored to the graph class of the layers. More specifically, we investigate the complexity of TEMPORAL (s, z)-SEPARATION on temporal graphs where every layer is a *split graph* or every layer is a *permutation graph*.

Temporal Split Graphs. In a split graph, the vertex set can be partitioned into a clique and an independent set. Split graphs can be used to model an idealized form of core-periphery structures, in which there exists a densely connected core and a periphery that only has connections to that core [4]. Core-periphery structures can be observed in social contact networks in which one group of people meets at some location, forming a fully connected core. Meanwhile, other people associated with the group may have contact with some of its members, but otherwise do not have any interactions relevant to the observed network. Such applications are naturally subject to change over time, for example due to vertices entering or leaving the network.[2] This temporal aspect is captured by *temporal split graphs*, where each layer has a separate core-periphery split.

We prove that TEMPORAL (s, z)-SEPARATION remains NP-complete on temporal split graphs, even with only four layers. On the positive side, we show that TEMPORAL (s, z)-SEPARATION is solvable in polynomial time on temporal graphs where the partition of the vertices into a clique and an independent set stays the same in every layer. We use this as a basis for a "distance-to-triviality"-parameterization [14,19], also motivated by the assumption that in application cases the core-periphery structure of a network will roughly stay intact over time with only few changes. Intuitively, we parameterize on how many vertices may switch between the two parts over time and use this parameterization to obtain fixed-parameter tractability results. Formally, we show fixed-parameter tractability for the combined parameter "number of vertices different from s and z that switch between the clique and the independent set at some point in time" and the lifetime τ.

Temporal Permutation Graphs. A permutation graph on an ordered set of vertices is defined by a permutation of that ordering. Two vertices are connected by an edge if their relative order is inverted by the permutation. They were introduced by Even et al. [11] and appear in integrated circuit design [20], memory layout optimization [11], and other applications [13]. In a *temporal permutation graph*, the edges of each layer are given by a separate permutation. We prove that TEMPORAL (s, z)-SEPARATION remains NP-complete on temporal permutation graphs. We then parameterize on how much the permutation changes over

---

[2] While the vertex set of a temporal graph formally remains unchanged, isolated vertices are equivalent to non-existing vertices as far as separators are concerned.

time to obtain fixed-parameter tractability results. We use the *Kendall tau* distance [16] to measure the dissimilarity of the permutations. More precisely, we obtain fixed-parameter tractability for the combined parameter "sum of Kendall tau distances between consecutive permutations" and the separator size k. We remark that in a similar context, the Kendall tau distance has also been used by Fluschnik et al. [12] to measure the amount of change over time in a temporal graph.

We remark that most of our results are not tight in the sense that our fixed-parameter tractability results use combined parameters and we currently cannot exclude fixed-parameter tractability for all single parameters. We point out open questions in the conclusion. Some proofs (marked by $\star$) are deferred to a full version [17].

2   Split Graphs

In this section, we study the computational complexity of finding temporal separators in *temporal split graphs*. Split graphs represent an idealized model of core-periphery structures with a well-connected core and a periphery only connected to that core [4]. In terms of numbers, they also constitute the majority of all chordal graphs [1].

Formally, a graph $G = (V, E)$ is called a *split graph* if V can be partitioned into two sets C, I such that C induces a clique and I induces an independent set. Then, (C, I) is called a *split partition* of G. In general, a split graph may admit multiple split partitions. A *temporal split graph* is a temporal graph $\mathcal{G}$ of which every layer is a split graph. A *temporal split partition* $(C_t, I_t)_{t=1}^\tau$ then contains a split partition of every layer of $\mathcal{G}$.

2.1   Hardness Results

The fact that TEMPORAL (s, z)-SEPARATION on temporal split graphs is NP-hard can be derived from a result of Fluschnik et al. [12] stating that TEMPORAL (s, z)-SEPARATION is hard on temporal graphs containing a single edge per layer. This is due to the fact that a graph with a single edge is clearly a split graph.

We now strengthen this result, showing that TEMPORAL (s, z)-SEPARATION is NP-hard on temporal split graphs with only a constant number of layers. We do this by building on a reduction by Zschoche et al. [21] showing NP-hardness of TEMPORAL (s, z)-SEPARATION on general temporal graphs.

Here and in the following we use "separator" as a shorthand for "temporal (s, z)-separator" when no ambiguity arises.

Theorem 1 ($\star$). TEMPORAL (s, z)-SEPARATION *is NP-hard on temporal split graphs with four layers.*

Theorem 1 shows that TEMPORAL (s, z)-SEPARATION on temporal split graphs is NP-hard when $\tau \geq 4$. Evidently, TEMPORAL (s, z)-SEPARATION is polynomial-time solvable for $\tau = 1$. The computational complexity for the cases $\tau = 2$ and $\tau = 3$ remains open.

2.2   Fixed-Parameter Tractability Results

The defining characteristic of temporal split graphs is that for each layer t they can be split into a clique C_t and an independent set I_t. While temporal split graphs allow for changes of this partition, for several application scenarios described in the introduction it is reasonable to assume that only a few of these changes occur while most vertices retain their role throughout all layers.

We will prove that TEMPORAL (s, z)-SEPARATION can be solved efficiently in this setting, that is, if the number of *switching vertices* $\bigcup_{t,t'} C_t \cap I_{t'}$ is low. Note that the set of switching vertices depends on the choice of partitions (C_t, I_t)—we will subsequently show how to compute these partitions.

We start by proving that TEMPORAL (s, z)-SEPARATION is polynomial-time solvable if the only switching vertices are s and z.

Lemma 2. *Let* $\mathcal{G} = (V \cup \{s, z\}, \mathcal{E}, \tau)$ *be a temporal split graph with a given temporal split partition having no switching vertices except possibly* s *and* z. *Then all minimal temporal* (s, z)*-separators in* $\mathcal{G}$ *can be found in* $\mathcal{O}(|\mathcal{G}| \cdot \tau)$ *time.*

Proof. We assume that there is never an edge between s and z, otherwise the problem is trivial. Let the given temporal split partition be $(C_t, I_t)_t$. Let $C := C_t \setminus \{s, z\}$ and $I := I_t \setminus \{s, z\}$ (for some and thus all layers t). We show that all minimal temporal (s, z)-separators are given by the set

$$\mathcal{S} := \left\{ \bigcup_{0 < t \leq i} (N_{G_t}(s) \cap C) \cup \bigcup_{i < t \leq \tau} (N_{G_t}(z) \cap C) \cup T \;\middle|\; 0 \leq i \leq \tau \right\}, \text{ where}$$

$$T := \bigcup_{1 \leq t \leq t' \leq \tau} N_{G_t}(s) \cap N_{G_{t'}}(z).$$

The set $\mathcal{S}$ can be constructed in $\mathcal{O}(|\mathcal{G}| \cdot \tau)$ time and contains at most $\tau + 1$ elements. This proves the stated time bound. It remains to verify that $\mathcal{S}$ contains all minimal separators.

First, note that T contains exactly those vertices which form temporal (s, z)-paths of length 2. Thus T has to be contained in any separator.

So it only remains to consider temporal (s, z)-paths of length at least 3. If such a path P contains a vertex $v \in I$, then let $(\{u, v\}, t)$ and $(\{v, w\}, t')$ be the two time-edges of P containing v. Then we must have $u \in C$ or $w \in C$ and thus the above two time-edges can be replaced by either $(\{u, w\}, t)$ or $(\{u, w\}, t')$, shortening P by one. Consequently, if $\mathcal{G}$ contains a temporal (s, z)-path of length at least 3, then there is also a temporal (s, z)-path in $\mathcal{G} - I$.[3] So it suffices to consider temporal paths in $\mathcal{G} - I$.

Thus it becomes clear that each element of $\mathcal{S}$ is in fact a temporal (s, z)-separator. Now let S be an arbitrary temporal (s, z)-separator and i maximal with $S \supseteq \bigcup_{0 < t \leq i} N_{G_t}(s) \cap C$. Thus, S does not contain some vertex $v \in$

---

[3] We denote by $\mathcal{G} - X$ the temporal graph resulting from removing vertices in X from the vertex set of $\mathcal{G}$.

$N_{G_{i+1}}(s) \cap C$. Then, $\mathcal{G} - S$ contains a path from s to every vertex in $C \setminus S$ since v is connected to all other vertices in C. So starting from layer $i + 1$, all edges to z from a vertex in $C \setminus S$ would complete a temporal path from s to z. Hence, $S \supseteq \bigcup_{i<t\leq\tau} N_{G_t}(z) \cap C$, concluding the proof.   □

Next, we first show that this is still an NP-hard problem and then give an FPT-algorithm for the solution size parameter.

Proposition 3 (⋆). *For temporal split graphs, it is NP-hard to compute a minimum-size set of switching vertices.*

Proposition 4 (⋆). *For a temporal split graph $\mathcal{G} = (V, \mathcal{E}, \tau)$, one can find a temporal split partition minimizing the number of switching vertices in $\mathcal{O}(|\mathcal{E}| + \tau \cdot |V| + |V|^2 \cdot (1.2738^p + p \cdot |V|))$ time, where p is the minimum number of switching vertices.*

Based on Lemma 2 and Proposition 4, we can now develop a fixed-parameter algorithm for the parameter lifetime τ combined with the parameter number p of switching vertices apart from s, z.

Theorem 5. *Let $\mathcal{G}$ be a temporal split graph with at most p switching vertices apart from s and z. Then TEMPORAL (s, z)-SEPARATION on $\mathcal{G}$ can be solved in $\mathcal{O}\left((\tau + 1)^{3^p(p+1)}|\mathcal{G}| + 1.2738^p \cdot |V|^2 + p \cdot |V|^3\right)$ time.*

Proof. We begin with using Proposition 4 to compute a temporal split partition of $\mathcal{G}$. We then use induction to show that for all values of p there are at most $(\tau + 1)^{3^p(p+1)}$ minimal temporal (s, z)-separators, which can all be found in $D(\tau + 1)^{3^p(p+1)}|\mathcal{G}|$ time for some constant D.

For the case $p = 0$, by Lemma 2, we can find all minimal separators of which there are at most $\tau + 1$, in $\mathcal{O}(|\mathcal{G}| \cdot \tau)$ time.

Now for the induction step suppose that our claim holds whenever the number of switching vertices (apart from s, z) is at most $p - 1$. We choose a switching vertex v from $\mathcal{G}$ ($v \notin \{s, z\}$). The subgraph $\mathcal{G} - v$ then contains $p - 1$ switching vertices apart from s, z, therefore we can find all its minimal separators in $D(\tau + 1)^{3^{p-1}p}|\mathcal{G}|$ time. Since a separator of $\mathcal{G}$ is also a separator of $\mathcal{G} - v$, every minimal separator of $\mathcal{G}$ must contain a separator of $\mathcal{G} - v$. We will base our separators for $\mathcal{G}$ on the minimal separators of $\mathcal{G} - v$, henceforth called *base separators*, by finding all possible combinations of vertices that can be added to turn them into minimal separators of $\mathcal{G}$.

Because we only added v, all temporal (s, z)-paths left in $\mathcal{G}$ after removing a separator of $\mathcal{G} - v$ must pass through v. Thus any separator of $\mathcal{G}$ must either contain v or some other set of vertices that cuts all these paths. To do the latter, it has to ensure that all remaining temporal (s, v)-paths arrive after the latest layer in which a temporal (v, z)-path can begin. In other words, such a separator of $\mathcal{G}$ needs to contain a temporal (s, v)-separator for the layers from 1 to some layer t, and a temporal (v, z)-separator from $t + 1$ to τ. We can compute all minimal separators of this form by applying the induction hypothesis to enumerate all temporal (s, v)-separators in layers 1 through t of $\mathcal{G} - \{z\}$ and all temporal (v, z)-separators in layers $t + 1$ through τ of $\mathcal{G} - \{s\}$. Note that both of these are temporal split graphs with at most $p - 1$ switching vertices (not counting s, v, and z).

So for any given t, there are no more than $((\tau + 1)^{3^{p-1}p})^2$ possible separator combinations. Additionally we have the option of simply taking v. As there are $\tau + 1$ options for t and $(\tau + 1)^{3^{p-1}p}$ base separators to choose from, the overall number of minimal temporal (s, z)-separators is thus upper-bounded by

$$(\tau + 1)^{3^{p-1}p}\left(\left((\tau + 1)^{3^{p-1}p}\right)^2(\tau + 1) + 1\right) \leq (\tau + 1)^{3^p(p+1)}.$$

We require $D(\tau + 1)^{3^{p-1}p}|\mathcal{G}|$ time to find all base separators. In addition, for each $t \in \{0, \ldots, \tau\}$ we need $2D(\tau + 1)^{3^{p-1}p}|\mathcal{G}|$ time to compute all minimal (s, v)- and (v, z)-separators. Afterwards we need $\mathcal{O}(|V|)$ time to output each found separator. The overall time required thus is

$$D(\tau + 1)^{3^{p-1}p}|\mathcal{G}| \cdot (1 + 2(\tau + 1)) + D(\tau + 1)^{3^{p-1}2p}|V|$$
$$\leq D(\tau + 1)^{3^p(p+1)}|\mathcal{G}|$$

where we assumed $\tau \geq 3$. This completes the induction. Together with the time required for Proposition 4, we obtain an overall upper time bound of

$$\mathcal{O}\left((\tau + 1)^{3^p(p+1)}|\mathcal{G}| + |\mathcal{E}| + \tau \cdot |V| + |V|^2 \cdot (1.2738^p + p \cdot |V|)\right)$$
$$\subseteq \mathcal{O}\left((\tau + 1)^{3^p(p+1)}|\mathcal{G}| + 1.2738^p \cdot |V|^2 + p \cdot |V|^3\right).$$

$\square$

We leave open whether TEMPORAL (s, z)-SEPARATION is fixed-parameter tractable for the single parameter number p of switching vertices.

3 Permutation Graphs

In this section, we investigate the complexity of finding temporal separators in *temporal permutation graphs*. A permutation graph is defined by a ordered set of vertices (say $1, \ldots, n$) and any permutation of that vertex set. Two vertices are connected by an edge if and only if their relative order is inverted by the permutation.

Formally, we call a graph $G = (V, E)$ with vertex set $V = [n] := \{1, 2, \ldots, n\}$ a *permutation graph*, if there exists a permutation $\pi : V \to V$ of the vertices such that for any two vertices $v < w$, we have $\{v, w\} \in E$ if and only if $\pi(w) < \pi(v)$. Clearly, π is then uniquely determined by G. Note that we follow the definition of Even et al. [11] in which the vertices are already labeled; some authors also define a permutation graph as any (unlabeled) graph for which such a labeling can be found [13].

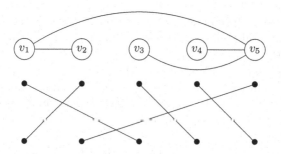

Fig. 1. An example of a permutation graph and the corresponding matching diagram.

We call a temporal graph $\mathcal{G} = ([n], \mathcal{E}, \tau)$ a *temporal permutation graph* if there exist τ permutations $\pi_1, \ldots, \pi_\tau$ of the vertices such that, for any two vertices $v < w$, we have $(\{v, w\}, t) \in \mathcal{E}$ if and only if $\pi_t(w) < \pi_t(v)$.[4]

One can visualize a permutation with a *matching diagram* [11]. A matching diagram for a given permutation graph $G = ([n], E)$ with permutation π can be constructed by drawing n points on a horizontal line and another n points on a parallel line below it. Then each vertex i is represented by a straight line segment connecting the i-th point on the top line to the $\pi(i)$-th point on the bottom line. Observe that two vertices share an edge in G if and only if their corresponding line segments cross in the matching diagram. Figure 1 provides an example for a permutation graph and the matching diagram of its underlying permutation.

First, we prove in Sect. 3.1 that TEMPORAL (s, z)-SEPARATION on temporal permutation graphs is NP-complete. In Sect. 3.2, we use the *Kendall tau* distance [16] to measure the dissimilarity of the permutations. Recall that the Kendall tau distance is a metric that counts the number of pairwise disagreements between two total orderings; it is also known as "bubble sort distance" since it measures the number of swaps needed to transform one permutation into the other. We show that TEMPORAL (s, z)-SEPARATION becomes fixed-parameter tractable when parameterized by the combined parameter "sum of Kendall tau distances between consecutive permutations" and the separator size k.

3.1   Hardness Results

As is the case for split graphs, the class of permutation graphs contains all graphs with only one edge. This means that TEMPORAL (s, z)-SEPARATION for temporal graphs in which every layer is a permutation graph is NP-hard [12]. We can also show that TEMPORAL (s, z)-SEPARATION remains NP-hard when restricted to the temporal permutation graphs (note that this means that the vertex ordering is the same for all layers).

---

[4] Note that it is not sufficient for each layer to be isomorphic to a permutation graph.

Theorem 6 ($\star$). TEMPORAL (s, z)-SEPARATION *is NP-complete on temporal permutation graphs.*

We remark that the reduction we use to obtain Theorem 6 uses an unbounded number of time steps and also the maximum Kendall tau distance between any two consecutive permutations is unbounded. However, by introducing additional layers, one can decrease the Kendall tau distance between any two consecutive layers to one. The main idea is to gradually change a layer to an independent set and then gradually to the next layer. This can be done in a way that does not introduce any new temporal paths.

3.2 Fixed-Parameter Tractability Results

In this section, we examine the effect of limiting how much the permutations of the layers of the temporal permutation graph change. We can do so by measuring the Kendall tau distance $\tau_K(\pi, \pi')$ between the permutations π, π' of two consecutive layers, which is defined as

$$\tau_K(\pi, \pi') = |\{(i, j) \mid i < j \wedge (\pi(i) - \pi(j))(\pi'(i) - \pi'(j)) < 0\}|.$$

We will show that TEMPORAL (s, z)-SEPARATION on temporal permutation graphs is fixed-parameter tractable with respect to the sum of Kendall tau distances between all pairs of consecutive permutations plus separator size k. For this we need to demonstrate that these parameters limit the number of (s, z)-separators that a layer does not have in common with another layer.

First, we introduce the concept of *scanlines*. A scanline is any line segment in the matching diagram of a permutation with one end on each horizontal line. If s lies on one side of the scanline and z on the other, then the set of all vertices whose line segments cross the scanline is an (s, z)-separator. We call such a separator a *scanline separator*. Bodlaender et al. [3] have shown that every minimal separator in a permutation graph is a scanline separator and that there are at most $(n-1)(2k+3)$ scanline separators of size at most k [3, Proof of Lemma 3.6], where n is the number of vertices.

Lemma 7. *Let $G_1 = ([n], E_1)$ and $G_2 = ([n], E_2)$ be two permutation graphs. If the two corresponding permutations have Kendall tau distance d, then the number of scanline separators of size at most k in G_2 that are not also scanline separators in G_1 is at most $d \cdot (2k + 1)$.*

Proof. We begin by showing that for every endpoint on the bottom line, there are at most $2k + 1$ separators of size at most k defined by a scanline with that endpoint. For a given bottom endpoint, we look at the leftmost endpoint on the top line that defines such a separator. Moving right, every time we pass an element of the permutation, its line is added to those crossed by the scanline if it was completely to the right of it before, and removed from them if it crossed the scanline before. This corresponds to adding or removing one vertex from the defined separator.

Since our scanline can pass every element only once, the vertices that were added in this process will not be removed again. If we have already passed more than $2k$ points on the top line, then more than k of these represent vertices which were not present in the initial separator and were thus added. Therefore the resulting separator contains more than k vertices. Therefore, all separators of size at most k must be produced in the first $2k$ steps of this process. Together with the separator defined by the initial scanline, this gives us at most $2k + 1$ scanline separators as claimed.

Upon moving from G_1 to G_2, the set of lines crossing any given scanline only changes, if any of the d swaps swapped the two points immediately to the left and right of the lower end of the scanline. By the above, this means at most $d \cdot (2k + 1)$ many scanlines produce different separators in G_1 and G_2. Hence, the number of new scanline separators in G_2 does not exceed $d \cdot (2k + 1)$.   □

Now we show that reachability in temporal permutation graphs follows the vertex order: If a vertex reaches another vertex, then it also reaches all vertices in between.

Lemma 8 ($\star$). *Let $G = ([n], E)$ be a permutation graph and $v, w, x \in [n]$ three vertices with $v < w < x$. If there exists a path from v to x, then there also exist paths from w to both, v and x.*

We now present a parameterized algorithm for solving TEMPORAL (s, z)-SEPARATION on a temporal permutation graph $\mathcal{G}$. For this, we introduce the parameter $d_{\Sigma} := \sum_{t=1}^{\tau-1} d_{\mathrm{Kt}}(\pi_t, \pi_{t+1})$, where d_{Kt} denotes the Kendall tau distance and π_t is the t-th permutation of $\mathcal{G}$. Note that taking the maximum instead of the sum does not provide a helpful parameter since the hardness reduction we used to obtain Theorem 6 can be modified such that the Kendall tau distance of any two consecutive layers is one.

Theorem 9. TEMPORAL (s, z)-SEPARATION *can be solved on a temporal permutation graph in* $\mathcal{O}((d_{\Sigma}(2k + 1))^k n \cdot |\mathcal{E}| + \tau n^2)$ *time.*

Proof. We present an algorithm that runs in $\mathcal{O}((d_{\Sigma}(2k+1))^k n \cdot |\mathcal{E}| + \tau n^2)$ time, which determines whether a given TEMPORAL (s, z)-SEPARATION-instance has a solution (see Algorithm 1).

Remember that the total number of scanline separators of size at most k in layer 1 is at most $(n - 1)(2k + 3)$. Furthermore, in all layers after layer 1, the number of minimal (i.e., scanline) separators which are not shared with the previous layer is at most $d_{\Sigma}(2k + 1)$ (see Algorithm 7). Hence the first call of GETSEPARATOR iterates at most $(n-1)(2k+3)+d_{\Sigma}(2k+1) \in \mathcal{O}(d_{\Sigma}(2k + 1) \cdot n)$ times and every recursive call iterates at most $d_{\Sigma}(2k + 1)$ times.

Due to the condition in Line 14, every time a recursive call is made, the set passed to S contains at least one vertex more than before, but never exceeds the size k. Thus the maximum recursion depth is k. In every call of GETSEPARATOR it is checked in $\mathcal{O}(|E|)$ time whether S is a temporal (s, z)-separator. This results in a running time of $\mathcal{O}\left((d_{\Sigma}(2k + 1))^k n \cdot |\mathcal{E}|\right)$ for the initial call of GETSEPARATOR.

Algorithm 1

Input: A TEMPORAL (s, z)-SEPARATION-instance $\mathcal{I} = (\mathcal{G} = ([n], \mathcal{E}, \tau), s, z, k)$, where $\mathcal{G}$ is a temporal permutation graph and $s, z \in [n]$

Output: true if $\mathcal{I}$ is a yes-instance, false otherwise

1: compute $\pi_1, \ldots, \pi_\tau$
2: let seplist be an empty list
3: append the set of scanline separators of size at most k in $G_1(\mathcal{G})$ to seplist
4: **for** $i \in \{2, ..., \tau\}$ **do**
5:     let seps be the set of scanline separators of size at most k in $G_i(\mathcal{G})$ that are not scanline separators in $G_{i-1}(\mathcal{G})$
6:     **if** seps is not empty **then**
7:         append seps to seplist
8: output GETSEPARATOR($\emptyset$, 1)

9: **function** GETSEPARATOR(S, i)
10:     **if** S is a temporal (s, z)-separator of $\mathcal{G}$ **then**
11:         **return** true
12:     **for** $j \in \{i, \ldots, \tau\}$ **do**
13:         **for** $S' \in$ seplist$[j]$ **do**
14:             **if** $|S| < |S \cup S'| \leq k$ **then**
15:                 **if** GETSEPARATOR($S \cup S'$, $j + 1$) **then**
16:                     **return** true
17:     **return** false

It remains to show that Lines 1 through 7 can be executed in $\mathcal{O}(\tau n^2)$ time. To construct some permutation π_i, we first iterate once through the edges of G_i and build, for each vertex $v \in [n]$, the set $I(v) := \{w \in [n] \mid w < v \text{ and } \{v, w\} \in E(G_i)\}$. Then we can incrementally construct π_i from an empty tuple by going through all vertices in ascending order and inserting each vertex v exactly to the left of all elements of $I(v)$. If $I(v)$ is implemented using a hash set, then this takes $\mathcal{O}(n^2)$ time for each layer. Afterwards, building seplist again takes $\mathcal{O}(n^2)$ time for each layer as there are n^2 potential scanlines and each only requires constant checking time if they are all iterated in order.

Correctness. It is easy to see that Algorithm 1 will never output true when $\mathcal{I}$ is a no-instance, as the condition in Line 10 can only evaluate to true if there exists a temporal (s, z)-separator.

It remains to be shown that Algorithm 1 will always output true when $\mathcal{I}$ is a yes-instance. Without loss of generality we assume that $s < z$. We also assume that there exists a minimal temporal (s, z)-separator S^* of size at most k in $\mathcal{G}$. Due to Lemma 8, every layer t in $\mathcal{G} - S^*$ has some farthest reachable vertex f_t between s and z such that until time t, s can reach all vertices v with $s \leq v \leq f_t$ via temporal paths but no vertex v with $v > f_t$. Clearly $f_t \leq f_{t+1}$. This means that in each layer t, S^* must contain a scanline separator that separates f_t from all vertices $v > f_t$. We denote this scanline separator by S_t^*. By minimality of S^*, $S^* = \bigcup_{t=1}^{\tau} S_t^*$.

Trivially, the following property holds for the initial call of GETSEPARATOR:

$$\bigcup_{t=1}^{i-1} S_t^* \subseteq S \subseteq S^* \tag{$*$}$$

We next show that whenever ($*$) holds for a call of GETSEPARATOR, then either $S = S^*$ or ($*$) also holds for some recursive call. This implies that some recursive call will eventually produce S^*.

So assume now ($*$) holds with $S \neq S^*$. Since S is then not a temporal (s, z)-separator, GETSEPARATOR(S, i) iterates through all scanline separators in seplist$[i]$ through seplist$[\tau]$, one of which must be the first scanline separator S_t^* which is not already contained in S. When it gets to that separator it makes a recursive call GETSEPARATOR$(S \cup S_t^*, i + 1)$. This recursive call then again satisfies ($*$).   □

We leave open whether TEMPORAL (s, z)-SEPARATION is fixed-parameter tractable or becomes W[1]-hard when parameterized by either only the separator size k or only the sum d_Σ of Kendall tau distances of permutations of consecutive layers.

4   Conclusion

We showed that TEMPORAL (s, z)-SEPARATION remains NP-complete on temporal split graphs even when there are only $\tau \geq 4$ layers, but it becomes fixed-parameter tractable when parameterized by the lifetime τ combined with the number p of "switching vertices", that is, vertices that switch between the independent set and the clique. We leave open whether one can obtain fixed-parameter tractability when only parameterizing by p. Another natural restriction we can place on temporal split graphs is limiting the size of the independent set for all layers. We conjecture that TEMPORAL (s, z)-SEPARATION is fixed-parameter tractable with respect to the maximum size of the independent set.

We also showed that TEMPORAL (s, z)-SEPARATION remains NP-complete on temporal permutation graphs, but becomes fixed-parameter tractable with respect to the separator size k plus the sum d_Σ of Kendall tau distances of permutations of consecutive layers. We left open the complexity of TEMPORAL (s, z)-SEPARATION on temporal permutation graphs when parameterized by the lifetime τ. Whether the problem stays fixed-parameter tractable or becomes W[1]-hard when parameterized by either only the separator size k or only the sum d_Σ of Kendall tau distances of permutations of consecutive layers remains open as well.

Lastly, we leave for future research whether our results also hold in the *strict* case, that is, when the temporal paths that are to be destroyed by the separator have strictly increasing time labels. Most of our algorithms heavily rely on the fact that temporal paths may use several time edges with the same label, and hence they can presumably not be adapted to the strict setting in a straightforward way.

References

1. Bender, E.A., Richmond, L.B., Wormald, N.C.: Almost all chordal graphs split. **38**(2), 214–221 (1985). https://doi.org/10.1017/S1446788700023077
2. Bentert, M., Himmel, A.-S., Nichterlein, A., Niedermeier, R.: Efficient computation of optimal temporal walks under waiting-time constraints. Appl. Netw. Sci. **5**(1), 73 (2020). https://doi.org/10.1007/s41109-020-00311-0
3. Bodlaender, H.L., Kloks, T., Kratsch, D.: Treewidth and pathwidth of permutation graphs. **8**, 606–616 (1995). https://doi.org/10.1137/S089548019223992X
4. Borgatti, S.P., Everett, M.G.: Models of core/periphery structures. **21**(4), 375–395 (2000). https://doi.org/10.1016/S0378-8733(99)00019-2
5. Brandstädt, A., Le, V.B., Spinrad, J.P.: Graph Classes–A Survey (1999). https://doi.org/10.1137/1.9780898719796
6. Xuan, B.B., Ferreira, A., Jarry, A.: Computing shortest, fastest, and foremost journeys in dynamic networks. **14**(02), 267–285 (2003). https://doi.org/10.1142/S0129054103001728
7. Buß, S., Molter, H., Niedermeier, R., Rymar, M.: Algorithmic aspects of temporal betweenness, pp. 2084–2092 (2020). https://doi.org/10.1145/3394486.3403259
8. Enright, J., Meeks, K., Mertzios, G.B., Zamaraev, V.: Deleting edges to restrict the size of an epidemic in temporal networks, **119**, 60–77 (2021) https://doi.org/10.1016/j.jcss.2021.01.007
9. Enright, J., Meeks, K., Skerman, F.: Assigning times to minimise reachability in temporal graphs, **115**, 169–186 (2021). https://doi.org/10.1016/j.jcss.2020.08.001
10. Erlebach, T., Hoffmann, M., Kammer, F.: On temporal graph exploration. **119**, 1–18 (2021). https://doi.org/10.1016/j.jcss.2021.01.005
11. Even, S., Pnueli, A., Lempel, A.: Permutation graphs and transitive graphs. **19**(3), 400–410 (1972). https://doi.org/10.1145/321707.321710
12. Fluschnik, T., Molter, H., Niedermeier, R., Renken, M., Zschoche, P.: Temporal graph classes: a view through temporal separators. **806**, 197–218 (2020). https://doi.org/10.1016/j.tcs.2019.03.031
13. Golumbic, M.C.: Algorithmic Graph Theory and Perfect Graphs (2004). ISBN 978-0-444-51530-8
14. Guo, J., Hüffner, F., Niedermeier, R.: A structural view on parameterizing problems: distance from triviality. In: Downey, R., Fellows, M., Dehne, F. (eds.) IWPEC 2004. LNCS, vol. 3162, pp. 162–173. Springer, Heidelberg (2004). https://doi.org/10.1007/978-3-540-28639-4_15
15. Kempe, D., Kleinberg, J., Kumar, A.: Connectivity and inference problems for temporal networks. **64**(4), 820–842 (2002). https://doi.org/10.1006/jcss.2002.1829
16. Kendall, M.G.: A new measure of rank correlation. **30**(1/2), 81–93 (1938). https://doi.org/10.2307/2332226
17. Maack, N., Molter, H., Niedermeier, R., Renken, M.: On finding separators in temporal split and permutation graphs (2021). http://arxiv.org/abs/2105.12003
18. Mertzios, G.B., Michail, O., Chatzigiannakis, I., Spirakis, P.G.: Temporal network optimization subject to connectivity constraints. In: Fomin, F.V., Freivalds, R., Kwiatkowska, M., Peleg, D. (eds.) ICALP 2013. LNCS, vol. 7966, pp. 657–668. Springer, Heidelberg (2013). https://doi.org/10.1007/978-3-642-39212-2_57

19. Niedermeier, R.: Invitation to fixed-parameter algorithms. (2006). https://doi.org/10.1093/ACPROF:OSO/9780198566076.001.0001
20. Sen, A., Deng, H., Guha, S.: On a graph partition problem with application to VLSI layout. **43**(2), 87–94 (1992). https://doi.org/10.1016/0020-0190(92)90017-P
21. Zschoche, P., Fluschnik, T., Molter, H., Niedermeier, R.: The complexity of finding small separators in temporal graphs. **107**, 72–92 (2020). https://doi.org/10.1016/j.jcss.2019.07.006

The Possible Winner Problem with Uncertain Weights Revisited

Marc Neveling, Jörg Rothe[iD], and Robin Weishaupt[✉]

Institut für Informatik, Heinrich-Heine-Universität Düsseldorf, Düsseldorf, Germany
{marc.neveling,rothe,robin.weishaupt}@hhu.de

Abstract. Baumeister et al. [8] introduced the possible winner with uncertain weights problem which, given a weighted election with the weights of some voters as yet unspecified, asks whether one can assign weights to these voters such that a distinguished candidate wins. Solving all questions they specifically left open *for nonnegative integer weights*, we show that two variants of this problem for 3-approval and four variants for plurality with runoff can be solved efficiently. In addition, we study variants of this problem for Borda, k-veto, and veto with runoff in terms of their computational complexity. Finally, we also prove that the problem of constructive control by adding voters in succinct representation belongs to P for plurality with runoff and veto with runoff.

1 Introduction

Over the previous two decades, computational social choice—with its many applications to collective decision making—has evolved into a central subarea of artificial intelligence and, in particular, multiagent systems. Looking into the textbook edited by Brandt et al. [12], one of the most intensively studied problems in computational social choice alongside manipulation, control, and bribery is the *possible winner problem* that Konczak and Lang [24] were the first to study. Generalizing the (unweighted) coalitional manipulation problem [13] and being a special case of swap bribery [14], in this problem we are given an election with only partial (not total) preferences over the candidates and a designated candidate c, and we ask whether one can extend the partial preferences to total ones to make c a winner. This problem and variations thereof as well as its companion, the *necessary winner problem* [24,33], have been studied by many authors for many voting rules—see, e.g., the very recent survey by Lang [26] and the references cited therein. The idea underlying the possible and necessary winner problems (to determine the winners in the presence of incomplete preferences for *some* or for *all* extensions to total preferences) is so fundamental that it has been applied successfully to many other areas, including *judgment aggregation* [3,4], *fair division* [2,25], *hedonic games* [23], and *abstract argumentation* [5,6,28,32].

While most work on the possible winner problem is concerned with an unweighted variant of the problem, Baumeister et al. [7] were the first to consider its *weighted* variant. Baumeister et al. [8] also introduced and studied another variant of the weighted possible winner problem, the *possible winner with uncertain weights problem*, where the uncertainty concerns the voters' weights instead of their preferences: Given a

© Springer Nature Switzerland AG 2021
E. Bampis and A. Pagourtzis (Eds.): FCT 2021, LNCS 12867, pp. 399–412, 2021.
https://doi.org/10.1007/978-3-030-86593-1_28

weighted election with (total preferences over the candidates and) the weights of some voters as yet unspecified, can one assign weights to these voters such that a designated candidate wins?

Based on the ever-increasing exchange of—almost real-time—data in our modern society, the possible winner with uncertain weights model can be applied to all kinds of elections today better than ever. For almost every election taking place, some pre-election or some polls are done and the results are published. These results can be translated by suitable probability-based methods into ranges for likely weights, upper bounds on the total weight, etc., so that the results presented here, together with the previous results by Baumeister et al. [8], provide powerful means to improve election forecasts in efficient ways.

Furthermore, with blockchain-related technologies advancing and experiencing more and more mainstream adoption, some elections start to take place on-chain entirely. Characteristic for these elections is that they satisfy full transparency, i.e., everyone can check who is eligible to vote, who has voted and how, who has not voted, etc. In this setting, the possible winner with uncertain weights problem can be applied quite beneficially: If there is an efficient algorithm available to determine, given uncertainty about the voters' weights, whether some distinguished candidate has a chance of winning or whether there is no hope, such an algorithm could be run periodically to make a good prediction.

Baumeister et al. [8] introduced a general framework for the possible winner with uncertain weights problem, both for nonnegative integer and nonnegative rational weights, with and without upper bounds on the total weight to be distributed, and with and without ranges to choose the weights from, and they studied the resulting problems in terms of their computational complexity for scoring protocols such as k-approval, for plurality with runoff, Copeland, ranked pairs, and (a simplified variant of) Bucklin and fallback voting.

Continuing this line of research, we solve all questions they specifically left open *for nonnegative integer weights*:

(a) with no restriction whatsoever (apart from nonnegative integer weights);
(b) with both an upper bound on the total weight to be distributed and with ranges to choose the weights from;
(c) with only an upper bound on the total weight; and
(d) with only ranges to choose the weights from.

Specifically, we provide polynomial-time algorithms for the variants (b) and (c) of this problem for 3-approval (Sect. 3) and for all four variants (a)–(d) for plurality with runoff (Sect. 4). Furthermore, we study the complexity of the same four variants of this problem for veto with runoff (Sect. 5) and for other prominent scoring protocols—namely, k-veto (Sect. 6) and Borda (Sect. 7)—and establish both NP-completeness results (for Borda in cases (a)–(d) and for k-veto in cases (b) and (c) for $k \geq 3$) and polynomial-time algorithms (in all other cases). Relatedly, in Sects. 4 and 5, we also prove that the problem of constructive control by adding voters in succinct representation can be solved in polynomial time for plurality with runoff and veto with runoff. Finally, in Sect. 8, we summarize our results in Table 1.

2   Preliminaries

An *election* (C,V) is given by its set C of candidates and its list V of votes expressing the voters' preferences over the candidates. We assume that each vote is represented by a (strict) linear preference order. For example, if there are four candidates in $C = \{a,b,c,d\}$ and a voter prefers b to c, c to d, and d to a, we write this vote as $b > c > d > a$. When possible, we represent elections *succinctly*, i.e., identical votes are not listed one by one but just once along with a number in binary representation giving the multiplicity of this vote.

A *voting rule* determines the winner(s) of a given election. Most voting rules we consider are *scoring protocols*: For m candidates, a *scoring vector* $(\alpha_1, \alpha_2, \ldots, \alpha_m)$ of integers with $\alpha_1 \geq \alpha_2 \geq \cdots \geq \alpha_m$ specifies the points the candidates receive from each vote based on their position in it, i.e., a candidate in the ith position of a vote receives α_i points from it, and the *score of candidate $c \in C$ in election* (C,V), denoted by $score_V(c)$, is the sum of the points c receives from the votes in V. Specifically, we consider the prominent scoring protocols (for m candidates) that are based on the following scoring vectors:

k-**approval:** $(1, \ldots, 1, 0, \ldots, 0)$, where the first $k \leq m$ entries are ones (1-approval is also known as *plurality*);
k-**veto:** $(1, \ldots, 1, 0, \ldots, 0)$, where the last $k \leq m$ entries are zeros (1-veto is also simply known as *veto*); and
Borda: $(m-1, m-2, \ldots, 0)$.

In addition to these scoring protocols, we consider the following rules:

Plurality with runoff: proceeds in two stages. In the first stage, all candidates except the two candidates with the highest and the second highest plurality score are eliminated. In the second stage (the runoff), among the two remaining candidates and with votes restricted to these, the candidate with the highest plurality score wins. In both stages, we use some predefined tie-breaking rule to determine the two candidates that proceed to the runoff and the overall winner in case there are ties.
Veto with runoff: works just like plurality with runoff, except it uses veto scores in both stages to determine who proceeds to the runoff and who is the overall winner.

In an unweighted election, each vote has unit weight. We will consider *weighted* elections where each vote comes with a weight. For example, if a vote has weight two, the scores that the candidates receive from this vote are doubled. Note further that we consider the *nonunique-winner model*, which means that it is enough for the distinguished candidate to be one among possibly several candidates with the highest (possibly weighted) score to win the election.

In our proofs, we make use of the following notation. Let $S \subseteq C$ be a set of candidates. When $\overrightarrow{S}$ appears in a vote, the candidates from S are ranked in any fixed (e.g., the lexicographical) order; when $\overleftarrow{S}$ appears in a vote, the candidates from S are ranked in the reverse order; when S appears in a vote (without an arrow on top), the order in which the candidates from S are ranked here does not matter for our argument; and when $\cdots$ appears in a vote, the order in which the remaining candidates occur does not matter.

For example, if $C = \{a,b,c,d\}$, $S = \{b,c\}$ and we use a lexicographical order, then $\overrightarrow{S} > d > a$ means $b > c > d > a$; $\overleftarrow{S} > d > a$ means $c > b > d > a$; and both $S > d > a$ and $\cdots > d > a$ indicate any one of $b > c > d > a$ and $c > b > d > a$. Further, for an election (C,V) and two candidates $c,d \in C$, we use $diff_{(C,V)}(c,d) = score_V(c) - score_V(d)$ to denote the difference of the scores of c and d in (C,V). By $diff_{(\{c,d\},V)}(c,d)$ we denote the difference of the scores of c and d in the head-to-head contest which is $(\{c,d\},V)$ with the votes in V being tacitly reduced to only c and d.

We now recall the definition of the problems introduced by Baumeister et al. [8] that we are interested in. For a given voting rule $\mathcal{E}$, define the problem:[1]

$\mathcal{E}$-POSSIBLE-WINNER-WITH-UNCERTAIN-WEIGHTS-$\mathbb{N}$ ($\mathcal{E}$-PWUW-$\mathbb{N}$)

Given: A set C of candidates, a list V_1 of unit-weight votes over C, a list V_0 of votes over C with unspecified weights, and a distinguished candidate c.

Question: Is there an assignment of weights $w_i \in \mathbb{N}$ for all $v_i \in V_0$, $1 \leq i \leq |V_0|$, such that c wins the weighted election $(C, V_1 \cup V_0)$ under $\mathcal{E}$?

Baumeister et al. [8] also introduced the following problem variants:

- $\mathcal{E}$-PWUW-RW-$\mathbb{N}$: In addition to the components of a $\mathcal{E}$-PWUW-$\mathbb{N}$ instance, we are given a set $R = \{R_1, \ldots, R_{|V_0|}\}$ of *regions* (or, intervals) $R_i = [l_i, r_i] \subseteq \mathbb{N}$, and the question is the same as for $\mathcal{E}$-PWUW-$\mathbb{N}$, except that each weight w_i is additionally required to be chosen from R_i.
- $\mathcal{E}$-PWUW-BW-$\mathbb{N}$: In addition to the components of a $\mathcal{E}$-PWUW-$\mathbb{N}$ instance, we are given a bound B, and the question is the same as for $\mathcal{E}$-PWUW-$\mathbb{N}$, except that the total weight of the votes in V_0 is additionally required to not exceed B, i.e., $\sum_{i=1}^{|V_0|} w_i \leq B$.
- $\mathcal{E}$-PWUW-BW-RW-$\mathbb{N}$ incorporates both the restrictions of $\mathcal{E}$-PWUW-RW-$\mathbb{N}$ and $\mathcal{E}$-PWUW-BW-$\mathbb{N}$.

Baumeister et al. [8] comprehensively discuss these definitions and justify their choices (for example, why votes in V_1 have unit weight and why weight-zero votes in V_0 are allowed and that this corresponds to modeling *control by deleting voters* as defined by Bartholdi et al. [1] and Hemaspaandra et al. [21]), and we refer the reader to this discussion.

We assume the reader to be familiar with the basic concepts of computational complexity, such as the complexity classes P and NP and the notions of NP-*hardness* and NP-*completeness*, based on *polynomial-time many-one reductions*, denoted by $A \leq_m^p B$. For more background on complexity theory, we refer to the textbooks by Garey and Johnson [18], Papadimitriou [29], and Rothe [30].

---

[1] Baumeister et al. [8] define and study these problems for nonnegative integer and nonnegative rational weights, and accordingly append either $\mathbb{N}$ or $\mathbb{Q}^+$ to their problems. We will consider the case of nonnegative integer weights only, but for clarity and consistency with the literature, we append the suffix "-$\mathbb{N}$" to our problems.

3   3-Approval

Baumeister et al. [8] have shown that PWUW-RW-$\mathbb{N}$ and PWUW-$\mathbb{N}$ are in P for 3-approval but left it open how hard it is to solve PWUW-BW-$\mathbb{N}$ or PWUW-BW-RW-$\mathbb{N}$ for this voting rule. Solving these open questions, we show that these two problems are in P as well. Our proof also shows that 3-approval-PWUW-RW-$\mathbb{N}$ is in P, establishing an alternative proof for this already known result. We first define the problem GENERALIZED-WEIGHTED-B-EDGE-MATCHING (GWBEM), which belongs to P and which we will use to prove membership of 3-approval-PWUW-BW-RW-$\mathbb{N}$ in P.[2]

GENERALIZED-WEIGHTED-B-EDGE-MATCHING (GWBEM)
Given: An undirected multigraph $G = (N, E)$ without loops, capacity-bounding functions $a_\ell, a_u \colon E \to \mathbb{N}$ and $b_\ell, b_u \colon N \to \mathbb{N}$, a weight function $w \colon E \to \mathbb{N}$, and a target integer $r \in \mathbb{N}$.
Question: Does there exist a function $x \colon E \to \mathbb{N}$ with $\sum_{e \in E} w(e)x(e) \geq r$ such that for every edge $e \in E$ it holds that $a_\ell(e) \leq x(e) \leq a_u(e)$ and for every node $z \in N$ it holds that $b_\ell(z) \leq \sum_{e \in \delta(z)} x(e) \leq b_u(z)$, where $\delta(z)$ is the set of edges incident to node z?

Next, we define a useful notation to simplify our proofs.

Definition 1. *Let* $(C, V_1 \cup V_0)$ *be a weighted election for a scoring protocol. For each* $c \in C$, *$score_{V_1}(c)$ denotes the score of c according to the (unit-weight) votes in V_1, and* $score_{V_0}(c)$ *denotes the score of c according to the weighted votes in V_0 once they are specified.*

We now present two lemmas (their proofs and some other proofs are omitted due to space limitations) that enable us to make generic assumptions about the problem instances used later on. By the first lemma, we may assume that, w.l.o.g., all intervals $R_i = [l_i, r_i] \in R$ fulfill $l_i = 0$ for every $\mathcal{E}$-PWUW-BW-RW-$\mathbb{N}$ instance $I = (C, V_1, V_0, R, B, c)$.

Lemma 1. *Let* $I = (C, V_1, V_0, R, B, c)$ *be an* $\mathcal{E}$-PWUW-BW-RW-$\mathbb{N}$ *instance for a scoring protocol* $\mathcal{E}$. *Then there exists an instance* $I' = (C, V_1', V_0, R', B', c)$ *with* $l_i' = 0$ *for all* $R_i' = [l_i', r_i'] \in R'$ *such that* $I \in \mathcal{E}$-PWUW-BW-RW-$\mathbb{N}$ *if and only if* $I' \in \mathcal{E}$-PWUW-BW-RW-$\mathbb{N}$.

Next, we present a lemma that allows us to make a statement about which weights for which votes in V_0 are greater than 0.

Lemma 2. *Let C, c, V_1, and V_0 be elements of a* $\mathcal{E}$-PWUW *instance for some scoring protocol* $\mathcal{E}$. *If there exist weights* $w_i \in \mathbb{N}$ *for* $v_i \in V_0$, $1 \leq i \leq |V_0|$, *such that c wins the weighted* $\mathcal{E}$ *election* $(C, V_1 \cup V_0)$, *then there exists an alternative weight assignment, where c still wins the election, with weights* $w_i' \in \mathbb{N}$ *for the votes in V_0, such that* $w_i' > 0$ *holds if and only if v_i assigns a positive score to c.*

---

[2] Formally, Gabow [17] and Grötschel et al. [20, p. 259] define a maximization variant of GWBEM and show its polynomial-time solvability, which immediately implies that GWBEM is in P. The same problem was first used by Lin [27] in the context of voting and later on also, for example, by Erdélyi et al. [15].

With Lemma 2 we may assume that, without loss of generality, for any 3-approval-PWUW-BW-RW-$\mathbb{N}$ instance $I = (C, V_1, V_0, R, B, c)$ where c wins, only votes in V_0 with c among the top three candidates have positive weight.

We are now ready to prove that 3-approval-PWUW-BW-RW-$\mathbb{N}$ is efficiently solvable. We do so by reducing this problem to GWBEM.

Theorem 1. *3-approval*-PWUW-BW-RW-$\mathbb{N}$ *is in* P.

Proof. In order to prove that 3-approval-PWUW-BW-RW-$\mathbb{N}$ is in P, we show that 3-approval-PWUW-BW-RW-$\mathbb{N}$ $\leq_m^P$ GWBEM. Let $I = (C, V_1, V_0, R, B, c)$ be a 3-approval-PWUW-BW-RW-$\mathbb{N}$ instance. According to Lemma 1, we assume for all $v_i \in V_0$ that $R_i = [0, r_i]$ holds. We can also assume $\sum_{i=1}^{|V_0|} r_i \geq B$, since the sum over the r_i provides an upper bound for the overall weight distributed among the votes in V_0.[3] We construct a GWBEM instance $I' = (G, a_\ell, a_u, b_\ell, b_u, w, r)$ with multigraph $G = (C', E)$ whose set of nodes $C' = C \setminus \{c\}$ consists of all candidates except c. Furthermore, denote the subset of votes from V_0 where candidate c is among the top three positions by $V_0' = \{x_1 > x_2 > x_3 > \cdots \in V_0 \mid c \in \{x_1, x_2, x_3\}\}$ and define the set of edges for G as

$$E = \{\{x_1, x_2, x_3\} \setminus \{c\} \mid x_1 > x_2 > x_3 > \cdots \in V_0'\}.$$

That is, for every vote from V_0 with c among the top three positions we add an edge between the vote's remaining two candidates. This can result in a multigraph, of course, but is in line with the problem definition. We set the target integer $r = B$ and, for every edge $e \in E$ linked to $v_i \in V_0$, we define the edge capacity bounds by $a_\ell(e) = 0$ and $a_u(e) = r_i$ and the weight function by $w(e) = 1$ for the corresponding interval $R_i = [0, r_i]$. For every $c' \in C'$, we define the node capacity bounds by $b_\ell(c') = 0$ and $b_u(c') = score_{V_1}(c) + B - score_{V_1}(c')$. We can assume for every candidate $c' \in C'$ (which is, recall, a node in G) that $b_u(c') \geq 0$ holds, as otherwise I would be a trivial NO-instance. That is the case, since $b_u(c') < 0$ implies $score_{V_1}(c') - score_{V_1}(c) > B$, which makes it impossible for c to beat c' with votes from V_0 having a total weight of at most B. Obviously, the construction of I' can be realized in time polynomial in $|I|$. We now provide our intuition on how to prove that $I \in$ 3-approval-PWUW-BW-RW-$\mathbb{N}$ if and only if $I' \in$ GWBEM. The full proof of correctness is omitted due to space constraints.

From left to right, we know that there exist weights $w_i \in \mathbb{N}$ such that c wins the weighted election and we can assume that the sum of these weights equals B. By Lemma 2, we know that only votes having c among its top three candidates have positive weight. Assigning these weights to the edges of G according to the corresponding votes allows us to obtain a weight allocation for I'. One can now validate that this weight assignment satisfies all requirements of I'. That the sum of the edges' weights in G equals r is easy to see. Furthermore, the edges' weights satisfy their upper and lower limits as these limits correspond to the corresponding intervals from I. Finally, every node in G satisfies its upper and lower limits because of the facts that c is a winner of the weighted election and c's overall score equals to $score_{V_1}(v) + B$. Hence, I' is a YES-instance.

---

[3] This assumption does make sense indeed: Otherwise, the parameter B would be meaningless and our instance becomes a PWUW-RW-$\mathbb{N}$ instance which, as mentioned earlier, is efficiently solvable.

From right to left, we know that there exists a weight assignment for the edges in G satisfying all requirements. By transferring these weights to the votes in V_0' and setting the weights of all remaining votes in $V_0 \backslash V_0'$ to 0, we obtain a weight assignment for I. All weights are within their intervals since the edges' weights fulfill their upper and lower limits, too. From the fact that all vertices in G satisfy their upper and lower limits, one can conclude by a simple calculation that c is a winner of the weighted election. Finally, we can calculate that c has a point lead over all other candidates, such that we can reduce some of the positively weighted weights in order to have the overall sum of weights assigned to the votes in V_0 to equal B. Consequently, all requirements of I are satisfied and therefore, I is a YES-instance.   □

From this result it immediately follows that the remaining two problem variants belong to P, too. The proof of Corollary 1 (omitted here due to space constraints) uses the reductions given by Baumeister et al. [8] and can be analogously applied to other rules as well, as we will do later on.[4]

Corollary 1. *For 3-approval, the problems* PWUW-BW-$\mathbb{N}$ *and* PWUW-RW-$\mathbb{N}$ *are in* P.

Let $\mathcal{E}$ be a scoring protocol. We say that $\mathcal{E}$ is a *binary scoring protocol* if all entries of its score vector are from $\{0, 1\}$. For example, k-veto and k-approval are binary scoring protocols for all $k \in \mathbb{N}$.

Theorem 2. *For any binary scoring protocol* $\mathcal{E}$, *the problems* $\mathcal{E}$-PWUW-RW-$\mathbb{N}$ *and* $\mathcal{E}$-PWUW-$\mathbb{N}$ *are in* P.

The proof of Theorem 2 (omitted due to space constraints) is also used later on in the proof that $\mathcal{E}$-PWUW-RW-$\mathbb{N}$ or -PWUW-$\mathbb{N}$ for some voting rule $\mathcal{E}$ is in P. Note that this result also provides an alternative proof that 3-approval-PWUW-RW-$\mathbb{N}$ and 3-approval-PWUW-$\mathbb{N}$ are in P, as 3-approval is a binary voting protocol.

4   Plurality with Runoff

For plurality with runoff, Baumeister et al. [8] showed that all four possible winner problems with uncertain weights are in P when the weights can be rational. For nonnegative integer weights, however, they left the complexity of these four problems open. We solve these open questions.

Baumeister et al. [8] showed that PWUW-BW-RW-$\mathbb{N}$ $\leq_m^p$ CCAV for every $\mathcal{E}$ if CCAV is used in succinct representation (which, recall, means that identical votes are

---

[4] As pointed out by Zack Fitzsimmons and Edith Hemaspaandra, for 3-approval, that PWUW-BW-$\mathbb{N}$ is (and, possibly, other of our problems are) in P also follows immediately from the results of Baumeister et al. [8] and Fitzsimmons and Hemaspaandra [16]. Even more, they note that since the dichotomy for the problem CCAV (the definition of which is recalled in Footnote 5 in the next section) shown by Hemaspaandra et al. [22] is exactly the same as for its succinct variant CCAV$_{succinct}$ [16], it follows that the same dichotomy holds for all problems X such that CCAV $\leq_m^p$ X $\leq_m^p$ CCAV$_{SUCCINCT}$. Hence, this immediately gives the same dichotomy for PWUW-BW-$\mathbb{N}$ (both its general and its succinct variant), for example.

not listed one by one but just once along with a binary number giving the multiplicity of this vote).[5] Together with the facts that PWUW-BW-N $\leq_{m}^{p}$ PWUW-BW-RW-N and PWUW-RW-N $\leq_{m}^{p}$ PWUW-BW-RW-N it follows that if CCAV for plurality with runoff (in succinct representation) is in P, each of PWUW-BW-RW-N, PWUW-BW-N, and PWUW-RW-N is in P for plurality with runoff as well. Erdélyi et al. [15] showed that CCAV in classical representation is in P for plurality with runoff. Alas, their approach cannot be easily adapted for succinct representation as their algorithm iterates over all possible values $\ell' \leq \ell$ and ℓ is not polynomially bounded in succinct representation. Instead, after some preprocessing steps we will solve the problem with an integer linear program (ILP) similarly to how Fitzsimmons and Hemaspaandra [16] have handled election problems in succinct representation. Regarding tie-breaking, we assume that whenever a tie occurs we can freely choose how it is broken. The technically rather involved proof of Theorem 3 is omitted due to space constraints.

Theorem 3. *Plurality-with-runoff-CCAV in succinct representation is in P.*

Corollary 2. *For plurality with runoff, the problems* PWUW-BW-RW-N, PWUW-RW-N, *and* PWUW-BW-N *are in P.*

Lastly, we consider PWUW-N for this rule, for which we can use Corollary 2.

Theorem 4. *For plurality with runoff,* PWUW-N *is in P.*

Proof. For plurality with runoff, a given instance $I = (C, V_1, V_0, c)$ of PWUW-N can be solved in polynomial time as follows. If there is a vote in V_0 with c on top, we have a YES-instance, as we can simply assign this vote a large enough weight (e.g., $|V_1|$) and other votes a weight of zero for c to win the runoff. Otherwise, we define a PWUW-RW-N instance $I' = (C, V_1, V_0, R, c)$ (for plurality with runoff), with each region $R_i \in R$ ranging from zero to $|V_0||V_1|$. We can solve this instance in polynomial time (see Corollary 2) and then use that I is a YES-instance of PWUW-N if and only if I' is a YES-instance of PWUW-RW-N. To see this, we show that we can never assign a weight greater than $|V_0||V_1|$ to a vote in V_0 if c wins. Assume for a contradiction that a vote $v \in V_0$ was given a weight greater than $|V_0||V_1|$ and c won the runoff. Let d be the candidate on top of v. We know that none of the votes in V_0 has c on top; otherwise, we would already be done. So c has a score of at most $|V_1|$ in the first round. As c reaches the runoff and the score of d is greater than c's score, all other candidates $C \setminus \{c, d\}$ have at most the same score as c. Thus the weight of each other vote in $V_0 \setminus \{v\}$ that does not have d on top is at most $|V_1|$ (i.e., the upper bound on c's score). Then the score c gains in the runoff from the eliminated candidates is at most $(|V_0| - 1)|V_1|$ which sums up to an upper bound of $|V_0||V_1|$ of c's score in the runoff. But d's score is greater than $|V_0||V_1|$, since the vote v with weight greater than $|V_0||V_1|$ has d on top, so c loses the

---

[5] As a reminder, CCAV is a shorthand for the problem CONSTRUCTIVE-CONTROL-BY-ADDING-VOTERS introduced by Bartholdi et al. [1]. The input of $\mathcal{E}$-CCAV consists of a set C of candidates with a distinguished candidate $c \in C$, a list V of registered (unit-weight) votes over C, an additional list W of as yet (unit-weight) unregistered votes, and a positive integer ℓ. The question is whether it is possible to make c win the election under $\mathcal{E}$ by adding at most ℓ votes from W to the election.

runoff, which is a contradiction. Note that there might be other votes in V_0 that have d on top, but their weight is irrelevant for our argument as d already wins the first round and the runoff even without additional points from them.   □

5   Veto with Runoff

Next, we turn to veto with runoff and solve all four cases. Again, we assume that if any ties occur, we can freely choose how to break these. Similarly to the previous section we will use the result of Baumeister et al. [8] that PWUW-BW-RW-ℕ, PWUW-BW-ℕ, and PWUW-RW-ℕ reduce to CCAV for any voting rule $\mathcal{E}$ if CCAV is in succinct representation. For veto with runoff, CCAV in classical representation was shown to be in P by Erdélyi et al. [15] as well, but we run into the same issue as in the previous section when trying to adapt their proof assuming succinct representation. Therefore, we prove the following theorem with a different approach; the proof is again omitted here.

Theorem 5. *For veto with runoff, CCAV in succinct representation is in* P.

Corollary 3. *For veto with runoff, the problems* PWUW-RW-ℕ, PWUW-BW-ℕ, *and* PWUW-BW-RW-ℕ *are in* P.

Lastly, we consider PWUW-ℕ for veto with runoff, omitting the proof of Theorem 6 as well.

Theorem 6. *For veto with runoff,* PWUW-ℕ *is in* P.

6 k-Veto

In this section, we focus on k-veto. For $k = 1$ (i.e., veto), we can give polynomial-time algorithms for all four problem variants, omitting the proof of Theorem 7.

Theorem 7. *For veto, the problems* PWUW-BW-RW-ℕ *and* PWUW-BW-ℕ *are in* P.

Using Theorem 2 from Sect. 3, we obtain the following corollary, as veto is a binary scoring protocol.

Corollary 4. *For veto, the problems* PWUW-RW-ℕ *and* PWUW-ℕ *are in* P.

In order to prove that our problems are in P for 2-veto as well, we introduce another variant of the earlier presented polynomial-time solvable GWBEM problem. According to Gabow [17] and Grötschel et al. [20, p. 259], the following variant of this problem is in P, too.[6]

---

[6] Again, Gabow [17] and Grötschel et al. [20, p. 259] formalize this problem variant as a minimization problem. Since this problem is polynomial-time solvable, the decision problem variant that we introduce is in P as well.

GENERALIZED-B-EDGE-COVER (GBEC)

Given: An undirected multigraph $G = (N, E)$ without loops, capacity-bounding functions $a_\ell, a_u \colon E \to \mathbb{N}$ and $b_\ell, b_u \colon N \to \mathbb{N}$, and a target integer $r \in \mathbb{N}$.

Question: Is there a function $x \colon E \to \mathbb{N}$ with $\sum_{e \in E} x(e) \le r$, such that for every edge $e \in E$ it holds that $a_\ell(e) \le x(e) \le a_u(e)$ and for every node $z \in N$ it holds that $b_\ell(z) \le \sum_{e \in \delta(z)} x(e) \le b_u(z)$, where $\delta(z)$ is the set of edges incident to node z?

The difference to the earlier introduced GWBEM problem is that this time the weights assigned to the edges of G are *not weighted* and we want the sum to be *at most r* instead of *at least r*. Especially the last difference seems to be a bit subtle but is crucial for the ensuing proof, as this time we work with a veto rule and, thus, construct the corresponding graph in such a way that positively weighted edges in the graph cause candidates to *not* obtain points.

Theorem 8. *For 2-veto, each of* PWUW-BW-RW-$\mathbb{N}$, PWUW-BW-$\mathbb{N}$, PWUW-RW-$\mathbb{N}$, *and* PWUW-$\mathbb{N}$ *is in* P.

Proof. In order to prove that 2-veto-PWUW-BW-RW-$\mathbb{N}$ is in P, we reduce this problem to GBEC. Let $I = (C, V_1, V_0, R, B, c)$ be a PWUW-BW-RW-$\mathbb{N}$ instance for 2-veto with $R_i = [0, r_i]$ for $1 \le i \le |V_0|$, according to Lemma 1. We construct a GBEC instance $I' = (G, a_\ell, a_u, b_\ell, b_u, r)$ similar to the GWBEM instance in the proof of Theorem 1. We define the multigraph $G = (C \setminus \{c\}, E)$, where in order to specify E we first define the set

$$V_0' = \{\cdots > x_1 > x_2 \in V_0 \mid \{x_1, x_2\} \cap \{c\} = \emptyset\}$$

consisting of all votes from V_0 with c not being ranked in one of the last two positions. Then we define the edge set of G as

$$E = \{\{x_1, x_2\} \mid \cdots > x_1 > x_2 \in V_0'\}.$$

Doing so, every edge in the graph corresponds to some vote in V_0, i.e., when we write e_i, we implicitly refer to the corresponding vote v_i from V_0 (re-indexing the indices as needed). For every edge $e_i \in E$, we define $a_\ell(e_i) = 0$ and $a_u(e_i) = r_i$ for $R_i = [0, r_i]$. For every node $d \in C \setminus \{c\}$, we define

$$b_\ell(d) = \max\{0, score_{V_1}(d) - score_{V_1}(c)\}$$

and $b_u(d) = B$. Lastly, we define $r = B$. This completes the construction of I', which can be realized in time polynomial in $|I|$. The proof to show that $I \in$ 2-veto-PWUW-BW-RW-$\mathbb{N}$ holds if and only if $I' \in$ GBEC is true is almost the same as the one for Theorem 1, adjusted for the different voting rule. We simply refer to the earlier proof; the reader can easily fill in the necessary details.

Using the same approach as already used in Corollary 1, it immediately follows that PWUW-BW-$\mathbb{N}$ and PWUW-RW-$\mathbb{N}$ are in P, too. Finally, to see that PWUW-$\mathbb{N}$ is in P, we can once again refer to Theorem 2, since 2-veto is a binary scoring protocol as well. $\qquad\square$

Turning to k-veto for $k \geq 3$, we show that PWUW-N and PWUW-RW-N are in P, while PWUW-BW-RW-N and PWUW-BW-N are NP-complete. For the first two problems, PWUW-N and PWUW-RW-N, we refer to Theorem 2, since 3-veto is a binary scoring protocol.

Corollary 5. *For each $k \geq 3$, k-veto-*PWUW-N *and k-veto-*PWUW-RW-N *are in* P.

For the other two problems, PWUW-BW-N and PWUW-BW-RW-N, we have NP-completeness results. The proof of Theorem 9 is again omitted.

Theorem 9. *For each $k \geq 3$, k-veto-*PWUW-BW-N *is* NP-*complete.*

By the reduction CCAV $\leq^P_m$ PWUW-BW-RW-N of Baumeister et al. [8], we immediately obtain the following result, as Lin [27] has shown that CCAV is NP-hard for k-veto.

Corollary 6. *For each $k \geq 3$, k-veto-*PWUW-BW-RW-N *is* NP-*complete.*

7 Borda

Finally, we turn to the voting rule due to Borda [11], which perhaps is the most famous scoring protocol and has been intensively studied in social choice theory and in computational social choice (see the survey by Rothe [31]). As for k-veto, $k \geq 3$, we establish hardness results for Borda, but now even for all four of our problem variants.

Theorem 10. *For Borda, each of* PWUW-BW-N*,* PWUW-RW-N*,* PWUW-BW-RW-N*, and* PWUW-N *is* NP-*complete.*

Proof. Membership in NP is obvious for all problems but Borda-PWUW-N. For Borda-PWUW-N, it is not trivial to show that a solution that can be used as a witness is polynomial in the input size. But in this case we can construct in polynomial time an integer linear program that solves the problem similarly to the linear programs that were constructed by Baumeister et al. [8] for the variants with rational weights. Then membership in NP of Borda-PWUW-N follows from this reduction from Borda-PWUW-N to the NP-complete problem INTEGER-PROGRAMMING-FEASIBILITY. We note in passing that with this technique we can show NP-membership of $\mathcal{E}$-PWUW-N for every scoring protocol $\mathcal{E}$.

Regarding the four NP-hardness results, we will prove them for Borda-PWUW-N first and then describe how the reduction can be altered to show NP-hardness for the other cases. So, in order to prove that Borda-PWUW-N is NP-hard, from X3C. Let $I = (X, \mathcal{S})$ be a given X3C instance with $X = \{x_1, \ldots, x_{3q}\}$ and $\mathcal{S} = \{S_1, \ldots, S_m\}$. We construct a Borda-PWUW-N instance $I' = (C, V_1, V_0, c)$ as follows. Define $C = \{c, d, d'\} \cup X \cup B$ with B being a set of $9q^2$ buffer candidates. For each $x_i \in X$ and $b_j \in B$, we add the two votes

$$\overrightarrow{C \setminus \{x_i, b_j\}} > x_i > b_j \quad \text{and} \quad x_i > b_j > \overleftarrow{C \setminus \{x_i, b_j\}}$$

and q times two votes of the form

$$\overrightarrow{C\setminus\{d,d'\}} > d' > d \quad \text{and} \quad d' > d > \overleftarrow{C\setminus\{d,d'\}}$$

to V_1. Doing so, we construct the following point distances between c and each of the other candidates in the election (C,V_1): $diff_{(C,V_1)}(c,d) = q$, $diff_{(C,V_1)}(c,d') = -q$, $diff_{(C,V_1)}(c,x_i) = -|B|$ for every $x_i \in X$, and $diff_{(C,V_1)}(c,b_j) = 3q$ for every $b_j \in B$. For each i, $1 \le i \le m$, we add a vote of the form $v_i = d > c > X \setminus S_i > B > S_i > d'$ to V_0. Obviously, this construction of I' is possible in time polynomial in $|I|$.

The proof of correctness for this reduction for Borda-PWUW-$\mathbb{N}$ is again omitted, as well as NP-hardness of the other three problem variants, which follows by augmenting the above construction appropriately. □

8   Conclusions and Open Questions

Table 1. Overview of complexity results. Our results are in boldface. Known results of Baumeister et al. [8] (who also settle all cases of k-approval, $k \neq 3$) are in gray. "NP-c." stands for NP-complete.

PWUW-	3-App.	Plurality/Veto with runoff	k-Veto for $k \in \{1,2\}$	k-Veto for $k \ge 3$	Borda
$\mathbb{N}$	P	**P**	**P**	**P**	**NP-c.**
BW-RW-$\mathbb{N}$	**P**	**P**	**P**	**NP-c.**	**NP-c.**
BW-$\mathbb{N}$	**P**	**P**	**P**	**NP-c.**	**NP-c.**
RW-$\mathbb{N}$	P	**P**	**P**	**P**	**NP-c.**

We have continued the research initiated by Baumeister et al. [8] regarding the computational complexity of four variants of the *possible winner with uncertain weights problem*, solving all of their open problems *for nonnegative integer weights*. In addition, we have established results for other very prominent voting rules, namely Borda, k-veto, and veto with runoff. In the course of our research we also further completed the landscape of results with respect to CCAV *in succinct representation* by showing that for both, plurality and veto with runoff, this problem can be solved in P. Table 1 presents a summary of previously known and our new results.

As to open problems, like for the *possible winner problem* studied by Betzler and Dorn [10] and Baumeister and Rothe [9], it is desirable to have a dichotomy result for our problems with respect to *all* scoring protocols.

.

Acknowledgments. This work was supported in part by Deutsche Forschungsgemeinschaft under grants RO 1202/14-2 and RO 1202/21-1.

References

1. Bartholdi III, J., Tovey, C., Trick, M.: How hard is it to control an election? Math. Comput. Model. **16**(8/9), 27–40 (1992)
2. Baumeister, D., et al.: Positional scoring-based allocation of indivisible goods. J. Auton. Agents Multi-Agent Syst. **31**(3), 628–655 (2017). https://doi.org/10.1007/s10458-016-9340-x
3. Baumeister, D., Erdélyi, G., Erdélyi, O., Rothe, J.: Complexity of manipulation and bribery in judgment aggregation for uniform premise-based quota rules. Math. Soc. Sci. **76**, 19–30 (2015)
4. Baumeister, D., Erdélyi, G., Erdélyi, O., Rothe, J., Selker, A.: Complexity of control in judgment aggregation for uniform premise-based quota rules. J. Comput. Syst. Sci. **112**, 13–33 (2020)
5. Baumeister, D., Järvisalo, M., Neugebauer, D., Niskanen, A., Rothe, J.: Acceptance in incomplete argumentation frameworks. Artif. Intell. **295** (2021). Article No. 103470
6. Baumeister, D., Neugebauer, D., Rothe, J., Schadrack, H.: Verification in incomplete argumentation frameworks. Artif. Intell. **264**, 1–26 (2018)
7. Baumeister, D., Roos, M., Rothe, J.: Computational complexity of two variants of the possible winner problem. In: Proceedings of the 10th International Conference on Autonomous Agents and Multiagent Systems, pp. 853–860. IFAAMAS (2011)
8. Baumeister, D., Roos, M., Rothe, J., Schend, L., Xia, L.: The possible winner problem with uncertain weights. In: Proceedings of the 20th European Conference on Artificial Intelligence, pp. 133–138. IOS Press (2012)
9. Baumeister, D., Rothe, J.: Taking the final step to a full dichotomy of the possible winner problem in pure scoring rules. Inf. Process. Lett. **112**(5), 186–190 (2012)
10. Betzler, N., Dorn, B.: Towards a dichotomy for the possible winner problem in elections based on scoring rules. J. Comput. Syst. Sci. **76**(8), 812–836 (2010)
11. Borda, J.: Mémoire sur les élections au scrutin. Histoire de L'Académie Royale des Sciences, Paris (1781). English translation appears in the paper by de Grazia [19]
12. Brandt, F., Conitzer, V., Endriss, U., Lang, J., Procaccia, A. (eds.): Handbook of Computational Social Choice. Cambridge University Press, Cambridge (2016)
13. Conitzer, V., Sandholm, T., Lang, J.: When are elections with few candidates hard to manipulate? J. ACM **54**(3), Article 14 (2007)
14. Elkind, E., Faliszewski, P., Slinko, A.: Swap bribery. In: Mavronicolas, M., Papadopoulou, V.G. (eds.) SAGT 2009. LNCS, vol. 5814, pp. 299–310. Springer, Heidelberg (2009). https://doi.org/10.1007/978-3-642-04645-2_27
15. Erdélyi, G., Reger, C., Yang, Y.: Towards completing the puzzle: Solving open problems for control in elections. In: Proceedings of the 18th International Conference on Autonomous Agents and Multiagent Systems, IFAAMAS, pp. 846–854 (2019)
16. Fitzsimmons, Z., Hemaspaandra, E.: High-multiplicity election problems. Auton. Agent. Multi-Agent Syst. **33**(4), 383–402 (2019). https://doi.org/10.1007/s10458-019-09410-4
17. Gabow, H.: An efficient reduction technique for degree-constrained subgraph and bidirected network flow problems. In: Proceedings of the 15th ACM Symposium on Theory of Computing, pp. 448–456. ACM Press (1983)
18. Garey, M., Johnson, D.: Computers and Intractability: A Guide to the Theory of NP-Completeness. W. H Freeman and Company, New York (1979)
19. de Grazia, A.: Mathematical deviation of an election system. Isis **44**(1–2), 41–51 (1953)
20. Grötschel, M., Lovász, L., Schrijver, A.: Geometric Algorithms and Combinatorial Optimization. Springer, Heidelberg (1988). https://doi.org/10.1007/978-3-642-97881-4

21. Hemaspaandra, E., Hemaspaandra, L., Rothe, J.: Anyone but him: the complexity of precluding an alternative. Artif. Intell. **171**(5–6), 255–285 (2007)
22. Hemaspaandra, E., Hemaspaandra, L., Schnoor, H.: A control dichotomy for pure scoring rules. In: Proceedings of the 28th AAAI Conference on Artificial Intelligence, pp. 712–720. AAAI Press (2014)
23. Kerkmann, A., Lang, J., Rey, A., Rothe, J., Schadrack, H., Schend, L.: Hedonic games with ordinal preferences and thresholds. J. Artif. Intell. Res. **67**, 705–756 (2020)
24. Konczak, K., Lang, J.: Voting procedures with incomplete preferences. In: Proceedings of the Multidisciplinary IJCAI-05 Workshop on Advances in Preference Handling, pp. 124–129 (2005)
25. Kuckuck, B., Rothe, J.: Duplication monotonicity in the allocation of indivisible goods. AI Commun. **32**(4), 253–270 (2019)
26. Lang, J.: Collective decision making under incomplete knowledge: Possible and necessary solutions. In: Proceedings of the 29th International Joint Conference on Artificial Intelligence, pp. 4885–4891. ijcai.org (2020)
27. Lin, A.: Solving hard problems in election systems. Ph.D. thesis, Rochester Institute of Technology, Rochester, NY, USA (2012)
28. Niskanen, A., Neugebauer, D., Järvisalo, M., Rothe, J.: Deciding acceptance in incomplete argumentation frameworks. In: Proceedings of the 34th AAAI Conference on Artificial Intelligence, pp. 2942–2949. AAAI Press (2020)
29. Papadimitriou, C.: Computational Complexity, 2nd edn. Addison-Wesley, Boston (1995)
30. Rothe, J.: Complexity Theory and Cryptology. An Introduction to Cryptocomplexity. EATCS Texts in Theoretical Computer Science, Springer, Heidelberg (2005). https://doi.org/10.1007/3-540-28520-2
31. Rothe, J.: Borda count in collective decision making: a summary of recent results. In: Proceedings of the 33rd AAAI Conference on Artificial Intelligence, pp. 9830–9836. AAAI Press (2019)
32. Skiba, K., Neugebauer, D., Rothe, J.: Complexity of nonempty existence problems in incomplete argumentation frameworks. IEEE Intell. Syst. **36**(2), 13–24 (2021)
33. Xia, L., Conitzer, V.: Determining possible and necessary winners given partial orders. J. Artif. Intell. Res. **41**, 25–67 (2011)

Streaming Deletion Problems
Parameterized by Vertex Cover

Jelle J. Oostveen[✉] and Erik Jan van Leeuwen

Department of Information and Computing Sciences,
Utrecht University, Utrecht, The Netherlands
{j.j.oostveen,e.j.vanleeuwen}@uu.nl

Abstract. Streaming is a model where an input graph is provided one edge at a time, instead of being able to inspect it at will. In this work, we take a parameterized approach by assuming a vertex cover of the graph is given, building on work of Bishnu et al. [COCOON 2020]. We show the further potency of combining this parameter with the Adjacency List streaming model to obtain results for vertex deletion problems. This includes kernels, parameterized algorithms, and lower bounds for the problems of Π-FREE DELETION, H-FREE DELETION, and the more specific forms of CLUSTER VERTEX DELETION and ODD CYCLE TRANSVERSAL. We focus on the complexity in terms of the number of passes over the input stream, and the memory used. This leads to a pass/memory trade-off, where a different algorithm might be favourable depending on the context and instance. We also discuss implications for parameterized complexity in the non-streaming setting.

1 Introduction

Streaming is an algorithmic paradigm to deal with data sets that are too large to fit into main memory [22]. Instead, elements of the data set are inspected in a fixed order[1] and aggregate data is maintained in a small amount of memory (much smaller than the total size of the data set). It is possible to make multiple passes over the data set. The goal is to design algorithms that analyze the data set while minimizing the combination of the number of passes and the required memory. We note that computation time is not measured in this paradigm. Streaming has proved very successful and is extensively studied in many diverse contexts [27,29]. In this work, we focus on the case where the data sets are graphs and the streamed elements are the edges of the graph.

---

[1] We consider insertion-only streams throughout this paper.

This work is based on the master thesis "Parameterized Algorithms in a Streaming Setting" by the first author. This work is partially supported by the NWO grant OCENW.KLEIN.114 (PACAN).

E. Bampis and A. Pagourtzis (Eds.): FCT 2021, LNCS 12867, pp. 413–426, 2021.
https://doi.org/10.1007/978-3-030-86593-1_29

A significant body of work on graph streaming works in the semi-streaming model, where $\tilde{\mathcal{O}}(n)$ memory[2] is allowed, with the aim of limiting the number of necessary passes to one or two. This memory requirement might still be too much for the largest of networks. Unfortunately, many basic problems in graphs require $\Omega(n)$ or even worse space [18,19] to compute in a constant number of passes. Therefore, Fafianie and Kratsch [17] and Chitnis et al. [13] introduced concepts and analysis from parameterized complexity [16] to the streaming paradigm. For example, it can be decided whether a graph has a vertex cover of size at most K using one pass and $\tilde{\mathcal{O}}(K^2)$ space, which is optimal. This led to various further works [4,6,11] and the first systematic study by Chitnis and Cormode [10].

Our work continues this line of research and follows up on recent work by Bishnu et al. [5,6][3]. They made two important conceptual contributions. First, they analyzed the complexity of parameterized streaming algorithms in three models that prescribe the order in which the edges arrive in the stream and that are commonly studied in the literature [6,14,27,28]. The *Edge Arrival* (EA) model prescribes some permutation of all the edges of the graph. The *Vertex Arrival* (VA) requires that the edges appear per vertex: if we have seen the vertices $V' \subseteq V$ already, and the next vertex is w, then the stream contains the edges between w and the vertices in V'. Finally, the *Adjacency List* (AL) gives the most information, as it requires the edges to arrive per vertex, but when vertex v appears in the stream, we also see all edges incident to v. This means we effectively see every edge twice in a single pass, once for both of its endpoints.

The second and more important contribution of Bishnu et al. [5] was to study the size K of a vertex cover in the graph as a parameter. This has been broadly studied in parameterized complexity (see e.g. the PhD thesis of Jansen [25]). They showed that the very general $\mathcal{F}$-SUBGRAPH DELETION and $\mathcal{F}$-MINOR DELETION problems all admit one pass, $\tilde{\mathcal{O}}(\Delta(\mathcal{F}) \cdot K^{\Delta(\mathcal{F})+1})$ space streaming algorithms in the AL model, by computing small kernels to which then a straightforward exhaustive algorithm is applied. On the other hand, such generic streaming algorithms are not possible in the EA and VA models, as then (super-) linear lower bounds exist even if the size of a smallest vertex cover is constant [5].

We focus on the induced subgraph version of the vertex deletion problem, parameterized by the size of a vertex cover. Here, Π is a collection of graphs.

Π-FREE DELETION [VC]
Input: A graph G with a vertex cover X, and an integer $\ell \geq 1$.
Parameter: The size $K := |X|$ of a vertex cover.
Question: Is there a set $S \subseteq V(G)$ of size at most ℓ such that $G[V(G) \backslash S]$ does not contain a graph in Π as an induced subgraph?

---

[2] Throughout this paper, memory is measured in bits. The $\tilde{\mathcal{O}}$ notation hides factors polylogarithmic in n. Note that $\mathcal{O}(\log n)$ bits is the space required to store (the identifier of) a single vertex or edge.

[3] As the Arxiv version contains more results, we refer to this version from here on.

To avoid triviality[4], we assume every graph in Π is edgeless or $K \geq \ell$. We assume the vertex cover is given as input; if only the size is given, we can use one pass and $\tilde{\mathcal{O}}(K^2)$ space or 2^K passes and $\tilde{\mathcal{O}}(K)$ space to obtain it [10,13] (this does not meaningfully impact our results). The unparameterized version of this problem is well known to be NP-hard [26] for any nontrivial and hereditary property Π. It has also been well studied in the parameterized setting (see e.g. [9, 20,31]). When parameterized by the vertex cover number, it has been studied from the perspective of kernelization: while a polynomial kernel cannot exist in general [8,21], polynomial kernels exist for broad classes of families Π [21,24]. As far as we are aware, parameterized algorithms for this parameterization have not been explicitly studied.

In the streaming setting, Chitnis et al. [11] showed for the unparameterized version of this problem in the EA model that any p-pass algorithm needs $\Omega(n/p)$ space if Π satisfies a mild technical constraint. For some Π-FREE DELETION [VC] problems, the results by Bishnu et al. [5] imply single-pass, poly(K) space streaming algorithms (through their kernel for $\mathcal{F}$-SUBGRAPH DELETION [VC]) in the AL model and lower bounds in the EA/VA model. They also provide an explicit kernel for CLUSTER VERTEX DELETION [VC] in the AL/EA/VA models. However, this still leaves the streaming complexity of many cases of the Π-FREE DELETION [VC] problem open.

Our Contributions. We determine the streaming complexity of the general Π-FREE DELETION [VC] problem. Our main positive result is a unified approach to a single-pass polynomial kernel for Π-FREE DELETION [VC] for a broad class of families Π. In particular, we show that the kernelization algorithms by Fomin et al. [21] and Jansen and Kroon [24] can be adapted to the streaming setting. The kernels of Fomin et al. [21] consider the case when Π can be characterized by few adjacencies, which intuitively means that for any vertex of any member of Π, adding or deleting edges between all but a few (say at most c_Π) distinguished other vertices does not change membership of Π. The exponent of the polynomial kernels depends on c_Π. Jansen and Kroon [24] considered even more general families Π. We show that these kernels can be computed in the AL model using a single pass and polynomial space (where the exponent depends on c_Π). This generalizes the previous results by Bishnu et al. [5] as well as their kernel for $\mathcal{F}$-SUBGRAPH DELETION [VC].

To complement the kernels, we take a direct approach to find more memory-efficient algorithms, at the cost of using many passes. We show novel parameterized streaming algorithms that require $\tilde{\mathcal{O}}(K^2)$ space and $\mathcal{O}(K)^{\mathcal{O}(K)}$ passes. Here, all hidden constants depend on c_Π. Crucially, however, the exponent of the space usage of these algorithms does not, which provides an advantage over computing the kernel. We also provide explicit streaming algorithms for CLUSTER VERTEX DELETION [VC] and ODD CYCLE TRANSVERSAL [VC] that require $\tilde{\mathcal{O}}(K)$ space (both) and $2^K K^2$ and 3^K passes respectively, as well as streaming algorithms for Π-FREE DELETION [VC,$|V(H)|$] when $\Pi = \{H\}$ and the problem is parameterized by K and $|V(H)|$. A crucial ingredient to these algorithms is a

---

[4] Otherwise, removing the entire vertex cover is a trivial solution.

streaming algorithm that finds induced subgraphs isomorphic to a given graph H. Further details are provided in Sect. 3.

The above results provide a trade-off in the number of passes and memory complexity of the algorithm used. However, we should justify using both the AL model and the parameter vertex cover. To this end, in Sect. 4, we investigate lower bounds for streaming algorithms for Π-FREE DELETION [VC]. The (unparameterized) linear lower bound of Chitnis et al. [11] in the EA model requires that Π contains a graph H for which $|E(H)| \geq 2$ and no proper subgraph is a member of Π. We prove that the lower bound extends to both the VA and AL models, with only small adjustments. Hence, parameterization is necessary to obtain sublinear passes and memory for most Π. Since VERTEX COVER is one of the few natural graph parameters that has efficient parameterized streaming algorithms [13,17], this justifies the use of the vertex cover parameter. We also extend the reductions by Bishnu et al. [5] to general hardness results for Π-FREE DELETION in the VA and EA model when the size of the vertex cover is a constant (dependent on Π), justifying the use of the AL model for most Π.

We also consider the parameterized complexity of H-FREE DELETION [VC] in the non-streaming setting. While polynomial kernels were known in the non-streaming setting [21], we are unaware of any investigation into explicit parameterized algorithms for these problems. We give a general $2^{O(K^2)}\text{poly}(n,|V(H)|)$ time algorithm. This contrasts the situation for H-FREE DELETION parameterized by the treewidth t of the graph, where a $2^{o(t^{|V(H)|-2})}\text{poly}(n,|V(H)|)$ time lower bound is known under the Exponential Time Hypothesis (ETH) [31]. We also construct a graph property Π for which we provide a lower bound of $2^{o(K \log K)}\text{poly}(n,|V(H)|)$ for Π-FREE DELETION [VC] under ETH. Further details are provided in Sect. 3.

Preliminaries. We work on undirected graphs $G = (V, E)$, where $|V| = n, |E| = m$. We denote an edge $e \in E$ between $v \in V$ and $u \in V$ with $uv \in E$. For a set of vertices $V' \subseteq V$, denote the subgraph induced by V' as $G[V']$. Denote the neighbourhood of a vertex v with $N(v)$ and for a set S denote $N(S)$ as $\bigcup_{v \in S} N(v)$. We write $N[v]$ for $N(v)$ including v, so $N[v] = N(v) \cup \{v\}$.

We denote the parameters of a problem in [·] brackets, a problem A parameterized by vertex cover number and solution size is denoted by A [VC, ℓ].

2   Adapting Existing Kernels

We first show that very general kernels for vertex cover parameterization admit straightforward adaptations to the AL streaming model. The kernels considered are those by Fomin et al. [21] and by Jansen and Kroon [24]. Fomin et al. [21] provide general kernelization theorems that make extensive use of a single property, namely that some graph properties can be characterized by few adjacencies.

Definition 1. ([21, Definition 3]) *A graph property Π is characterized by $c_\Pi \in \mathbb{N}$ adjacencies if for all graphs $G \in \Pi$, for every vertex $v \in V(G)$, there is a set $D \subseteq V(G) \setminus \{v\}$ of size at most c_Π such that all graphs G' that are obtained from G by adding or removing edges between v and vertices in $V(G) \setminus D$, are also contained in Π.*

Fomin et al. show that graph problems such as Π-FREE DELETION [VC], can be solved efficiently through kernelization when Π is characterized by few adjacencies (and meets some other demands), by making heavy use of the REDUCE algorithm they provide. The idea behind the REDUCE algorithm is to save *enough* vertices with specific adjacencies in the vertex cover, and those vertices that we forget have equivalent vertices saved to replace them. The sets of adjacencies we have to consider can be reduced by making use of the characterization by few adjacencies, as more than c_Π adjacencies are not relevant. The number of vertices we retain is ultimately dependent on $\ell \leq K$.

In the AL streaming model, we have enough information to compute this kernel, by careful memory management in counting adjacencies towards specific subsets of the vertex cover. The following theorem then shows how this algorithm leads to streaming kernels for Π-FREE DELETION [VC] as an adaptation of [21, Theorem 2]. We call a graph G *vertex-minimal* with respect to Π if $G \in \Pi$ and for all $S \subsetneq V(G)$, $G[S] \notin \Pi$.

Theorem 1 ($\clubsuit^5$). *If Π is a graph property such that:*

(i) Π *is characterized by c_Π adjacencies,*
(ii) *every graph in Π contains at least one edge, and*
(iii) *there is a non-decreasing polynomial $p : \mathbb{N} \to \mathbb{N}$ such that all graphs G that are vertex-minimal with respect to Π satisfy $|V(G)| \leq p(K)$,*

then Π-FREE DELETION [VC] admits a kernel on $\mathcal{O}((K + p(K))K^{c_\Pi})$ vertices in the AL streaming model using one pass and $\mathcal{O}((K + K^{c_\Pi}) \log(n))$ space.

We note that the theorem applies to $\mathcal{F}$-SUBGRAPH DELETION [VC] when $\Delta(\mathcal{F})$ (the maximum degree) is bounded as well as to CLUSTER VERTEX DELETION [VC]. As such, our streaming kernels generalize the kernels of Bishnu et al. [5] for these problems, while the memory requirements and kernel sizes are fairly comparable. A discussion and further implications for several general problems, following Fomin et al. [21], appear in the full version of the paper.

We also give an adaptation ($\clubsuit$) of a more recent kernel by Jansen and Kroon [24], which has another broad range of implications for streaming kernels. This kernel uses a different characterization of the graph family, however, the adaptation to the AL streaming model is very similar. We observe that the adaptation of this kernel leads to a streaming algorithms for problems like PERFECT DELETION [VC], AT-FREE DELETION [VC], INTERVAL DELETION [VC], and WHEEL-FREE DELETION [VC].

3   A Direct FPT Approach

In this section, we give direct FPT streaming algorithms for Π-FREE DELETION [VC] for the same cases as Theorem 1. This is motivated by the fact that Chitnis and Cormode [10] found a direct FPT algorithm for VERTEX COVER

---

[5] Further discussions and proofs for results marked with $\clubsuit$ appear in the full online version of the paper.

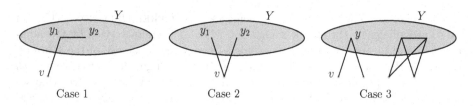

Fig. 1. The different cases how a P_3 can exists with respect to Y, part of the vertex cover. Notice that the case where the entire P_3 is contained in Y is not included here. Case 3 assumes there are no Case 1 or Case 2 P_3's in the graph anymore.

using $\mathcal{O}(2^k)$ passes and only $\tilde{\mathcal{O}}(k)$ space in contrast to the kernel of Chitnis et al. [11] using one pass and $\tilde{\mathcal{O}}(k^2)$ space. Therefore, we aim to explore the pass/memory trade-off for Π-FREE DELETION [VC] as well.

3.1 P_3-free Deletion

We start with the scenario where $\Pi = \{P_3\}$, which means we consider the problem CLUSTER VERTEX DELETION [VC]. The general idea of the algorithm is to branch on what part of the given vertex cover should be in the solution. For managing the branching correctly, we use a black box enumeration technique also used by Chitnis and Cormode [10]. In a branch, we first check whether the 'deletion-free' part of the vertex cover (Y) contains a P_3, which invalidates a branch. Otherwise, what remains is some case analysis where either one or two vertices of a P_3 lie outside the vertex cover, for which we deterministically know which vertices have to be removed to make the graph P_3-free. We illustrate this step in Fig. 1. Case 1 and 2 have only one option for removal of a vertex. After Case 1 and 2 no longer occur, we can find Case 3 occurrences and show that we can delete all but one of the vertices in such an occurrence. So, if this process can be executed in a limited number of passes, the algorithm works correctly.

To limit the number of passes, the use of the AL model is crucial. Notice that for every pair of vertices y_1, y_2 in the vertex cover, we can identify a Case 1 or 2 P_3 of Fig. 1, or these cases but with v in the vertex cover as well, in a constant number of passes. This is because we can first use a pass to check the presence of an edge between y_1 and y_2, and afterwards use a pass to check the edges of every other vertex towards y_1 and y_2 (which are given together because of the AL model). This means we can find P_3's contained in the vertex cover or corresponding to Case 1 or 2 P_3's in $\mathcal{O}(K^2)$ passes total. The remaining Case 3 can be handled in $\mathcal{O}(K)$ passes from the viewpoint of each $y \in Y$. So this algorithm takes $\mathcal{O}(2^K K^2)$ passes (including branching).

Theorem 2 (♣). *We can solve* CLUSTER VERTEX DELETION [VC] *in the* AL *streaming model using* $\mathcal{O}(2^K K^2)$ *passes and* $\mathcal{O}(K \log n)$ *space.*

Let us stress some details. The use of the AL model is crucial, as it allows us to locally inspect the neighbourhood of a vertex when it appears in the stream. The same approach would require more memory or more passes in other models

---

Algorithm 1. The procedure FINDH.

1: **function** FINDH(solution set S, forbidden set $Y \subseteq X$, integer i)
2:   **for each** Set O of i vertices of H that can be outside X **do** ▷ Check non-edges
3:     Denote $H' = H \setminus O$
4:     **for each** Occurrence of H' in Y **do**   ▷ Check $\mathcal{O}(\binom{|X|}{|H|-i})(|H|-i)!)$ options
5:       $S' \leftarrow \emptyset, O' \leftarrow O$
6:       **for each** Vertex $v \in V \setminus (S \cup X)$ **do**
7:         Check the edges/non-edges towards $H' \in Y$
8:         **if** v is equivalent to some $w \in O'$ for H' **then**
9:           $S' \leftarrow S' \cup \{v\}, O' \leftarrow O' \setminus \{w\}$
10:          **if** $O' = \emptyset$ **then return** S' ▷ We found an occurrence of H
11:   **return** $\emptyset$ ▷ No occurrence of H found

---

to accomplish this result. Also note that we could implement this algorithm in a normal setting (the graph is in memory, and not a stream) to get an algorithm for CLUSTER VERTEX DELETION [VC] with a running time of $\mathcal{O}(2^K \cdot K^2 \cdot (n+m))$.

3.2 H-free Deletion

We now consider a more generalized form of Π-FREE DELETION [VC], where $\Pi = \{H\}$, a single graph. Unfortunately, the approach when $H = P_3$ does not seem to carry over to this case, because the structure of a P_3 is simple and local.

Theorem 3 (♣). *We can solve H-FREE DELETION [VC] in $2^{O(K^2)} \text{poly}(n, | V(H)|)$ time, where H contains at least one edge and K is the size of the vertex cover.*

In the proof, we rely on the assumption that $\ell < K$ and use that the vertices outside the vertex cover can be partitioned into at most 2^K equivalence classes. Moreover, we use the algorithm implied by the work of Abu-Khzam [1] to find occurrences of H in G.

In order to analyze the complexity with respect to H more precisely and to obtain a streaming algorithm, we present a different algorithm that works off a simple idea. We branch on the vertex cover, and then try to find occurrences of H of which we have to remove a vertex outside the vertex cover. We branch on these removals as well, and repeat this find-and-branch procedure. In an attempt to keep the second branching complexity low, we start by searching for occurrences of H such that only one vertex lies outside the vertex cover, and increase this number as we find no occurrences. For briefness, we only present the occurrence detection part of the algorithm here, a procedure we call FINDH. Note that this is not (yet) a streaming algorithm.

Lemma 1 (♣). *Given a graph G with vertex cover X, graph H with at least one edge, and sets S, $Y \subseteq X$, and integer i, Algorithm 1 finds an occurrence of H in G that contains no vertices in S and $X \setminus Y$ and contains $|V(H)| - i$ vertices in Y. It runs in $\mathcal{O}\left(\binom{h}{i}[i^2 + \binom{K}{h-i}(h-i)!((h-i)^2 + Kn + (h-i)in)]\right)$ time, where $|V(H)| = h$ and $|X| = K$.*

FINDH is adaptable to the streaming setting, as is the complete algorithm. All the actions FINDH takes are local inspection of edges, and many enumeration actions, which lend itself well to usage of the AL streaming model. The number of passes of the streaming version is closely related to the running time of the non-streaming algorithm. This then leads to the full find-and-branch procedure.

Theorem 4 (♣). *We can solve H-FREE DELETION [VC] in the AL model, where H contains at least one edge, using $\mathcal{O}(2^K h^{K+2} K^h h!)$ or alternatively $\mathcal{O}(2^K h^{K+2} K! h!)$ passes and $\mathcal{O}((K + h^2) \log n)$ space, where $|V(H)| = h$.*

3.3   Towards Π-free Deletion

An issue with extending the previous approach to the general Π-FREE DELE-TION problem is the dependence on the maximum size h of the graphs $H \in \Pi$. Without further analysis, we have no bound on h. However, we can look to the preconditions used by Fomin et al. [21] on Π in e.g. Theorem 1 to remove this dependence.

The first precondition is that the set $\Pi' \subseteq \Pi$ of graphs that are vertex-minimal with respect to Π have size bounded by a function in K, the size of the vertex cover. That is, for these graphs $H \in \Pi'$ we have that $|V(H)| \leq p(K)$, where $p(K)$ is some function. We can prove (♣) that it suffices to only remove vertex-minimal elements of Π to solve Π-FREE DELETION. Note that Fomin et al. [21] require that this is a polynomial, we have no need to demand this. If we also assume that we know the set Π', we obtain the following result.

Theorem 5 (♣). *If Π is a graph property such that:*

(i) *we have explicit knowledge of $\Pi' \subseteq \Pi$, which is the subset of q graphs that are vertex-minimal with respect to Π, and*

(ii) *there is a non-decreasing function $p : \mathbb{N} \to \mathbb{N}$ such that all graphs $G \in \Pi'$ satisfy $|V(G)| \leq p(K)$, and*

(iii) *every graph in Π contains at least one edge,*

then Π-FREE DELETION [VC] can be solved using $\mathcal{O}(q \cdot 2^K \cdot p(K)^K \cdot K! \cdot K \cdot p(K)! \cdot p(K)^2 \cdot n)$ time.

We argue this algorithm is essentially tight, under the Exponential Time Hypothesis (ETH) [23], by augmenting a reduction by Abu-Khzam et al. [2].

Theorem 6 (♣). *There is a graph property Π for which we cannot solve Π-FREE DELETION [VC] in $2^{o(K \log K)} \mathrm{poly}(n)$ time, unless ETH fails, where K is the vertex cover number of G, even if each graph that has property Π has size quadratic in its vertex cover number.*

Next, we look to further improve the bound of Theorem 5. Note that so far, we have made no use at all of the characterization by few adjacencies of Π, as in Theorem 1. We now argue that there may be graphs in Π that cannot occur in G simply because it would not fit with the vertex cover.

Lemma 2 (♣). *If Π is a graph property such that*

(i) every graph in Π is connected and contains at least one edge, and
(ii) Π is characterized by c_Π adjacencies,

and G is some graph with vertex cover X, $|X| = K$, and $S \subseteq V(G)$ some vertex set. Then $G[V(G) \setminus S]$ is Π-free if and only if $G[V(G) \setminus S]$ is Π'-free, where $\Pi' \subseteq \Pi$ contains only those graphs in Π with $\leq (c_\Pi + 1)K$ vertices.

The precondition that every graph in Π is connected is necessary to obtain this result. We can use Lemma 2 in combination with Theorem 5 to obtain a new result. Alternatively, using a streaming version of the algorithm instead of the non-streaming one, immediately also provides a streaming result.

Theorem 7 (♣). *Given a graph G with vertex cover X, $|X| = K$, if Π is a graph property such that*

(i) every graph in Π is connected and contains at least one edge, and
(ii) Π is characterized by c_Π adjacencies, and
(iii) we have explicit knowledge of $\Pi' \subseteq \Pi$, which is the subset of q graphs of at most size $(c_\Pi + 1)K$ that are vertex-minimal with respect to Π,

then Π-FREE DELETION [VC] can be solved using $\mathcal{O}(q \cdot 2^K \cdot ((c_\Pi + 1)K)^K \cdot K! \cdot K \cdot ((c_\Pi + 1)K)! \cdot ((c_\Pi + 1)K)^2 \cdot n)$ time. Assuming $c_\Pi \geq 1$ this can be simplified to $\mathcal{O}(q \cdot 2^K \cdot c_\Pi^K \cdot K^{K+3} \cdot K! \cdot (c_\Pi K)! \cdot n)$ time. In the streaming setting, Π-FREE DELETION [VC] can be solved using $\mathcal{O}(q \cdot 2^K \cdot c_\Pi^K \cdot K^{K+2} \cdot K! \cdot (c_\Pi K)!)$ passes in the AL streaming model, using $\tilde{\mathcal{O}}((c_\Pi K)^2 + q \cdot (c_\Pi + 1)K)$ space.

The required explicit knowledge of Π' might give memory problems. That is, we have to store Π' somewhere to make this algorithm work, which takes $\tilde{\mathcal{O}}(q \cdot (c_\Pi + 1)K)$ space. Note that q can range up to $K^{\mathcal{O}(K)}$. We adapt the streaming algorithm to the case when we have oracle access to Π in ♣.

3.4 Odd Cycle Transversal

Specific forms of Π-FREE DELETION [VC] allow for improvement over Theorem 7, which we illustrate for the problem of ODD CYCLE TRANSVERSAL [VC]. Note that odd cycle-free and induced odd cycle-free are equivalent.

ODD CYCLE TRANSVERSAL [VC]
Input: A graph G with a vertex cover X, and an integer ℓ.
Parameter: The size $K := |X|$ of the vertex cover.
Question: Is there a set $S \subseteq V(G)$ of size at most ℓ such that $G[V(G) \setminus S]$ contains no induced odd cycles?

The interest in this problem comes from the FPT algorithm using iterative compression provided in [15, Section 4.4], based on work by Reed et al. [30]. Although Chitnis and Cormode [10] have shown how iterative compression can be used in the streaming setting, adapting the algorithm out of Reed et al. seems difficult. The main cause for this is the use of a maximum-flow algorithm, which does not seem to lend itself well to the streaming setting because of its memory requirements. Instead, we present the following approach.

It is well known that a graph without odd cycles is a bipartite graph (and thus 2-colourable) and vice versa. In the algorithm, we guess what part of the vertex cover is in the solution, and then we guess the colouring of the remaining part. Then vertices outside the vertex cover for which not all neighbours have the same colour must be deleted. This step can be done in one pass if we use the AL streaming model. In the same pass, we can also check if the colouring is valid within the vertex cover. If the number of deletions does not exceed the solution size and the colouring is valid within the vertex cover, then the resulting graph is bipartite and thus odd cycle free.

The total number of guesses comes down to $\mathcal{O}(3^K)$ options, as any vertex in the vertex cover is either in the solution, coloured with colour 1 or coloured with colour 2. This directly corresponds to the number of passes, as only one pass is needed per guessed colouring.

Theorem 8 (♣). *Given a graph G given as an AL stream with vertex cover X, $|X| = K$, we can solve* ODD CYCLE TRANSVERSAL [VC] *using $\mathcal{O}(3^K)$ passes and $\mathcal{O}(K \log n)$ space.*

If we think about this algorithm, we can notice that often the colouring we guess on the vertex cover is invalid. An alternative approach (♣) follows by noting that a connected component within the vertex cover can only have two possible valid colourings. We can exploit this to decrease the number of passes when the number of connected components in the vertex cover is low. This comes at a price: to easily find components of the vertex cover, we store it in memory, which increases the memory complexity. Alternatively, we can use $\mathcal{O}(K)$ passes to find the connected components of the vertex cover in every branch.

4   Lower Bounds

We show lower bounds for Π-FREE DELETION. To prove lower bounds for streaming, we can show reductions from problems in communication complexity, as first shown by Henzinger et al. [22]. An example of such a problem is DISJOINTNESS.

DISJOINTNESS
Input: Alice has a string $x \in \{0,1\}^n$ given by $x_1 x_2 \ldots x_n$. Bob has a string $y \in \{0,1\}^n$ given by $y_1 y_2 \ldots y_n$.
Question: Bob wants to check if $\exists 1 \leq i \leq n$ such that $x_i = y_i = 1$. (Formally, the answer is NO if this is the case.)

The following proposition is given and used by Bishnu et al. [5], and gives us one important consequence of reductions from a problem in communication complexity to a problem for streaming algorithms.

Proposition 1. *(Rephrasing of item (ii) of [5, Proposition 5.6]) If we can show a reduction from* DISJOINTNESS *to problem* Π *in streaming model* $\mathcal{M}$ *such that the reduction uses a 1-pass streaming algorithm of* Π *as a subroutine, then any streaming algorithm working in the model* $\mathcal{M}$ *for* Π *that uses p passes requires* $\Omega(n/p)$ *bits of memory, for any* $p \in \mathbb{N}$ *[3,7,12].*

The structure of these reductions is relatively simple: have Alice and Bob construct the input for a streaming algorithm depending on their input to DIS-JOINTNESS. If we do this in such a manner that the solution the streaming algorithm outputs gives us exactly the answer to DISJOINTNESS, we can conclude that the streaming algorithm must abide the lower bound of DISJOINTNESS.

Chitnis et al. [11, Theorem 6.3] prove hardness for many Π, those that obide to a small precondition. However, Chitnis et al. do not describe in their reduction how Alice and Bob give their 'input' as a stream to the algorithm for Π-FREE DELETION, and thus it would apply only to the EA streaming model. However, if we observe the proof closely, we can see it extends to the VA model.

We would also like it to extend to the AL model. However, this requires a slightly stronger precondition on the graph class Π.

Theorem 9. *If* Π *is a set of graphs such that each graph in* Π *is connected, and there is a graph* $H \in \Pi$ *such that*

– *H is a minimal element of* Π *under the operation of taking subgraphs, i.e., no proper subgraph of H is in* Π, *and*
– *H has at least two **disjoint** edges,*

then any p-pass (randomized) streaming algorithm working on the AL *streaming model for* Π-FREE DELETION *[ℓ] needs* $\Omega(n/p)$ *bits of space.*

Proof. We add onto the proof of [11, Theorem 6.3], by specifying how Alice and Bob provide the input to the p-pass streaming algorithm.

Let H be a minimal graph in Π which has at least two disjoint edges, say e_1 and e_2. Let $H' := H \setminus \{e_1, e_2\}$. Create as an input for the streaming algorithm n copies of H', where in copy i we add the edges e_1 and e_2 if and only if the input of DISJOINTNESS has a 1 for index i for Alice and Bob respectively.

As e_1 and e_2 are disjoint, e_2 is incident on two vertices v, w which are not incident to e_1. For every pass the algorithm requires, we do the following. We provide all the copies of H as input to the streaming algorithm by letting Alice input all vertices $V(H) \setminus \{v, w\}$ as an AL stream. Note that Alice has enough information to do this, as the vertices incident on the edge e_2 in each copy of H is never included in this part of the stream. Then Alice passes the memory of the streaming algorithm to Bob, who inputs the edges incident to the vertices v, w for each copy of H (which includes e_2 if and only if the respective bit in the input of DISJOINTNESS is 1). This ends a pass of the stream.

Note that Alice and Bob have input the exact specification of a graph as described by Chitnis et al. [11, Theorem 6.3], but now as a AL stream. Hence, the correctness follows. □

Theorem 9 provides a lower bound for, for example, EVEN CYCLE TRANSVERSAL [ℓ] (where Π is the set of all graphs that contain a $C_4, C_6, \ldots$), and similarly ODD CYCLE TRANSVERSAL [ℓ] and FEEDBACK VERTEX SET [ℓ]. Theorem 9 does not hold for the scenario where Π contains only stars.

Notice that the lower bound proof makes a construction with a vertex cover size linear in n. Therefore, these bounds do not hold when the vertex cover size is bounded. We can prove lower bounds with constant vertex cover size for H-FREE DELETION with specific requirements on H, for the EA and VA models. These results follow by adapting the known lower bound construction by Bishnu et al. [5]. Here, we give a summarizing theorem for these lower bounds.

Theorem 10. (♣). *If H is such that either:*

1. *H is a connected graph with at least 3 edges and a vertex of degree 2, or,*
2. *H is a graph with a vertex of degree at least 2 for which every neighbour has an equal or larger degree,*

then any algorithm for solving H-FREE DELETION [VC] on a graph G with $K \geq |VC(H)| + 1$ requires $\Omega(n/p)$ bits when using p passes in the VA/EA models, even when the solution size $\ell = 0$.

Theorem 10 proves lower bounds for ODD CYCLE TRANSVERSAL [VC], EVEN CYCLE TRANSVERSAL [VC], FEEDBACK VERTEX SET [VC], and COGRAPH DELETION [VC]. Examples for which Theorem 10 does not give a lower bound include CLUSTER VERTEX DELETION [VC] (indeed, then a kernel is known [5]), or more generally, H-FREE DELETION [VC] when H is a star.

5   Conclusion

We have seen different streaming algorithms and lower bounds for Π-FREE DELETION and its more specific forms, making use of the minimum vertex cover as a parameter. We have seen the potency of the AL streaming model in combination with the vertex cover, where in other streaming models lower bounds arise. It is interesting that for very local structures like a P_3, this combination works effortlessly, giving a very efficient memory-optimal algorithm. For more general structures troubles arise, but nonetheless, we can solve the more general problems with a many-pass, low-memory approach. Alternatively, the adaptations of kernels gives rise to a few-pass, high-memory algorithm, which provides a possible trade-off when choosing an algorithm.

We also propose the following open problems. Can lower bounds be found expressing a pass/memory trade-off in the vertex cover size for the Π-FREE DELETION [VC] problem? Or alternatively, can we find an upper bound for Π-FREE DELETION [VC] using $\mathcal{O}(K \log n)$ bits of memory but only a polynomial in K number of passes? Essentially, here we ask whether or not our algorithm is reasonably tight, or can be improved to only use a polynomial number of passes in K. A lower bound expressing a trade-off in terms of the vertex cover size is a standalone interesting question, as most lower bound statements about streaming algorithms express a trade-off in terms of n.

We also ask about the unparameterized streaming complexity of CLUSTER VERTEX DELETION in the AL model. While lower bounds for most other Π-FREE DELETION problems in the AL model follow from our work (Theorem 9) and earlier work of Bishnu et al. [5], this appears an intriguing open case.

Finally, we ask if there is a $2^{o(K \log K)}$ lower bound for Π-FREE DELETION [VC] when Π is characterized by few adjacencies?

References

1. Abu-Khzam, F.N.: Maximum common induced subgraph parameterized by vertex cover. Inf. Process. Lett. **114**(3), 99–103 (2014)
2. Abu-Khzam, F.N., Bonnet, É., Sikora, F.: On the complexity of various parameterizations of common induced subgraph isomorphism. Theor. Comput. Sci. **697**, 69–78 (2017)
3. Agarwal, D., McGregor, A., Phillips, J.M., Venkatasubramanian, S., Zhu, Z.: Spatial scan statistics: approximations and performance study. In: Proceedings of SIGKDD 2006, pp. 24–33. ACM (2006)
4. Agrawal, A., et al.: Parameterized streaming algorithms for min-ones d-sat. In: Proceedings of FSTTCS 2019. LIPIcs, vol. 150, pp. 8:1–8:20. Schloss Dagstuhl - Leibniz-Zentrum für Informatik (2019)
5. Bishnu, A., Ghosh, A., Kolay, S., Mishra, G., Saurabh, S.: Fixed-parameter tractability of graph deletion problems over data streams. CoRR abs/1906.05458 (2019)
6. Bishnu, A., Ghosh, A., Kolay, S., Mishra, G., Saurabh, S.: Fixed parameter tractability of graph deletion problems over data streams. In: Kim, D., Uma, R.N., Cai, Z., Lee, D.H. (eds.) COCOON 2020. LNCS, vol. 12273, pp. 652–663. Springer, Cham (2020). https://doi.org/10.1007/978-3-030-58150-3_53
7. Bishnu, A., Ghosh, A., Mishra, G., Sen, S.: On the streaming complexity of fundamental geometric problems. CoRR abs/1803.06875 (2018)
8. Bodlaender, H.L., Jansen, B.M.P., Kratsch, S.: Kernelization lower bounds by cross-composition. SIAM J. Discret. Math. **28**(1), 277–305 (2014)
9. Cai, L.: Fixed-parameter tractability of graph modification problems for hereditary properties. Inf. Process. Lett. **58**(4), 171–176 (1996)
10. Chitnis, R., Cormode, G.: Towards a theory of parameterized streaming algorithms. In: Proc. IPEC 2019. LIPIcs, vol. 148, pp. 7:1–7:15. Schloss Dagstuhl - Leibniz-Zentrum für Informatik (2019)
11. Chitnis, R., et al.: Kernelization via sampling with applications to finding matchings and related problems in dynamic graph streams. In: Proceedings of SODA 2016, pp. 1326–1344. SIAM (2016)
12. Chitnis, R.H., Cormode, G., Esfandiari, H., Hajiaghayi, M., Monemizadeh, M.: Brief announcement: New streaming algorithms for parameterized maximal matching & beyond. In: Proceedings of SPAA 2015, pp. 56–58. ACM (2015)
13. Chitnis, R.H., Cormode, G., Hajiaghayi, M.T., Monemizadeh, M.: Parameterized streaming: Maximal matching and vertex cover. In: Proceedings of SODA 2015, pp. 1234–1251. SIAM (2015)

14. Cormode, G., Dark, J., Konrad, C.: Independent sets in vertex-arrival streams. In: Proceedings of ICALP 2019. LIPIcs, vol. 132, pp. 45:1–45:14. Schloss Dagstuhl - Leibniz-Zentrum für Informatik (2019)
15. Cygan, M., et al.: Parameterized Algorithms. Springer, Heidelberg (2015). https://doi.org/10.1007/978-3-319-21275-3
16. Downey, R.G., Fellows, M.R.: Parameterized Complexity. Monographs in Computer Science, Springer, Heidelberg (1999). https://doi.org/10.1007/978-1-4612-0515-9
17. Fafianie, S., Kratsch, S.: Streaming kernelization. In: Csuhaj-Varjú, E., Dietzfelbinger, M., Ésik, Z. (eds.) MFCS 2014. LNCS, vol. 8635, pp. 275–286. Springer, Heidelberg (2014). https://doi.org/10.1007/978-3-662-44465-8_24
18. Feigenbaum, J., Kannan, S., McGregor, A., Suri, S., Zhang, J.: Graph distances in the streaming model: the value of space. In: Proceedings of SODA 2005, pp. 745–754. SIAM (2005)
19. Feigenbaum, J., Kannan, S., McGregor, A., Suri, S., Zhang, J.: On graph problems in a semi-streaming model. Theor. Comput. Sci. 348(2–3), 207–216 (2005)
20. Fomin, F.V., Golovach, P.A., Thilikos, D.M.: On the parameterized complexity of graph modification to first-order logic properties. Theory Comput. Syst. 64(2), 251–271 (2020). https://doi.org/10.1007/s00224-019-09938-8
21. Fomin, F.V., Jansen, B.M.P., Pilipczuk, M.: Preprocessing subgraph and minor problems: when does a small vertex cover help? J. Comput. Syst. Sci. 80(2), 468–495 (2014)
22. Henzinger, M.R., Raghavan, P., Rajagopalan, S.: Computing on data streams. In: Abello, J.M., Vitter, J.S. (eds.) Proceedings of DIMACS 1998. DIMACS, vol. 50, pp. 107–118. DIMACS/AMS (1998)
23. Impagliazzo, R., Paturi, R.: On the complexity of k-SAT. J. Comput. Syst. Sci. 62(2), 367–375 (2001)
24. Jansen, B.M.P., de Kroon, J.J.H.: Preprocessing vertex-deletion problems: characterizing graph properties by low-rank adjacencies. In: Proceedings of SWAT 2020, LIPIcs, vol. 162, pp. 27:1–27:15. Schloss Dagstuhl - Leibniz-Zentrum für Informatik (2020)
25. Jansen, B.M.: The power of data reduction: Kernels for fundamental graph problems. Ph.D. thesis, Utrecht University (2013)
26. Lewis, J.M., Yannakakis, M.: The node-deletion problem for hereditary properties is NP-complete. J. Comput. Syst. Sci. 20(2), 219–230 (1980)
27. McGregor, A.: Graph stream algorithms: a survey. SIGMOD Rec. 43(1), 9–20 (2014)
28. McGregor, A., Vorotnikova, S., Vu, H.T.: Better algorithms for counting triangles in data streams. In: Proceedings of PODS 2016, pp. 401–411. ACM (2016)
29. Muthukrishnan, S.: Data streams: algorithms and applications. Found. Trends Theor. Comput. Sci. 1(2) (2005)
30. Reed, B.A., Smith, K., Vetta, A.: Finding odd cycle transversals. Oper. Res. Lett. 32(4), 299–301 (2004)
31. Sau, I., dos Santos Souza, U.: Hitting forbidden induced subgraphs on bounded treewidth graphs. In: Proceedings of MFCS 2020. LIPIcs, vol. 170, pp. 82:1–82:15. Schloss Dagstuhl - Leibniz-Zentrum für Informatik (2020)

On the Hardness of the Determinant: Sum of Regular Set-Multilinear Circuits

S. Raja[✉] and G. V. Sumukha Bharadwaj

IIT Tirupati, Tirupati, India
{raja,cs21d003}@iittp.ac.in

Abstract. In this paper, we study the computational complexity of the commutative determinant polynomial computed by a class of set-multilinear circuits which we call *regular set-multilinear circuits*. Regular set-multilinear circuits are commutative circuits with a restriction on the order in which they can compute polynomials. A regular circuit can be seen as the commutative analogue of the *ordered circuit* defined by Hrubes, Wigderson and Yehudayoff [5]. We show that if the commutative determinant polynomial has small representation in the sum of constantly many regular set-multilinear circuits, then the commutative permanent polynomial also has a small arithmetic circuit.

Keywords: Hardness of determinant · Set-multilinear circuits · Regular set-multilinear circuits

1 Introduction

Arithmetic circuit complexity studies the complexity of computing polynomials using arithmetic operations. Arithmetic circuits are a natural computational model for computing and describing polynomials. Arithmetic circuit is a directed acyclic graph with internal nodes labeled by + or ×, and leaves labeled by either variables or elements from a underlying field $\mathbb{F}$. The complexity measures associated with arithmetic circuits are size, which measures number of gates in the circuit, and depth, which measures length of the longest path from a leaf to the output gate in the circuit. Two important examples of polynomial family are the determinant and the permanent polynomials. The determinant polynomial is ubiquitous in linear algebra, and it can be computed by polynomial-sized arithmetic circuits (see e.g., [3]). On the other hand, the permanent of 0/1 matrices is #P-complete [10], where #P corresponds to the counting class in the world of Boolean complexity classes. Thus, it is believed that, over fields of characteristic different from 2, the permanent $PERM = (PERM_n)$ polynomial family cannot be computed by any polynomial-sized circuit family. A central open problem of the field is proving super-polynomial size lower bounds for arithmetic circuits that compute the permanent polynomial $PERM_n$. Motivated by this problem, Valiant, in his seminal work [9], defined the arithmetic analogues of P and NP: denoted by VP and VNP. Informally, VP consists of multivariate (commutative)

© Springer Nature Switzerland AG 2021
E. Bampis and A. Pagourtzis (Eds.): FCT 2021, LNCS 12867, pp. 427–439, 2021.
https://doi.org/10.1007/978-3-030-86593-1_30

polynomials that have polynomial size circuits. Valiant showed that $PERM$ is VNP-complete w.r.t. projection reductions. Thus, $VP \neq VNP$ iff $PERM_n$ requires arithmetic circuits of size super-polynomial in n.

Set-multilinear circuits are introduced in the work of [7]. Let $\mathbb{F}$ be a field and $X = X_1 \sqcup X_2 \sqcup \cdots \sqcup X_d$ be a partition of the variable set X. A *set-multilinear polynomial* $f \in \mathbb{F}[X]$ w.r.t. this partition is a homogeneous degree d multilinear polynomial such that every nonzero monomial of f has exactly one variable from X_i, for all $1 \leq i \leq d$. Some of the well-known polynomial families like the permanent $PERM_n$ and the determinant DET_n, are set-multilinear. The variable set is $X = \{x_{ij}\}_{1 \leq i,j \leq n}$ and the partition can be taken as the row-wise partition of the variable set. I.e. $X_i = \{x_{ij} \mid 1 \leq j \leq n\}$ for $1 \leq i \leq n$. In this work, we study the set-multilinear circuit complexity of the determinant polynomial DET_n. A *set-multilinear arithmetic circuit* C computing f w.r.t. the above partition of X, is a directed acyclic graph such that each in-degree 0 node of the graph is labeled with an element from $X \cup \mathbb{F}$. Each internal node v of C is of in-degree 2, and is either a $+$ gate or $\times$ gate. With each gate v we can associate a subset of indices $I_v \subseteq [d]$ and the polynomial f_v computed by the circuit at v is set-multilinear over the variable partition $\bigsqcup_{i \in I_v} X_i$. If v is a $+$ gate then for each input u of v, $I_u = I_v$. If v is a $\times$ gate with inputs v_1 and v_2 then $I_v = I_{v_1} \sqcup I_{v_2}$. Clearly, in a set-multilinear circuit every gate computes a set-multilinear polynomial (in a syntactic sense). The output gate of C computes the polynomial f, which is set-multilinear over the variable partition $\bigsqcup_{i \in [d]} X_i$. The *size* of C is the number of gates in it and its *depth* is the length of the longest path from an input gate to the output gate of C. Additionally, a set-multilinear circuit C is called a *set-multilinear formula* if out-degree of every gate is bounded by 1.

Set-multilinear arithmetic circuits are a natural model for computing set-multilinear polynomials. It can be seen that each set-multilinear polynomial can be computed by a set-multilinear arithmetic circuit. For set-multilinear formulas, super-polynomial size lower bounds are known [8]. Super-polynomial lower bounds for a class of set-multilinear ABPs computing the determinant DET_n is shown in [1]. It is known that proving super-polynomial lower bound result for general set-multilinear circuits computing the permanent polynomial $PERM_n$ would imply that $PERM_n$ requires super-polynomial size non-commutative arithmetic circuits, and this is an open problem for over three decades. Non-commutative circuits are a restriction on the computational power of circuits. Though non-commutative circuits compute non-commutative polynomials, one can study what is the power of commutativity in computing the DET_n polynomial. Noncommutative arithmetic circuit models are well studied, see e.g., [2,5,6]. In [2], it was shown that computing the non-commutative determinant polynomial is as hard as computing the commutative permanent polynomial.

1.1 Our Results

To explain our results, we first define the computational model that we study. Let S_n denote the set of all permutations over the set $\{1, 2..., n\}$.

Definition 1 (Regular Set-Multilinear Circuits). *Let $X = X_1 \sqcup X_2 \sqcup \cdots \sqcup X_d$ be a partition of the variable set X. Let $\sigma \in S_d$. A set-multilinear circuit C that computes a set-multilinear polynomial $f \in F[X]$ w.r.t the above partition is called* regular set-multilinear circuit w.r.t $\sigma \in S_d$, *if every gate v in C is associated with an interval I_v w.r.t $\sigma \in S_d$. In other words, $\sigma \in S_d$ defines an ordering $(\sigma(1), \sigma(2), \cdots, \sigma(d))$ and every gate v in C is associated with an interval I_v w.r.t σ-ordering $(\sigma(1), \sigma(2), \cdots, \sigma(d))$.*

Let C be a regular set-multilinear circuit w.r.t σ computing a commutative polynomial f of degree d. Let v be a gate in C computing the polynomial f_v of degree k. By definition, f_v is a set-multilinear polynomial w.r.t $I_v = [\sigma(i), \sigma(i+1), \cdots, \sigma(i+k)]$, where $i <= d-k$. Let $order(f_v) = I_v = (\sigma(i), \sigma(i+1), \cdots, \sigma(i+k))$.

Since for each gate v in C, I_v can be viewed as an interval w.r.t $\sigma \in S_d$, the two children u and w of v can be designated as left and right child. In particular, for each product gate v with children u and w such that $I_v = I_u \sqcup I_w$, we refer to u as the left child of v, and w as the right child of v.

We make the following observations about regular set-multilinear circuits:

- If v is an input gate (leaf node) labeled by a field constant, then $order(f_v) = ()$, where $()$ is the empty sequence. If v is an input gate labeled by a variable $x_{i,j}$, then $order(f_v) = (i)$.
- If v is an product gate, then $order(f_v) = order(f_u) \sqcup order(f_w)$, where the interval $order(f_v)$ is obtained by appending $order(f_u)$ with $order(f_w)$.
- If v is a sum gate, then $order(f_v) = order(f_u) = order(f_w)$.

One can define several versions of non-commutative DET_n polynomial. Non-commutative circuits computing the DET_n polynomial, where the first index of the variables in each monomial is in increasing order, can be seen as regular set-multilinear w.r.t the identity permutation. In [2], it was shown that computing the non-commutative determinant polynomial is as hard as computing the commutative permanent polynomial. A natural next step is to find the set-multilinear circuit complexity of the commutative determinant polynomial.

We study the computational complexity of the commutative determinant polynomial DET_n computed by a sum of regular set-multilinear circuits. We show that if the determinant polynomial DET_n is computed by a circuit C of size s, where C is a sum of constantly-many regular set-multilinear circuits, then we can modify C to compute the permanent polynomial $PERM_{n^\epsilon}$, where $\epsilon > 0$, such that the new circuit size is polynomially related to the size of C. We remark that in our result, there is no restriction on the number of different parse tree types/shapes (see e.g., [1]) allowed in each regular circuits.

One can view this as a generalization of the result shown in [2] to a class of set-multilinear circuits computing the determinant polynomial DET_n. We obtain our result by carefully combining *Erdös-Szekeres theorem* [4] and some properties that we prove about regular set-multilinear circuits and the result of [2].

2   Preliminaries

2.1   Determinant and Permanent

Definition 2 *(Commutative Determinant and Permanent). Given the set of variables $X = \{x_{i,j} \mid 1 \leq i, j \leq n\}$, the $n \times n$ commutative determinant and the $n \times n$ commutative permanent over X, denoted by $DET_n(X)$ and $PERM_n(X)$ respectively, are n^2-variate polynomials of degree n given by:*

$$DET_n(X) = \sum_{\sigma \in S_n} sgn(\sigma) \prod_{i=1}^{n} x_{i,\sigma(i)}$$

$$PERM_n(X) = \sum_{\sigma \in S_n} \prod_{i=1}^{n} x_{i,\sigma(i)},$$

Non-commutative determinant can be defined in various ways depending on the order in which variables are multiplied. One natural type of non-commutative determinant, called the *Cayley determinant $CDET_n$*, is one where the order of multiplication is the identity permutation w.r.t first index of the variable.

2.2   Erdös-Szekeres Theorem

Theorem 1 (Erdös-Szekeres Theorem, [4]). *Let n be a positive integer. Let S be a sequence of distinct integers of length at least $n^2 + 1$. Then, there exists a monotonically increasing subsequence of S of length $n + 1$, or a monotonically decreasing subsequence of S of length $n + 1$.*

Let A, B be two $n \times n$ matrices. The following are known facts about the determinant and permutations.

Fact 1: $det(A \times B) = det(A) \times det(B)$.
Fact 2: The determinant of a permutation matrix is either $+1$ or -1.
Fact 3: Let $\tau, \sigma \in S_n$. Then $sign(\tau \circ \sigma) = sign(\tau) \times sign(\sigma)$.

For $n \in \mathbb{N}$, let $[n] = \{1, 2, \cdots, n\}$.

3   Hardness of the Determinant: Sum of Two Regular Set-Multilinear Circuits

In this section, we show that if the determinant polynomial is computed by a sum of two regular set-multilinear circuits then the permanent polynomial can also be represented as a regular set-multilinear circuit. This result involves all the techniques which will be used in the main result and it is easy to explain in this sum of two regular circuits model. In the next section, we will prove the result for sum of constantly many regular set-multilinear circuits. We note that all our polynomials are commutative. For the purpose of readability, we sometimes ignore the floor operation.

Let $X = \{x_{i,j} \mid 1 \leq i, j \leq n\}$ be the set of variables. Let $X_i = \{x_{ij} \mid 1 \leq j \leq n\}$ for $1 \leq i \leq n$. Our aim is to show that if $C = C_1^{\sigma 1} + C_2^{\sigma 2}$ computing the determinant polynomial $DET_n(X) \in \mathbb{F}[X]$, where the circuits $C_1^{\sigma 1}$, $C_2^{\sigma 2}$ are *regular set-multilinear circuits* w.r.t $\sigma 1, \sigma 2 \in S_n$ respectively, then there is an efficient transformation that converts the given circuit C to another circuit C' computing the permanent polynomial of degree $\sqrt{n}/2$. Given $C = C_1^{\sigma 1} + C_2^{\sigma 2}$ computing $DET_n(X)$, if $\sigma 1 = \sigma 2$ then we can directly adapt the result of [2] and get a circuit C' computing the permanent polynomial of degree $n/2$. If n is not even then we can substitute variables in the set X_n suitably from $\{0, 1\}$ such that C computes $DET_{n-1}(X)$ before using the result of [2].

The case of $\sigma 1 \neq \sigma 2$ needs more work that we explain now. The idea is to use the well known Erdös-Szekeres Theorem [4] that guarantees that any sequence of n distinct integers contains a subsequence of length at least $\sqrt{n}$ that is either monotonically increasing or decreasing. By viewing $\sigma = (\sigma(1), \sigma(2), \cdots, \sigma(n))$ as a sequence of integers, we apply the above result to permutations $\sigma 1, \sigma 2 \in S_n$. We first apply it to $\sigma 1 = (\sigma 1(1), \sigma 1(2), ..., \sigma 1(n))$ and let $A = \{i_1, i_2, \cdots, i_{\sqrt{n}}\}$ be the set of indices that appear in this monotone subsequence. If the subsequence is monotonically increasing then we do substitutions in $DET_n(X)$ so that it computes the determinant polynomial of $\sqrt{n} \times \sqrt{n}$ matrix whose rows and columns are labeled by the elements of set A. This is done by making suitable substitutions to the variables in X from $X \cup \{0, 1\}$ in the given circuit C. After this we get a circuit C' from C that computes $DET_{\sqrt{n}}(X')$ where $X' = \bigsqcup_{i \in A} X_i$.

We note that $C' = C_1^{\sigma 1'} + C_2^{\sigma 2'}$ where $\sigma 1', \sigma 2' \in S_{\sqrt{n}}$ and $\sigma 1' = (\sigma 1'(1), \sigma 1'(2), \cdots, \sigma 1'(\sqrt{n}))$ is in increasing order. If $\sigma 1' = \sigma 2'$, then we can use [2] and get the permanent of degree $\sqrt{n}/2$. Otherwise, we apply Erdös-Szekeres Theorem to permutations $\sigma 1', \sigma 2'$. In particular, this will give us a monotone subsequence in $\sigma 2' = (\sigma 2'(1), \sigma 2'(2), \cdots, \sigma 2'(\sqrt{n}))$ with length at least $n^{1/4}$. If this sequence is increasing, then the same subsequence is also increasing in $\sigma 1'$ as we already noted that it is in increasing order. Let $A_1 = \{j_1, j_2, \cdots, j_{n^{1/4}}\}$ be the set of indices that appear in this monotone subsequence. Now we project, as before so that it computes the determinant polynomial of a $n^{1/4} \times n^{1/4}$ matrix whose rows and columns are labeled by the elements in set A_1. After substituting from $X' \cup \{0, 1\}$ for each variable in the given circuit C', we get a regular circuit C'' that computes $DET_{n^{1/4}}(X'')$, where $X'' = \bigsqcup_{i \in A_1} X_i$.

The important thing to note here is that in the new circuit $C'' = C_1^{\sigma 1''} + C_2^{\sigma 2''}$, where $\sigma 1'', \sigma 2'' \in S_{n^{1/4}}$, both $\sigma 1''$ and $\sigma 2''$ are the same, i.e., $\sigma 1'' = \sigma 2''$. We can rename the variable sets in $X'' = \bigsqcup_{i \in A_1} X_i$ to $X_1, X_2, \cdots, X_{n^{1/4}}$. For example, if $i_1 \in A_1$ is the lowest index then we can rename X_{i_1} to X_1, and for all j, rename $X_{i_1, j}$ to $X_{1, j}$. Similarly, the k-th lowest index is modified. After these modifications, we can assume that $\hat{X} = \bigsqcup_{i \in [n^{1/4}]} X_i$.

As we noted before, any non-commutative circuit computing DET_n, where the first index of the variables in each monomial is in increasing order, can be seen as *regular set-multilinear w.r.t identity permutation*. Now we can apply the following theorem (Theorem 10 from [2]) to get our result.

Theorem 2 *(Theorem 10, [2]). For any $n \in \mathbb{N}$, if there is a non-commutative circuit C of size s computing the Cayley determinant $DET_{2n}(X)$ then there is a circuit C' of size polynomial in s and n that computes the Cayley permanent $PERM_n(Y)$.*

If $n' = \lfloor n^{1/4} \rfloor$ is not an even number then we ignore the $X_{n'}$ variable set in $\hat{X}$ by following substitutions: $X_{n',n'} = 1$ and for all $j \in [n'-1]$, $X_{n',j} = 0$ and $X_{j,n'} = 0$. After this substitutions, we have a circuit that computes the determinant $DET_{n'-1}$ polynomial. Now applying the above theorem we get a circuit $\hat{C}$ that computes the permanent polynomial of degree $\frac{n'-1}{2}$.

We now explain how to handle if Erdös-Szekeres Theorem guarantees only monotonically decreasing sequence. For that we define the *reverse* of a regular set-multilinear circuit C w.r.t $\sigma \in S_n$ computing a polynomial f. This results in a regular set-multilinear circuit C^{rev} w.r.t $\sigma^{rev} \in S_n$, where $\sigma^{rev} = (\sigma(n), \sigma(n-1), ..., \sigma(1))$, *computing the same commutative polynomial f as circuit C.* We note that if σ has monotonically decreasing subsequence of length k then σ^{rev} has a monotonically increasing subsequence of *same* length k. We obtain C^{rev} by interchanging the left and right children of product gates in C. This is proved in the following lemma.

Lemma 1 (Reversal Lemma). *Let $X = \{x_{i,j} \mid 1 \le i, j \le n\}$ be a set of variables and $X = X_1 \sqcup X_2 \sqcup ... \sqcup X_n$ be a partition of X, where for all $1 \le i \le n$, $X_i = \{x_{i,1}, x_{i,2}, ..., x_{i,n}\}$. Let C be a regular set-multilinear circuit w.r.t a permutation $\sigma \in S_n$ computing the polynomial $f \in F[X]$. Then, there exists a regular set-multilinear circuit C^{rev} w.r.t $\sigma^{rev} \in S_n$ where $\sigma^{rev} = (\sigma(n), \sigma(n-1), ..., \sigma(1))$ computing the same commutative polynomial f as circuit C. Moreover, the size of C^{rev} is same as that of C.*

Proof. First, we describe the construction of the circuit C^{rev}, and then prove its correctness. Let v be a gate in C. As C is a regular set-multilinear circuit w.r.t $\sigma \in S_n$, we have an interval I_v w.r.t the permutation σ associated with the gate v.

Construction of C^{rev}: Starting with the product gates at the bottom of C and gradually moving up level-by-level, swap the left and right children of each product gate.

Correctness: We show by induction on depth d of C that both circuits C and C^{rev} compute the same polynomial $f \in F[X]$ and C^{rev} is a regular set-multilinear circuit w.r.t $\sigma^{rev} \in S_n$, where $\sigma^{rev} = (\sigma(n), \sigma(n-1), ..., \sigma(1))$. Let f_v and f_v^{rev} denote the polynomials computed at any node v in C and C^{rev}, respectively. Let $order(f_v) = I_v$. We will show that f_v and f_v^{rev} are the same polynomial and the only difference is in their orders. That is, $order(f_v^{rev}) = rev(order(f_v))$, where $rev(order(f_v))$ is $order(f_v)$ written in reverse (i.e., the interval I_v is reversed).

The proof is by induction on the depth d of the circuit C^{rev}. Let f^{rev} denote the polynomial computed by C^{rev}.

Base Case: The base case is any node at depth 0, i.e., a leaf node. Consider any leaf node l. Then f_l, the polynomial computed at l, is either a variable or a field constant in F. If f_l is a field constant, then $order(f_l) = ()$. Therefore, $order(f_l^{rev}) = ()$. If f_l is a variable $x_{i,j}, 1 \leq i,j \leq n$, then $order(f_l) = (i)$. Therefore, the $order(f_l^{rev}) = (i)$. In both cases, $f_l^{rev} = f_l$ and $order(f_l^{rev}) = rev(order(f_l))$.

Induction Hypothesis: Assume for any node u at depth d', $1 \leq d' \leq d - 1$, that $f_u^{rev} = f_u$ and $order(f_u^{rev}) = rev(order(f_u))$.

Induction Step: Consider any node v at depth $d' + 1$, with v_L and v_R as its left and right children, respectively. By induction hypothesis, $f_{v_L}^{rev} = f_{v_L}$ and $order(f_{v_L}^{rev}) = rev(order(f_{v_L}))$. Similarly, $f_{v_R}^{rev} = f_{v_R}$ and $order(f_{v_R}^{rev}) = rev(order(f_{v_R}))$.

If v is a product gate, then $f_v^{rev} = f_{v_R}^{rev} \times f_{v_L}^{rev}$, which is equivalent to $f_{v_R} \times f_{v_L} = f_v$ by induction hypothesis. By induction hypothesis, $order(f_v^{rev})$ is $order(f_{v_R}^{rev})$ appended with $order(f_{v_L}^{rev})$. The $order(f_{v_L}^{rev}) = rev(order(f_{v_L}))$, and $order(f_{v_R}^{rev}) = rev(order(f_{v_R}))$. Therefore, $order(f_v^{rev}) = rev(order(f_v))$.

If v is a sum gate, then $f_v^{rev} = f_{v_L}^{rev} + f_{v_R}^{rev}$, which is equivalent to $f_{v_L} + f_{v_R} = f_v$ by induction hypothesis. As v is a sum gate, $order(f_v) = order(f_{v_L}) = order(f_{v_R})$. As $order(f_{v_L}^{rev}) = rev(order(f_{v_L}))$ by induction hypothesis, we have that $order(f_v^{rev}) = rev(order(f_v))$ and $order(f_{v_R}^{rev}) = rev(order(f_{v_R}))$. Thus, $order(f_v^{rev}) = order(f_{v_L}^{rev}) = order(f_{v_R}^{rev})$.

The size of C^{rev} is same as that of C because the only modification we are doing to C is swapping the children of product gates. This completes proof of the lemma.

Using Lemma 1, we can handle the monotonically decreasing sequence without modifying the polynomial computed by a regular set-multilinear circuit. This gives us a circuit $\widehat{C}$ that computes the permanent polynomial of degree $\frac{\sqrt[4]{n}}{2}$. We remark that Lemma 1 can be adapted for non-commutative circuits as well.

We now explain how to get the permanent polynomial of degree $\frac{\sqrt{n}}{2}$ instead of $\frac{\sqrt[4]{n}}{2}$. This gives us quadratic improvement in the degree of the permanent polynomial. This is based on the observation that if C is a regular set-multilinear circuit w.r.t a permutation $\sigma \in S_n$ computing the determinant polynomial $DET_n(X)$, then for any permutation $\tau \in S_n$, there is another regular set-multilinear circuit C' w.r.t $\tau \circ \sigma \in S_n$ computing the same determinant polynomial $DET_n(X)$. Moreover, the size of C' is at most one more than the size of C.

In other words, composition of permutations can be efficiently carried out for regular set-multilinear circuits computing the determinant polynomial $DET_n(X)$.

Lemma 2 (Composition Lemma). *Let $C = C_1 + C_2$ be the sum of two regular set-multilinear circuits computing the determinant polynomial $DET_n(X)$, where the circuits C_1, C_2 are regular set-multilinear circuits w.r.t $\sigma_1, \sigma_2 \in S_n$ respectively. Then for any permutation $\tau \in S_n$, there exists another circuit C' that computes $DET_n(X)$. C' is also a sum of two regular set-multilinear circuits* *(regular set-multilinear w.r.t $\tau \circ \sigma1, \tau \circ \sigma2 \in S_n$). Moreover, the size of C' is at most one more than the size of C.*

Proof. First, we describe the construction of the circuit $C' = C_1' + C_2'$ and then prove its correctness.

Construction of C': For every variable $x_{i,j}$ in $C = C_1 + C_2$, substitute the variable $x_{\tau(i),j}$. Let $\widehat{C}$ be this modified circuit. If $sgn(\tau)$ is -1, then add a leaf node labeled -1 and multiply the root node of $\widehat{C}$ with this leaf node. Let C' be this modified circuit. The size of C' is at most one more than the size of C.

Correctness: Now we will prove that C' computes $DET_n(X)$. Let m_1 and m_2 be any two monomials in $DET_n(X)$ computed by C. Let m_1' and m_2' be the monomials obtained by applying τ to the first index of each of the variables in m_1 and m_2 respectively. The permutations corresponding to m_1' and m_2' are $\tau \circ \sigma_1$ and $\tau \circ \sigma_2$ respectively.

- Case 1: $m_1 = m_2$. We show that $m_1' = m_2'$ in C'. We note that m_1 and m_2 could be computed by circuits C_1 and C_2 respectively. Thus, the order of variables appearing in m_1 and m_2 could be different in general. By construction of C', $x_{i,j}$ is substituted by the variable $x_{\tau(i),j}$. Since $m_1 = m_2$, we get $m_1' = m_2'$.
- Case 2: $m_1 \neq m_2$. We show that $m_1' \neq m_2'$ in C'. Since $m_1 \neq m_2$, there exists a variable x_{i_1,j_1} in m_1 and a variable x_{i_2,j_2} in m_2 such that $x_{i_1,j_1} \neq x_{i_2,j_2}$. Suppose $j_1 = j_2$, then $i_1 \neq i_2$. Then, $x_{\tau(i_1),j_1} \neq x_{\tau(i_2),j_2}$. This implies $m_1' \neq m_2'$. Suppose $j_1 \neq j_2$, then $x_{\tau(i_1),j_1} \neq x_{\tau(i_2),j_2}$, which again implies that $m_1' \neq m_2'$.

By construction of C', we note that coefficients of monomials are not affected. Now we will prove that C' computes $DET_n(X)$. Let A_X be a $n \times n$ matrix where row i contains all variables of the set X_i. In other words, the entry of i-th row and j-th column of the matrix A_X is $x_{i,j}$. Let $\beta \in S_n$. By changing $x_{i,j}$ to $x_{\beta(i),j}$, in effect it permutes the rows of A_X. In other words, the determinant is equal to the determinant of $P_\beta \times A_X$, where P_β is the $n \times n$ permutation matrix. The entry of i-th and j-th column of P_β is 1 iff $j = \beta(i)$ and 0 otherwise. By Fact 1 and 2, we have $det(P_\beta \times A_X) = det(P_\beta) \times det(A_X) = sign(\beta) \times det(A_X)$.

Thus, composing the permutation τ with σ_1, σ_2 maps different monomials to different monomials and in effect does not change the determinant computed except that the sign changes. Note that $sgn(\tau \circ \beta) = sgn(\tau).sgn(\beta)$ (by Fact 3). Therefore, if $sgn(\tau) = -1$, then the coefficients of m_1' and m_2' are the negatives of the coefficients of m_1 and m_2 respectively. Therefore, if $sgn(\tau) = -1$, C' computes $DET_n(X)$, as the leaf gate labeled -1 multiplied to the output gate ensures that C' computes $DET_n(X)$. However, the coefficients of m_1' and m_2' are the same as the coefficients of m_1 and m_2 respectively, if $sgn(\tau) = +1$. In the case that $sgn(\tau) = +1$, there is no need of this leaf gate. In both cases, the polynomial computed by C' is $DET_n(X)$.

Now we will show that $order(C_j) = (\tau(\sigma_j(1)), \tau(\sigma_j(2)), ..., \tau(\sigma_j(n)))$, $j \in \{1, 2\}$. The proof is by induction on the depth d of the circuit. We will prove it for C_1. The proof is similar for the circuit C_2. Recall that C_1 is regular set-multilinear circuit w.r.t σ_1. Let v be a gate in the circuit. We denote polynomial computed at v in C and C' by f_v and f'_v respectively.

Base Case: The base case is any node at depth 0, i.e., a leaf node. Let ℓ be any leaf node. Then f_ℓ is either a field constant or a variable $x_{i,j}$. If $f_\ell \in F$, then the $order(f_\ell)$ is the empty sequence $()$. As there is no variable in f_ℓ, there is no change to be made. Therefore, $order(f'_\ell) = ()$, and therefore the claim trivially holds. If f_ℓ is a variable $x_{i,j}$, then $order(f_\ell) = (i) = (\sigma_1(k))$, for some $k \in \{1, 2, ..., n\}$. We change $x_{i,j}$ to $x_{\tau(i),j}$, which means $order(f'_\ell) = (\tau(\sigma_1(k)))$.

Induction Hypothesis: Suppose the claim holds for any node at depth $d', 1 \leq d' < d$.

Induction Step: Consider any node v at depth $d' + 1$. Let u and w be its left and right children with degrees d_u, d_w respectively.

- Case 1: v is a sum gate. Thus, $f'_v = f'_u + f'_w$. Then $order(f'_u) = order(f'_w) = order(f'_v)$.
- Case 2: v is a product gate. Thus, $f'_v = f'_u \times f'_w$. Let $0 \leq a \leq n - d_u - d_w$, where d_u, d_w denote degrees of f_u, f_w respectively.
  Let $order(f_u) = (\sigma_1(a + 1), \sigma_1(a + 2), \cdots, \sigma_1(a + d_u))$ and $order(f_w) = (\sigma_1(a + d_u + 1), \sigma_1(a + d_u + 2), \cdots, \sigma_1(a + d_u + d_v))$. By IH, $order(f'_u) = (\tau(\sigma_1(a+1)), \tau(\sigma_1(a+2)), ..., \tau(\sigma_1(a+d_u)))$, and let $order(f'_w) = (\tau(\sigma_1(a+d_u+1)), \tau(\sigma_1(a+d_u+2)), ..., \tau(\sigma_1(a+d_u+d_v)))$. Then $order(f'_v) = (\tau(\sigma_1(a + 1)), \cdots, \tau(\sigma_1(a + d_u)), \tau(\sigma_1(a + d_u + 1)), \cdots, \tau(\sigma_1(a + d_u + d_v)))$.

Thus, in both cases, the claim holds. This completes the proof of the lemma.

Unlike Lemma 1, we note that in general this composition operation may not hold for any polynomial f computed by a regular circuit. For example, if C is a regular set-multilinear circuit computing the polynomial $f = x_{1,1}x_{2,0}x_{3,0}x_{4,1}$ then by swapping the 3rd and 4th indices, we get a different polynomial $f' = x_{1,1}x_{2,0}x_{4,0}x_{3,1}$. Now we have all results needed to the case where the determinant polynomial is computed by a sum of two regular set-multilinear circuits.

Theorem 3. *Let* $X = \{x_{i,j}\}_{i=1,j=1}^n$. *If the determinant polynomial over* X *is computed by a circuit* C *of size* s, *where* C *is the sum of two regular set-multilinear circuits, then the permanent polynomial of degree* $\sqrt{n}/2$ *can be computed by a regular set-multilinear circuit* C' *of size polynomial in* n *and* s.

Proof. Let $C = C_1^{\sigma_1} + C_2^{\sigma_2}$, where the circuits $C_1^{\sigma_1}, C_2^{\sigma_2}$ are *regular set-multilinear circuits* w.r.t $\sigma_1, \sigma_2 \in S_n$ respectively. We show that there is an efficient transformation that converts the given circuit C to another circuit C' computing the permanent polynomial of degree $\sqrt{n}/2$.

Without loss of generality, we can assume that σ_1 is the identity permutation. This is because otherwise by Lemma 2 we can get a new circuit $\hat{C} = C_1^{\sigma_1^{-1} \circ \sigma_1} +$

$C_2^{\sigma_1^{-1} \circ \sigma_2}$ with $\sigma_1^{-1} \circ \sigma_1, \sigma_1^{-1} \circ \sigma_2 \in S_n$ as the two permutations used. This does not increase the circuit size. By the Erdös-Szekeres Theorem, there is a monotone subsequence of length $\sqrt{n}$. Let A be the set of all such indices.

- Case 1: Subsequence is increasing. As σ_1 is the identity, the same subsequence of indices in σ_1 is also increasing. We do the following substitutions. For all $i \notin A$, set $x_{j,j} = 1$ and for all $i \in [n]$ and $i \neq j$, set $x_{j,i} = 0$ and $x_{i,j} = 0$. After this substitutions, the circuit computes the determinant polynomial over $A' = \bigsqcup_{i \in A} X_i$ and the order of the subsequence in both C_1 and C_2 are the same. We rename the variable sets in A' as follows: if $i_1 \in A_1$ is the j-th lowest index in the subsequence then we rename X_{i_1} to X_j, and for all k, rename $X_{i_1,k}$ to $X_{j,k}$. The modified circuit C' computes the determinant polynomial over $\hat{X} = \bigsqcup_{i \in [n^{1/2}]} X_i$ and it is regular w.r.t the identity permutation in $S_{\sqrt{n}}$.
- Case 2: Subsequence is decreasing. Then by Lemma 1, we modify the circuit $C_2^{\sigma_2}$ to get a new circuit computing the same polynomial as computed by the circuit $C_2^{\sigma_2}$ but the new circuit is regular set-multilinear w.r.t the permutation $\sigma_2^{rev} = (\sigma_2(n), \sigma_2(n-1), \cdots, \sigma_2(1))$. We note that, by applying Lemma 1, no sign change occurs to the determinant polynomial. In this modified (second) circuit, the corresponding subsequence now becomes increasing. This reduces this case to case 1.

Thus, after this modifications we have a new regular circuit C', that computes the determinant polynomial of degree $\sqrt{n}$, w.r.t the identity permutation. If $\lfloor \sqrt{n} \rfloor$ is not an even number then we substitute variables in $X_{\sqrt{n}}$ as explained before. Thus, C' computes the determinant polynomial of even degree. Now by the result of [2], we can compute the permanent polynomial of degree $\frac{\sqrt{n}}{2}$ by a circuit of size polynomial in s and n. This completes the proof of the theorem.

4 Hardness of the Determinant: Sum of Constantly-Many Regular Set-Multilinear Circuits

In this section, we show that if the determinant polynomial $DET_n(X)$ is computed by a sum of constantly many regular set-multilinear circuits then the permanent polynomial $PERM_{n^\epsilon/2}(X)$, $\epsilon > 0$ depends on k, computed a regular circuit. The proof of the following lemma is omitted due to lack of space. This is a generalization of the (composition) Lemma 2 but idea of the proof is similar.

Lemma 3. *Let $C = C_1 + C_2 + \cdots + C_k$ be a sum of k regular set-multilinear circuits such that C computes $DET_n(X)$. Let $C_1, C_2, ..., C_k$ be regular set-multilinear w.r.t $\sigma_1, \sigma_2, ..., \sigma_k$ respectively, where each $\sigma_i \in S_n$. For any $\tau \in S_n$, let $C_1^{\tau(\sigma_1)}, C_2^{\tau(\sigma_2)}, ..., C_k^{\tau(\sigma_k)}$ be the circuits obtained by substituting $x_{\tau(i),j}$ for each variable in $x_{i,j}$ in each of the k circuits. Let C' be the sum of $C_1^{\tau(\sigma_1)}, C_2^{\tau(\sigma_2)}, ..., C_k^{\tau(\sigma_k)}$. Then C' also computes $DET_n(X)$. Moreover, the size of C' is at most one more than the size of C.*

Without loss of generality we can assume that for each $i \neq j \in [k]$, $\sigma_i \neq \sigma_j$. Otherwise, we can combine all C_i's which use same σ_i into a single C_i using addition gates and get a circuit C that is a sum of k' regular set-multilinear circuits, where $k' < k$. Therefore, C is the sum of k' regular set-multilinear circuits such that no two permutations used by any two of these k' circuits is same. We call such a circuit C as k'-*regular circuit*.

Theorem 4. *Let C be the sum of k-many regular set-multilinear circuits, of size s, computing the determinant polynomial $DET_n(X)$. Then there exists a regular set-multilinear circuit whose size is at most $s+1$ that computes the determinant polynomial $DET_{n^\epsilon}(X')$, where $X' = \{x_{i,j}\}_{i=1,j=1}^{n^\epsilon}$ and $\epsilon \geq 1/2^{k-1}$.*

Proof. Let $C = C_1^{\sigma_1} + C_2^{\sigma_2} + \cdots + C_k^{\sigma_k}$, where the circuits $C_i^{\sigma_i}$ are regular set-multilinear circuits w.r.t $\sigma_i \in S_n$, $i \in [k]$. We show that there is an efficient transformation that converts the given circuit C to another circuit C' computing the determinant polynomial of degree n^ϵ, $\epsilon = 1/2^{k-1}$.

Without loss of generality, we can assume that σ_1 is the identity permutation. This is because otherwise by Lemma 2 we can get a new circuit $\hat{C} = \hat{C}_1 + \hat{C}_2 + \cdots + \hat{C}_{k'}$ where $\hat{C}_i$ is a regular set-multilinear circuit w.r.t the permutation $\sigma_1^{-1} \circ \sigma_i \in S_n$, where $i \in [k]$. We note that $\hat{C}$ computes the same polynomial as circuit C and both circuits have the same size.

Denote by $C^{(\ell)}$ the circuit obtained after the ℓ-th iteration, where $0 \leq \ell < k$. We will show that $C^{(\ell)}$ computes the determinant polynomial of degree $n^{1/2^\ell}$ and $C^{(\ell)}$ is a $(k - \ell)$-regular circuit.

At iteration 0, this condition holds, as $C^{(0)} = C$ computes the determinant polynomial over X and $C^{(0)}$ is a k-regular circuit.

Suppose the condition is true for some m, where $0 \leq m < k$. We will show that $C^{(m+1)}$ computes the determinant polynomial of degree $n^{1/2^{m+1}}$ and $C^{(m+1)}$ is a $k - (m + 1)$-regular circuit. Note that $C_1, C_2, \cdots, C_k$ have been modified during the first m iterations. Let us denote these modified circuits at the end of the m-th iteration by $C_1', C_2', \cdots, C_k'$. Thus, $C^{(m)} = C_1' + C_2' + \cdots + C_k'$.

Without loss of generality, we will assume that each variable in the determinant computed by $C^{(m)}$ has both its indices in $X^{(m)} = \{1, 2, \cdots, k_m\}$, where $k_m = n^{\frac{1}{2^m}}$. We note that the first m regular set-multilinear circuits $C_1', C_2', \cdots, C_m'$ are regular w.r.t identity permutation $id \in S_{k_m}$. As noted before, we can combine all C_i''s which has same σ_i as single C_i using addition gates. By Erdös-Szekeres Theorem [4], in σ_{m+1}', there is a monotone subsequence of length $n^{\frac{1}{2^{m+1}}}$. There are two cases to handle based on whether the subsequence is increasing or decreasing.

- Case 1: Suppose σ_{m+1}' has an increasing subsequence. Let $S^{(m+1)} = \{i_1, i_2, \cdots, i_{k_{m+1}}\}$ be the set of indices in this increasing subsequence, where $k_{m+1} = n^{\frac{1}{2^{m+1}}}$. We do the following substitutions. For all $j \notin S^{(m+1)}$, set $x_{j,j} = 1$ and for all $i \in [k_m]$ and $i \neq j$, set $x_{j,i} = 0$ and $x_{i,j} = 0$. After these substitutions, the circuit computes the determinant polynomial over $A' = \bigsqcup_{i \in S^{(m+1)}} X_i$. We rename the variable sets in A' as follows: if

$i_1 \in S^{(m+1)}$ is the j-th lowest index in the subsequence then we rename X_{i_1} to X_j, and for all k, rename $x_{i_1,k}$ to $x_{j,k}$. The modified circuit $C^{(m+1)}$ computes the determinant polynomial over $\hat{X} = \bigsqcup_{i \in [k_{m+1}]} X_i$. It is clear that $\sigma'_1 = \sigma'_2 = \cdots = \sigma'_m = \sigma'_{m+1} = identity$. This shows that $C^{(m+1)}$ is a $k - (m+1)$-regular circuit.

- Case 2: Suppose σ'_{m+1} has only a decreasing subsequence, then, we modify the sub-circuit C'_{m+1} by Lemma 1 to get a new circuit computing the same polynomial as computed by the $(m+1)$-th sub-circuit in the previous iteration but the new circuit is regular set-multilinear w.r.t the permutation $\sigma^{rev}_{m+1} = (\sigma'_{m+1}(k_m), \sigma'_{m+1}(k_m - 1), \cdots, \sigma'_{m+1}(1))$. Note that after reversal operation, Lemma 1 guarantees that the polynomial computed by the circuit C'_{m+1} does not change. In σ^{rev}_{m+1}, the corresponding subsequence now becomes increasing. It is clear that the same sequence of indices in $\sigma'_1, \sigma'_2, \cdots, \sigma'_m$ are also increasing. This reduces this case to case 1.

Clearly, $C^{(m+1)}$, obtained at the end of the $(m+1)$-th iteration, computes the determinant over $X^{(m+1)} = \{x_{i,j} \mid i, j \in S^{(m+1)}\}$. This implies that at the end of $(k-1)$-th iteration, $C^{(k-1)}$ computes the determinant of degree n^ϵ over $X^{(k-1)}$, where $\epsilon = 1/2^{k-1}$. Moreover, $C^{(k-1)}$ is a 1-regular set-multilinear circuit. This completes the proof of the theorem.

Let d be the degree of the determinant polynomial computed by the circuit $C^{(k-1)}$ in the above theorem. Clearly, $d \geq n^{\frac{1}{2^{k-1}}}$. If $\lfloor d \rfloor$ is not an even number then like before we substitute variables in the set $X_{\lfloor d \rfloor}$ such that the modified circuit computes the determinant of even degree $\lfloor d \rfloor - 1$. Now by the result of [2], we can compute the permanent polynomial of degree $d/2$ by a circuit of size polynomial in s and n. Thus, we get the following main result as a corollary.

Corollary 1. *Let C be the sum of k-many regular set-multilinear circuits computing the determinant polynomial $DET_n(X)$. Let s denote the size of the circuit C. Then there exists a regular set-multilinear circuit $\hat{C}$ computing the permanent polynomial $PERM_{n^\epsilon/2}$, where $\epsilon = 1/2^{k-1}$. Moreover, the size of $\hat{C}$ is polynomial in s and n.*

We note that to compute the permanent polynomial of degree n, we need to consider the determinant polynomial of degree $n^{2^{k-1}}$ computed by a k-regular circuit. So, our methods need k to be a constant.

5 Discussion

In this paper we studied the complexity of computing the determinant polynomial using sum of constant number of regular set-multilinear circuits. We showed that computing the determinant in this model is at least as hard as computing the commutative permanent polynomial. An interesting open question is whether our results can be extended to the sum of a non-constant (some function of the degree of the determinant) number of regular set-multilinear circuits. Another question is: What is the complexity of computing the determinant polynomial using set-multilinear circuits?. This question was also raised in [1].

References

1. Arvind, V., Raja, S.: Some lower bound results for set-multilinear arithmetic computations. Chicago J. Theor. Comput. Sci. **2016**(6)
2. Arvind, V., Srinivasan, S.: On the hardness of the noncommutative determinant. In: Proceedings of the 42nd ACM Symposium on Theory of Computing, STOC 2010, Cambridge, Massachusetts, USA, 5–8 June 2010, pp. 677–686 (2010)
3. Berkowitz, S.J.: On computing the determinant in small parallel time using a small number of processors. Inf. Process. Lett. **18**(3), 147–150 (1984)
4. Erdös, P., Szekeres, G.: A combinatorial problem in geometry. Compos. Math. **2**, 463–470 (1935)
5. Hrubes, P., Wigderson, A., Yehudayoff, A.: Non-commutative circuits and the sum-of-squares problem. In: Proceedings of the 42nd ACM Symposium on Theory of Computing, STOC 2010, Cambridge, Massachusetts, USA, 5–8 June 2010, pp. 667–676 (2010)
6. Nisan, N.: Lower bounds for non-commutative computation (extended abstract). In: STOC, pp. 410–418 (1991)
7. Nisan, N., Wigderson, A.: Lower bounds for arithmetic circuits via partial derivatives (preliminary version). In: 36th Annual Symposium on Foundations of Computer Science, Milwaukee, Wisconsin, 23–25 October 1995, pp. 16–25 (1995)
8. Raz, R.: Multi-linear formulas for permanent and determinant are of super-polynomial size. J. ACM **56**(2), 1–17 (2009)
9. Valiant, L.G.: Completeness classes in algebra. In: Proceedings of the 11h Annual ACM Symposium on Theory of Computing, 30 April–2 May 1979, Atlanta, Georgia, USA, pp. 249–261 (1979)
10. Valiant, L.G.: The complexity of computing the permanent. Theor. Comput. Sci. **8**, 189–201 (1979)

Concentration of the Collision Estimator

Maciej Skorski[✉]

University of Luxembourg, Luxembourg City, Luxembourg

Abstract. This work establishes strong concentration properties of the natural collision estimator, which counts the number of co-occurrences within a random sample from a discrete distribution.

Although this estimator has a wide range of applications, including uniformity testing and entropy assessment, prior works don't give any insights into its stochastic properties beyond the variance bounds.

We circumvent technical difficulties, applying elegant techniques to bound higher moments and conclude concentration properties. Remarkably, this shows that the simple unbiased estimator in many settings achieves desired accuracy on its own, with no need for confidence boosting.

Keywords: Entropy estimation · Collision estimation · Birthday paradox

1 Introduction

1.1 Background

In a number of important applications, such as derivation of cryptographic keys and security analysis [11,37], randomness extraction [19,20], property testing [4, 14] and others [1] one needs to estimate the *collision probability* of an unknown distribution X [7]:

$$P_2 \triangleq \mathbb{P}[X' = X''], \quad X', X'' \sim^{iid} X. \tag{1}$$

Given a sample $X_1, \ldots, X_n \sim^{iid} X$ of size n, the "natural" collision estimator is

$$\hat{P}_2 \triangleq \frac{1}{\binom{n}{2}} \sum_{1 \leqslant i < j \leqslant n} \mathbb{I}(X_i = X_j). \tag{2}$$

This estimator is intuitive, equal to the average number of collisions within a sample. It is unbiased, that is $\mathbb{E}\hat{P}_2 = P_2$. Although the sum consists of *dependent terms*, correlations are weak (each term is correlated to at most $2n$ out of $\binom{n}{2}$ possible) so the weak law of large numbers holds: $\hat{P}_2 \to P_2$ in probability when $n \to \infty$. In fact, with more powerful machinery it is possible to establish the

The work was supported by the FNR grant C17/IS/11613923. The full version is available at https://arxiv.org/abs/2006.07366.

E. Bampis and A. Pagourtzis (Eds.): FCT 2021, LNCS 12867, pp. 440–456, 2021.
https://doi.org/10.1007/978-3-030-86593-1_31

asymptotic normality [18]. We want to go beyond that and established *strong concentration for a finite sample*, namely a bound of the following form

$$\mathbb{P}[|\hat{P}_2 - P_2| > \epsilon] \leqslant \delta \tag{3}$$

for possibly small accuracy ϵ and error probability δ.

The motivation behind this work is *poor understanding of concentration properties* of this estimator. The problem has been touched by the research on property testing, Rényi entropy estimation; in principle it could be also attacked by concentration bounds for quadratic forms. However, none of these approaches gives a satisfactory answer to the posed problem. Works on property testing [10,15,34] and Rényi entropy estimation [2,31] established only weak bounds on (2) and used it rather as a building block of other, biased, estimators.

To close this gap, this work establishes *sharp concentration bounds* in form of (3) for the collision estimator, along with applications.

1.2   Main Result

Below we give exponentially strong concentration for the collision estimator (2) under *any distribution* X, and quantifies the error term δ in (3).

Theorem 1 (Collision Estimator Tails). *The concentration holds with the error probability*

$$\mathbb{P}[|\hat{P}_2 - P_2| > \epsilon] = O(1) \exp(-\Omega(\min(\epsilon^2/v^2, n\sqrt{\epsilon}))) \tag{4}$$

for any ϵ, absolute constants under $O(\cdot)$ and $\Omega(\cdot)$, and the factor v^2 defined as

$$v^2 \triangleq \frac{1}{n^2} \sum_k \mathbb{P}[X = k]^2 + \frac{1}{n} \sum_k \mathbb{P}[X = k]^3. \tag{5}$$

Below we discuss the obtained estimate in more detail.

Remark 1 (Subgaussian behavior). For sufficiently small ϵ the tail behaves like $\exp(-\Omega(\epsilon^2/v^2))$, e.g. as gaussian with variance $\Theta(v^2)$; in fact $\mathbf{Var}[\hat{P}_2] = O(v^2)$.

Remark 2 (Heavy-tail behavior). For bigger values of ϵ the tail behavior is like $\exp(-\Omega(n\sqrt{\epsilon}))$. This dependency is necessary, and is caused by small probability masses of X. We will clarify this phenomena later, when discussing applications.

Remark 3 (Sharp tails). The bound in Theorem 1 is sharp in the following sense: to have a gaussian-like tail for small ϵ the variance proxy has to satisfy $v^2 = \Omega(\mathbf{Var}[\hat{P}_2])$; but for a "typical" distribution we have $v^2 = O(\mathbf{Var}[\hat{P}_2])$ and then the exponent is optimal up to a constant. Specifically, from [10] we get

$$\mathbf{Var}[\hat{P}_2] = \Theta\left(n^{-2} \sum_x \mathbb{P}[X = x]^2 + n^{-1}(\sum_x \mathbb{P}[X = k]^3 - (\sum_x \mathbb{P}[X = k]^2)^2)\right)$$

and comparing this with our v^2 we see that $\mathbb{V}[\hat{P}_2] = \Theta(v^2)$ when $(\sum_k \mathbb{P}[X = k]^2)^2 < 0.99 \cdot \sum_k \mathbb{P}[X = k]^3$. Jensen's inequality, applied to $u \to u^{\frac{1}{2}}$, gives $(\sum_k \mathbb{P}[X = k]^2)^2 < \sum_x \mathbb{P}[X = k]^3$ for any non-uniform X; thus the condition holds when X is not "too close" to the uniform distribution.

1.3 Related Work

To the best author's knowledge, *there has been no works on concentration* of the collision estimator (2). Although its variance has been studied extensively in the context of uniformity testing [4,10,14,15,34], and Rényi entropy estimation [1,2,31], we lack of understanding of the higher moments and concentration properties. The techniques used to study the variance in prior works are merely manipulation of algebraic expressions with some combinatorics to carry out term cancellations, and do not scale to higher moments.

Outline of Techniques. Our findings are valuable not only because of strong quantitative guarantees (this becomes more apparent when discussing applications) but also because of elegant proof techniques that are of broader interest. We elaborate on that below, presenting the proof roadmap.

Handling Correlations: Negative Dependence. Negative dependency of random variables, roughly speaking, captures the property that one of them increases others are more likely to decrease. This property is a very strong form of negative correlation, known to imply concentration bounds comparable to those for independent random variables; essentially (in the context of concentration bounds) one works with negatively dependent random variables as if they were independent. Best-known from applications to balls and bin problems, the theory of negative dependence has been extensively investigated in [25] and [12].

We leverage the negative dependence to address the problem of correlations among estimatorcontributions. We group estimator's contributions $\tilde{Q}_k$ (thought as loads) on possible outcomes of the samples x (thought as bins), obtaining

$$\hat{P}_2 = \frac{1}{n(n-1)} \sum_k \tilde{Q}_k, \quad \tilde{Q}_k \triangleq \sum_{i \neq j} \mathbb{I}(X_i = X_j = k) \tag{6}$$

and then we prove that $\tilde{Q}_k$ (indexed by k) are *negatively dependent*. This reduces the problem to studying sums of *independent* random variables distributed as $\tilde{Q}_k$. We note that this trick can be used to simplify a bulk of complicated computations studied in higher-order Rényi entropy estimators [1,2,31].

Estimating Quadratic Forms. In order to understand the behavior of $\hat{P}_2(X)$, we have to establish the concentration properties of the components contributing to this sum. The contribution from the bin k given by $\tilde{Q}_k = \sum_{i \neq j} \mathbb{I}(X_i = X_j = k)$ can be seen as a quadratic form in $\tilde{Q}_k = \sum_{i \neq j} \xi_{k,i}\xi_{k,j}$ where $\xi_{k,i} = \mathbb{P}[X_i = k]$. In order to apply the probability machinery, we need to *center* these forms, that is to express them in zero-mean random variables. Specifically, fixing the k-th bin and denoting $p_k = \mathbb{P}[X = k]$ we have the following decomposition:

$$\tilde{Q}_k - \mathbb{E}\tilde{Q}_k = U_2 + 2(n-1)p_k \cdot U_1 \tag{7}$$

where U_1 and U_2 are the elementary symmetric polynomials in $\xi_i' = \xi_{k,i} - p_k$:

$$U_1 = \sum_i \xi_i', \quad U_2 = \sum_{i \neq j} \xi_i' \xi_j', \quad . \tag{8}$$

We use this decomposition to derive the final bound. While the linear term U_1 is easy to handle, the quadratic one U_2 requires a subtle treatment, involving *decoupling* (reducing quadratic to bilinear forms), *symmetrization* (replacing components by symmetrized versions), and accurate moment estimates.

The state-of-the art moment bounds for quadratic forms are essentially established by the two groups of results: a) *General Hypercontractivity* [32], which extends the celebrated result of Bonami [6] the cornerstone technique in research on boolean functions and b) *Hanson-Wright* inequality [17] and its refinements [5,36], better known the theoretical statisticians. Although these bounds are very powerful in general, they require much of work to be applied in our setup. The hypercontractivity does not produce good bounds because it yields sub-exponential type bounds, while we need more flexibility (our bounds are much of sub-gamma type, depending on two parameters). In turn, the Hanson-Wright inequality was developed for the symmetric case (sub-gaussian random variables), so that an accurate reduction from our case requires much of work and auxiliary results (Fig. 1).

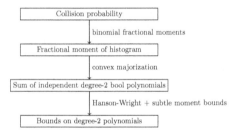

Fig. 1. The proof roadmap.

Estimation of Sum of Independent Random Variables. Most of concentration results in TCS papers are obtained by a black-box application of Chernoff-like bounds, but sometimes these inequalities fail to produce good results. The concentration of collision estimator is such an interesting use-case. The problem is that the estimator may be heavy tailed, with d-th moments growing faster than $O(d)^d$, so that the moment generating function does not exist. Furthermore, the sum terms are of different magnitude (when the distribution of X is "dispersed"); classical inequalities are not well-suited to such heterogenic setups.

What to do when standard concentration inequalities fail? The solution are subtle moment methods. The corollary below slightly extends the state-of-art result of Latala [26], which estimates moments of sums by *controlling moments of the individual components* (such bounds are called Rosenthal-type [35]).

Corollary 1. *Let* $(W_x)_x$ *be zero-mean and independent random variables (indexed by* x*),* d *be even and positive, and define the auxiliary function* $\phi(u) \triangleq \frac{(1+u)^d + (1-u)^d}{2}$*. Then*

$$\|\sum_x W_x\|_d = \Theta(1) \cdot \inf\{t : \sum_x \log \mathbb{E}\phi(W_x/t) \leqslant d\}.$$

The bounds can be used when classical concentration bounds are not sharp enough, for example the missing mass problem [33] or random projections [13].

Applications

Binomial Majorization. Recall the notation of *stochastic convex order*: a distribution Y majorizes X, denoted by $X \prec^{cvx} Y$ iff $\mathbb{E}f(X) \leqslant C \cdot \mathbb{E}f(Y)$ for all convex functions f and some absolute constant C. Below we link the performance of the estimator to an explicit binomial-like random variable.

Theorem 2 (Binomial Majorization). *The following bound holds for the collision estimator*

$$\hat{P}_2 \prec^{cvx} \binom{n}{2}^{-1} \sum_k \binom{S_k}{2}, \quad S_k \sim^{iid} \text{Binom}(n, \mathbb{P}[X = k]). \tag{9}$$

where k *runs over the alphabet of* X*.*

Remark 4 (Implications). What the majorization really means is that $\hat{P}_2$ has, roughly speaking, same or better concentration properties as the right-hand side of (9). This is because virtually all existing tail bounds are derived by an application of Markov's inequality to a carefully chosen function f which happen to be convex (this is the case of the moment or exponential moment method, in particular all classical probabilistic inequalities).

Remark 5 (Evaluation). The inequality provides us with a framework to estimate the right-hand side, which is a sum of *independent* random variables, with *explicit* distributions. As demonstrated later, under some structural assumption on X one can get bounds in a more closed-form or other interesting insights.

Remark 6 (Optimality). The result is sharp in the sense that both sides would be identically distributed if the terms in the estimator sum were truly independent (we know that they are weakly correlated).

Understanding Limitations to High-Confidence Estimation. There has been recently a trend to re-examine state-of-art results in property testing with the intent to further improve bounds in high-confidence regimes (see [9] and follow-up works). Such constructions are quite involved, much beyond the simplicity of intuitive estimators. This motivates the question about the performance of (2) if one wants small δ. The best we could get in (3) is the sample size

$n = \Theta_{\epsilon,v}(\log^{1/2}(1/\delta))$ by benchmarking with the CLT. We now show that in certain regimes the best we can get is only $n = \Theta_{\epsilon,v}(\log(1/\delta))$.

We know that for $S \sim \text{Binom}(n,p)$ the tail $\mathbb{P}[S > t]$ is only like $e^{-O(t \log t)}$ when p is small $t \gg np$ (this follows from sharp binomial tails obtained in [28,29], see also the survey [23]). Now in Theorem 2 we have terms $S^2 - S$ which are necessarily heavy tailed, because $\mathbb{P}[S^2 - S > t]$ for large t behaves like $e^{-O(\sqrt{t} \log t)}$. Using $t = n(n-1)\epsilon$ we obtain $\mathbb{P}[\binom{n}{2}^{-1}\binom{S}{2} > \epsilon] = \exp(-\tilde{O}(n\epsilon^{1/2}))$ (logarithmic terms hidden under $\tilde{O}$) and indeed we need $n = \Omega_\epsilon(\log(1/\delta))$, thus worse in terms of δ by a factor $\log^{1/2}(1/\delta)$ than suggested by the CLT. Note that already one x with too small probability $\mathbb{P}[X = x] \ll t/n$ makes $\hat{P}_2$ heavy-tailed.

High-Entropy Estimation with High-Confidence. We now discuss an important use-case where high-confidence estimation, that is $n = \Theta_\epsilon(\log^{1/2}(1/\delta))$ (matching the CLT), is actually possible. As opposed to the negative example before, here we will fall in the regime of "moderate" deviations and "light" tails.

For simplicity consider $X \in \{0,1\}^m$ and recall that min-entropy and collision-entropy are defined, respectively, as $H_\infty(X) \triangleq -\log \max_x \mathbb{P}[x = x]$ and $H_2(X) \triangleq -\log \sum_x \mathbb{P}[X = x]^2$. The previous section shows that estimation of $H_2(X)$ within error of ϵ and with confidence $1 - \delta$ costs $n = O(\log(1/\delta)2^{\frac{1}{2}H_2(X)}\epsilon^{-2})$ samples. The dependency on $H_2(X)$ and ϵ are in general optimal, but the factor $\log(1/\delta)$ can be improved in certain regimes. One example is when the min-entropy is slightly more than one half of the maximal value.

Corollary 2 (High-Confidence High-Entropy Estimation). *With notation as above, we can estimate $H_2(X)$ within an additive error of ϵ and correctness probability $1 - \delta$ given*

$$n = O(\log^{1/2}(1/\delta)2^{\frac{1}{2}H_2(X)}/\epsilon^2)$$

samples, provided that $H_\infty(X) \geqslant \frac{m + \log\log(1/\delta)}{2}$ and that $\epsilon = O(\log^{-1}(1/\delta))$.

Remark 7 (On improvement). The improvement in Corollary 2, due to the restriction on $H_\infty(X)$, is by the factor of $\log^{1/2}(1/\delta)$, which is significant when very high confidence is desired, e.g. in cryptography. The restriction on $H_\infty(X)$ in cryptography is considered mild, for example randomness sources are often required to have at least half of full min-entropy.

Optimal Rényi Entropy Estimation. Consider the problem of *relative* estimation, where $\epsilon := \epsilon Q$, $Q = P_2$. This can be seen as estimation of collision entropy $H_2(X) \triangleq -\log Q$ within an additive error of ϵ [2]. We prove that the estimator achieves high-probability guarantee on its own, without parallel runs.

Corollary 3 (Collision Estimation). *We have*

$$\mathbb{P}[|\hat{P}_2 - Q| > \epsilon Q] \leqslant O(1)\exp(-\Omega(n\epsilon^2 Q^{1/2})), \quad 0 \leqslant \epsilon \leqslant 1.$$

In particular the estimator in Listing 1.1 achieves relative error of ϵ and correctness probability of $1 - \delta$ when the number of samples is:

$$n = O(\log(1/\delta)Q^{-1/2}/\epsilon^2)$$

Remark 8 (Comparison with [2,31]). When a close bound on Q is known, we obtain the same bound on n as in [31]. When no bound on Q is known, we have $n = O(m^{1/2}/\epsilon^2)$ where m is the size of the domain of X, matching the bound from [2]. Our novelty is to show that the simple estimator $\tilde{Q}$ suffices, while previous works used it only as a building block [2,31].

Remark 9 (Difference from Uniformity Testing: Phase Transition). The bounds for collision estimation are sharp when $Q = \Omega(1/m)$, however for uniformity testing one considers the extreme regime of $Q = 1/m \cdot (1 + o(1))$, so the lower bounds [2] no longer apply. Indeed, uniformity testing allows a better dependency on ϵ that suggested by the general collision estimation. This is an interesting phenomena that could be seen as a "phase transition".

Beating Median-of-Means Trick. The multiplicative penalty of $\log(1/\delta)$ in the sample bound is typical for the "median trick" [3,24,30] and appears in a number of works on various estimators. The novelty of this work is in providing interesting examples where the median trick can be avoided, or even significantly improved (as in Corollary 2), by the natural concentration.

Optimal and Unbiased Collision Entropy Estimation. Another consequence of the lack of concentration bounds was that collision estimators in prior works were biased (due to confidence boosting). This motivates the question if a biased estimator is necessary to improve the accuracy. Interestingly, we show that one can achieve at the same time optimal efficiency and unbiased estimation.

Estimation of Quadratic Forms. In a follow-up work we show how the techniques developed here can be used to simplify to a great extent the analysis of sparse random projections. The bottleneck in prior works was the problem with estimation of quadratic forms, which our bounds can handle (these forms were found many times [13,21,22] not working well with existing variants of Hanson-Wright lemma). The resulting analysis is simpler and quantitatively stronger.

Sublinear Uniformity Testing. In uniformity testing one wants to know, based on random samples, how close is some unknown distribution to the uniform one. An appealing idea is to relate the closeness to the collision probability Q: for X over m elements the smallest value of Q is $1/m$ realized by the uniform distribution U_m, and (intuitively) the closer is Q to $1/m$, the closer is X to U_m. This test was studied in many works [4,10,14,15,34] with optimal bounds found in [10]. Although more sophisticated constructions may perform bit better in certain regimes [9], it is of interest to know the performance of the natural "birthday" estimator $\hat{P}_2$, which fits the long line of research gradually improving its variance properties [10,15,34]. We analyze the test using our concentration bounds.

```
def l2closeness_to_uniform (X ,n ,ε ):
    m ← |dom(X)|  /* domain  size */
    x[1]...x[n] ← X  /* get  iid  samples */
    Q ← #{(i,j) : x[i] = x[j], i ≠ j}  /* count  collisions */
    Q ← Q/n(n − 1)  /* normalize */
    if  Q > (1 + ε)/m :
        return False
    elif  Q < (1 + ε)/m :
        return True
```

Listing 1.1. Uniformity Tester for Discrete Distributions.

Corollary 4 (Optimal Sublinear Collision Tester). *If X is distributed over m elements, with*

$$n = O(\log(1/\delta)m^{1/2}/\epsilon)$$

samples the algorithm in Listing 1.1 distinguishes with probability $1 - \delta$ between a) $\|\mathbf{P}_X − U_m\|_2^2 \leqslant \frac{\epsilon}{2m}$ and b) $\|\mathbf{P}_X − U_m\|_2^2 \geqslant \frac{2\epsilon}{m}$, when $1/\sqrt{m} \leqslant \epsilon \leqslant 1$.

Remark 10. (Comparison with [10]). Our analysis is optimal in terms of dependency on the alphabet m and the error ϵ under the mild assumption that $n = O(m)$, that is that the sample size is *sublinear* with respect to the alphabet, which is what we usually want in applications (this means $\epsilon = \Omega(m^{-1/2}) = \Omega(2^{-d/2})$ where $d = \lceil \log m \rceil$ is the alphabet length; accuracy exponentially small in the alphabet length is more than enough, say, for security applications). In this regime we improve upon [10] by proving that there is no need for confidence boosting (parallel runs) of $\hat{P}_2$, as the estimator concentrates well on its own.

Sharp Binomial Estimates. To our best knowledge, the problem of determining *closed-form sharp bounds for binomial moments* has not been fully solved in the literature. The demand for such bounds comes from applications where the subtle moment calculations are needed to approximate tail probabilities, for example in analyses of random projections [21] or missing mass [33].

Binomial moments cannot be estimated accurately and easily from, much better understood, tail bounds such as the Chernoff bound (with exponent sharp up to a constant for the binomial case [8]) or the strong result of Littlewood [28,29] (with exponent sharp up to a sub-constant error); the difficulty is in handling complicated integrals that appear in the tail-expectation formula. Sharp binomial moments in closed-form are known in certain regimes, for example a) under the assumptions of the Central Limit Theorem, b) for the very special case of the symmetric binomial, by Khintchine's Inequaltiy [16]; they are also combinatorial formulas, but not of closed-form [39]. As pointed out in [21], no simple formula covering all the regimes is known. We close this gap with the following result:

Theorem 3 (Binomial Moments). *For $S \sim \mathsf{Binom}(p)$, $p \leqslant \frac{1}{2}$ and $d \geqslant 2$:*

$$\|S - \mathbf{E}S\|_d = \Theta(1) \begin{cases} (dnp)^{1/2} & \log(d/np) < \max(2, d/n) \\ d/\log(d/np) & \max(2, d/n) \leqslant \log(d/np) \leqslant d \\ (np)^{1/d} & d < \log(d/np). \end{cases} \quad (10)$$

1.4   Organization

Preliminaries are presented in Sect. 2, the main result is proved in Section 3. Section 4 concludes the work. Some details appear in the full version.

2   Preliminaries

2.1   Negative Dependence

For a thoughtful discussion of negatively dependent variables we refer the reader to [12,25]. Below we overview basic properties used in this paper.

Definition 1 (Negatively Dependent Random Variables [12,25]). *A collection of real-valued random variables $(Z_x)_{x \in \mathcal{X}}$ is negatively dependent when for any two disjoint subsets $\mathcal{X}_1, \mathcal{X}_2$ of $\mathcal{X}$ and any two real functions $f_i : \mathbb{R}^{|\mathcal{X}_i|} \to \mathbb{R}$ for $i = 1, 2$ both increasing or both decreasing in each coordinate, we have*

$$\mathbf{Cov}[f_1(Z_x : x \in \mathcal{X}_1), f_2(Z_x : x \in \mathcal{X}_2)] \leqslant 0.$$

Proposition 1 (Zero-One Principle (cf Lemma 8 in [12])). *0–1 valued random variables $(W_x)_x$ such that $\sum_x W_x = 1$, are negatively dependent.*

Proposition 2 (Augmentation Property (cf Proposition 7 in [12])). *If $(W_x)_x$ are negatively dependent and $(Z_x)_x$ are negatively dependent and independent of $(W_x)_x$, then the collection $(W_x)_x \cup (Z_x)_x$ is negatively dependent.*

Proposition 3 (Aggregation by Monotone Functions (cf Proposition 7 in [12])). *If $(W_x)_x$ are negatively dependent, $\mathcal{X}_1, \ldots, \mathcal{X}_k$ are non-overlapping sets and $f_1, \ldots, f_k$ are functions all increasing or decreasing then the random variables $f_i(W_x : x \in \mathcal{X}_i)$ are negatively dependent.*

Proposition 4 (Majorization by IID variables (cf main result of [38] or Section 11.6 in [27])). *If $(W_x)_x$ are negatively dependent and $(W'_x)_x$ are distributed as $(W_x)_x$ but independent, then $\mathbb{E}f((W_x)_x) \leqslant \mathbb{E}f((W'_x)_x)$ for any convex function f. In other words $\sum_x W_x \prec^{cvx} \sum_x W'_x$.*

2.2   Symmetrization and Decoupling

We need the following decoupling inequality ([40,41]), which essentially allows for replacing quadratic forms by bi-linear forms (much easier to handle).

Proposition 5 (Decoupling for Quadratic Forms). *Let $\xi = (\xi_1, \ldots, \xi_n)$ be a random vector with zero-mean independent components, A be a $n \times n$ matrix with zero diagonal, ξ' be an independent copy of ξ and f be convex. Then*

$$\mathbb{E}f(\xi^T A \xi) \leqslant \mathbb{E}f4(\xi^T A \xi').$$

We also need the following fact on symmetrization (cf. Lemma 6.1.2 in [41])

Proposition 6 (Symmetrization). *Let Y, Z be independent and $\mathbb{E}Z = 0$, then $\mathbb{E}f(Y) \leqslant \mathbb{E}f(Y + Z)$ for any convex f.*

3 Proofs

3.1 Proof of Theorem 1

Negative Dependency and Decomposition. Denote $p_k = \Pr[X = k]$ and $\xi_{i,k} = \mathbb{I}(X_i = k)$. Then we can write

$$\hat{P}_2 = \frac{2}{n(n-1)} \sum_k Q_k, \quad Q_k \triangleq \sum_{i \neq j} \xi_{i,k} \xi_{j,k}. \tag{11}$$

Let $\tilde{Q}_k$ be independent with same marginal distributions as Q_k. We claim that

$$\mathbf{E}f(\hat{P}_2) \leqslant \mathbf{E}f\left(\frac{2}{n(n-1)} \sum_k \tilde{Q}_k\right), \quad \tilde{Q}_k \sim^{iid} \sum_{i \neq j} \xi_i \xi_j, \quad \xi_i \sim^{iid} \mathsf{Bern}(p_k) \tag{12}$$

for any convex f. In particular $\frac{2}{n(n-1)} \sum_k \mathbf{E}[\tilde{Q}_k] = \mathbf{E}\hat{P}_2$. Thus the claim implies:

$$\|\hat{P}_2 - \mathbf{E}\hat{P}_2\|_d \leqslant O(n^{-2}) \cdot \|\sum_k [\tilde{Q}_k - \mathbf{E}\tilde{Q}_k]\|_d. \tag{13}$$

To prove the claim, we need prove that $\tilde{Q}_k$ are *negatively dependent*. Then the inequality follows by Proposition 4. Let $S_k = \sum_i \xi_i$, then we have $\tilde{Q}_k = S_k^2 - S_k$ For any fixed i the random variables $\mathbb{I}(X_i = k)$, indexed by k, are negatively dependent because they are boolean and add up to one (see Proposition 1) Since X_i for different i are independent, we obtain that $(\mathbb{I}(X_i = k))_{i,k}$ indexed by *both* i and x are negatively dependent (by augmentation property, see Proposition 2). Observe that $S_k^2 - S_k = f((\mathbb{I}(X_i = k))_i)$ with $f(u_1, \ldots, u_n) = \sum_{i \neq j} u_i u_j$ increasing in each u_i when $u_i \geqslant 0$. Applying increasing functions to non-overlapping subsets of negatively dependent variables produces negatively dependent variables (see Proposition 3), so $S_k^2 - S_k$ are negatively dependent. Same holds for $\tilde{Q}_k$ (differ by a scaling factor). This proves the negative dependency.

Observe that this property combined with Proposition 4 implies $\sum_k \tilde{Q}_k \prec^{cvx} \sum_x (S_k^2 - S_k)$ where $S_x \sim \mathsf{Binom}(n, \mathbb{P}[X = x])$. This proves Theorem 2, we need only to rescale by the constant $n(n-1)$ (which preserves the convex order).

Observe that we can decompose

$$\tilde{Q}_k - \mathbf{E}\tilde{Q}_k \sim \sum_{i \neq j} \xi_i' \xi_j' + 2np_k \sum_i \xi_i', \quad \xi_i' \sim^{iid} \xi' = \mathrm{Bern}(p_k) - p_k, \quad (14)$$

by plugging $\xi_i = \xi_i' + p$ into $\sum_{i \neq j} \xi_i \xi_j$, expanding and grouping linear and second-order terms separately. Therefore we obtain:

$$\sum_k [\tilde{Q}_k - \mathbf{E}\tilde{Q}_k] = \sum_k \sum_{i \neq j} \xi_{k,i}' \xi_{k,j}' + 2n \sum_k \sum_i p_k \xi_{k,i}', \quad \xi_{k,i}' \sim^{iid} \mathrm{Bern}(p_k) - p_k. \quad (15)$$

We shall estimate separately moments for the second-order and the linear terms.

Bounding Second-Order Chaos. Fix an even integer $d > 1$. We first prove

$$\left\| \sum_k \sum_{i \neq j} \xi_{k,i}' \xi_{k,j}' \right\|_d \leqslant 4 \left\| \sum_k \sum_{i \neq j} \xi_{k,i}' \xi_{k,j}'' \right\|_d, \quad (16)$$

where $\xi_{k,j}''$ are distributed as $\xi_{k,j}'$ but independent. This follows by the decoupling theorem (Proposition 5), applied to the family $\xi_{k,i}'$ indexed by the *tuples* (k, i). The matrix A has rows indexed by (k, i), columns indexed by (k', j) and has entries 1 when $k = k', i \neq j$ and 0 otherwise; it is clearly off-diagonal.

In the next step we prove the *Rademacher symmetrization* inequality:

$$\left\| \sum_k \sum_{i \neq j} \xi_{k,i}' \xi_{k,j}'' \right\|_d \leqslant 4 \left\| \sum_k \sum_{i \neq j} \xi_{k,i}' \xi_{k,j}'' r_{k,i}' r_{k,j}'' \right\|_d, \quad r_{k,i}', r_{k,j}'' \sim^{iid} \{-1, 1\}. \quad (17)$$

We achieve this in the following two steps: we apply the symmetrization (Proposition 6) to $f(\sum_k \sum_{i \neq j} \xi_{k,i}' \xi_{k,j}'')$ but *conditionally* on values of $\xi_{k,i}'$. With these values fixed, the function argument is linear in $\xi_{k,j}''$ and thus the theorem for sums of random variables applies. With $f(u) = |u|^d$ we have $\mathbf{E}[f(\sum_k \sum_{i \neq j} \xi_{k,i}' \xi_{k,j}'') | \xi_{k,i}'] \leqslant 2^d \mathbf{E}[f(\sum_k \sum_{i \neq j} \xi_{k,i}' \xi_{k,j}'' r_{k,j}'') | \xi_{k,i}']$ and thus also uncoditionally: $\mathbf{E} f(\sum_k \sum_{i \neq j} \xi_{k,i}' \xi_{k,j}'') \leqslant 2^d \mathbf{E} f(\sum_k \sum_{i \neq j} \xi_{k,i}' \xi_{k,j}'' r_{k,j}'')$. In the second step we proceed analogously, but conditioning on all values of $\xi_{k,j}''$ *and also* $r_{k,j}''$; this gives $\mathbf{E} f(\sum_k \sum_{i \neq j} \xi_{k,i}' \xi_{k,j}'' r_{k,j}'') \leqslant 2^d \mathbf{E} f(\sum_k \sum_{i \neq j} \xi_{k,i}' \xi_{k,j}'' r_{k,i}' r_{k,j}'')$ which proves the claim.

Applying the Hanson-Wright inequality [36] to the right-hand side of (17), conditionally on fixed values of ξ_i, ξ_j', we obtain the following inequality:

$$\sum_k \sum_{i \neq j} \xi_{k,i}' \xi_{k,j}' \leqslant O(\sqrt{d})^d \cdot \mathbf{E}_{(\xi_i'), (\xi_j'')} [\|A(\xi_{k,i}', \xi_{k',j}'')_{(k,i),(k',j)}\|_F^d]$$

$$+ O(d)^d \cdot \mathbf{E}_{(\xi_i'), (\xi_j'')} [A(\xi_{k,i}', \xi_{k',j}'')_{(k,i),(k',j)}\|_2^d], \quad (18)$$

for the following matrix (note it is indexed by tuples!):

$$A(\xi_{k',i}', \xi_{k',j}'')_{(k,i),(k',j)} = \mathbb{I}(k = k', i \neq j) \xi_{k,i}' \xi_{k,j}''. \quad (19)$$

It remains to bound these matrix norms. We start with the $\| \cdot \|_F$ term:

$$\|A(\xi'_{k,i}, \xi''_{k',j})_{(k,i),(k',j)}\|_F = \left(\sum_{k,i \neq j} \xi'^2_{k,i} \xi''^2_{k,j} \right)^{1/2}. \tag{20}$$

We know claim the following majorization by Bernoulli variables, for *even* d:

$$\left(\sum_{k,i \neq j} \xi'^2_{k,i} \xi''^2_{k,j} \right)^{d/2} \leqslant \left(\sum_{k,i \neq j} \eta'_{k,i} \eta''_{k,j} \right)^{d/2}, \quad \eta'_{k,i}, \eta''_{k,j} \sim^{iid} \mathsf{Bern}(p_k). \tag{21}$$

Indeed, since d is even, this follows by the multinomial expansion, independence and that $|\xi| \leqslant 1$ implies $\mathbf{E}\xi^{2\ell} \leqslant \mathbf{E}\xi$ for integer $\ell \geqslant 1$ (applied to $\xi = \xi'_{k,i}$ and $\xi''_{k,j}$, this yields $\xi'^{2\ell}_{k,i} \leqslant p \leqslant \eta'^{\ell}_{k,i}$ and $\xi''^{2\ell}_{k,i} \leqslant p \leqslant \eta''^{\ell}_{k,i}$ accordingly). This gives:

$$\mathbf{E}\|A(\xi'_{k,i}, \xi''_{k',j})_{(k,i),(k',j)}\|_F^d \leqslant \mathbf{E} \left(\sum_{k,i \neq j} \eta'_{k,i} \eta''_{k,j} \right)^{d/2}. \tag{22}$$

By adding extra non-negative terms for $i = j$ and observing that we have $\sum_{k,i} \eta'_{k,i} \sim \mathsf{Binom}(p_k), \sum_{k,i} \eta''_{k,i} \sim \mathsf{Binom}(p_k)$ we can further bound this as:

$$\mathbf{E}\|A(\xi'_{k,i}, \xi''_{k',j})_{(k,i),(k',j)}\|_F^d \leqslant \mathbf{E} \left(\sum_k S'_k S''_k \right)^{d/2}, \quad S'_k, S''_k \sim^{iid} \mathsf{Binom}(n, p_k). \tag{23}$$

We now move to the $\| \cdot \|_2$ term. Instead of estimating it directly, we use the matrix norm inequality $\|A\|_2 \leqslant \|A\|_F$. This, for even d, gives:

$$\mathbf{E}\|A(\xi'_{k,i}, \xi''_{k',j})_{(k,i),(k',j)}\|_F^d \leqslant \mathbf{E} \left(\sum_k S'_k S''_k \right)^{d/2}. \tag{24}$$

Plugging (24) and (23) back in (18), and using the elementary bound $(a+b)^{1/d} < a^{1/d} + b^{1/d}$ for $a, b > 0$ we conclude the bound in terms of binomial moments:

$$\| \sum_k \sum_{i \neq j} \xi'_{k,i} \xi'_{k,j} \|_d \leqslant O(d) \cdot \| \sum_k S'_k S''_k \|_{d/2}^{1/2}. \tag{25}$$

In the final part, we estimate the expression with binomials. Denote $W_k = S'_k S''_k$, we are interested in moments of $\sum_k W_k$. By the moment bound from [26]:

$$\sum_k \log \mathbf{E}(1 + S'_k S''_k / t)^d) = \sum_k \log \left(1 + (np_k)^2 / t \cdot \sum_{\ell \geqslant 2} \binom{d}{\ell} / t^{\ell-1} \right). \tag{26}$$

Assuming that $t \geqslant d$ and using $\log(1+u) \leqslant u$ we obtain:

$$\sum_k \log \mathbf{E}(1 + S_k' S_k''/t)^d \leqslant d/t \cdot n^2 \sum_k p_k^2, \tag{27}$$

and we see that this is smaller than $c \cdot d$ when $t \geqslant n^2 \sum_k p_k^2/c$. Thus:

$$\| \sum_k \sum_{i \neq j} \xi_{k,i}' \xi_{k,j}' \|_d \leqslant O(d + n^2 \sum_k p_k^2), \tag{28}$$

and we finally get the desired bound on the quadratic contributions:

$$\| \sum_k \sum_{i \neq j} \xi_{k,i}' \xi_{k,j}' \|_d \leqslant O(d^{1/2}) \cdot (n^2 \sum_k p_k^2)^{1/2} + O(d)^{3/2}. \tag{29}$$

Bounding First-Order Chaos. By symmetrization (Proposition 6) we obtain:

$$\| \sum_i \xi_{k,i}' \|_d = \Theta(1) \| \sum_i \xi_{k,i}' r_i' \|_d, \quad r_i' \sim^{iid} \{-1, 1\}. \tag{30}$$

Observe that the LHS equals $\|S - \|ES\|_d$, for $S \sim \text{Binom}(n, p)$ with $p = p_k$. We start by deriving the binomial estimates in Theorem 3.

Latala pointed out [26] that for symmetric and IID r.vs. Z_i, the moment framework gives the following more handy estimate:

$$\| \sum_{i=1}^n Z_i \|_d = \Theta(1) \sup\{d/q \cdot (n/d)^{1/q} \cdot \|Z\|_q : \max(2, d/n) \leqslant q \leqslant d\}.$$

Let $\xi_i, \xi_i' \sim \text{Bern}(p)$ with $p \leqslant 1/2$, and let $Z_i = \xi_i - \xi_i'$. Since $\|Z\|_q = \Theta(p^{1/q})$:

$$\|S - S'\|_d = \Theta(1) \sup\{d/q \cdot (np/d)^{1/q} : \max(2, d/n) \leqslant q \leqslant d\}.$$

Now it all boils down to the auxiliary function $g(q) = 1/q \cdot a^{1/q}$ (in our case $a = np/d$). When $a > 1$ the function is decreasing for $q > 0$; when $a < 1$ it achieves its global maximum at $q = \log(1/a)$ with the value $\frac{1}{e \log(1/a)}$ (by the derivative test). By comparing the global maximum with the interval $\max(2, d/n) \leqslant q \leqslant d$,

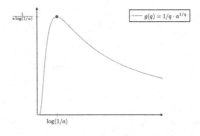

Fig. 2. The function $g(q) = 1/q \cdot a^{1/q}$ appearing in binomial moment estimates ($a < 1$).

we obtain the bounds for $\|S - S'\|_d$ with the same formula as in Theorem 3, and thus for $\|S - \mathbf{E}S\|_d$ because $\|S - S'\|_d \leqslant \|S - \mathbf{E}S\|_d + \|S' - \mathbf{E}S'\|_d = 2\|S - \mathbf{E}S\|_d$ and $\|S - \mathbf{E}S\|_d = \|S - \mathbf{E}S'\|_d \leqslant \|S - S'\|_d$ respectively by the triangle and Jensen's inequality. The proof works for also *real* d (Fig. 2).

Let $S_k, S_k' \sim^{iid} \mathsf{Binom}(n, p_k)$. Then:

$$\|2n \sum_k \sum_i p_k \xi_{k,i}'\|_d = 2n\| \sum_k p_k (S_k - \mathbf{E}S_k)\|. \tag{31}$$

Let $W_k = np_k(S_k - \mathbf{E}S_k)$. The bound in Theorem 3 implies $\|S_k - \mathbf{E}S_k\|_d \leqslant O(1) \max\{\sqrt{dnp_k}, d\}$, so $\|W_k\|_d \leqslant np_k \max\{\sqrt{dnp_k}, d\}$.

$$\log \mathbf{E}(1 + W_k/t)^d \leqslant \log\left(1 + \sum_{q \geqslant 2, q \text{ even}} \binom{d}{q}(np_k)^2 q^{2q-2}/t^q + q^{q/2}((n^3 p_k^3/t^2)^{1/2})^q\right). \tag{32}$$

We will make use of the following inequalities:

$$\sum_{q \geqslant 2, q \text{ even}} \binom{d}{q}\sqrt{qu}^q \leqslant \cosh(O(du^{1/2})) - 1 \leqslant \exp(O(d^2 u)), \quad u > 0$$

$$\sum_{q \geqslant 2, q \text{ even}} \binom{d}{q}(q^2 u)^q \leqslant 1 + O(1), \quad \frac{1}{2d^2} > u > 0. \tag{33}$$

The first follows by Taylor's expansion of $\exp(\cdot)$, the bound $\binom{d}{q} = O(d/q)^q$ and Stirling's approximation $q! = \Theta(q)^q$, and the second by the formula on the geometric progression. We apply the first one to $u = n^3 p_k^3/t^2$ and the second one to $u = 1/t$. Since $\log(1 + a + b) < \log(1 + a) + \log(1 + b)$ for positive a, b (equivalent to $1 + a + b < (1 + a)(1 + b)$), we obtain for $t > 2d^2$:

$$\log \mathbf{E}(1 + W_k/t)^d \leqslant O(d^2(n^3 p_k^3)/t^2) + O(n^2 p_k^2/t^2). \tag{34}$$

This gives us the desired bound:

$$\|2n \sum_k \sum_i p_k \xi_{k,i}'\|_d = 2\| \sum_k W_k\|_d \leqslant O(d^{1/2}) \cdot (n^3 \sum_k p_k^3 + n^2 \sum_k p_k^2)^{1/2} + O(d^2). \tag{35}$$

Putting Bounds Together. By the results of the two previous sections and (14) we obtain:

$$\| \sum_k [\tilde{Q}_k - \mathbf{E}\tilde{Q}_k]\|_d \leqslant O(vd^{1/2}) + O(d^2), \quad v^2 \triangleq n^2 \sum_k p_k^2 + n^3 \sum_k p_k^3.$$

By Markov's inequality, for some constant $C > 0$, we obtain:

$$\mathbb{P}[| \sum_k (\tilde{Q}_k - \mathbb{E}\tilde{Q}_k)| > \epsilon] \leqslant (Cd^2/\epsilon + C\sqrt{d}v/\epsilon)^d.$$

Setting d so that a) $d^2 \leqslant \epsilon/4C$ and b) $d \leqslant \epsilon^2/v^2/16C^2$, gives the tail of 2^{-d}. Since d has to be even and at least 2, if $\epsilon/4C \geqslant 4$ and $\epsilon^2/v^2/16C^2 \geqslant 2$ we obtain the tail of $2^{-\Omega(\min(\epsilon^2/v^2, \epsilon^{1/2}))}$. Otherwise $\epsilon = O(1)$ or $\epsilon = O(v)$ so that $2^{-\Omega(\min(\epsilon^2/v^2, \epsilon^{1/2}))} = 2^{-O(1)} = \Omega(1)$, and the claimed bound holds trivially, because $B \cdot 2^{-\Omega(\min(\epsilon^2/v^2, \epsilon^{1/2}))} \geqslant 1$ for an appropriate constant B.

The result in Theorem 1 follows because $\|\hat{P}_2 - P_2\|_d = \frac{1}{n(n-1)} \sum_k (\tilde{Q}_k - \mathbb{E}\tilde{Q}_k)$, so it suffices to change $\epsilon := n(n-1)\epsilon$ and scale the formula for v accordingly.

4   Conclusion

We have obtained, for the first time, strong concentration guarantees for the collision estimator. This subsumes variance bounds from previous works. Such concentration bounds, for example, eliminate the need for the median trick.

References

1. Acharya, J., Orlitsky, A., Suresh, A.T., Tyagi, H.: The complexity of estimating rényi entropy. In: Proceedings of the Twenty-Sixth Annual ACM-SIAM Symposium on Discrete Algorithms, pp. 1855–1869. SIAM (2014)
2. Acharya, J., Orlitsky, A., Suresh, A.T., Tyagi, H.: Estimating rényi entropy of discrete distributions. IEEE Trans. Inf. Theory **63**(1), 38–56 (2016)
3. Alon, N., Matias, Y., Szegedy, M.: The space complexity of approximating the frequency moments. J. Comput. Syst. Sci. **58**(1), 137–147 (1999)
4. Batu, T., Fischer, E., Fortnow, L., Kumar, R., Rubinfeld, R., White, P.: Testing random variables for independence and identity. In: Proceedings 42nd IEEE Symposium on Foundations of Computer Science, pp. 442–451. IEEE (2001)
5. Bellec, P.C.: Concentration of quadratic forms under a Bernstein moment assumption. arXiv preprint arXiv:1901.08736 (2019)
6. Bonami, A.: Étude des coefficients de fourier des fonctions de $lp(g)$. In: Annales de l'institut Fourier, vol. 20, pp. 335–402 (1970)
7. Cachin, C.: Entropy measures and unconditional security in cryptography. Ph.D. thesis, ETH Zurich (1997)
8. Csiszár, I.: The method of types [information theory]. IEEE Trans. Inf. Theory **44**(6), 2505–2523 (1998)
9. Diakonikolas, I., Gouleakis, T., Peebles, J., Price, E.: Optimal identity testing with high probability. arXiv preprint arXiv:1708.02728 (2017)
10. Diakonikolas, I., Gouleakis, T., Peebles, J., Price, E.: Collision-based testers are optimal for uniformity and closeness. Chicago J. Theor. Comput. Sci. **1**, 1–21 (2019)
11. Dodis, Y., Yu, Yu.: Overcoming weak expectations. In: Sahai, A. (ed.) TCC 2013. LNCS, vol. 7785, pp. 1–22. Springer, Heidelberg (2013). https://doi.org/10.1007/978-3-642-36594-2_1
12. Dubhashi, D.P., Ranjan, D.: Balls and bins: a study in negative dependence. BRICS Rep. Ser. **3**(25) (1996). https://www.brics.dk/RS/96/25/BRICS-RS-96-25.pdf
13. Freksen, C.B., Kamma, L., Larsen, K.G.: Fully understanding the hashing trick. In: Advances in Neural Information Processing Systems, pp. 5389–5399 (2018)

14. Goldreich, O.: Introduction to Property Testing. Cambridge University Press, Cambridge (2017)
15. Goldreich, O., Ron, D.: On testing expansion in bounded-degree graphs. In: Goldreich, O. (ed.) Studies in Complexity and Cryptography. Miscellanea on the Interplay between Randomness and Computation. LNCS, vol. 6650, pp. 68–75. Springer, Heidelberg (2011). https://doi.org/10.1007/978-3-642-22670-0_9
16. Haagerup, U.: The best constants in the khintchine inequality. Stud. Math. **70**, 231–283 (1981)
17. Hanson, D.L., Wright, F.T.: A bound on tail probabilities for quadratic forms in independent random variables. Ann. Math. Stat. **42**(3), 1079–1083 (1971)
18. Hoeffding, W., et al.: A class of statistics with asymptotically normal distribution. Ann. Math. Stat. **19**(3), 293–325 (1948)
19. Impagliazzo, R., Levin, L.A., Luby, M.: Pseudo-random generation from one-way functions. In: Proceedings of the Twenty-First Annual ACM Symposium on Theory of Computing, pp. 12–24 (1989)
20. Impagliazzo, R., Zuckerman, D.: How to recycle random bits. FOCS **30**, 248–253 (1989)
21. Jagadeesan, M.: Simple analysis of sparse, sign-consistent JL. In: APPROX/RANDOM 2019. Schloss Dagstuhl-Leibniz-Zentrum fuer Informatik (2019)
22. Jagadeesan, M.: Understanding sparse JL for feature hashing. In: Advances in Neural Information Processing Systems, pp. 15203–15213 (2019)
23. Janson, S.: Large deviation inequalities for sums of indicator variables. arXiv preprint arXiv:1609.00533 (2016)
24. Jerrum, M.R., Valiant, L.G., Vazirani, V.V.: Random generation of combinatorial structures from a uniform distribution. Theoret. Comput. Sci. **43**, 169–188 (1986)
25. Joag-Dev, K., Proschan, F.: Negative association of random variables with applications. Ann. Stat., 286–295 (1983)
26. Latała, R., et al.: Estimation of moments of sums of independent real random variables. Ann. Probab. **25**(3), 1502–1513 (1997)
27. Lin, Z., Bai, Z.: Probability Inequalities. Springer, Heidelberg (2011). https://doi.org/10.1007/978-3-642-05261-3
28. Littlewood, J.E.: On the probability in the tail of a binomial distribution. Adv. Appl. Probab. **1**(1), 43–72 (1969)
29. McKay, B.D.: On littlewood's estimate for the binomial distribution. Adv. Appl. Probab. **21**(2), 475–478 (1989)
30. Nemirovsky, A.S., Yudin, D.B.: Problem complexity and method efficiency in optimization (1983)
31. Obremski, M., Skorski, M.: Renyi entropy estimation revisited. In: APPROX-/RANDOM 2017, volume 81 of Leibniz International Proceedings in Informatics (LIPIcs), pp. 20:1–20:15 (2017)
32. O'Donnell, R.: Analysis of Boolean Functions. Cambridge University Press, Cambridge (2014)
33. Ortiz, L.E., McAllester, D.A.: Concentration inequalities for the missing mass and for histogram rule error. In: Advances in Neural Information Processing Systems, pp. 367–374 (2003)
34. Paninski, L.: A coincidence-based test for uniformity given very sparsely sampled discrete data. IEEE Trans. Inf. Theory **54**(10), 4750–4755 (2008)
35. Rosenthal, H.P.: On the subspaces of L^p (p > 2) spanned by sequences of independent random variables. Israel J. Math. **8**(3), 273–303 (1970). https://doi.org/10.1007/BF02771562

36. Rudelson, M., Vershynin, R., et al.: Hanson-wright inequality and sub-Gaussian concentration. Electron. Commun. Probab. **18**, 1–9 (2013)
37. Schaub, A., Rioul, O., Boutros, J.J.: Entropy estimation of physically unclonable functions via chow parameters. In: 57th Annual Allerton Conference on Communication, Control, and Computing, pp. 698–704. IEEE (2019)
38. Shao, Q.-M.: A comparison theorem on moment inequalities between negatively associated and independent random variables. J. Theor. Probab. **13**(2), 343–356 (2000). https://doi.org/10.1023/A:1007849609234
39. Skorski, M.: Handy formulas for binomial moments. arXiv preprint arXiv:2012.06270 (2020)
40. Vershynin, R.: A simple decoupling inequality in probability theory (2011)
41. Vershynin, R.: High-Dimensional Probability: An Introduction with Applications in Data Science, vol. 47. Cambridge University Press, Cambridge (2018)

Valency-Based Consensus Under Message Adversaries Without Limit-Closure

Kyrill Winkler[1]([✉]) [iD], Ulrich Schmid[2][iD], and Thomas Nowak[3][iD]

[1] University of Vienna, Vienna, Austria
[2] TU Wien, Vienna, Austria
[3] Université Paris-Saclay, CNRS, Orsay, France

Abstract. We introduce a novel two-step approach for developing a distributed consensus algorithm, which does not require the designer to identify and exploit intricacies of the underlying system model explicitly. In a first step, which is typically done off-line only once, labels representing valid decision values (valencies) are assigned to suitable prefixes of all possible runs. The challenge here is to assign them consistently for indistinguishable runs. The second step consists in deploying a simple generic distributed consensus algorithm, which just uses the previously computed labeling. If it observes that all runs that may lead to a local state that is indistinguishable from the current local state have the same label, it decides on the value determined by this label, otherwise it has to keep on checking. We demonstate the power of our approach by developing a new and asymptotically optimal consensus algorithm for dynamic networks under eventually stabilizing message adversaries for arbitrary system sizes.

1 Introduction

The focus of this paper is on the classic deterministic distributed consensus problem [11], where each process starts with an input value and has to eventually, deterministically and irrevocably select ("decide on") an output value that was the input of some process, such that the same value is selected at every process. The literature on consensus is abundant: besides impossibility and lower bound results, many different consensus algorithms have specifically been tailored to a wide variety of system models.

In sharp contrast to the traditional "tailoring" design paradigm, we propose a *generic* approach for developing consensus algorithms, which consists of two steps: (i) the centralized problem of assigning *labels* to the tree of (prefixes of) admissible runs of a message-passing protocol, in a way that is consistent for indistinguishable runs, and (ii) a simple distributed observation routine that uses this labeling for determining a decision value in the current run. Labels

K. Winkler—Supported by the Vienna Science and Technology Fund (WWTF), grant number ICT19-045 (WHATIF), 2020–2024, and by the Austrian Science Fund (FWF) under project ADynNet (P28182).

ⓒ Springer Nature Switzerland AG 2021
E. Bampis and A. Pagourtzis (Eds.): FCT 2021, LNCS 12867, pp. 457–474, 2021.
https://doi.org/10.1007/978-3-030-86593-1_32

are typically, but not necessarily, sets of processes, whose input values are valid decision values, i.e., values that everyone has learned in the run. Needless to say, consensus algorithms designed via our approach are very different from any possibly existing tailored solution.

In essence, our approach thus reduces the task of finding a distributed consensus algorithm for a given model to labeling the tree of prefixes of admissible runs, which is, fundamentally, an off-line problem.[1] Alternatively, however, processes could also iteratively compute the labels of those finite prefixes that they actually need to check in the observation routine at runtime. The main advantages of our approach are its genericity and hence its wide applicability, as well as the unconventional view of the runs of a distributed algorithm, which may be of independent theoretical interest.

In this paper, we will specialize our approach to the case of full-information protocols[2] in directed dynamic networks, i.e., lock-step synchronous message-passing systems consisting of a possibly unknown number of processes, where the communication in every round is controlled by a message adversary [1]. The solution power of our approach will be demonstrated by providing the first asymptotically optimal algorithm for an eventually stabilizing message adversary.

Main Contributions and Paper Organization. After introducing some related work below, and our formal model in Sect. 2, we define our deterministic labeling function $\Delta(.)$ in Sect. 3. It acts on prefixes of the communication graph sequences that can be produced by a given message adversary: In round r of a given run, it applies a permanent label to the prefix consisting of the first r rounds of the corresponding communication graph sequence. This is done in a way that guarantees that all communication graph sequences extending this prefix keep this label, such that indistinguishable runs will never receive inconsistent labels. This labeling is used by the generic distributed algorithm presented as Algorithm 1.

In Sect. 4, we apply our approach to the non-limit-closed eventually stable message adversary $\Diamond\mathsf{STABLE}_n(n)$ from [13]. This results in the first consensus algorithm with asymptotically optimal termination time under this message adversary. We round off our paper by some conclusions in Sect. 5.

Related Work. Our approach has been stimulated by our paper [10], where it was proved that the solvability of consensus in the message-adversary model corresponds to a connectivity property of a suitable topological space defined on the set of infinite runs. The constructive side of this result, i.e., a consensus algorithm, relies on the existence of a partition of the (usually uncountable) set of runs. Unfortunately, this cannot be considered an "operational" algorithm

---

[1] It may require infinite space, however, for storing the system specification required for constructing the admissible runs.

[2] Full-information protocols, where everyone stores and forwards its entire view to everybody all the time, simplify the presentation and align perfectly with our goal of designing optimal algorithms.

in general, as it is not clear how a process with its bounded resources could maintain/construct the sets of this partition. Also considering its topological basis, the approach [10] must be considered far away from being practical.

Interestingly, however, for the restricted class of limit-closed message adversaries (where the limits of growing sequences of prefixes of admissible executions are also admissible), which are compact in the topological sense, a purely combinatorial treatment and algorithmic operationalization has been provided by Winkler, Schmid and Moses in [12]. We showed that there is some "uniform" prefix length, that is, some round r, by which *all* admissible runs can be labeled by the kernel (the set of processes that have reached everyone) in the equivalence class of all indistinguishable runs. Since such a "uniform" prefix length (which also allows all processes to decide simultaneously) does not exist for non-limit-closed message adversaries like the eventually stabilizing message adversary from [13], however, the approach of [12] cannot be applied in our context.

The same is true for all the related work on combinatorial topology in distributed consensus we are aware of, in particular, [2,5]: Whereas studying the indistinguishability relation of prefixes of runs is closely connected to connectivity properties of the r-round protocol complex, we need to go beyond a uniform prefix length r that is inherently assumed here. Indeed, the models considered in topological studies of consensus like [2,5] are all limit-closed.

By contrast, our paper addresses the considerably more complex case of general non-limit-closed message adversaries, albeit without any explicit topological machinery. In more detail, our approach is based on an algorithmic assignment of labels to finite prefixes of infinite runs. This assignment is done in a way that guarantees consistent labels even for longer indistinguishable prefixes later on. Our assignment procedure effectively guarantees that all runs that are indistinguishable for some process will receive the same label. Note that, in the spirit of [2], our labels can be seen as sets of *valencies* for the consensus valency task.

2   Model and Notations

Dynamic networks consist of a finite set Π of synchronous processes, modeled as state machines, which communicate by exchanging messages over an unreliable network. Conceptually, it is assumed that the communication environment is under the control of an omniscient, malevolent *message adversary* [1] that may suppress certain messages. In particular, message adversaries allow modeling of dynamic networks with time-varying communication, ranging from systems with transient link failures due to, e.g., wireless interference phenomena, to systems where an attacker has gained the ability to control the communication links. Obviously, in order to solve a non-trivial distributed computing problem, the message adversary needs to be restricted in some way.

We assume that the processes in Π are always correct and operate in lock-step synchronous rounds, where each round $r = 1, 2, \ldots$ consists of a phase of message exchange according to a directed communication graph (see Fig. 1), followed by a phase of local, deterministic and instantaneous computations of

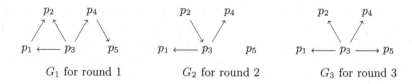

$$G_1 \text{ for round } 1 \qquad G_2 \text{ for round } 2 \qquad G_3 \text{ for round } 3$$

Fig. 1. Communication graphs G_1, G_2, G_3 for 3 rounds. Self-loops are assumed but not shown.

all processes that depend only on the current state and the messages received. Rounds are communication-closed, i.e., messages are only delivered in the same round or dropped altogether. We use the convenient convention that time t is precisely the moment where round t ends (i.e., after the round t computation), and the next round $t + 1$ (i.e., sending the round $t + 1$ messages) begins.

We will adopt the terminology used in most existing papers, including [6–8, 13], which relies on the round r *communication graph* G_r as the crucial structure of a message adversary. Every communication graph is a directed graph, where every process of Π is represented by a node (see Fig. 1). An edge (p, q) in the communication graph G_r represents the fact that q receives the message from p in round r. By convention, it is assumed that processes always receive their own messages, so $(p, p) \in G_r$ for every $p \in \Pi$.

A *communication pattern* is then a dynamic graph, that is, a sequence of communication graphs $G_1, G_2, \ldots$, and a message adversary is just a set of infinite communication patterns. Note that the message adversary has no control over the input values of the processes (see below). The communication patterns that are in a message adversary are called *admissible*. Given some distributed algorithm, an *admissible execution* or *admissible run* is a sequence of global states, or *configurations* $C_0, C_1, \ldots$, which is constructed as follows: Given some initial configuration C^0 and an admissible communication pattern σ, for $i > 0$, C_i results from exchanging the messages according to the round i communication graph of σ and performing the deterministic computations according to the algorithm based on C^{i-1}.

Perhaps the most crucial insight for this model is that the execution of a deterministic distributed algorithm under a message adversary is completely determined by the initial configuration and the communication pattern. It thus makes sense to denote an execution as $\langle C_0, \sigma \rangle$, and a configuration C' as $C' = \langle C, \rho \rangle$, where C is a configuration of an admissible execution and C' can be reached from C via ρ. That is, ρ is a graph sequence that, when starting from C and exchanging the messages according to the communication graphs of ρ, as well as performing the state transitions according to the algorithm and the received messages, results in C'. Since most of our runs will start from the same initial configuration C^0, we will usually use executions and corresponding graph sequences interchangably.

Consensus. We study deterministic *consensus*, a classic distributed computing problem [9,11], where each process $p \in \Pi$ initially holds an input value $x_p \in \mathcal{I}$

from a set $\mathcal{I}$, which we assume w.l.o.g. to be a subset of integers, and has an output or decision variable y_p, which is initially $y_p = \bot$ and may be written to only once. An algorithm solves consensus if it satisfies, for all $p, q \in \Pi$,

(i) eventually $y_p \neq \bot$ (termination),
(ii) if $y_p \neq \bot$ then $y_p = x_k$ for some $k \in \Pi$ (validity), and
(iii) if $y_p \neq \bot$ and $y_q \neq \bot$ then $y_p = y_q$ (agreement).

The special case where $\mathcal{I} = \{0, 1\}$ is called binary consensus.

Local Views. In addition to the basic model outlined above, we introduce some essential terminology, most of which is taken from [6–8,13]. We use lower case Greek letters $\rho, \sigma, \tau, \ldots$ to denote finite or infinite communication patterns and $\sigma(r)$ to denote the round-r communication graph of σ. We call the communication pattern that consists of the first r communication graphs of a communication pattern σ the round-r prefix of σ, which is denoted by $\sigma|_r := \sigma(1), \ldots, \sigma(r)$. By convention, for $r < 1$, we let $\sigma|_r$ be the empty sequence.

At the heart of our proofs lies the *indistinguishability* of two different execution prefixes $\varepsilon, \varepsilon'$ for some process p, denoted by $\varepsilon \sim_p \varepsilon'$, which denotes that $p \in \Pi$ is in the same state, i.e., has the same local view, in each round r configuration of ε and ε'. We will use $\varepsilon \sim \varepsilon'$ if $\varepsilon \sim_p \varepsilon'$ holds for some $p \in \Pi$.

Most of the time, however, we will use the corresponding indistinguishability relation $\sigma \sim_p \sigma'$ defined on the communication patterns σ resp. σ' leading to ε resp. ε' when starting from the same initial configuration C^0. As already argued, $\sigma \sim_p \sigma'$ and $\varepsilon \sim_p \varepsilon'$ are equivalent here, since our algorithms are deterministic and full-information. Formally, this indistinguishability can be defined as follows: For a single communication graph G, the view of a process p is simply the set of its in-neighbors $\text{In}_G(p)$. For evolving views, we use the convenient notion of a *process-time graph* (c.f. [3]) of a communication pattern. For a finite communication pattern σ that consists of r communication graphs with nodes Π, the *process-time graph*[3] PT_σ is defined as

$$\text{PT}_\sigma := \langle V, E \rangle : \quad V = \Pi \times \{0, \ldots, r\},$$
$$((p, r'), (q, r'')) \in E \Leftrightarrow (r'' = r' + 1) \wedge (p, q) \in \sigma(r'').$$

We say (p, r') influences (q, r''), written as $(p, r') \rightsquigarrow_\sigma (q, r'')$, if there is a path from (p, r') to (q, r'') in PT_σ and denote the set of processes that managed to reach some process p in σ as p's heard-of set [6]

$$\text{HO}_\sigma(p) := \{q : (q, r') \rightsquigarrow_\sigma (p, r)\}.$$

Those processes that managed to reach everyone in a finite communication pattern σ are called the *kernel* of σ. It is of central importance, as the decision must be on an input value of some process in the kernel: If a decision were made on the input value of a process that is not in the kernel, some process would need

---

[3] Note that, in contrast to [10], our process time graphs do not incorporate initial values, since we usually start from the same initial configuration.

to decide on a value of which it cannot be sure that it really was the input value of another process, thus violating validity. A more formal argument for this can be found in the proof of Theorem 1. The kernel is denoted by

$$\text{Ker}(\sigma) := \bigcap_{p \in \Pi} \text{HO}_\sigma(p). \tag{1}$$

For an infinite communication pattern τ, we write

$$\text{Ker}(\tau) = \bigcup_{r \geq 1} \text{Ker}(\tau|_r) = \lim_{r \to \infty} \text{Ker}(\tau|_r)$$

Using this notation, the *view* of a process p in a finite communication pattern σ that consists of r communication graphs can be formally defined as the subgraph of PT_σ that is induced by the vertex-set $\{(q, r') : (q, r') \leadsto_\sigma (p, r)\}$ and denoted as $\text{PT}_\sigma(p)$. Two finite communication patterns ρ, σ are thus indistinguishable for process p, written as $\rho \sim_p \sigma$ if $\text{PT}_\rho(p) = \text{PT}_\sigma(p)$. We write $\rho \sim \sigma$ if $\rho \sim_p \sigma$ for some $p \in \Pi$ and $\rho \not\sim \sigma$ if $\nexists p : \rho \sim_p \sigma$.

3   A Generic Consensus Algorithm

The Function Δ

Our algorithm will rely on a labeling function $\Delta(.)$, which is a member of the class of Δ-functions introduced in the following Definition 1. Fixing some admissible communication pattern σ of a message adversary, $\Delta(\sigma|_r)$ assigns a label to every round r prefix $\sigma|_r$ of σ as follows: Going from shorter prefixes to longer ones, the label is initially $\emptyset$ until it becomes a fixed, non-empty subset of the kernel of σ. As this subset consists of processes of which, by definition of the kernel, every other process has heard of eventually in σ, the input value of each process from this subset is a valid decision value. Note that we do not a priori restrict the range of $\Delta(.)$ to one of these processes to provide some flexibility of choice here. It is in this sense that $\Delta(\sigma|_r)$ assigns a valency to $\sigma|_r$, as the decision value of all extensions of $\sigma|_r$ can be fixed to the input value of some deterministically chosen process from $\Delta(\sigma|_r)$.

Definition 1. *Let* MA *be a message adversary. The class of Δ-functions for* MA *is made up by all functions*

$$\Delta : \bigcup_{r \geq 1} \{\sigma|_r : \sigma \in \text{MA}\} \to 2^\Pi$$

such that, for every admissible communication pattern $\sigma \in$ MA, there is a round $s \geq 0$ and a set $K \subseteq \text{Ker}(\sigma)$ with $K \neq \emptyset$ such that

$$\Delta(\sigma|_r) = \begin{cases} \emptyset, & \text{for } r \leq s, \\ K & \text{for } r > s. \end{cases}$$

As shown in Theorem 1 below, already the solvability of binary consensus implies the existence of a Δ-function, thus the non-emptiness of the class of Δ-functions (for a given MA) is ensured if consensus (for any non-trivial set of input values) is solvable (under MA).

Theorem 1. *If binary consensus is solvable under* MA, *then there exists a function* $\Delta(.)$ *such that* $\Delta(.)$ *is a member of the class of Δ-functions for* MA.

Proof. First, we show that if binary consensus is solvable under MA, then every $\sigma \in$ MA satisfies $\mathrm{Ker}(\sigma) \neq \emptyset$. Suppose that some algorithm A solves consensus in a communication pattern σ in spite of $\mathrm{Ker}(\sigma) = \emptyset$. By termination, for every input configuration C, there is a round t such that all processes have decided when executing A in $\langle C, \sigma|_t \rangle$. In fact, since there are only finitely many processes and hence input assignments, for every $\sigma \in$ MA there exists a round t' such that all processes have decided when running A in $\langle C, \sigma|_{t'} \rangle$, irrespective of the input value assignment in the initial configuration C. Assuming that the set of input values is $\mathcal{I} = \{0, 1\}$, by validity, when all processes start with input 0, they must also decide 0, and when all processes start with 1, they must decide 1. Starting from the input configuration where all processes start with 0 and flipping, one by one, the input values to 1, reveals that there exist two input configurations C', C'' that differ in the input value assignment of a single process p but all processes decided 0 in $\langle C', \sigma|_{t'} \rangle$, whereas all processes decided 1 in $\langle C'', \sigma|_{t'} \rangle$. Since $\mathrm{Ker}(\sigma|_{t'}) = \emptyset$ by our supposition, there is a process q such that $p \notin \mathrm{HO}_{\sigma|_{t'}}(q)$. But then $\langle C', \sigma|_{t'} \rangle \sim_q \langle C'', \sigma|_{t'} \rangle$ and q decides the same in both executions, a contradiction.

We conclude the proof by showing that the existence of $\Delta(.)$ follows if every $\sigma \in$ MA has $\mathrm{Ker}(\sigma) \neq \emptyset$, which implies that for every $\sigma \in$ MA there is a round r such that $\mathrm{Ker}(\sigma|_r) \neq \emptyset$. Given $\sigma \in$ MA, let s be the smallest such round where $K = \mathrm{Ker}(\sigma|_s) \neq \emptyset$. Then

$$\Delta(\sigma|_r) = \begin{cases} \emptyset, & \text{for } r \leq s \\ K & \text{for } r > s \end{cases}$$

is a Δ-function according to Definition 1: Because $\mathrm{Ker}(\sigma|_s) = K \neq \emptyset$ and the kernel of a communication pattern prefix monotonically increases with increasing prefix length, i.e., if $r \leq r'$ then $\mathrm{Ker}(\sigma|_r) \subseteq \mathrm{Ker}(\sigma|_{r'})$, every $\rho \in$ MA with $\rho|_s = \sigma|_s$ satisfies $K \subseteq \mathrm{Ker}(\rho)$. $\qquad\square$

One might argue that the core statement of Theorem 1 is not very strong, as the definition of $\Delta(.)$ is almost equivalent to the consensus problem specification itself. However, the point is not that this formulation is very deep, but rather that it is a very useful abstraction: We show in the remainder of our paper how, by means of encapsulating the model (i.e., MA) in the computation of $\Delta(.)$, consensus can be solved quickly and using relatively simple algorithms. So rather than tailoring a consensus algorithm to the intricacies of a particular model, we abstract the model into $\Delta(.)$ and use a simple and generic consensus algorithm on top of it.

Solving Consensus with Δ

From the contraposition of Theorem 1, it immediately follows that the existence of a Δ-function $\Delta(.)$ is necessary for solving consensus. At a first glance, it might seem that the existence of $\Delta(.)$ is also sufficient for consensus, i.e., that Theorem 1 can actually be extended to an equivalence, thus providing an efficient consensus solvability characterization. The reason why this is not the case, however, is that the local uncertainty, i.e., the incomplete view a process has of the actual communication pattern, makes it unsure of the latter. Thus, the mere existence of $\Delta(\sigma|_r)$ is not enough, since process p may be uncertain of the actual $\sigma|_r$: The actual communication pattern may in fact be ρ, with $\sigma|_r \sim_p \rho|_r$, but still $\Delta(\rho|_r) = \emptyset$. In order to make consensus solvable, we thus require an additional consistency condition, stated in Assumption 1, which guarantees that the label that will eventually be assigned to ρ *later on* is the same as $\Delta(\sigma|_r)$:

Assumption 1. $\forall \sigma \in$ MA $\exists r \in \mathbb{N} \ \forall \sigma' \in$ MA: $\sigma'|_r \sim \sigma|_r \Rightarrow \Delta(\sigma'|_r) = \Delta(\sigma|_r) \neq \emptyset$.

Assumption 1 states that every admissible communication pattern has a round r after which the values of $\Delta(.)$ are the same for every indistinguishable prefix. Note carefully, however, that this neither implies (i) that σ and σ' must get their labels in the same round (before or at r) nor (ii) that there is a uniform round r that holds for all $\sigma \in$ MA.

We do not know whether Assumption 1 is necessary for solving consensus. The following Algorithm 1 will show that it is sufficient, however. Our algorithm works as follows: Every process p waits until it detects that all prefixes ρ that it still considers possible, i.e., those prefixes that do not contradict what p observed so far, have the same value $\Delta(\rho) = K \neq \emptyset$. When this occurs, p waits until it has heard from the process $p \in K$ with the largest identifier and decides on p's input value. Obviously, the processes also need to know the input values of all processes that they have heard of so far, which is inherently the case wenn using a full-information algorithm.[4]

We note that the algorithm uses both (i) the pre-computed Δ-functions and (ii) the set P of communication pattern prefixes process p considers possible in the current round r, which can be determined from its view, given the message adversary specification MA (for example, via the *finite* sets of admissible r-prefixes, for $r = 1, 2, \dots$).

Note carefully that a way to compute $\Delta(.)$, together with Algorithm 1, gives us a purely on-line solution algorithm for consensus, which operates in two phases per round: In a run with actual communication pattern σ, in its round r computation step, every process p first pre-computes $\Delta(\rho)$ for all communication pattern prefixes ρ that are compatible with its view, i.e., all $\rho \in P$, where P is the set of round r prefixes that p considers possible, as in Line 1 of Algorithm 1. Process p then proceeds to run Algorithm 1, using these pre-computed values instead of actual calls to $\Delta(.)$.

---

[4] Obviously, just communicating and keeping track of the input values received so far would also suffice.

Algorithm 1: Consensus algorithm, given $\Delta(.)$ for process p and round r when run under communication pattern σ of message adversary MA.

---

1 Let $P = \{\rho|_r : \rho \in \text{MA} \wedge \rho|_r \sim_p \sigma|_r\}$ be the prefixes that are indistinguishable for p from the actual prefix

2 **if** $y_p = \bot$ *and* $\exists K \subseteq \Pi \; \forall \rho \in P : \Delta(\rho) = K \neq \emptyset$ **then**

3 $\quad$ Let $q \in \Pi$ be the process in K with the largest identifier

4 $\quad$ **if** *heard from q* **then** $y_p \leftarrow x_q$

The correctness of Algorithm 1 is stated by the following Theorem 2.

Theorem 2. *Algorithm 1 solves consensus under a message adversary* MA, *given that $\Delta(.)$ is a Δ-Function for* MA *that satisfies Assumption 1.*

Proof. We show that the consensus properties are satisfied by Algorithm 1 when it is run in an admissible communication pattern $\sigma \in$ MA. In the proof, we use v_p^r to denote the value of variable v at process p at time r (i.e., at the end of round r).

Validity follows immediately, since every decision in Line 4 is on some other process's input value.

Termination follows because, when run under a communication pattern $\sigma \in$ MA, Assumption 1 guarantees that there is a set $K \neq \emptyset$ and a round r such that $\Delta(\rho) = K$ for all ρ with $\rho \sim \sigma|_r$. Furthermore, $K \subseteq \text{Ker}(\sigma)$, thus the guard of Line 4 will eventually be passed.

In order to show agreement, let r be the round where some process, say p, passes the guard in Line 2 for the first time. This implies that $K_p^r \neq \emptyset$ and p has $\Delta(\rho) = K_p^r$ for all $\rho \sim_p \sigma|_r$ so, in particular, $\Delta(\sigma|_r) = K_p^r$. Definition 1 implies that for all $s \geq r$, $\Delta(\sigma|_s) = K_p^r$ as well. Thus, if some process p' passes Line 2 again in round s, because it clearly has $\sigma|_s \in P_{p'}^s$, every decision via Line 4 must be on the input of the process with the largest identifier in $K_{p'}^s = \Delta(\sigma|_s) = K_p^r$. $\square$

Whereas Algorithm 1 is generic, i.e., independent of MA, this is not the case for the computation of $\Delta(.)$, which depends heavily on the MA specification. In Sect. 4, we will show how $\Delta(.)$ can be computed under the message adversaries from [13].

4   Computing $\Delta(.)$ for the Eventually Stable Message Adversary

One of the most severe restrictions of limit-closed message adversaries is that they cannot express any guarantees of the form "eventually something good will happen", that is, guarantees that hold only after a finite but unbounded number of rounds. Limit-closed message adversaries cannot hence be used for modeling systems with chaotic boot-up behavior, where it takes an unknown number of rounds until reasonably reliable synchronous communication could be established, as well as systems with massive transient faults during periods of unknown duration.

The Message Adversary $\lozenge\text{STABLE}_n(x)$

One instance of a non-closed message adversary is the eventually stable message adversary $\lozenge\text{STABLE}_n(n)$ from [13], which we will now briefly describe.

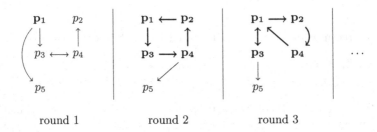

round 1   round 2   round 3

Fig. 2. The root component of round 1 consists of the node p_1. In round 2 and 3 there exists a vertex-stable root component consisting of the vertex set $\{p_1, p_2, p_3, p_4\}$.

It is based on the notion of a *root component* (see Fig. 2), a strongly connected component, where no node has an incoming edge from a node outside of the component. A root component is called *vertex-stable* if it consists of the same set of nodes for multiple consecutive rounds of a communication pattern. We note that the connectivity inside of a vertex-stable root component may vary significantly, as long as it remains a root component of the respective communication graphs.

The message adversary $\lozenge\text{STABLE}_n(x)$ from [13], as illustrated in Fig. 3, consists of message patterns with n processes and is subject to the following two conditions[5]:

(i) Every communication graph is *rooted*, i.e., has precisely one root component and

(ii) every admissible communication pattern $\sigma \in \lozenge\text{STABLE}_n(x)$ has a root component that remains vertex stable for at least x rounds.

In our analysis, we will denote the round where the first such vertex-stable root component occurs as $r_{stab}(\sigma)$. Furthermore, we denote the member nodes of this vertex-stable root component as $\text{Root}(\sigma)$ and use $\text{Root}(\sigma|_r)$ to denote the vertex-stable root component of σ if the stability phase already occurred by round r, i.e., $\text{Root}(\sigma|_r) = \text{Root}(\sigma)$ if $r \geq r_{stab}(\sigma) + n - 1$ and $\text{Root}(\sigma|_r) = \emptyset$

---

[5] Technically, the message adversary from [13] has an additional parameter, the dynamic diameter D, which we will neglect here for simplicity. Since it was shown in [4, Corollary 1] that $D < n$, we will just conservatively assume $D = n - 1$. This has the downside of the lower bound established in Theorem 15 being formally weaker in those cases where $D = o(n)$.

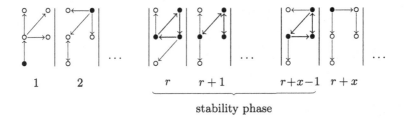

stability phase

Fig. 3. An admissible communication pattern σ of the message adversary $\Diamond\mathsf{STABLE}_n(x)$. The vertex-stable root component (highlighted in black) occurs in round $r = r_{stab}(\sigma)$ and lasts until round $r + x - 1$.

otherwise. Note that the processes have no a priori knowledge of it in a given execution $\langle C, \sigma \rangle$.

The central result of [13] was a consensus solvability characterization wrt. the parameter x, i.e., the duration of the stability phase, rephrased in Theorem 3.

Theorem 3 ([13, Theorem 3 and 4]). *Consensus is solvable under* $\Diamond\mathsf{STABLE}_n(x)$ *if and only if* $x \geq n$.

The consensus algorithm of [13, Algorithm 2] is able to decide $\Theta(n^2)$ rounds after $r_{stab}(\sigma)$ in $\sigma \in \Diamond\mathsf{STABLE}_n(n)$ in the worst case. In the next section, we will present a computation for $\Delta(.)$ that, combined with Algorithm 1, leads to a consensus algorithm that terminates already after $r_{stab}(\sigma) + \Theta(n)$ rounds, for which we also provide a matching lower bound.

An Algorithm for Computing $\Delta(.)$

We now introduce Algorithm 2 to compute Δ under the message adversary $\Diamond\mathsf{STABLE}_n(n)$. Starting from round $r = 1$, it operates by assigning labels to selected round-r prefixes $\sigma|_r$. In order to satisfy monotonicity of the labels, the algorithm first preserves the labels from the previous round (Line 4). Subsequently, it checks whether there is a prefix that was not assigned a label yet and whose vertex-stable root component lasted from round $r - 3n + 1$ to (at least) round $r - 2n$ (Line 9). If such a "mature" prefix $\sigma|_r$ is found, and there is an indistinguishable prefix that already had its label assigned, $\Delta(\sigma|_r)$ adopts this label (Line 10). If, on the other hand, there is no such indistinguishable prefix, $\Delta(\sigma|_r)$ is set to the vertex-stable root component of $\sigma|_r$ (Line 12). Finally, a prefix that was not assigned a label in the current round adopts the label of an indistinguishable prefix that was "mature" already n rounds ago (Line 7).

Correctness Proof

We first establish some important basic properties of vertex-stable root components and rooted communication graphs. We will use the notation from Sect. 2

Algorithm 2: Computing Δ for each r-prefix $\sigma|_r$, $r = 0, 1, 2, \ldots$, for a given $\sigma \in \Diamond\mathsf{STABLE}_n(n)$.

---

1 Initially, let $\Delta(\sigma|_0) = \emptyset$
2 **for** $r = 1, 2, \ldots$ **do**
3   **foreach** $\sigma|_r$ **do**
4          $\Delta(\sigma|_r) \leftarrow \Delta(\sigma|_{r-1})$
5   **foreach** $\sigma|_r$ with $\Delta(\upsilon|_r) = \emptyset$ **do**
6   **if** $\exists \rho|_r \sim \sigma|_r$ with $\Delta(\rho|_r) \neq \emptyset$ and $r_{stab}(\rho) \leq r - 4n$ **then**
7             $\Delta(\sigma|_r) \leftarrow \Delta(\rho|_r)$
8   **foreach** $\sigma|_r$ with $r_{stab}(\sigma) \leq r - 3n$ and $\Delta(\sigma|_r) = \emptyset$ **do**
9   **if** $\exists \rho|_r \sim \sigma|_r$ with $\Delta(\rho|_r) \neq \emptyset$ **then**
10            $\Delta(\sigma|_r) \leftarrow \Delta(\rho|_r)$
11   **else**
12            $\Delta(\sigma|_r) \leftarrow \mathrm{Root}(\sigma)$

$(p, r) \leadsto_\sigma (q, r')$, or simply $(p, r) \leadsto (q, r')$ if σ is understood, to denote that there is a path from (p, r) to (q, r') in the process-time graph of σ, i.e., that p at round r is in the causal past $\mathrm{CP}_q^{r'} r$ of q at round r'. In the following proofs, let ρ, σ denote two arbitrary communication patterns of $\Diamond\mathsf{STABLE}_n(n)$ and r be an arbitrary round.

We will rely on the following generic result from [13]:

Theorem 4 (Root propagation [13, Theorem 1]). *Let $G = \{G^{r_1}, \ldots, G^{r_n}\}$ be an ordered set of rooted communication graphs on the same vertex set Π where $|\Pi| = n > 1$. Pick an arbitrary mapping $f \colon [1, n] \mapsto \Pi$ s.t. $f(i) \in \mathrm{Root}(G^{r_i})$. Then $\forall p \in \Pi \setminus \{f(n)\}$, $\exists i \in [1, n-1]\colon f(i) \in \mathrm{CP}_p^{r_n}(r_i)$.*

Corollary 5. *For all $q \in \mathrm{Root}(\sigma)$, for all $p \in \Pi$, we have that $(q, r_{stab}(\sigma)) \leadsto (p, r_{stab}(\sigma) + n - 1)$.*

Proof. For every $q \in \mathrm{Root}(\sigma)$, just take $G = \{G^{r_{stab}(\sigma)}, \ldots, G^{r_{stab}(\sigma)+n-1}\}$ and pick the mapping $f(i) = q$ for all i in Theorem 4 to show that $(q, r_{stab}(\sigma)) \leadsto (p, r_{stab}(\sigma) + n - 1)$ for every $p \in \Pi$. □

Corollary 6. *If $\rho|_r \sim \sigma|_r$ and $r_{stab}(\sigma) = r - n + 1$ then for all $p \in \mathrm{Root}(\sigma|_r)$ we have $\rho|_{r-n+1} \sim_p \sigma|_{r-n+1}$.*

Proof. Suppose $\rho|_{r-n+1} \not\sim_p \sigma|_{r-n+1}$ for some $p \in \mathrm{Root}(\sigma|_r)$. Since $(p, r - n + 1) \leadsto (q, r)$ for every process q by Corollary 5, we have $\rho|_r \not\sim \sigma|_r$, a contradiction. □

Corollary 7. *Let $r \geq n$ and $\tau \in \{\rho|_r, \sigma|_r\}$. If $\rho|_r \sim_p \sigma|_r$ and $\rho|_{r-n+1} \sim_P \sigma|_{r-n+1}$ for the maximal set P containing p, then for every $q \in \Pi$ there is some $p' \in P$ such that $(p', r - n + 1) \leadsto_\tau (q, r)$.*

Proof. For $k \in [r-n+1, r]$, consider the set of processes P_k such that $\rho|_k \sim_{P_k} \sigma|_k$, and the set of processes Q_k that did not hear from at least one process in P by round k in τ. Clearly, $P_{r-n+1} = P$ and $Q_{r-n+1} \subseteq \Pi \backslash P$. Given P_k and Q_k, if the root component R_{k+1} of round $k+1$ in τ contains only processes in P_k, then at least one process in Q_k hears from a member of R_{k+1} and thus leaves, i.e., $Q_{k+1} \subset Q_k$. Similarly, if $R_{k+1} \subseteq Q_k$, then at least one process $p' \in P_k$ can now distinguish $\rho|_{k+1} \not\sim_{p'} \sigma|_{k+1}$, so $P_{k+1} \subset P_k$. If R_{k+1} contains both processes from P_k and Q_k, at least one of P_{k+1} and Q_{k+1} shrinks. So, after round r, at least one of P_r and Q_r must be empty. Since we know that $|P_r| \geq 1$ as $\rho|_r \sim_p \sigma|_r$, we must have $Q_r = \emptyset$, which proves Corollary 7. $\qquad\square$

Corollary 8. *If $\sigma|_r \sim \rho|_r$ and $r \geq r_{stab}(\sigma) + n - 1$ then for every $q \in \mathrm{Root}(\sigma)$, for every $p \in \Pi$ we have $(q, r_{stab}(\sigma)) \rightsquigarrow_\rho (p, r)$ and thus the root component of ρ in round $r_{stab}(\sigma)$ is $\mathrm{Root}(\sigma)$.*

Proof. Due to Corollary 5, for all $q \in \mathrm{Root}(\sigma)$ and all $p \in \Pi$, $(q, r_{stab}(\sigma)) \rightsquigarrow_\sigma (p, r_{stab}(\sigma) + n - 1)$. And since $\sigma|_r \sim_p \rho|_r$ for some p, we must also have $(q, r_{stab}(\sigma)) \rightsquigarrow_\rho (p, r_{stab}(\sigma) + n - 1)$. This implies that the members of $\mathrm{Root}(\sigma)$ must also form a strongly connected component without incoming edges in round $r_{stab}(\sigma)$ in run ρ. Since all graphs in ρ are single-rooted, $\mathrm{Root}(\sigma)$ must hence be the root component of this round. $\qquad\square$

This leads to our first instrumental result, namely, that indistinguishable communication patterns cannot have overlapping stability phases:

Theorem 9. *Let $s = r_{stab}(\sigma)$ and $s' = r_{stab}(\rho)$. If $\sigma|_{s+n-1} \sim \rho|_{s+n-1}$ then $s = s'$ or $s' \leq s - n$ or $s' \geq s + n$.*

Proof. Let us suppose that $\sigma|_{s+n-1} \sim \rho|_{s+n-1}$ and $s' \in [s - n + 1, s - 1] \cup [s + 1, s + n - 1]$. By Corollary 8, we thus have that the root component of round s in ρ is $\mathrm{Root}(\sigma) = R$, which implies $\mathrm{Root}(\sigma) = \mathrm{Root}(\rho) = R$. If, w.l.o.g. $s' < s$, the root component of σ in all rounds $k \in [s', s]$ consists of the processes of R as well: By Corollary 5, we have $(q, s) \rightsquigarrow_\sigma (p, s + n - 1)$ for all $q \in R, p \in \Pi$, i.e., everyone has learned the in-neighborhood of q for all rounds $\leq s$. Since $\sigma|_{s+n-1} \sim \rho|_{s+n-1}$ implies $\sigma|_k \sim \rho|_k$, the in-neighborhood of every such q must be the same in σ and in ρ in round k. But then the stability phase of σ actually starts in round s' (or before) and we have $s' \geq r_{stab}(\sigma) \neq s$, a contradiction.

The most important prerequisite for our correctness proof is Theorem 10, which says that that transitive indistinguishability in round r implies direct indistinguishability at round $r - n + 1$ for at least one process. Note carefully that this implies that, by going $(k - 2)n$ rounds up in a chain of transient indistinguishability $\sigma^1|_r \sim \sigma^2|_r \sim \cdots \sim \sigma^k|_r$ guarantees $\sigma^i|_{r-(k-1)n} \sim_p \sigma^j|_{r-(k-1)n}$ for some process p, for any choice of i, j.

Theorem 10. *Suppose the round r prefixes $\rho|_r, \sigma|_r, \tau|_r$ with $r \geq n$ satisfy $\rho|_r \not\sim \tau|_r$ and $\rho|_r \sim_p \sigma|_r \sim_q \tau|_r$. Let P and Q be the maximal sets guaranteeing $\rho|_{r-n+1} \sim_P \sigma|_{r-n+1}$ resp. $\sigma|_{r-n+1} \sim_Q \tau|_{r-n+1}$. Then, either $\rho|_{r-n+1} \sim_p \tau|_{r-n+1}$ and $\forall s \in \Pi \; \exists p' \in P$ such that $(p', r - n + 1) \rightsquigarrow_\sigma (s, r)$, or $\rho|_{r-n+1} \sim_q \tau|_{r-n+1}$, and and $\forall s \in \Pi \; \exists q' \in Q$ such that $(q', r - n + 1) \rightsquigarrow_\sigma (s, r)$.*

Proof. For a contradiction, suppose that $(\rho|_{r-n+1} \not\sim_p \tau|_{r-n+1}) \wedge (\rho|_{r-n+1} \not\sim_q \tau|_{r-n+1})$ for any $p \in P$ and any $q \in Q$. This assumption prohibits any feasible choice $p \in P \cup Q$, since it also implies $(\rho|_{r-n+1} \not\sim_q \sigma|_{r-n+1}) \wedge (\sigma|_{r-n+1} \not\sim_p \tau|_{r-n+1})$.

Since $\rho|_r \not\sim \tau|_r$ and hence $\rho|_{r-n+1} \not\sim \tau|_{r-n+1}$ must hold due to $\rho|_r \not\sim \tau|_r$, our assumption implies that $\sigma|_{r-n+1} \not\sim_P \tau|_{r+n-1}$ and $\rho|_{r-n+1} \not\sim_Q \sigma|_{r-n+1}$. Consequently, $P \subseteq \Pi \backslash Q$ and $Q \subseteq \Pi \backslash P$, $P \cap Q = \emptyset$ and $|P| + |Q| \leq n$.

Now consider the sets of processes P_k and Q_k, such that $\rho|_k \sim_{P_k} \sigma|_k$ resp. $\sigma|_k \sim_{Q_k} \tau|_k$. Clearly, $P_{r-n+1} = P$ and $Q_{r-n+1} = Q$, and given P_k and Q_k, if the root component R_{k+1} of round $k + 1$ in σ contains only processes in P_k, then at least one process in Q_k hears from a member of R_{k+1} and thus leaves, i.e., $Q_{k+1} \subset Q_k$. Similarly, if $R_{k+1} \subseteq Q_k$, then $P_{k+1} \subset P_k$. If R_{k+1} contains both processes from P_k and Q_k or none of those, at least one of P_{k+1} and Q_{k+1} shrinks. So, after round r, at least one of P_r and Q_r must be empty, which contradicts $(\rho|_r \sim_p \sigma|_r) \wedge (\sigma|_r \sim_q \tau|_r)$.

Corollary 7 applied to the "winning" set, say, P where $p \in P$ satisfies $\rho|_{r-n+1} \sim_p \tau|_{r-n+1}$ reveals that every $s \in \Pi$ hears from some $p' \in P$ such that $(p', r - n + 1) \rightsquigarrow_\sigma (s, r)$ in σ (and in τ).   □

The following corollary, in conjunction with Corollary 6, shows that $\rho|_r \sim \sigma|_r \sim \tau|_r$ in round $r = r_{stab}(\sigma|_r) + n - 1$ (at the end of the VSRC stability) implies that $\rho|_{r-n} \sim_{\text{Root}(\sigma)} \sigma|_{r-n} \sim_{\text{Root}(\sigma)} \tau|_{r-n}$ (at the beginning of VSRC stability).

Corollary 11. *Suppose the round r prefixes $\rho|_r, \sigma|_r, \tau|_r$ with $r \geq n$ satisfy $\rho|_r \sim_p \sigma|_r \sim_q \tau|_r$ and $(p, r - n + 1) \rightsquigarrow_\sigma (q, r)$. Then, $(p, r - n + 1) \rightsquigarrow_\tau (q, r)$ and $\rho|_{r-n+1} \sim_p \sigma|_{r-n+1} \sim_p \tau|_{r-n+1}$.*

Proof. Since $\sigma|_r \sim_q \tau|_r$, $(p, r-n+1) \rightsquigarrow_\tau (q, r)$ follows immediately. If $\sigma|_{r-n+1} \not\sim_p \tau|_{r-n+1}$ would hold, then $\sigma|_r \not\sim_q \tau|_r$ since (q, r) would be different in $\sigma|_r$ and $\tau|_r$.   □

We can now prove our main result:

Theorem 12. *The function $\Delta(\sigma)$ computed by Algorithm 2 outputs $\emptyset$ until it eventually outputs $\Delta(\sigma) \neq \emptyset$ forever, which happens by round $r_{stab}(\sigma)+3n$, where every process has heard from every member of $\text{Root}(\sigma)$ already. Furthermore, $\Delta(\sigma)$ satisfies Assumption 1 with $r = r_{stab}(\sigma) + 4n$.*

Proof. It follows immediately from the code of Algorithm 2 that every σ gets its label by $r \leq r_{stab}(\sigma) + 3n$. Moreover, if ρ is assigned a label $R \neq \emptyset$ in round r, then every σ that satisfies $\sigma|_{r+n} \sim \rho|_{r+n}$ also gets the label R in round $r + n$, which secures Assumption 1.

We still need to prove that the labels are correctly assigned, however. First of all, in Algorithm 2, $\Delta(\sigma|_0)$ gets initialized to $\emptyset$ in Line 1. Once a non-empty label is assigned, it is never modified again, since each assignment, except the one in Line 4, may only be performed if the label was still $\emptyset$. In accordance with

Assumption 1, we must hence show that if a label $\Delta(\sigma|_r) \leftarrow R \neq \emptyset$ is assigned to a round r prefix $\sigma|_r$, then every indistinguishable prefix $\rho|_r \sim \sigma|_r$ has either $\Delta(\rho|_r) = \Delta(\sigma|_r)$ or $\Delta(\rho|_r) = \emptyset$.

We prove this by induction on $r = 0, 1, \ldots$. The base case is the "virtual" round $r = 0$, where $\Delta(\sigma|_0)$ gets initialized to $\emptyset$ in Line 1 and our statement is vacuously true.

For the induction step from $r - 1$ to r, assume by hypothesis that, for all rounds $k < r$ prefixes that already have some label $R \neq \emptyset$ assigned, all their indistinguishable prefixes have label R or $\emptyset$.

For the purpose of deriving a contradiction, suppose that two r-round prefixes $\rho|_r \sim \sigma|_r$ end up with non-empty labels $\Delta(\rho|_r) = R \neq R' = \Delta(\sigma|_r)$. They can be assigned via Line 4, Line 12, Line 10 or Line 7. Note that we must have $r \geq 3n + 1$ here, as no non-empty labels are assigned before.

If both $\rho|_r$ and $\sigma|_r$ get their label via Line 4, the induction hypothesis guarantees $R = R'$.

If both $\rho|_r$ and $\sigma|_r$ get their label via Line 12, Theorem 9 implies that both $r_{stab}(\rho) = r_{stab}(\sigma)$ and $\text{Root}(\rho) = \text{Root}(\sigma)$ and hence $R = R'$, which provides a contradiction. However, we can show an even stronger result: Consider a chain of transitively indistinguishable r-round prefixes $\omega|_r \sim \rho|_r \sim \sigma|_r \sim \tau|_r$. Applying Theorem 10 yields $\rho|_{r-n} \sim_p \sigma|_{r-n} \sim_p \sigma|_{r-n}$ for some process p, and hence $\omega|_{r-n} \sim \sigma|_{r-n} \sim \tau|_{r-n}$. Another application of Theorem 10 then provides $\tau|_{r-2n} \sim \omega|_{r-2n}$. Consequently, if both $\omega|_r$ and $\tau|_r$ get their label via Line 12, Theorem 9 shows that even $r_{stab}(\omega) = r_{stab}(\tau)$ and $\text{Root}(\omega) = \text{Root}(\tau)$ holds. In addition, if $\rho|_r$ and $\sigma|_r$ get their labels $\Delta(\rho|_{r+n}) = \Delta(\omega|_{r+n})$ and $\Delta(\sigma|_{r+n}) = \Delta(\tau|_{r+n})$ via Line 7, n rounds later, we end up with $R = R'$ also in this case.

If $\omega|_r$ and $\tau|_{r'}$ get different labels R and R' via Line 12 in different iterations $r' < r$, the above arguments show that this requires $r' \leq r - n$. Line 7 causes $\Delta(\sigma|_{r'+n}) = \Delta(\tau|_{r'+n})$, as well as $\Delta(\rho|_{r+n}) = \Delta(\omega|_{r+n})$, and applying Theorem 10 to the chain $\rho|_{r'+n} \sim \sigma|_{r'+n} \sim \tau|_{r'+n}$ reveals $\rho|_{r'} \sim \tau|_{r'}$. Consequently, Line 10 is first used to set the label $\Delta(\rho|_{r'}) = \Delta(\tau|_{r'}) = R'$ and also to set $\Delta(\omega|_r) = \Delta(\rho|_r) = R'$, so $R = R'$ also here. Note that this scenario also covers the simpler case where $\sigma|_r$ gets its label via Line 4. Moreover, it also covers the case where $\sigma|_{r'}$ gets its label directly via Line 12, and/or when $\rho|_r$ gets its label directly via Line 12.

However, if $\rho|_r$ and $\sigma|_r$ get their labels via "forwarding" via Line 10 and/or Line 7, we also need to consider the possibility that the "exporting" prefixes $\omega|_r$ and $\tau|_r$ may have got non-empty labels $R = \Delta(\omega|_r) \neq \Delta(\tau|_r) = R'$ via Line 10 or Line 7. To deal with these cases, we need the following technical lemma:

Lemma 13. *Algorithm 2 can forward a label R over a chain of indistinguishable prefixes with root components different from R, via Line 10 or Line 7, at most once every n iterations.*

Proof. Since Line 10 can only import a label to $\sigma|_r$ with $r_{stab}(\sigma) \leq r - 3n$ whereas Line 7 can only export a label from $\rho|_r$ with $r_{stab}(\rho) \leq r - 4n$, it follows from

Theorem 9 that a label R that is different from the root component $\text{Root}(\sigma)$ of the importing prefix $\sigma|_r$ must originate from an exporting prefix $\rho|_r$ with $|r_{stab}(\sigma) - r_{stab}(\rho)| \geq n$. This immediately implies our lemma.   □

So if $\omega|_r$ and $\tau|_{r'}$ have got labels $R \neq \text{Root}(\omega)$ and $R' \neq \text{Root}(\tau)$ either in Line 12 and/or Line 7, Lemma 13 reveals that they may at most originate from some $\omega'|_r \sim \omega|_r$ and $\tau'|_{r'} \sim \tau|_{r'}$, but may not be further apart w.r.t. transitive indistinguishability. On the other hand, applying Theorem 10 to the resulting 5-chain reduces it to the 3-chain $\omega'|_{r-n} \sim \rho|_{r-n} \sim \sigma|_{r-n} \sim \tau'|_{r-n}$. We can repeat this argument iteratively until we reach the 3-chain $\omega^k|_{r-kn} \sim \rho|_{r-kn} \sim \sigma|_{r-kn} \sim \tau^k|_{r-kn}$ where $\omega^k|_{r-kn}$ and $\tau^k|_{r-kn}$ got their labels R and R' assigned in one of the cases analyzed above, i.e., do not get them via forwarding. Since we have shown that $R = R'$ here, we obtain the required contradiction also for this case.

Finally, we must show that every process in ρ has heard from every member of its label R, which need not be equal to $\text{Root}(\rho)$, of course: Ultimately, R must have been generated via Line 12 in iteration $r = r_{stab}(\sigma) + 3n$ for some prefix $\sigma|_{r+3n}$, but it could have been forwarded to some indistinguishable prefix $\rho|_{r+3n} \sim \sigma|_{r+3n}$. However, Corollary 8 implies that every process both in $\rho|_{r_{stab}(\sigma)+n-1}$ and $\sigma|_{r_{stab}(\sigma)+n-1}$ has already heard from $\text{Root}(\sigma)$. If this forwarding of R happened transitively, we can again use Lemma 13 to show that the latter also holds in this case.

This finally concludes the proof of Theorem 12.   □

Corollary 14. *The consensus Algorithm 1 when run with $\Delta(.)$, as computed by Algorithm 2, terminates for $\sigma \in \Diamond\text{STABLE}_n$ by round $r_{stab}(\sigma) + 4n$.*

Finally, we provide a lower bound that matches asymptotically the termination time of Algorithm 1 when run with $\Delta(.)$, as computed by Algorithm 2.

Theorem 15. *For $n \geq 3$, solving binary consensus under $\sigma \in \Diamond\text{STABLE}_n$ takes $r_{stab}(\sigma) + n + \Omega(n)$ rounds.*

Proof. Fix an arbitrary input assignment C^0 and suppose some binary consensus algorithm A terminates in round $r_{stab}(\sigma) + 2n - 3$ in every run $\langle C^0, \sigma \rangle$. Let σ be the communication pattern where each communication graph consists of the same cycle, thus $r_{stab}(\sigma) = 1$. Starting from the initial configuration, where all input values are 0 and toggling, one at a time, the input assignments to 1 until we arrive at the initial configuration where all inputs are 1, the validity condition shows that there are two initial configurations C, C' that differ only in the input value of a single process p, yet all processes decide 0 in $\langle C, \sigma|_{2n-3}\rangle$ and 1 in $\langle C', \sigma|_{2n-3}\rangle$.

Consider the graph sequence ρ that is $\sigma|_{n-2}$, followed by at least $n-1$ repetitions of a directed line graph that starts from the in-neighbor p' of p in the cycle, i.e., from p', such that (p', p) is in the cycle (see Fig. 4). It is not hard to see that $\rho|_{2n-3} \sim_q \sigma|_{2n-3}$, where q is the process at the end of the line graph.

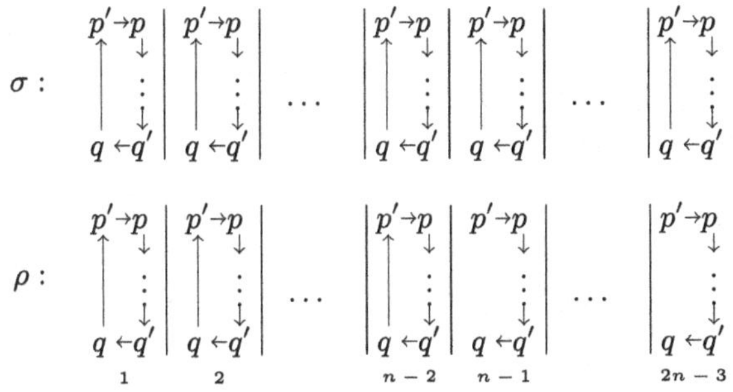

Fig. 4. Communication patterns ρ, σ for Theorem 15. Note the missing in-edge of p' in the last few rounds of ρ.

Thus q, and by agreement, all processes, decide 0 in $\varepsilon = \langle C, \rho \rangle$ and 1 in $\varepsilon' = \langle C', \rho \rangle$. This is a contradiction, however, since $\varepsilon \sim_{p'} \varepsilon'$.   □

5 Conclusions

We presented a new generic algorithmic technique for solving consensus in dynamic networks controlled by a message adversary, and demonstrated its power by devising the first algorithm for the message adversary $\Diamond\mathsf{STABLE}_n$ with assymptotically optimal termination time. The simplicity arguably stems mainly from dividing the task of solving consensus into two separate tasks: A centralized task that assigns labels to the admissible communication pattern prefixes of a given message adversary, and a simple generic distributed algorithm Algorithm 1 that uses this labeling to compute a valid decision value. Future work will be devoted to applying our labeling approach also to other distributed computing models.

References

1. Afek, Y., Gafni, E.: Asynchrony from synchrony. In: Frey, D., Raynal, M., Sarkar, S., Shyamasundar, R.K., Sinha, P. (eds.) ICDCN 2013. LNCS, vol. 7730, pp. 225–239. Springer, Heidelberg (2013). https://doi.org/10.1007/978-3-642-35668-1_16
2. Attiya, H., Castañeda, A., Rajsbaum, S.: Locally solvable tasks and the limitations of valency arguments. In: Bramas, Q., Oshman, R., Romano, P. (eds.) 24th International Conference on Principles of Distributed Systems, OPODIS 2020, 14–16 December 2020, Strasbourg, France (Virtual Conference). LIPIcs, vol. 184, pp. 18:1–18:16. Schloss Dagstuhl - Leibniz-Zentrum für Informatik (2020). https://doi.org/10.4230/LIPIcs.OPODIS.2020.18
3. Ben-Zvi, I., Moses, Y.: Beyond Lamport's happened-before: on time bounds and the ordering of events in distributed systems. J. ACM **61**(2), 13:1–13:26 (2014)
4. Biely, M., Robinson, P., Schmid, U., Schwarz, M., Winkler, K.: Gracefully degrading consensus and k-set agreement in directed dynamic networks. Theor. Comput. Sci. **726**, 41–77 (2018)

5. Castañeda, A., Fraigniaud, P., Paz, A., Rajsbaum, S., Roy, M., Travers, C.: Synchronous t-resilient consensus in arbitrary graphs. In: Ghaffari, M., Nesterenko, M., Tixeuil, S., Tucci, S., Yamauchi, Y. (eds.) SSS 2019. LNCS, vol. 11914, pp. 53–68. Springer, Cham (2019). https://doi.org/10.1007/978-3-030-34992-9_5

6. Charron-Bost, B., Schiper, A.: The heard-of model: computing in distributed systems with benign faults. Distrib. Comput. **22**(1), 49–71 (2009)

7. Coulouma, É., Godard, E., Peters, J.G.: A characterization of oblivious message adversaries for which consensus is solvable. Theor. Comput. Sci. **584**, 80–90 (2015)

8. Fevat, T., Godard, E.: Minimal obstructions for the coordinated attack problem and beyond. In: Proceedings of IPDPS 2011, pp. 1001–1011 (2011)

9. Fischer, M.J., Lynch, N.A., Paterson, M.S.: Impossibility of distributed consensus with one faulty process. J. ACM **32**(2), 374–382 (1985)

10. Nowak, T., Schmid, U., Winkler, K.: Topological characterization of consensus under general message adversaries. In: Proceedings of PODC 2019, pp. 218–227. ACM (2019)

11. Pease, M., Shostak, R., Lamport, L.: Reaching agreement in the presence of faults. J. ACM **27**(2), 228–234 (1980)

12. Winkler, K., Schmid, U., Moses, Y.: A characterization of consensus solvability for closed message adversaries. In: Proceedings of OPODIS 2019, pp. 17:1–17:16. LIPIcs (2019)

13. Winkler, K., Schwarz, M., Schmid, U.: Consensus in rooted dynamic networks with short-lived stability. Distrib. Comput. **32**(5), 443–458 (2019). https://doi.org/10.1007/s00446-019-00348-0

Author Index

Printed in the United States
by Baker & Taylor Publisher Services